SOIL SCIENCE:
Methods and Applications

SOIL SCIENCE:
Methods and Applications

David L Rowell
Department of Soil Science,
University of Reading

Longman
Scientific &
Technical

Longman Scientific & Technical,
Longman Group Limited,
Longman House, Burnt Mill, Harlow,
Essex, CM20 2JE, England
and Associated Companies, throughout the world.

Copublished in the United States with
John Wiley & Sons, Inc., 605 Third Avenue, New York, NY 10158

© Longman Group UK Limited 1994

First published 1994
Reprinted 1995

ISBN 0 582 087848

British Library Cataloguing in Publication Data
A CIP record for this book is available from the British Library

Library of Congress Cataloging-in-Publication Data
Rowell, David L., 1937–
  Soil science: methods and applications/by David L.
  Rowell,
    p.   cm.
  Includes bibliographical references and index.
  ISBN 0–470–22141–0 (US). – ISBN 0–582–08784–8
(UK)
  1. Soil science. I. Title.
S591.R68   1993
631.4 – dc20   93–12830
                              CIP

Set in Plantin by 3
Produced by Longman Singapore Publishers (Pte) Ltd
Printed in Singapore

Contents

Acknowledgements

It is a pleasure to acknowledge the contributions made by many soil scientists to the writing of this book. Donald Payne (Reading University) has been my advisor-in-chief helping with content, presentation and proof checking. His time, and friendly, constructive criticism have been given generously over the four years that we have worked on the manuscript. I have also received specialist help with each chapter and would like to thank those listed below for their input of ideas and for sparing time to work through my manuscript and to discuss many issues with me. One of the pleasures of writing this book has been the interaction with these scientists as together we have attempted to sieve out what is important and to present the material in what I hope is a clear and understandable manner.

Chapter 1   Dr. S. Nortcliff, Reading University

Chapter 2   Dr. D. A. Jenkins, University College of North Wales, Bangor and Dr. A. A. Jones, Reading University

Chapter 3   Professor D. S. Jenkinson, Rothamsted Experimental Station

Chapter 4   Mr. D. Payne, Reading University and Dr. D. L. O. Smith, Silsoe Research Institute

Chapter 5   Dr. L. P. Simmonds, Reading University

Chapter 6   Mr. D. Payne, Reading University

Chapter 7   Dr. S. D. Young, Nottingham University and Dr. E. Tipping, Institute of Freshwater Ecology, Ambleside

Chapter 8   Dr. R. A. Skeffington, National Power plc, Leatherhead and Mr. A. G. Chalmers, ADAS Reading

Chapter 9   Dr. K. W. T. Goulding, Rothamsted Experimental Station

Chapter 10   Dr. P. Le Mare, Reading University

Chapter 11   Dr. D. Barraclough, Reading University and Mr. A. E. Johnston, Rothamsted Experimental Station

Chapter 12   Dr. L. P. Simmonds, Reading University

Chapter 13   Mr. A. E. Johnston, Rothamsted Experimental Station

Chapter 14   Dr. R. Keren, The Volcani Center, Israel

Chapter 15   Mr. M. C. G. Lane, ICI Agrochemicals (now Zeneca plc), Jealott's Hill Research Station, Dr. B. J. Wilson, Long Ashton Research Station and Mr. C. Chumbley, ADAS Reading

The input from Lester Simmonds (Chapters 5 and 12) and from Mike Lane (Chapter 15) has been particularly important. They developed most of the ideas, did some of the writing and helped me to put the material together through many useful discussions. Lester Simmonds has also been responsible for the modelling and computing which appear in Chapters 11, 12 and 15. Others have responded willingly to my requests for information, photographs, clarification of data and in many other ways: W. Adams, S. Allen, J. Archer, A. Armstrong, C. Bishop, P. Brookes, K. Cameron, D. Campbell, B. Chambers, L. Chubb, M. Court, J. Decroux, T. Edwards, M. Froment, M. Goss, D. Greenland, P. Gregory, P. Harris, S. Heming, C. Henkens, M. Hornung, J. Irwin, D. Kinneburgh, H. Koyumdjisky, R. MacEwen, S. McGrath, W. McHardy, S. McKean, A. McNeill, P. Nye, W. Patefield, A. Parsons, L. Petersen, D. Powlson, S. Prasher, K. Ritchey, R. Smith, B. Soane, M. Stansfield, P. Stevens, R. Sylvester-Bradley, R. Tayler, R. Unwin, C. Vincent, G. Wadsworth, A. Walker, G. Warren, M. Wong, G. Wyn Jones.

Anne Dudley has checked non-standard experimental methods, Mike Lane's technical staff have checked some of the experiments in Chapter 15 and Anne Gillibrand has prepared and organized laboratory classes for many years, leading to simplification and clarification of some of the methods for student use.

The word processing has been done by Alice Doyle, Valerie Keane, Dorothea Fitzgerald and Sue Hawthorne in the department office and by my wife and daughters, with technical advice from my son. It has been a laborious and often boring task and I am most grateful to them all. The figures have been prepared by Heather Browning in the Geography Department drawing office.

Although we have attempted to eliminate errors it is inevitable that some will have crept through into the final copy. Please let me know of any that you find so that they can be corrected.

I am grateful to the University and to the members of staff here in the department for giving me so much time and freedom to produce this book. Even in the academic world these are now rare commodities. Other members of staff here carried much of my normal work load for nine months in 1990 when I had to leave to work on the first draft. In the subsequent three years my summer term has been kept relatively light giving almost six months each year to concentrate on the writing. I trust that the investment of time and effort will prove to be worthwhile in helping others to learn how to use the 'tools of the trade'.

Extensive use has been made of the standard texts dealing with methods of soil analysis. In most cases, methods have developed over the years with input of ideas from many scientists, and so it is not possible to acknowledge the sources in detail. Similarly, it is not possible to acknowledge individually the input of students, many from overseas. Both at Oxford and here at Reading, their questions, ideas and experience have contributed much to my understanding and during the writing of the book they have read, used and criticised sections of the manuscript. It has been a privilege to work with them all.

Acknowledgements for illustrative material

We are grateful to the following for permission to reproduce copyright material.

For line work: American Society of Agronomy, Inc for Figures 12.4 (Denmead and Shaw, 1962) & 13.2 (Ramig and Rhoades, 1963); Blackwell Scientific Publications for Figures 4.10 (Smith, 1987), 8.2 (Goulding, McGrath and Johnston, 1989), 13.7 & 13.8 (Johnston, 1986); Cambridge University Press for Figures 11.3 & 11.4 (Gregory, Crawford and McGowan, 1979); Cranfield Institute of Technology for Fig. 1.3 (Hodgson, 1974); Elsevier Science Publishers BV (Hamblin, 1981) for Fig. 5.10; the Controller of Her Majesty's Stationery Office for Figures 4.11, 6.3, 10.5 & 11.10 © Crown Copyright; Longman Group UK Limited and the author, Dr. R. S. Russell for Fig. 3.6 (Wild, 1988); Dr. R. S. Russell for Fig. 3.1 (Russell, 1977); Scottish Centre of Agricultural Engineering for Fig. 4.4.

For black and white photography: ADAS Aerial Photography, Cambridge for Plate 1.2 © Crown Copyright; American Phytopathological Society and the author, R. C. Foster for Plate 2.2 (Foster, Rovira and Cock, 1983); DLG-Verlags-GmbH for Plate 3.1 (Kutschera, 1960); Prof. Dr. Graf von Reichenbach for Plate 9.1 (Smart and Tovey, 1981); for Plate 1.1 © Crown Copyright 1993/MOD reproduced with the permission of the Controller of Her Majesty's Stationery Office; W. J. McHardy and The Macaulay Land Use Research Institute for Plate 2.3 (Smart and Tovey, 1981); New Zealand Society of Soil Science and Mallinson Rendel Publishers, Wellington for Plate 2.4 (Molloy, 1988); Scottish Centre of Agricultural Engineering for Plate 4.1; John Wiley & Sons Ltd for Plate 6.4 (Emerson, Bond and Dexter, 1978).

For colour photography: Holt Studios International and Nigel Cattlin for Colour Plate 6; Rothamsted Experimental Station for Colour Plate 19; Scottish Centre of Agricultural Engineering for Colour Plates 3 & 7; Warren Spring Laboratory for Colour Plates 13, 16 & 17.

Copyright material has been acknowledged where appropriate. The quotations in the Prologue and Epilogue are from the Good News Bible published by The Bible Society/Harper Collins Publishers Ltd., UK©, American Bible Society 1966, 1971, 1976, 1992 with permission. The quotations in Section 13.5 from Nye, P. H. and Greenland, D. J. 1960 *The Soil Under Shifting Cultivation* are included by permission of CAB International.

David L. Rowell
Reading, July 1993

Prologue

The parable of the sower

Once there was a man who went out to sow some corn. As he scattered the seed in the field, some of it fell along the path and the birds came and ate it up. Some of it fell on rocky ground where there was little soil. The seeds soon sprouted because the soil wasn't deep. But when the sun came up, it burnt the young plants; and because the roots had not grown deep enough, the plants soon dried up. Some of the seed fell among thorn bushes, which grew up and choked the plants. But some seeds fell in good soil, and the plants produced corn; some produced a hundred grains, others sixty and others thirty.

And Jesus concluded, 'Listen, then, if you have ears!'

(*c.* AD 30)

CHAPTER I

Soils in the Field

INTRODUCTION

The earth's surface is covered by three materials in various combinations; water (including snow), rock and soil. Soil results from the *weathering* of rock materials, which involves both the physical breakdown of rock into smaller particles and chemical alteration of its composition (Ch. 2). There are, however, many other *processes* which together produce the distinctive features of the material we call soil and organize this material into soils on the surface of the land. Of primary importance are the processes associated with plants, animals and micro-organisms which colonize the soil (Ch. 3). The rates at which these processes occur depend primarily on climate, resulting in the wide range of soils and soil properties which are observed around the world.

The soil is always an important component in the system comprising the lithosphere, the atmosphere and the biosphere. Soil properties reflect the varying nature of the interactions within this system. Soil is essential for many human activities, and we can only successfully and sustainably undertake these activities if we understand how soil has been developed and how it is affected by changes in the system, particularly those in the biosphere caused by our manipulation of vegetation and soil.

SOIL DEVELOPMENT

During the second half of the nineteenth century Russian soil scientists led by Dokuchaiev observed that soil properties seemed to be influenced by various *soil-forming factors*, and expressed these relationships in a quasi-mathematical formula:

$$S = \text{a function of } (c\ell, p, r, v, o)_t \qquad [1.1]$$

where S represents the soil or some soil property, and the soil-forming factors are climate ($c\ell$), parent rock (p), topography (r), vegetation (v), soil organisms (o) and the time over which the soil has been forming (t).

The inclusion of time in Eq. 1.1 highlights the fact that soil is a dynamic system, in which the soil-forming factors influence the rates at which processes occur, with rate $\times$ time measuring the changes produced by the processes. The lengths of time involved in causing changes in properties vary from hours for the formation of an earthworm cast for example, to years for the formation and incorporation of humus and to millennia for the weathering of rocks and the formation of clay minerals. Thus the properties of a soil profile composed of several horizons and extending perhaps beyond a metre below the surface are the result of processes operating over very long periods of time. During this time changes in weather may have altered the topography, vegetation and organisms (the factors are not independent) and the rates at which processes have occurred have not been constant during the formation of the soil. Even without changes in climate, processes would not have operated at a constant rate during soil development. For example, changes in the composition of rock may have occurred initially at a rapid rate, slowing as the weathered material approached an equilibrium with its environment. Similarly as physical weathering of rock produces a deepening layer of soil material, so a succession of plants and organisms can colonize the area until a climax system is reached where little further change in soil properties may occur. Thus the rates at which processes operate today are not the same as the rates which have occurred in the past.

One further limitation of Eq. 1.1 is that the relative importance of the factors seems to vary depending on the scale at which the effects are being considered. For example, Dokuchaiev's original observations emphasized the importance of climate as the factor primarily responsible for the broad distribution of soils over the land of Russia. However within a climatic region differences in parent material may be the major factor causing soil differences. At an even more local scale, vegetation and topography may be the controlling factors if climate and parent material vary little over the area (Fig. 1.1 and Plate 1.1). At the scale mostly considered in this book, local variability within a field or a plot is the primary concern, and normally reflects variations in parent material, topography and the vegetation present before the land was brought into agricultural use (Plate 1.2).

Despite the limitations of Eq. 1.1 as a basis for understanding soil development, it does provide a

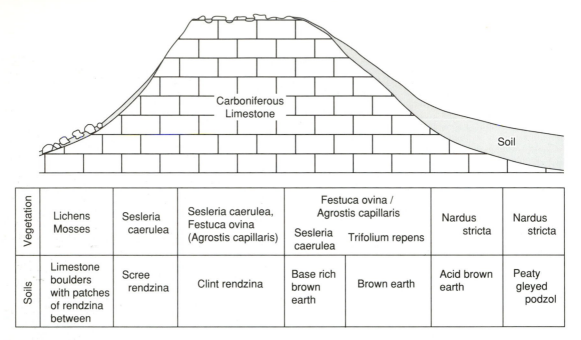

Figure 1.1 An idealized cross-section of the Malham area of North Yorkshire showing the patterns of soils developed over Carboniferous Limestone. Plate 1.1 is an aerial photograph of this area. There are two distinct sets of soils; those developed directly on the weathering residues of the limestone, and those on a non-calcareous drift overlying the limestone. The former are various forms of rendzinas, the main differences being the amount of limestone rubble in the soil and soil depth. The latter are base-rich brown earths where the drift is shallow, and more acidic brown earths where the drift thickens. At the foot of the slope acidic and poorly drained peaty gleyed podzols occur where the drift is several metres thick. From Bullock (1971). © Crown Copyright 1993/MOD reproduced with the permission of the Controller of HMSO.

logical framework within which soil properties and processes can be investigated and a century after its introduction is still the basis for understanding the spatial distribution of soils.

Of the many introductory books which deal with soils in the field, their development and distribution, Bridges (1978) and Molloy (1988) are recommended. The latter is unique in that it places scientific information in the context of an appreciation of the beauty of soils and landscape in New Zealand.

THE USE OF SOILS

Soil development viewed through Eq. 1.1. does not take direct account of the changes in soil properties which result from the use of soils for the production of food, fibre and timber. This is a major factor influencing soil properties in the inhabited regions of the world, the effects beginning after the Neolithic period with the shift from hunter–gatherer economies to the more sedentary agriculturally based communities (Hillel, 1992). Thus to Dokuchaiev's equation, management could be added as an extra factor. It would, however, have a different character, operating only over recent

time, in many instances counteracting natural processes of development and in very recent times adding agricultural chemicals and domestic and industrial wastes. An extension of Eq. 1.1 can therefore be written which takes account of these aspects of management as a soil-forming factor as follows:

$$S = \text{a function of } (c\ell, p, r, v, o)_{t_1} + (m)_{t_2}$$

where m is the management factor, t_1, is the total time the soil has been forming, and t_2 is the time since soil use began. This equation suggests an independence between management and the other factors which is not the case, since management has its effects primarily through alterations in vegetation, organisms and local climatic conditions. The possibility of global climate change resulting partly from the carbon dioxide (CO_2) produced by the burning of vegetation and loss of soil organic matter emphasizes the link between management and the other soil-forming factors.

The last few decades have seen a rapid increase in concern regarding the effects of our activities on the environment. Our effects on soils are highlighted by catastrophic events such as salinization (Ch. 14) resulting from poorly managed irrigation schemes, and

Plate 1.1 Limestone scenery north of Malham, North Yorkshire. The Cove is in the foreground and the Tarn in the distance. Limestone pavements and a dry valley dominate the centre of the photograph. Figure 1.1 shows the distribution of soils and vegetation in this area. Photograph by the University of Cambridge Committee for Aerial Photography. © Crown Copyright 1993/MOD reproduced with the permission of the Controller of HMSO.

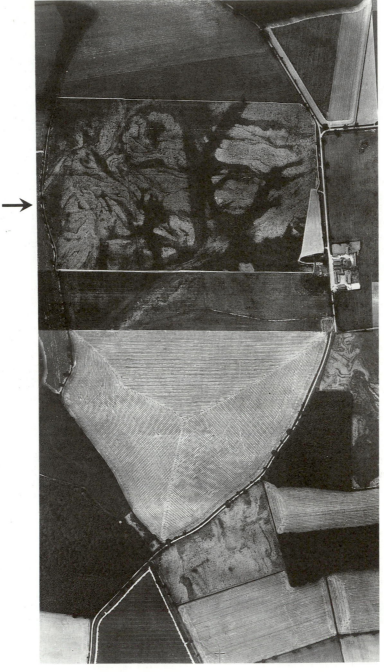

Plate 1.2 Soil variability on part of the Cromer Ridge, Norfolk. The pattern of grey tones across the field marked with an arrow shows the differential growth of the sugar beet crop in late August 1970. The lighter tones indicate poor crop growth, yellow foliage and in places crop failure, and the darker tones indicate a healthy green crop. The latter are on loams deeper than 90 cm, and the former are on loams less than 15 cm deep over sand. The deep loams have a larger water-holding capacity (Section 12.3) and sustain crop growth for longer periods during drought. The cereal crops in the adjacent fields had shown similar patterns before harvest, and with careful examination they can still be seen in the stubble. Similar patterns are seen where abandoned settlements have left building foundations near to the surface of the land (Banks and Stanley, 1990). Photography by ADAS Aerial Photography, Cambridge. Crown Copyright.

erosion resulting from the exposure of sensitive soils to wind and rain. For the most part, however, soils are able to accommodate changes in use, their properties being naturally well buffered (stable), changing only slowly towards a new equilibrium which depends on the altered soil-forming factors.

Apart from salinization and exposure of soil to erosion, the effects of the burning of vegetation and reduction in soil organic matter content are probably the most fundamental changes caused by cultivation (Ch. 13).

- Burning releases nutrient elements partly to the atmosphere and partly to the soil, the latter temporarily increasing nutrient availability and counteracting soil acidity which is the basis of shifting cultivation (Ch. 13). It is a natural part of nutrient cycling, but accelerates the processes which occur slowly during the decay of vegetation, and may increase the loss of nutrients.
- The effects of cultivation on organic matter in soils can be contrasted with natural conditions where organic matter content increases towards a stable maximum depending on the environment. Cultivation reverses this trend by reducing inputs of organic material from vegetation and increasing the rate of decomposition. As a result, the natural fertility of the soil is reduced through decreased availability of nutrients (Chs. 9–11) and deterioration of structural stability (Ch. 4). The latter increases the sensitivity of soils to erosion and to compaction by cultivation equipment and treading by animals. Cultivation then becomes more difficult, and root growth may be restricted both by the increased mechanical resistance of the soil and by reduction of aeration (Ch. 6). These changes are counteracted to some extent by the inclusion of leys (grasses and legumes for 1 or more years) in crop rotations and the use of organic manures, but very large inputs are required to prevent the organic matter content of soils from decreasing.

The use of organic manures, including domestic wastes and sewage, is part of the natural cycling of materials through soil, plant and animal. However, the natural cycle has been broken by the application of other materials to soils, which adds a new dimension to soil formation. This began with the ancient practice of liming soils with chalk. Although chalk is a naturally occurring material dug from beneath the soil and spread on the surface, its use, like cultivation, counteracts a natural trend in soil formation, in this case the development of soil acidity (Chs. 7 and 8). However, unlike cultivation, liming increases the fertility of soil by reducing aluminium and manganese toxicities and increasing nutrient availability.

More recently, manufactured chemicals have been added to soils. These can be divided into three groups:
1. *Fertilizers* which are compounds that release ions which are already present naturally in soils, and take their place in the natural cycling of nutrients between soil and plant.
2. *Organic chemicals* (pesticides and growth-controlling compounds) which do not occur naturally in soils. They are decomposed by soil organisms to produce compounds which do occur naturally, but their effects need to be understood (Ch. 15).
3. *Industrial wastes* which contain many inorganic and organic compounds. Some wastes are added to soil inadvertently through emissions into the atmosphere and these include metals and the acid producing oxides of sulphur and nitrogen. Some are applied through dumping, or are spread to make use of the nutrients they contain, such as industrially contaminated sewage. For the most part, there are no long-term effects, because soil is a remarkably effective system for the biodegradation of added organic materials into naturally occurring compounds and ions. Of more concern are the effects of increased acidity but at least within the soil these can be reversed by liming. However, metals which occur naturally in small concentrations may become toxic in larger concentrations (Ch. 15). They are not decomposed, and so cause irreversible changes in soil properties.

THE MEASUREMENT OF SOIL PROPERTIES

Because our use of soils changes their properties, influencing both their interaction with the environment and their ability to produce crops, it is important that these properties should be measured and the measurements understood. In the absence of exact information, there is little foundation either for discussions regarding environmental issues or for effective management for agricultural and other purposes. Our mistakes in the past show the dangers of working without this foundation.

Much useful information can be obtained by observation of soils in the field, but observation is to a large extent subjective. This book deals with the measurement of soil properties as a means of obtaining an objective understanding of our use of soils and the environmental implications.

There are three approaches to the measurement of soil properties:
1. Measurements in the field of properties which can be observed after digging a soil pit. These are normally semi-quantitative assessments of properties rather than direct measurements.
2. On-site measurements using equipment inserted

into soil, without significant disturbance of the soil. This approach applies particularly to soil water studies (Chs. 5 and 12).

3. Measurements made in the laboratory on soil samples taken from the field.

This chapter deals with soils in the field within the context of these three approaches. Section 1.1 gives guidance on the choice of site for a soil pit, digging procedure and sampling methods. Section 1.2 describes methods that can be used to assess texture, stone content and porosity. Section 1.3 discusses soil variability and describes procedures to obtain representative samples from fields or plots. Subsequent chapters deal with the measurement of basic soil properties (Chs. 2–8) and nutrient and water availability (Chs. 9–12). Building on information in these chapters, principles are then integrated in a discussion of soil fertility (Ch. 13). Problems resulting from our use of soils follow: soil salinity is both an ancient and modern pollution problem (Ch. 14) and pollution of soils by pesticides and metals presents us with the most recent challenges in soil management (Ch. 15). Ideally erosion should be included but we do not have sufficiently simple methods which can be used to give meaningful data, and so this topic has been omitted.

The following criteria underlie the selection of methods in this book. They:

- are basic methods which are generally accepted as the tools of soil science,
- can be understood without an advanced knowledge of the sciences,
- can be carried out without access to sophisticated analytical equipment, and
- measure fundamental soil properties and so provide data from which appropriate calculation gives information beyond that of the immediate measurement.

Based on these criteria some soil properties cannot yet be measured adequately, resulting in an uneven coverage of topics in this book.

Apart from erosion, this applies particularly to the measurement of certain aspects of soil structure, mechanical properties and aeration, and to soil biochemical properties where methods are at an early stage of development.

FURTHER STUDIES

Calculations are in Section 1.4.

Section 1.1 The use of soil pits

If soil samples are required from deeper than about 20 cm, either an auger can be used (Section 1.3) or a pit must be dug. The former will provide only limited information because the samples are taken below a small area of the surface (a few cm^2) and are disturbed. The exposure of a *soil profile* in a pit allows information to be obtained both on the vertical arrangement of soil material into *horizons* and their horizontal variability. Soil structure, porosity and other features, some of which would have been disturbed by an auger, are displayed for observation and measurement. The choice of sampling method therefore depends on the purpose of sampling.

HOW TO CHOOSE A SITE

Background information

Before going into the field to investigate or sample soils it is important to gain as much information about the area and its soils as possible. Books and maps produced by both local and national surveys may be available which deal with the topography, geology and soils of the area. For example in Britain, publications by the Ordnance Survey, the Soil Survey and Land Research Centre, the Macaulay Land Use Research Institute and the Geological Survey are valuable sources of information.

Addresses are given in Appendix 1.

Access to the site

Where land is privately owned permission to go on to the area and to carry out your work is required, In general landowners are generous in allowing access, but if permission is abused your current work may be halted, and future investigators refused permission.

Preliminary investigation and choice of site

Walk the area in order to view the site from more than one position.

An understanding of the landscape aids understanding of the distribution of soils. Refer back to the topography, geology and soil maps of the area. Use a screw auger to establish the general properties of the soils, making a number of borings to find the 'most representative' site for your purpose (Section 1.3). While it is impossible here to give guidance on where to sample because of the wide range of purposes, it is sensible to exclude some sites on the basis of excessive disturbance such as the positions of old roads, where old buildings once stood or where river or field drain dredgings have been dumped. For most purposes the following should also be avoided:

- areas close to gateways, paths and tracks;
- headlands of arable fields (the outer 10 m);

- sites where straw or fertilizer have been stored;
- sites used for localized burning of crop residues or hedge trimmings;
- old field boundaries where a hedge or bank has been removed and the land levelled.

Note that although the 'most representative' site may be chosen in terms of *spatial variability* over an area, *temporal variability* (changes with time) means that some observed and measured properties are only representative of the soil at the time of sampling. Most obviously, soil water content varies from day to day, and this and root distribution vary seasonally. Less obviously, biological activity varies seasonally, causing changes in aeration and the amounts of available nutrients. Agricultural management also causes seasonal changes particularly in the topsoil through cultivation and fertilization, and in the longer term liming and subsequent reacidification change soil pH values. Other properties are more permanent, for example horizon depth, texture, stone content and ion exchange capacity.

DIGGING A SOIL PIT

Equipment

Spade, auger, pickaxe, trowel, small knife, wooden shoring and props, polythene sheets, 1 inch paint brush.

Method

Observe and record features of the site and the soil surface following standard procedures, for example those given in the *Soil Survey Field Handbook* (Hodgson, 1974).

Pit size The areal extent of the pit depends on the required depth of observation and sampling. A pit to 1 m depth should be about 1 × 1.5 m in extent.

Pit orientation The face of the pit to be used for observation, photographing and sampling should face the sun.

Excavation Mark out the area of the pit. To preserve the characteristics of the upper few cm of soil avoid treading the soil on top of the face to be examined. Lay a polythene sheet alongside the pit. If the soil is covered with grass or other vegetation, cut square turfs and place them on the sheet maintaining their relative position.

Excavate the soil keeping the topsoil and subsoil in separate piles.

Record the conditions encountered when digging, for example dense or stony layers.

If the soil pit is deep (> about 1 m) it may be necessary to support the walls with shoring to prevent collapse. If there is a danger of collapse, there must always be two persons at the site during excavation. It may be necessary to dig out steps to facilitate entry to and exit from the pit.

Cleaning the profile The faces of the pit will have been cut and smeared by the spade. Using a trowel or knife pick soil from the face from the surface downwards to remove contamination and expose the features of the soil horizons. In a dry soil finally clean the face with a small brush.

Profile description Follow standard procedures (for example, Hodgson, 1974). Section 1.2 gives methods of assessing only those properties which are important in the context of this book. Sampling procedures are described below.

Leaving the site A pit should only be left unattended during the day if there is no possibility of people or animals falling into the hole.

A warning rope should be placed around the site. If the pit is to be left overnight it should be fenced or covered with boards. When work is complete the soil should be replaced in the correct sequence. Tread the soil occasionally to compact it into the hole. Finally replace the turfs and tread them into place, leaving the site as you found it.

SOIL SAMPLING

Equipment

Polythene bags, tie labels.

Method

The description of the soil profile will include information on horizons, distinguished on the basis of colour, texture, structure and other observable features. The boundaries between horizons may be distinct or merging, and the depth of a boundary may vary across the exposed face. Samples are normally taken to represent a horizon.

Record the depth of the horizon, and the nature of the boundaries.

Sample to give either a *bulked sample* for the horizon, or if required sample at various depths within the horizon. Using a trowel, prise soil from the profile face into a polythene bag, taking soil from a number of places within the horizon and mixing to give one representative sample known as the bulked sample. This can be taken from the four faces of the pit if it is intended to be representative of a larger area.

Special sampling methods are required for root measurements (Section 3.1), bulk density (Section 4.2) and mineral–nitrogen (Section 11.1).

Label the samples as follows:

- After placing the soil in the polythene bag, expel much of the air and tie the neck.
- Write a label using an indelible pen giving a sample number, the horizon (or depth), the pit number, the location, the name of the sampler and the date the sample was collected.
- Place the bagged sample in a second polythene bag and insert the label between the two bags. Expel the air and tie as before.
- Write a second label and attach to the second bag. Writing directly on the polythene bag with an indelible marker is not a reliable method.

The above procedure may seem unnecessarily tedious, but card labels placed in the soil bag quickly rot, and marking on the polythene bag is easily removed in transit.

Preparing the samples for analysis After returning to the laboratory spread the samples to dry in the air or dry them in a forced draught cabinet at 30 °C. Crush the soil using a pestle and mortar or a soil-grinding machine and pass through a 2 mm sieve, rejecting roots and stones to give the *fine earth* fraction. This is commonly referred to as a *<2 mm air-dry sample*. Clay soils can be very hard when dry. It is sometimes helpful to grind and sieve before drying is complete, and then to spread the sample for final drying. Clay subsoils are extremely difficult to grind even before they are completely dry. It is helpful to place the wet soil in a freezer for several days before allowing it to dry when it will break naturally into small pieces.

To obtain a subsample from a bulked sample, spread the soil on a polythene sheet. Divide into four quadrants. Take two opposite quadrants of soil, mix them together, spread and divide again. Repeat this process until a subsample of the required size is obtained.

SOIL VARIABILITY

It is normally assumed that by bulking soil from several places within a horizon, a representative sample is obtained. Grinding and sieving produce a homogenized sample from which a small subsample (e.g. 1 g) can be taken for analysis, and it is assumed that this subsample is then representative of the horizon. Table 1.1 gives data for samples taken from three horizons: a surface horizon (0–10 cm) in a permanent pasture and two horizons (0–5 and 5–10 cm) under woodland. The pasture and the woodland soils are both developed on an alluvial parent material adjacent to the River Loddon and are about 150 m apart. A 1 m length of each horizon was exposed, and across the length of each five samples were taken from within 20 cm sections. Approximately half of each sample was labelled and bagged separately and the remaining soil from each set of five samples was mixed together to give a bulked sample. The samples were air dried and sieved to give the fine earth. The bulked sample was then divided into five subsamples. Thus for each horizon there were five separate field samples, and five subsamples from the bulked sample.

The organic carbon (C) content (Section 3.4) of each sample was measured in triplicate to give the mean values shown in Table 1.1. The variability within each

Table 1.1 The variability of organic carbon contents (%, *m/m*) of samples taken from three horizons of soils at the University of Reading Shinfield Farm.

Sample No.	Pasture (0–10 cm)		Woodland			
			(0–5 cm)		(5–10 cm)	
	Separate	Bulked	Separate	Bulked	Separate	Bulked
1	5.18	4.84	6.80	8.81	4.18	4.00
2	5.02	4.76	7.51	8.18	3.01	3.55
3	5.02	4.96	9.98	7.43	3.19	3.61
4	5.70	5.18	9.18	9.07	3.61	3.79
5	4.99	4.83	8.76	8.43	3.18	3.38
Mean, $\bar{x}$	5.18	4.91	8.45	8.38	3.43	3.67
Standard deviation, s	0.299	0.165	1.28	0.634	0.472	0.237
95% confidence limits	±0.37	±0.21	±1.59	±0.79	±0.59	±0.29

Data by S. Nortcliff and A. Dudley, Reading University.

set of measurements is given in terms of the mean, the standard error and the 95 per cent confidence limits (Section 3.8).

These results lead to the following conclusions:

- The production of a bulked sample followed by sub-sampling reduces the variability between measurements.
- There is substantial variability in organic (C) content across a 1 m length of each horizon.
- The variability in the surface horizon under pasture is much less than in the woodland.

These results support the widely used practice of bulking a number of samples from across a horizon when endeavouring to characterize the horizon with one sample. The spatial variability over a plot or field is discussed in Section 1.3.

Section 1.2 Quantitative assessment of soil properties in the field

Many soil properties can be described once a soil profile has been exposed. Routine procedures have been established, some internationally standardized and others varying a little between countries. Some properties are assessed qualitatively, for example colour, using standard Munsell colour charts. Others are assessed semi-quantitatively: for example, the presence of roots may be recorded as 'many coarse roots, few fine roots' with 'many' and 'few' being based on the approximate number of roots per unit area of the exposed profile, and 'coarse' and 'fine' based on root diameter. The assessment of quantity is often aided by the use of standard charts: for example, the quantity of stones exposed in the profile can be compared to the charts in Fig. 1.3. Properties can also be measured in the field using simple equipment: for example, the stone content can be determined by sieving and weighing using a spring balance.

The methods given below are either semi-quantitative or quantitative. They satisfy the criteria given on p. 6 and are of importance in the context of the methods and applications which follow in subsequent chapters. Hodgson (1974) deals with other soil properties normally included in profile descriptions.

SOIL TEXTURE

Probably the most commonly used description of a soil is its texture which is a property of the fine earth (<2 mm) fraction which depends on the *particle-size distribution*. Particle size in this fraction varies from 2 mm diameter down to less than 0.1 μm, and the distribution of particles over this size range influences

many important soil properties such as ease of cultivation and water-holding characteristics. Chapter 2 develops this topic in more detail and Section 2.3 gives the laboratory method for determining particle-size distribution. For convenience, a soil is allocated to a textural class, depending on its content of *sand-, silt-* and *clay-sized* particles. In the field, the textural class can be determined subjectively from the feel of a moist soil moulded between the fingers and thumb because the particle-size distribution influences the mechanical properties of the material. With experience, a soil scientist comes to know the feel of each textural class and can accurately allocate a soil to a class. Over the years schemes have been developed to enable scientists to 'learn the trade', and have been brought together to give the guide shown in Fig. 1.2.

Various textural classifications are available. Figure 1.2 is based on the widely used USDA system which is shown in Fig. 2.6(a). Note also that Fig. 1.2 applies to mineral soils, i.e. light-textured soils with less than 6 per cent organic matter, medium-textured <8 per cent and heavy-textured <10 per cent. Soils with more organic matter have a fibrous or silky feel depending on the amount and degree of decomposition of the organic matter and may have to be classified as *organic soils* without a textural class. A silky feel is also given by silt and experience is required to distinguish the two materials (Hodgson, 1974).

Once a soil has been allocated to a textural class, its particle-size distribution has been determined within broad limits. Section 2.3 shows how these limits can be found from Fig. 2.6.

Finger assessment of soil texture for mineral soils

Equipment

Wash bottle, old cloth, trowel.

Method

As far as possible use only one hand for the soil sample, keeping the other hand clean or at least reasonably dry for writing down your results:

- Take about half a handful of soil from the profile face.
- Remove 'foreign bodies' such as roots, seeds and insects.
- Remove stones to leave the fine earth fraction. Although all particles >2 mm should be removed, in practice very small stones may remain.
- Add a little water from a wash bottle, allow the soil to absorb the water and then work (mould) the moist soil in the hand and then between the thumb and

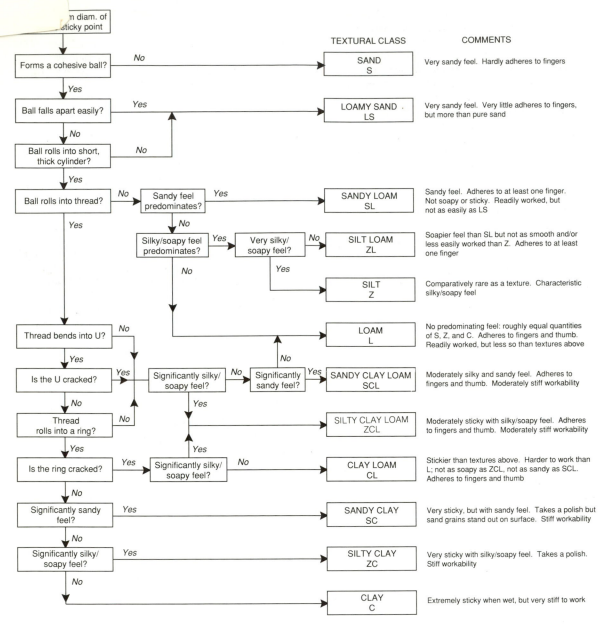

Figure 1.2 A guide to field assessment of texture for mineral soils. By S. Nortcliff, Reading University and J.R. Landon, Booker Agricultural International.

first two fingers until the soil is uniformly moist and has been broken down into its individual particles. Clay soils initially dry need much working to satisfy these requirements.

- Add more water or more soil, working the soil until the sample is at its *sticky point*, i.e. the condition in which the soil being wetted just begins to stick to the fingers. Clay soils may seem to become drier as they are worked due to the continued absorption of water. More water may need to be added until the condition of the soil is stable.
- Follow the guidelines in Fig. 1.2 and record a textural class.
- Wipe residual soil off the hand thoroughly before taking another sample of soil for texture assessment.

Note that the terms sand, silt and clay are used both for textural classes, and for particle-size classes (Ch. 2). Note also that there are no abrupt changes in feel between textural classes. Thus texture assessment is not so precise as is indicated by Fig. 1.2.

STONE CONTENT

Stones are for most purposes an inert component of the soil. Although of interest to the geologist and pedologist as an indication of the history of the site and soil, in the context of this book they affect soil fertility by taking up space which would otherwise be occupied by fine earth. Thus the ability of a given volume of soil to hold water (Section 5.1) and nutrients (Section 9.1) is reduced. Stones are also a hindrance to cultivation. For these reasons they are included in soil fertility and land-use capability classifications (Ch. 13 and Dent and Young, 1981).

Assessment of stone content

After cleaning the profile, estimate by eye the percentage stone content using the charts in Fig. 1.3. Essentially an area percentage is determined which is assumed to be equal to a volume percentage in the soil, this being the measure which is normally required for assessing the effects of stones on soil fertility.

Measurement of stone content

Normally the stone content is expressed as a percentage by volume. The stones in a soil sample taken from a measured volume in the field are weighed and their volume determined after measuring their density in the laboratory. Differences in stone porosity have to be taken into account. A percentage by mass can also be determined.

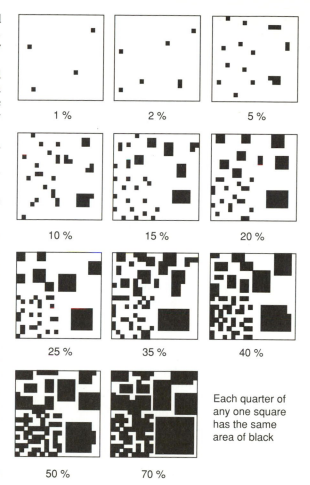

1 %   2 %   5 %

10 %   15 %   20 %

25 %   35 %   40 %

50 %   70 %

Each quarter of any one square has the same area of black

Figure 1.3 Charts for the assessment of stone quantity in a profile. From Hodgson (1974).

Equipment

Spade, spring balance (50 ± 0.2 kg), plastic bucket, polythene sheet, plywood board with a square hole (30 × 30 cm), sieve (10 mm square mesh), polythene measuring cylinder (2 ℓ), paraffin wax (congealing point about 49 °C).

Method

Sampling and field measurements Place the board on the soil surface or on the horizon of interest after excavating the superficial horizon(s). Dig out a hole of the required depth, often 20 cm for a cultivated topsoil, using the board as a template and placing the excavated soil on a polythene sheet. With care the hole can be dug with dimensions 30 × 30 × 20 cm, giving a volume of 18 000 cm³. Alternatively, the hole volume can be measured by the method given under 'Problem soils' in Section 4.2.

Except for wet or clay-rich soils, sieve the excavated soil through a 10 mm sieve (Note 1). Remove roots and other debris and weigh the stones. Place a subsample of the stones in a polythene bag, seal and take to the laboratory for the determination of density. If the stone content is to be determined as a percentage by mass (Note 2), weigh the sieved soil and place a subsample in a polythene bag, seal and take to the laboratory for the determination of water content.

If the soil is too wet or clay-rich to sieve in the field, take the whole sample to the laboratory, air-dry and weigh. Sieve out and weigh the stones. Take a sub-sample of stones for the determination of density, and if required a subsample of sieved soil for the determi-nation of water content.

Determination of water content Weigh the soil sub-sample. Dry at 105 °C and reweigh. Knowing the mass of moist sieved soil in the field (or of air-dry sieved soil in the laboratory) and these weighings, calculate the mass of oven-dry soil in the field sample (Section 3.3).

Stone density (non-porous) Provided the stones con-tain little iron and are non-porous, e.g. flint, quartzite, granite, basalt, a density of $2.65 \, g \, cm^{-3}$ can be assumed (see Table 4.4). The moist mass of stones in the field will not be significantly different from their dry mass. Thus the volume of stones can be calculated from the field mass and their density.

If required the density of non-porous stones can be determined as follows. Weigh a few stones. Put them into a 2 l polythene measuring cylinder containing water and note the increase in volume. Stone density = mass (g)/volume (cm^3).

Stone density (porous) Porous stones (limestones, chalks, sandstones) have densities less than $2.65 \, g \, cm^{-3}$, and contain water when sampled in the field. Proceed as follows: this method can also be used for soil aggregates as a means of determining their *bulk density* (Section 4.2).

Weigh the moist subsample of stones. Oven dry and reweigh. Take a representative stone, brush away loose soil, and tie round it a length of nylon thread. Weigh the dry stone + thread. Warm some paraffin wax in a beaker until it is just melted (about 55 °C) (Note 3). Dip the stone into the wax and immediately remove. Allow the wax coating to solidify. Inspect to ensure that there are no breaks in the coating. If necessary dip again. Weigh the stone + wax. Weigh a beaker of water. Leaving the beaker on the balance pan, hang the waxed stone on a stand so that the stone is suspended in the water and note the increase in weight.

Calculations

Example: non-porous stones

$$\text{Volume of hole} = 18\,000 \, cm^3$$
$$\text{Mass of stones} = 4500 \, g$$
$$\text{Mass of moist sieved soil} = 20\,250 \, g$$
$$\text{Mass of moist subsample} = 52.40 \, g$$
$$\text{Mass of oven-dry subsample} = 44.50 \, g$$

Stone content (*m/m*)
The mass of dry sieved soil
$$= 20\,250 \times 44.50/52.40$$
$$= 17\,197 \, g$$
Total mass of dry soil $= 17197 + 4500 = 21\,697 \, g$
Therefore the stone content (% *m/m*) $= 100 \times 4500/21\,697 = 21.$

Stone content (*v/v*) The volume of stones assuming a density of $2.65 \, g \, cm^{-3}$ is
$$4500 \, g/2.65 \, g \, cm^{-3} = 1698 \, cm^3$$
Therefore the stone content (% *v/v*) $= 100 \times 1698/18\,000 = 9.$

The use of similar data to calculate field bulk density is given in Section 4.2.

Example: porous stones

Mass of moist stones in the field $= 6255 \, g$
Mass of moist stone subsample $= 72.35 \, g$
Mass of oven-dry subsample $= 51.50 \, g$
Therefore, the mass of dry stones in the field $= 6255 \times 51.50/72.35 = 4452 \, g$.

$$\text{Mass of an oven-dry stone} = 32.43 \, g$$
$$\text{Mass of stone + wax} = 33.75 \, g$$
$$\text{Mass of wax} = 1.32 \, g$$

Increase in mass of beaker + water with the suspended stone + wax = 15.28 g. Therefore, the volume of the stone + wax = $15.28 \, cm^3$ (Note 4). The volume of wax coating assuming a wax density (Note 5) of $0.90 \, g \, cm^3$ is $1.32 \, g/0.9 \, cm^3 \, g^{-1} = 1.47 \, cm^3$.
Volume of the stone $= 15.28 - 1.47 = 13.81 \, cm^3$
Density of the dry stone $= 32.43 \, g/13.81 \, cm^3$
  $= 2.35 \, g \, cm^{-3}$ (Note 6).
Therefore, the volume of dry stones in the field is

$$4452 \, g/2.35 \, g \, cm^{-3} = 1894 \, cm^3$$

The calculation can then proceed as in the case of non-porous stones.

Note 1 Strictly the stones should be separated from the fine earth using a 2 mm sieve. This is not practicable in the field. Including the very small stones (2–10 mm) with the fine earth normally makes only small differ-ences to the measured values. If it is essential to use a 2 mm sieve, the field sample must be dried in the labo-ratory and crushed before sieving.

Note 2 Although the volume of soil need not be measured, the sample must represent the horizon from which it is taken. The same sampling procedure is recommended to ensure that the soil is sampled uniformly through the horizon.

Note 3 Paraffin wax can be replaced by Saran resin (Section 4.3).

Note 4 Use is made in the calculation of Archimedes' principle which states that when a solid is suspended in a liquid there is an upthrust on the body equal to the weight of liquid displaced. The weight of the body therefore decreases by an amount equal to the upthrust. However, there must also be a downthrust on the liquid equal in magnitude to the upthrust. In the experiment the weight of the beaker + water increases due to the suspended stone. The increase in weight (the downthrust) is equal to the mass of water displaced which is equal to the volume of water displaced if the density of water is 1 g cm^{-3}. Thus the volume of the stone + wax (cm^3) is numerically equal to the increase in weight (g). The purpose of the wax coating is to prevent water from entering the pores in the stone.

Note 5 The density of paraffin wax varies between about 0.90 and 0.93 g cm^{-3}. Provided the mass of wax is small compared to the mass of the stone, the assumption that the density is 0.90 g cm^{-3} causes negligible errors.

Note 6 The density of the stone (its bulk density) decreases as its porosity increases. Section 4.2 develops this principle. The solid particles within the stone have a density of about 2.65 g cm^{-3}. The density of a piece of chalk may be as low as 1.3 g cm^{-3} (porosity 0.5 cm^3 pores per cm^{-3} chalk) and sandstones have densities down to 1.9 g cm^{-3} (porosity $0.3 \text{ cm}^3 \text{ cm}^{-3}$).

POROSITY AND HYDRAULIC CONDUCTIVITY

Semi-quantitative assessments of porosity can be made from observation of profile characteristics. The method used in England and Wales by the Soil Survey and Land Research Centre is given in Hodgson (1974). Macroporosity (the volume of pores $>50 \mu\text{m}$ in diameter) can be estimated using charts similar to those for stones (Fig. 1.3). The estimates are however only broad indications of porosity and require experience in observing and describing soil structure. For this reason the methods are not included here. Quantitative methods are in Section 4.2.

Saturated hydraulic conductivity can be estimated from observed pore characteristics (McKeague *et al.*, 1982). A quantitative method is given in Section 5.5.

CALCIUM CARBONATE CONTENT: FIELD METHOD

The field estimate of calcium carbonate ($CaCO_3$) content is based on the reaction of soil with dilute acid giving both visible and audible effects. The method is not sensitive to differences in $CaCO_3$ contents above 10 per cent. The quantitative method is given in Section 2.4.

Table 1.2 Calcium carbonate contents and the reaction of soil with 10% hydrochloric acid

Field description	$CaCO_3$ %	Audible effects	Visible effects
Non−calcareous, < 0.5%	0.1	None	None
Very slightly calcareous, 0.5–1%	0.5	Faintly increasing to slightly audible	None
Slightly calcareous, 1–5%	1.0	Slightly increasing to moderately audible	Slight effervescence confined to individual particles, just visible
	2.0	Moderately to distinctly audible; heard away from the ear	Slightly more general effervescence visible on close inspection
Calcareous, 5–10%	5.0	Easily audible	Moderate effervescence; obvious bubbles up to 3 mm diameter
Very calcareous	10.0	Easily audible	General strong effervescence; ubiquitous bubbles up to 7 mm diameter; easily seen

From Hodgson (1974)

Reagent

Hydrochloric acid, 10 per cent HCl m/m. To 770 ml of water add 295 ml of concentrated HCl (36% m/m, specific gravity 1.18 g ml⁻¹) and mix. Use from a wash bottle.

Method

On to a piece of soil apply a few drops of 10 per cent HCl. Observe the effects. Hold the soil close to your ear and note the audible effects. Table 1.2 shows their meaning in terms of $CaCO_3$ content. The method is less reliable for dolomite.

Section 1.3 Sampling from a field or a plot

Frequently the aim of sampling a field or a plot is to obtain a 'representative' value for a soil property. Occasionally it may be important to obtain information about the nature of the spatial variability of that property. Both purposes can be achieved by taking a number of samples over the area. In the first case the samples are bulked and subsampled to give a representative sample, and in the second case the samples are analysed separately, in a similar manner to the characterization of a horizon in a soil pit (Section 1.1).

Because of the need to take many samples, it is not normally feasible to dig pits at each location, and a soil auger is therefore used.

Sampling using augers

Augers vary in size and design. Those commonly used are the screw auger, the Jarret (bucket) auger and the root sampling auger (Section 3.1) shown in Plate 1.3. These augers are rotated and pressed into the soil to take samples from depth increments of between 15 and 20 cm. The samples are inevitably 'disturbed' to varying degrees and so the observations which can be made on the samples will be restricted: colour, texture, mottles, stones, roots and horizon depth can be recorded but soil structure cannot. Special coring equipment is required to obtain 'undisturbed' samples (Sections 4.2, 5.5, 11.1 and 12.1).

To avoid contamination between successive vertical samples care is needed in placing the auger back into the hole and in extracting it and the auger should be cleaned between samplings.

Obtaining a representative sample

To obtain a representative sample from an area, a

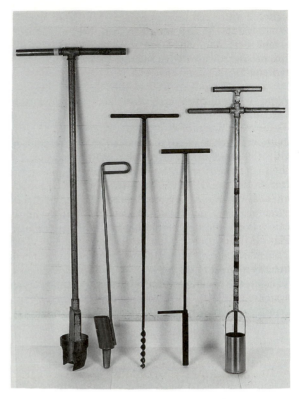

Plate 1.3 Augers and corers. Left to right: a Jarret auger, a grassland pot corer, a screw auger, a cheese-type corer and a root sampling auger.

widely used sampling strategy for agricultural purposes in the UK is to walk along a 'W' shaped path taking at least 25 samples which are bulked, as shown in Fig. 1.4 for a field. A smaller number of samples may be adequate. Sampling of arable fields can be by auger normally to 15 cm depth (screw or bucket depending on the amount of soil sample needed). Grassland should be

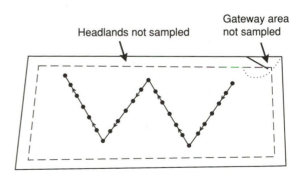

Figure 1.4 A soil sampling strategy for a field area.

sampled using a cheese-type corer or tubular corer to avoid losing the top few cm of soil (Plate 1.3), and samples are normally taken from the top 7.5 cm. In Britain, ADAS advise that samples should not generally represent more than about 4 ha.

Variability within a field or plot

There is no standard way of sampling to obtain information regarding variability over an area (Webster and Oliver, 1990) but a square grid is often used. Fig. 1.5 shows the organic carbon contents of 100 samples taken from 5 to 10 cm depth across a regular 30 × 30 m grid on (a) a long-established woodland with mature oak and sycamore trees and (b) an adjacent field under arable cultivation. The soils of both areas are developed on Plateau Gravel parent material in the Thames Valley. The values given are the means of triplicate measurements on each sample.

The data show the contrasting influence of soil management:

- The mean organic carbon (C) content in the arable area is less than under the woodland reflecting the effect of cultivation on the decomposition of soil organic matter (Sections 3.5 and 13.3).
- The variability of the data in the arable area is less than under the woodland, reflecting both the mixing of soil by cultivation and the more uniform inputs of plant material. The woodland plot has four mature trees and the extent of their canopies is shown by the dotted lines which enclose areas with larger organic C contents.

The variability of each set of data has been expressed through the standard deviation and the 95 per cent confidence limits (Section 3.8). In addition the *coefficient of variation* has been calculated. This is the standard deviation expressed as a percentage of the mean ($100\,s/\bar{x}$) and so allows variability to be compared for sets of data with very different means.

The two sets of data illustrate the complex patterns of variability which are often found in the field, and the need to take them into account when sampling for analysis. In the woodland site it may be that some form of structured sampling would be appropriate to investigate separately the soil properties below and outside the canopies.

4.73	4.82	3.81	3.95	3.26	3.14	3.92	4.74	4.63	4.69
4.72	4.69	4.94	3.71	3.43	3.30	4.06	4.92	4.82	4.91
4.32	4.64	4.89	3.26	3.25	3.25	4.27	4.54	4.76	4.97
4.91	4.52	4.79	3.24	3.43	4.14	4.01	4.66	4.94	4.83
4.71	4.89	4.40	3.01	3.71	2.95	3.95	4.21	4.54	4.74
3.21	3.43	3.11	3.15	3.87	3.14	3.71	3.14	3.83	4.03
3.14	3.11	3.17	3.96	4.26	4.14	3.23	3.11	3.25	3.13
3.08	3.42	3.93	4.76	4.43	4.52	3.14	3.06	3.81	3.91
3.34	3.86	4.41	4.53	4.76	4.63	3.70	3.68	4.96	4.94
3.49	3.94	4.89	4.93	4.94	4.86	3.92	3.93	4.74	4.65

(a)

1.84	1.65	1.59	1.77	1.75	1.63	1.96	1.93	1.52	2.06
1.63	1.91	1.48	2.06	2.03	1.94	1.84	1.65	1.63	1.67
1.62	1.60	1.53	1.77	1.69	1.66	1.89	1.79	1.59	1.96
1.66	1.55	1.63	2.06	1.66	1.85	1.57	2.09	1.73	2.11
2.19	1.74	1.92	1.90	1.66	1.75	2.13	1.83	2.09	1.66
1.58	1.69	1.78	2.06	1.74	1.87	1.76	1.64	2.00	1.97
1.56	1.49	1.98	1.68	1.86	1.77	1.78	1.76	1.81	1.88
1.72	2.03	1.62	1.96	2.11	2.08	2.10	1.45	1.99	2.08
2.02	1.95	1.48	1.67	1.65	2.00	1.75	1.79	1.92	1.68
1.64	2.02	1.53	1.82	1.86	1.63	1.98	2.04	1.80	1.75

(b)

Figure 1.5 The distribution of organic carbon in soils. (a) Under woodland: mean, $\bar{x} = 4.06$; standard deviation, $s = 0.668$; 95 per cent confidence limits = ± 0.133; coefficient of variation = 16.5 per cent. (b) Under arable cropping: mean, $\bar{x} = 1.80$; standard deviation, $s = 0.185$; 95 per cent confidence limits = ± 0.037; coefficient of variation = 10.3 per cent.

Section 1.4 Calculations

1. Using the data in Fig. 1.5(b), imagine that you were sampling the field to estimate its organic C content. You walk a 'W' path and take 25 separate samples and measure the organic C content of each. Draw on the figure your path and record the 25 values. Calculate the mean, the standard deviation,

the confidence limits and the coefficient of variability (Section 3.8). Repeat the calculation with 15 samples along the path. Compare the mean estimates and their reliability.

2. Using Fig. 1.5(a) calculate separately for the areas under the tree canopies and the remaining area the mean organic C contents, the standard deviations, the 95 per cent confidence limits and the coefficients of variability. Does the statistical analysis confirm that structured sampling would be appropriate for this area?

3. Using Table 1.1 compare the variability of the data on the bulk sample with that discussed in Section 3.8.

CHAPTER 2

Mineral Particles – the Soil Framework

An understanding of the materials present in soils begins at the simple level of what can be observed and felt. Observation can be aided using microscopes because soils are composed of material ranging from large stones to small clay particles. They feel different, particularly when moist with some feeling gritty and sandy, and others feeling smoother, sticky and plastic. These properties have been used in Section 1.2 as a basis for describing the texture of soils.

Soil is a mixture of inorganic and organic material with variable amounts of water and air. The inorganic material is in the form of mineral particles derived from the soil's *parent material*, which is the rock or sediment on which the soil has formed. This may be the material underlying the soil, or may have been brought from elsewhere by wind, water, ice or man. The weathering of rocks and the processes of soil formation alter rock minerals so that soil minerals are partly inherited from the parent material and partly developed in the soil (Dixon and Weed, 1989). Of particular significance is the development of altered or new minerals less than $2\,\mu m$ in size, known as the *clay-sized fraction*. This together with associated organic materials known as *soil organic matter*, controls to a large extent the physical and chemical properties of soils.

THE OBSERVATION OF SOIL PARTICLES

Without magnification the smallest particles that can be seen are about 0.5 mm in size. With a hand lens of $\times 10$ magnification, particles down to about 0.05 mm or $50\,\mu m$ can be seen. These and smaller particles can be observed if soil is placed on a microscope slide, water added and the soil broken up using a needle. When viewed using a microscope with $\times 100$ magnification particles down to about $5\,\mu m$ can be seen. Particles smaller than this give a cloudy appearance to the water. The larger particles have a crystalline appearance, and are usually the mineral quartz (silica) in soils of temperate regions. They are normally either angular or roughly rounded but occasionally more perfect crystals can be seen. Many of the particles appear to be 'dirty', having brown materials on their surfaces. This is an important characteristic of soils, where there is a close

association between the small mineral particles, the organic matter and the surfaces of larger mineral particles.

Note that the term 'mineral' is used in two ways, firstly as in geology where it applies to a specific crystalline inorganic particle, and secondly for the inorganic nutrients in soils which are taken up by plants, such as nitrate, calcium, etc.

The use of soil thin sections

The characteristics of mineral particles can be seen more effectively using thin slices of soil prepared and viewed with a microscope in a similar manner to thin sections of plant, animal or rock material. Soil thin sections, like soft plant and animal tissue have to be set (hardened) before they can be cut and mounted, and soil is impregnated with a hard setting resin for this purpose. The study of soil minerals using thin sections and a petrological microscope is described by Fitzpatrick (1980) and Bullock *et al.* (1985). Normally thin sections are about $30\,\mu m$ thick, but ultra-thin sections can be prepared (with great difficulty) which are about $0.1\,\mu m$ thick and allow soil material to be examined using transmission electron microscopy. Fracture surfaces can be examined using scanning electron microscopy. Smart and Tovey (1981) and Foster *et al.* (1983) include many excellent electron micrographs.

Parent materials

Colour Plate 1(a) shows a thin section of a dolerite rock composed of colourless prisms of feldspar, grey-green pyroxene and brown alteration products of olivine. Physical weathering causes disintegration into smaller particles, which can then be moved by gravity, ice, water and wind. Thus, soils may form in parent materials which are not derived directly from the underlying rock. An example is shown in Colour Plate 1(b) which is a *till* composed of a mixture of rock types moved by glaciation. The dolerite is both dense and resistant to weathering. In contrast some rocks are porous and easily weathered. Plate 2.1 shows a scanning electron micrograph of a chalk. In this case a

17

Plate 2.1 Scanning electron microscope photograph of an Upper Chalk fracture surface, Berkshire Downs, Southern England. Upper Chalk is a soft white porous limetone composed of almost pure calcite. Wheel-shaped microfossils known as coccoliths can be seen (about 5 μm diameter). The 1 to 2 μm particles are presumably pieces of broken coccolith. Photograph by K. Smith, Reading University. The bar represents 10 μm.

fractured surface is photographed, giving a three-dimensional impression of the material. It is easy to imagine physical breakdown releasing particles about 1 μm in size, and any larger pieces of chalk remaining in the soil will contain pores (spaces) of about 1 to 0.1 μm in size. The particles are the mineral calcite. The weathering of rocks and the development of soil minerals are discussed further in Section 2.1.

Soil materials

Soil thin sections are shown in Colour Plate 1(c) and (d). Several distinct features are apparent when compared to rock thin sections:

- The particle size in soils is generally much smaller than the crystal size in rocks. The minimum particle size visible at this magnification is about 2 μm.
- Brown material fills the spaces between the visible mineral particles. This is a complex mixture of clay *minerals* and *sesquioxides* (discussed below and in Section 2.2), and *humus* (Section 7.1). In Colour Plate 1(d) this material has been selectively accumulated in regions known as *clay skins* or *argillans* because of the apparent layering of the material on the surface of adjacent particles. Fig. 2.2 shows this in a diagrammatic form, and Plate 2.2 is an electron microscope photograph of an ultra-thin section which shows the organization of regions of clay.
- Pores are present, those visible occupying about 25 per cent of the field of view (in Colour Plate 1(c)). In addition many very small pores are present which cannot be seen at this magnification.

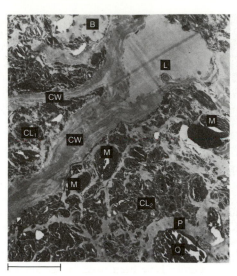

Plate 2.2 Transmission electron microscope (TEM) photograph of an ultra thin (50 nm) section of a clay soil adjacent to a wheat root. The section was cut tangentially to the root surface. L is a protruding epidermal cell which has been cut through. CW is cell wall material from collapsed cells, one of which has been filled with clay (CL_1). The clay occurs in regions (CL_2) and as silt-sized microaggregates (M) from which some clay has been lost during sectioning. A quartz crystal (Q) has been shattered by sectioning, and is surrounded by a clay skin and a pore (P). Pores are present throughout the soil, some of which may have been produced during preparation. Bacteria (B) can also be seen. The root diameter was about 200 μm. From Foster *et al.* (1983). The bar represents 10 μm.

These features are the result of the processes of rock weathering and soil formation which have operated over time to produce soil material. There are three main groups of processes:

1. *Physical processes.* Disintegration results from differential thermal expansion and contraction, the action of ice and the mechanical effects of particles rubbing against each other. These produce smaller particles, and form spaces.
2. *Chemical processes.* Resistant minerals remain relatively unchanged and others interact with natural waters (dilute carbonic acid) to produce altered or completely new minerals (Plate 2.3 and Section 2.1). The unchanged rock minerals are termed *inherited* or *primary minerals*, and the altered or new minerals are termed *pedogenic* or *secondary minerals* (pedogenesis = soil formation). Terminology causes problems here because primary and secondary are used by geologists for igneous and sedimentary rocks.
3. *Biological processes.* Plant roots grow into the soil, moving particles and leaving organic residues when

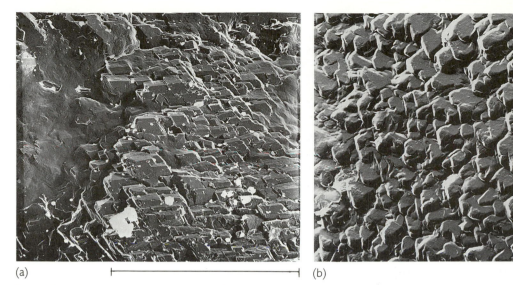

(a)　　　　　　　　　　　　　　　　　　　　(b)

Plate 2.3 Feldspar grains in granite – a TEM photograph using platinum shadowed carbon replicas: (a) from the core of a boulder and (b) from the weathered surface. The sharp edges and corners in (a) have been rounded by weathering in (b). Photograph by W. J. McHardy reproduced courtesy of the MLURI. From Smart and Tovey (1981). The bar represents 10 μm.

they die. Soil animals also create spaces and use plant materials as food leaving residues. Micro-organisms mostly live on plant and animal residues and their dead cells remain in the soil (Ch. 3). Metabolic products added to soil water by these organisms and their residues increase the rate of chemical reactions between minerals and water.

From a geologist's viewpoint rocks are degraded into soils, but from a soil scientist's viewpoint a material is built up which has new mineralogical, chemical, physical and biological characteristics which are vital for the life of organisms within the soil and for the growth of plants and production of crops.

PARTICLE-SIZE DISTRIBUTION AND SOIL TEXTURE

The sizes of the mineral particles profoundly affect the physical properties of soils: drainage, the ability to hold water for plant use and the ease with which they can be cultivated are examples. They also affect the chemical characteristics of soils because of the special properties of the surfaces of the very small particles. Thus we need a system for classifying particle size and for describing quantitatively the size distribution of these particles in soils.

There is no natural classification of particle size: the limits chosen are based on the contribution which particles of different sizes make to the physical and chemical properties of the soil. Table 2.1 gives the most commonly used system. Note that the names apply to

Table 2.1 Particle size classes

		Size (mm)
Stones		>2
Fine earth:	coarse sand	2–0.2 (2000–200 μm)
	fine sand	0.2–0.06 (200–60 μm)
	silt	0.06–0.002 (60–2 μm)
	clay	<0.002 (<2 μm)

The European system (including Britain) uses 60 μm as the limit between fine sand and silt: the United States Department of Agriculture uses 50 μm. Occasionally 20 μm is the limit. For many purposes coarse and fine sand are grouped together as sand.

particle-size classes and not to the types of minerals present in each class. We sometimes make this clear by talking about *sand-sized particles*, *clay-sized particles*, etc. However, certain minerals may predominate in the size groups: for example the sand-sized particles are commonly quartz and the clay-sized particles are often clay minerals. (Note that the term 'clay' has a third more popular meaning, i.e. any fine-grained plastic material.) The analysis of particles in terms of size classes (the *particle-size distribution*) initially separates the stones from the oven-dry soil. The remaining *fine earth* composed of particles less than 2 mm in diameter is divided in terms of a percentage by mass of each size fraction in the fine earth.

For convenience, soils are classified according to

texture, each textural class having a given range of particle-size distribution (Section 2.3). Thus there are small but generally unimportant variations in soil properties within each class, and significant differences between classes. The limits chosen do not indicate sudden changes in properties as composition changes.

Determination of particle-size distribution and soil texture

The field method for the determination of soil texture through the feel of a moist soil moulded between fingers and thumb is given in Section 1.2. By this method the texture of the soil is found in a qualitative way, and approximate values for particle-size distribution can be obtained from the triangular diagrams described in Section 2.3.

The standard analysis of particle-size distribution involves the dispersion of mineral particles after destroying the organic matter. The size classes are then separated using sieves and by sedimentation (Section 2.3) and the mass in each particle class is determined. The method also serves to separate the different size fractions for observation and analysis. The effects of sedimentation on the separation of soil particles in a field situation can be seen where soil has been puddled by cattle or machinery during a wet period around a drinking trough or in a gateway. After this disturbance the sand settles quickly followed by silt and then clay to form a layered skin when the soil subsequently dries out.

The significance of texture and particle-size distribution

Loams with a fairly even mixture of different sized particles generally have the best combination of physical and chemical properties in terms of cultivation and crop growth. When soils are dominated by a single particle-size class they are less suitable for crop production (Table 2.2) but become more easily managed and more fertile if the content of organic matter is increased.

The minerals present in the sand-, silt- and clay-sized fractions

Sand and silt

These fractions are dominated by resistant inherited minerals. Thus in temperate regions much quartz may be present unless the soil is formed on limestone in which case calcite and dolomite may dominate. In tropical regions iron and aluminium oxides and hydroxides dominate because more intense weathering may have dissolved away the quartz. In arid regions, inherited

Table 2.2 The characteristics of soils in relation to texture

Loams

Advantages:	easy drainage of excess water
	good retention of water for plant use
	easy cultivation over a wide range of water contents
	good supply of nutrients for plant use

Coarse sands

Advantages:	easy drainage
	easy cultivation
	warm up quickly in the spring
Problems:	poor ability to hold water for plant use – 'thirsty' soils
	poor supply of nutrients for plant use – 'hungry' soils
	poor ability to hold applied nutrients – leaching losses

Fine sands and silts

Advantages:	easy cultivation
Problems:	prone to erosion, compaction and capping†

Clays

Advantages:	good supply of nutrients for plant use
	good retention of nutrients against leaching
	good retention of water for plant use
Problems:	poor drainage of excess water – may become waterlogged
	high power requirement for cultivation
	easily puddled by animals or machinery when wet
	very hard when dry
	cultivation restricted to a narrow range of water contents
	warm up slowly in the spring

Stony soils

Problems:	drought because of the reduced volume of soil to hold water
	difficulty of cultivation and wear of machinery by abrasion
	increased leaching of nutrients

† Capping is the formation of a compact surface layer on the cultivated soil by rain. It can hinder germination and emergence of seedlings and cause water to run off the soil surface.

minerals dominate the whole soil because of the lack of water necessary for chemical and biological processes: the mineral particles are simply broken fragments of rock. Some pedogenic minerals such as gypsum and calcite are formed because although they are relatively soluble they crystallize on evaporation of solutions produced by the low intensity of chemical weathering.

The clay-sized fraction

The mineral components in this fraction are primarily clay minerals, sesquioxides and amorphous minerals in association with humus. Because of their small particle size and their electrical charge they have physical and chemical characteristics described as *colloidal*, and the clay-sized fraction is sometimes termed the *colloidal fraction*. These characteristics include the ability to absorb water with resultant swelling and the development of plasticity. They may also disperse in water as a result of the development of repulsive forces between the particles (Section 14.4).

This fraction has electrical charge which results in the particle surfaces holding ions, including important plant nutrients. In soils formed on limestone parent materials, the clay-sized fraction may include calcite as a major component.

The clay minerals These are crystalline hydroxy silicates containing aluminium, iron and magnesium with other metallic elements present in small amounts. They can be either inherited minerals if the soil is formed on a sedimentary material rich in clay, or pedogenic minerals when produced by weathering. They are generally plate-shaped, ranging in size down from about $2 \, \mu m$ (Plate 2.4). Clay minerals vary in crystalline structure, electrical charge, surface area and swelling characteristics (Section 2.2). Several types are permanently negatively charged and hold *exchangeable cations* including potassium, calcium and magnesium.

The identification and analysis of clay minerals require X-ray diffraction techniques and total elemental analysis which are beyond the scope of this book (Wilson, 1987). Soil clay minerals are not such perfect crystals as are those found in rocks, and are often present as intimate mixtures of two or more clay minerals with sesquioxides on their surfaces. Thus although these minerals can normally be identified, the quantities present can only be determined approximately ($\pm 5\%$ of the clay-sized fraction).

Allophane is an amorphous hydrous aluminium silicate, formed from volcanic ash and in other rapidly weathering systems, where it may be the dominating mineral. It has a very large surface area, carrying both positive and negative charge and is thus similar to the sesquioxides in some respects.

The sesquioxides These are also called iron and aluminium oxide clays, but this use of the term clay adds a fourth meaning to the word and for this reason *sesquioxide* or *hydrous oxide* is preferred. Various forms are produced by the weathering of silicates and other minerals, all being pedogenic minerals. They are mainly oxides or hydroxides of iron and aluminium including gibbsite, $Al(OH)_3$, goethite, $FeOOH$, and haematite, Fe_2O_3

with the poorly ordered ferrihydrite, $Fe_2O_3.nH_2O$, merging into a range of amorphous materials. They occur as discrete particles in the clay-sized fraction, and as very small particles (about 5 nm in diameter) and amorphous coatings on the surfaces of clay minerals (Plate 2.4). The iron oxides and hydroxides have red–brown colours, which combined with the black colour of humus gives the dark-brown colour of soil.

Sesquioxides act as binding agents between clay mineral particles, and as a source of electrical charge. This can be either positive or negative depending on soil pH, and is known as *pH-dependent charge*. Their surfaces often adsorb anions, including phosphate, nitrate and sulphate, which are plant nutrients. The clay mineral kaolinite also has this property but to a lesser extent. Sesquioxides and kaolinite dominate the clay-sized fractions of many soils of humid tropical regions, so that these soils have very different charge and anion adsorption characteristics from those of soils of temperate regions which are dominated by clay minerals having a permanent negative charge not dependent on pH.

The characteristics of the sesquioxides are discussed in Section 2.2. Their identification and analysis require special techniques (Wilson, 1987). Various methods are available to extract and measure the iron and aluminium in the sesquioxides by selective dissolution (Page, 1982).

Soil carbonates

The most common minerals are calcite, $CaCO_3$, magnesian calcites with up to 20 per cent $MgCO_3$ and dolomite, $CaMg(CO_3)_2$. Calcite occurs geologically both as soft white porous chalk and as hard limestone.

Carbonates may be found in all the particle-size classes of soil, being either inherited minerals in the stone, sand- and silt-sized fractions, or a mixture of inherited and pedogenic minerals in the clay-sized fraction. They may also occur in arid regions cementing large volumes of soil into hardened material termed *calcrete*. Carbonates maintain alkaline conditions in soils (Section 8.3) and influence the growth of plants through the direct effect of dissolved bicarbonate and the indirect effect of high pH on the solubility and availability of nutrients, particularly phosphorus, copper, zinc, iron and manganese.

The measurement of carbonate in soils is described in Section 2.4 and Section 2.5 deals with some chemical principles of this and other soil analyses.

FURTHER STUDIES

Ideas for projects are given in Section 2.6 and calculations in Section 2.7.

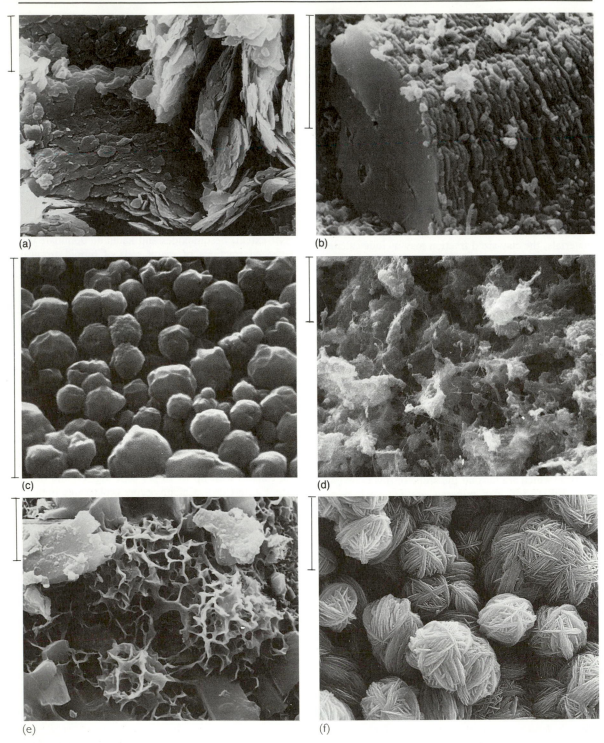

Plate 2.4 Scanning electron microscope photographs of clay minerals and sesquioxides. (a) Flakes of mica; (b) plates of kaolinite stacked together like pages of a book; (c) spheres of halloysite which is a 1:1 clay mineral similar to kaolinite; (d) allophane, teased out like threads of cotton wool; (e) grains of quartz, bound together by a lattice of iron oxide (probably ferrihydrite) in the iron pan of a podzol; (f) needles of goethite, wrapped together like skeins of wool. The lines represent 5 μm. From Molloy (1988).

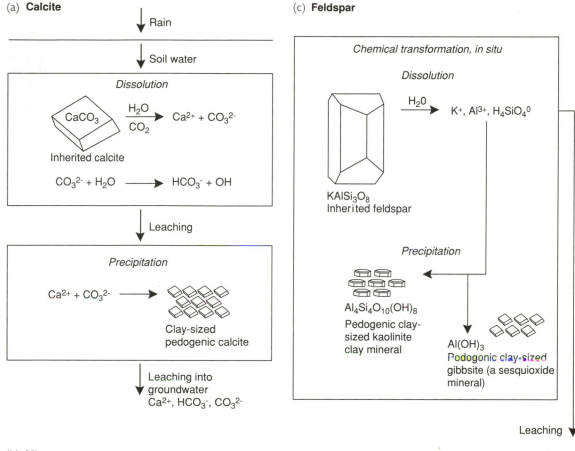

(a) **Calcite**

Rain

Soil water

Dissolution

$CaCO_3$

Inherited calcite

$$CaCO_3 \xrightarrow[CO_2]{H_2O} Ca^{2+} + CO_3^{2-}$$

$$CO_3^{2-} + H_2O \longrightarrow HCO_3^- + OH$$

Leaching

Precipitation

$$Ca^{2+} + CO_3^{2-} \longrightarrow$$

Clay-sized
pedogenic calcite

Leaching into
groundwater
Ca^{2+}, HCO_3^-, CO_3^{2-}

(c) **Feldspar**

Chemical transformation, in situ

Dissolution

$$\xrightarrow{H_2O} K^+, Al^{3+}, H_4SiO_4^0$$

$KAlSi_3O_8$
Inherited feldspar

Precipitation

$Al_4Si_4O_{10}(OH)_8$
Pedogenic clay-
sized kaolinite
clay mineral

$Al(OH)_3$
Pedogenic clay-sized
gibbsite (a sesquioxide
mineral)

Leaching

(b) **Mica**

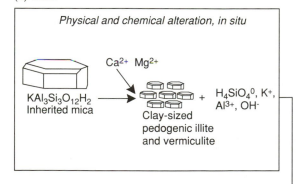

Physical and chemical alteration, in situ

Ca^{2+} Mg^{2+}

$KAl_3Si_3O_{12}H_2$
Inherited mica

$+$ $H_4SiO_4^0$, K^+,
Al^{3+}, OH^-

Clay-sized
pedogenic illite
and vermiculite

Leaching

Figure 2.1 The weathering of minerals. (a) Calcite (see also Section 8.3); (b) mica (Note: *in situ* means that the change is occurring 'in place' (without movement)); (c) feldspar.

Section 2.1 The weathering of rocks and the development of soil minerals

Minerals dissolve in the water moving through soils. The more soluble minerals, e.g. halite (common salt), $NaCl$, and gypsum, $CaSO_4.2H_2O$, are dissolved and leached away except in arid regions. With increasing leaching the less soluble minerals may also be dissolved and the solutes moved down the profile where they may be reprecipitated or leached away. Calcite behaves in this way (Fig. 2.1(a)). The removal of water by roots concentrates the Ca^{2+} and CO_3^{2-} ions causing precipitation. Similarly an increase in pH involves an increase in the CO_3^{2-} concentration so that the solution becomes supersaturated with respect to calcite. The precipitated material is likely to be in the form of clay-sized crystals.

Quartz is relatively resistant to weathering and persists in soils of temperate regions as the most common mineral in the sand and silt fractions. It is, however, slightly soluble in water and apparently more soluble in the presence of certain organic acids and sesquioxides. In the humid tropics the large amounts of water moving through the soil increase the rate of dissolution and may result in its complete loss.

Mica, a layered aluminium silicate mineral, undergoes alteration during weathering but retains its major characteristics so that there are similarities in crystal form between micas and the minerals which may be produced (Fig. 2.1(b) and Plate 9.1). For example, illite and vermiculite are moderately altered products and smectite more altered; all retain a layered form, and are usually clay-sized due to the combined effects of physical weathering and chemical alteration of the crystal. Because weathering occurs initially as a reaction at the surface of the particle, inherited minerals in soils may have a relatively unchanged core, but a changed surface. For example, the surface of quartz crystals may be altered to amorphous silica and the surface of calcite may become contaminated with calcium phosphate.

The ions and molecules dissolved from primary minerals may interact to form pedogenic minerals. A common weathering process occurring in tropical regions is shown in Fig. 2.1(c).

Some sedimentary parent materials (initially deposited by sedimentation in water) contain large amounts of clay minerals, and soils formed on these materials contain inherited clay minerals. These are normally altered to some extent by weathering and pedogenesis.

Section 2.2 The clay minerals and sesquioxides

THE CLAY MINERALS

The clay minerals are present primarily in the clay-sized fraction of soils. These particles separate when soil is dispersed and give the cloudiness which persists in soil suspensions. In thin sections of soil (Plates 2.3 and 2.4) clay occurs as areas of aggregated particles. The size and arrangement of these areas relative to other soil particles are illustrated in Fig. 2.2. *Domains* of clay are regions up to 10 μm in extent and *clay skins* may be up to 100 μm in extent. As an aid in visualizing the system, the analogy of a soil crumb being like a library is useful. The clay domains are like bookshelves, the clay particles like the books and the pages are the layered structure within the particles which is described below. The books can be taken off the bookshelf just as the particles can be dispersed in soil suspension.

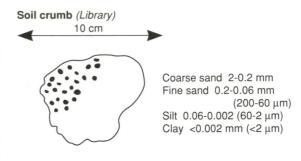

Soil crumb (Library)

10 cm

Coarse sand 2-0.2 mm
Fine sand 0.2-0.06 mm
(200-60 μm)
Silt 0.06-0.002 (60-2 μm)
Clay <0.002 mm (<2 μm)

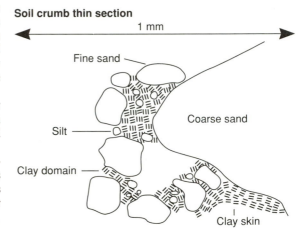

Soil crumb thin section

1 mm

Fine sand

Coarse sand

Silt

Clay domain

Clay skin

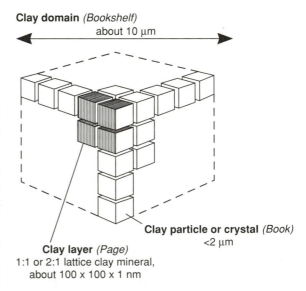

Clay domain (Bookshelf)
about 10 μm

Clay particle or crystal (Book)
<2 μm

Clay layer (Page)
1:1 or 2:1 lattice clay mineral,
about 100 x 100 x 1 nm

Figure 2.2 Clay minerals in soils – size and arrangement.

Clay particles or crystals

These are composed of successive planes of oxygen and hydroxyl ions bonded with silicon, aluminium, magnesium and other cations into tetrahedrally or octahedrally coordinated sheets extending in two dimensions:

- *The tetrahedral sheet* is composed of silicon bound to four oxygen atoms (Fig. 2.3(a)).
- *The octahedral sheet* is composed of aluminium or magnesium bound to six oxygen or hydroxyl ions (Fig. 2.3(b)).
- *The layers* are composed of associations of sheets giving three main types (Fig. 2.3(c)).
- *Clay crystals* are composed of associations of layers (Fig. 2.4). The *basal spacing*, which is the distance between similar faces of adjacent layers, can often characterize a mineral and indicate its properties.

The properties of the most important soil clays are given in Table 2.3.

When two layers come together such that a tetrahedral and octahedral sheet are adjacent as in 1:1 or 2:2 type minerals, a force resulting from hydrogen bonding between the oxygen associated with silicon in one layer and hydroxyl associated with the aluminium or magnesium in the adjacent layer ensures that the crystal is

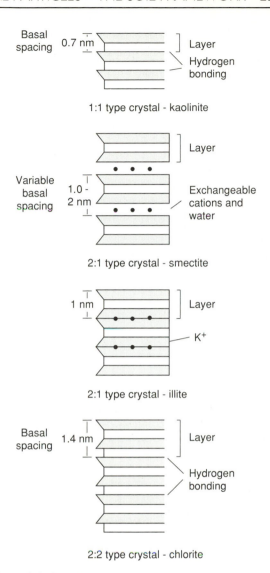

Figure 2.4 Crystal arrangement of clay minerals.

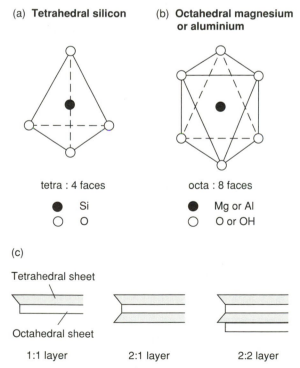

(a) **Tetrahedral silicon**
tetra : 4 faces
● Si
○ O

(b) **Octahedral magnesium or aluminium**
octa : 8 faces
● Mg or Al
○ O or OH

(c)
Tetrahedral sheet
Octahedral sheet
1:1 layer   2:1 layer   2:2 layer

Figure 2.3 Layer arrangement of clay minerals.

stable. When 2:1 layers come together the tetrahedral sheets in adjacent layers have no hydrogen atoms to form these bonds and the crystals are held together by electrostatic forces between the layers resulting from their electrical charge. In illite, as in mica, this is a strong force due to the presence of numerous potassium ions, but in smectite and vermiculite the layers can separate with entry of water between them and this increases the basal spacing. Usually the crystals remain as units in soils with only limited swelling. Strictly they are not crystals because of the variable basal spacing but for convenience the term is still used. They are often poorly ordered; 'crystals' of smectite, in particular, are

Table 2.3 The properties of the main groups of clay minerals

Clay	Layer structure	Layer thickness (nm)	Negative charge (cmol$_c$ kg^{-1})	Surface area (m^2 g^{-1})	Swelling properties of crystals
Kaolinite	1:1	0.7	Up to 10†	10	None
Allophane	1:1‡	0.7	20–50	700–900	None
Smectite	2:1	1.0	100	800	Extensive
Illite	2:1	1.0	25(250)§	20	None or very little
Vermiculite	2:1	1.4	150	400	Limited
Chlorite	2:2	1.4	10	10	None

† The negative charge on pure kaolinite is small (Fig. 7.7) and impurities of 2:1 clays probably give these higher values. Kaolinite may also carry positive charge (Section 7.6).

‡ The layer forms the wall of a hollow spherical particle about 4 nm in diameter. It also carries between 5 and 30 cmol of positive charge kg^{-1}.

§ The layers are bound together by potassium ions which partially enter the tetrahedral surface. This forms a strong bond between most of the layers, with no water entry, and so reduces the effective charge from 250 to about 25, i.e. 225 cmol$_c$ kg^{-1} is balanced by K$^+$ which is not exchangeable but can be slowly released.

composed of many small plate-like particles not perfectly aligned which disperse readily to form the very small (<0.1 μm) clay-size material.

Surface area

The small size of clay particles results in a large surface area per unit mass known as *specific surface area*. This can be calculated: spherical particles 2 mm in diameter and density 2.65 g cm^{-3} have a surface area of 11.3 cm^2 g^{-1}, whereas 2 μm particles have an area of 1.13 m^2 g^{-1}. However, the layered structure of clays gives much larger areas than spheres, and the swelling 2:1 clays have areas up to 800 m^2 g^{-1}.

Electrical charge

One further property distinguishes the main types of clay minerals. 'Perfect' clay crystals form with a balance of electrical charge between the electropositive ions and electronegative ions in the structure just as a crystal of common salt, Na$^+$Cl$^-$, is electrically neutral. For example, the ratio of ions in talc, a 2:1 clay mineral, can be expressed by the formula $Mg_6^{2+}Si_8^{4+}O_{20}^{2-}(OH^-)_4$; from this we can calculate the change balance as $+12 +32 -40 -4 = 0$. This is a simplification of the real situation because the bonding between the ions is partly ionic and partly covalent but we can assume that a 'perfect' crystal is electrically neutral. Soil clays, however, are not 'perfect' crystals: impurities are incorporated into the mineral during its formation in the form of ions substituted for either magnesium or aluminium in the octahedral layer or for silicon in the tetrahedral layer. Thus, if one silicon ion is substituted by an aluminium ion in the talc crystal unit we have

$Mg_6^{2+}Si_7^{4+}Al_3^{3+}O_{20}^{2-}(OH^-)_4$ with a charge of $+12 +28 +3 -40 -4 = -1$. The clay then carries a negative electrical charge per formula unit equal to that carried on one electron. The charge is termed a *permanent negative charge* because it is an intrinsic property of the crystal structure. Soil clays normally have substitutions spread through both tetrahedral and octahedral sheets.

Clays also carry pH-dependent charges although, except for kaolinite and allophane, the amounts are small compared to permanent charge. These charges develop in a similar way to that described for sesquioxides below.

Exchangeable cations

Charges on clay minerals are always balanced by ions on the crystal surfaces. The most common of these ions are aluminium, calcium, magnesium, potassium and hydrogen and they are interchangeable, being held by electrostatic forces between the negative clay charge and the positive ion charge. These ions are known as *exchangeable cations*, the process as *cation exchange* and the total charge as the *cation exchange capacity* (CEC) of the clay. It is measured as the quantity of charge on the cations held on the clay, and the units are mol (normally cmol) of charge kg^{-1} of clay where a mol of charge is that carried by 1 mol of hydrogen ions ($= 6.02 \times 10^{23}$ ions or electron charges). This concept and the meaning of the unit are developed further in Section 7.2.

The electrostatic forces which hold 2:1 layers together result from the presence of exchangeable cations: adjacent layers are attracted to the cations. In turn, water enters between the layers because it is attracted to the cations. The electrostatic attraction and the entry of water pushing the layers apart act in opposition, and the balance between these two forces determines the basal spacing of the crystal and the extent of swelling (Section 14.4). This spacing can be measured by X-ray diffraction analysis and is used to identify clay minerals.

Some common uses of clay minerals

Talc is mined to produce talcum powder. It is an uncharged clay mineral and its small plate-shaped particles sliding over each other give it a smooth feel and, lacking charge and attraction for water, do not become sticky on contact with water.

Other clay minerals are also found as geological deposits. One of the smectite group, montmorillonite, is used as 'fuller's earth' because its large surface area attracts organic molecules so removing stains from clothes. It is also used as a drilling mud in mineral exploration and for making moulds in the iron and steel industry. Kaolinite is also mined and used as a coating on paper to give a fine glaze. It is widely used for making high quality porcelain and china because on firing it readily transforms to other silicates forming a crystal network. 'Expanded vermiculite' is produced by rapidly heating the mineral so that the interlayer water is converted to a large volume of steam. It is used as an insulation material, and as a rooting medium for the growth of seedlings. It has spaces in the expanded crystals which can hold water and air, it provides good drainage through the spaces between the crystals and can hold nutrient ions in exchangeable form.

THE SESQUIOXIDES

This group of minerals consists of gibbsite, goethite and haematite which are well crystallized, ferrihydrite which is poorly crystallized and amorphous iron and aluminium hydroxides. The crystalline minerals are built up from sheets of oxygen or hydroxyl in octahedral arrangement with either aluminium or iron. The sheets are held together by hydrogen bonding as for the 1:1 and 2:2 clay minerals, so there is no expansion of the crystal (Fig. 2.5(a)).

There is no substitution by ions of different charge in the crystal and therefore no permanent negative charge. However at the broken edge of the crystal or on the surfaces of amorphous particles, exposed −AlOH or −FeOH groups have an amphoteric character being able to accept or lose hydrogen ions depending on the pH of the solution with which they are in contact (Fig. 2.5(b)). Thus a pH-dependent or variable charge is developed. Soils dominated by minerals carrying this type of charge are known as *variable charge soils*.

The charges are always balanced by ions attracted to the surface. Thus at high pH cations are held and the charge is part of the CEC of the soil along with the permanent negative charge of the 2:1 clay minerals and the negative charge associated with organic material (Section 7.1). At low pH anions are held and the soil has an *anion exchange capacity* (AEC) which is generally less than about 1 cmol charge kg^{-1} using the same units as for CEC. Positive charge develops by a similar mechanism on the edges of kaolinite and allophane particles. Further details on the surface charge characteristics of sesquioxides are given in Section 7.5.

(a) **The gibbsite crystal**

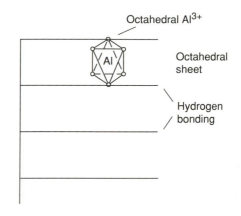

(b) **The development of charge**

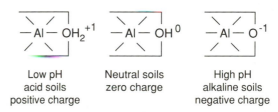

Low pH	Neutral soils	High pH
acid soils	zero charge	alkaline soils
positive charge		negative charge

Figure 2.5 Layer arrangement and charge of gibbsite. (a) The gibbsite crystal; (b) the development of charge.

Section 2.3 Soil texture and particle-size analysis

Particle-size distribution gives a broad indication of the physical and chemical properties of soils. However, particle shape and surface properties particularly of the clay-sized fraction significantly modify these properties. Thus absolute values of particle-size distribution are not important in this context. However, comparative values down a soil profile or between a profile and its parent material can be important evidence regarding the origin of, and the processes which have occurred in soils. A carefully standardized procedure for the analysis of particle-size distribution is needed for these comparisons to be meaningful.

TEXTURAL CLASSES

The classification of texture in terms of particle size-distribution is normally shown as a triangular diagram. Various classifications have been devised, the most commonly used being the USDA system (Fig. 2.6(a)). The UK classification is shown in Fig. 2.6(b). The three variables can be plotted in this way because they are related (sand + silt + clay = 100). The names used for the textural classes comprise a final noun indicating the main characteristic of the soil and an adjective and/ or a noun as a prefix indicating the way in which the main characteristic is modified.

Triangular diagrams can be used to determine texture if particle-size distribution is known, or to determine a range of particle-size distributions if texture is determined by the finger method (Section 1.2). However, there will not necessarily be a close correlation between texture determined by the two methods because of the influence of particle shape, size and constitution and organic matter on the finger texturing method. Soils with textural properties dominated by organic matter are termed *humic*.

Determination of approximate particle-size distribution if texture is known

Texture recorded in the literature or determined in the field places the soil in one area of the triangular diagram. For example, to determine the range of sand content in loams use the horizontal axis, and read from Fig. 2.6(a) using the sloping lines, following the direction indicated by the line joining the axis to the number. The ranges are 23–52 per cent sand, 6–27 per cent clay and 29–50 per cent silt.

Note that clay has a pronounced influence on texture: a soil with only 40 per cent clay-sized particles already has the extreme properties of a clay, whereas 87 per cent sand or 80 per cent silt is required to give the extreme properties of these classes.

Other terms are used to describe soil texture giving fewer broad textural classes:

- coarse = sands, medium = loams, fine = silts and clays, or
- light = sands, heavy = silts and clays.

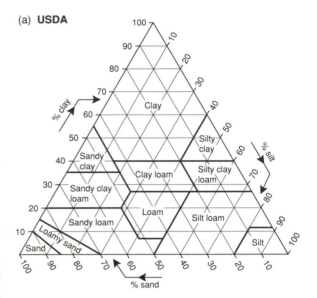

(a) **USDA**

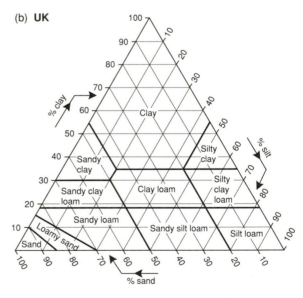

(b) **UK**

Figure 2.6 Soil textural classes and their particle-size distribution. (a) The USDA system; (b) The UK system. The size limit dividing fine sand and silt for the USDA system is 50 μm. The effect of using this limit instead of 60 μm is negligible for most purposes.

The names 'light' and 'heavy' originated from the difference in the work needed to cultivate them. For example, the Norfolk clays in Britain used to be known as four-horse land because four horses were needed to pull a single furrow plough. They do seem to have different weights because the clays hold more water under similar climatic conditions.

PARTICLE-SIZE ANALYSIS: THE STANDARD METHOD

Organic matter is destroyed by hydrogen peroxide and the remaining mineral soil dispersed by shaking in the presence of sodium hexametaphosphate; the soil is analysed by sieving and then by sedimentation using a pipette sampling technique (Avery and Bascomb, 1974). There are other methods, the most important differences being in the dispersion of the particles and the use of a hydrometer instead of pipette sampling (Page, 1982). Particularly for soils rich in sesquioxides, there are practical problems in dispersing the soil and it is difficult to know the extent to which particles should be broken into smaller ones in the laboratory, even though they may never break up under field conditions. Thus it should be noted that the answers obtained depend on the methods used to disperse the soil.

Reagents and equipment

Hydrogen peroxide, approximately 6 g of H_2O_2 per 100 ml of solution. Transfer approximately 200 ml of '100 vol. H_2O_2' (30 g H_2O_2 per ml) into a 1 litre measuring cylinder and make up to the mark. CARE: H_2O_2 is corrosive, oxidizing skin as well as soil organic matter. Use gloves and safety glasses.
Octan-2-ol.
Sodium hexametaphosphate.
Anhydrous sodium carbonate.
Sieves: 212 and 63 μm apertures mounted in 100 mm diameter frames complete with cover and receiver.

Soil pretreatment

Removal of organic matter

The organic matter is destroyed because (a) it binds mineral particles together, particularly clay-sized particles and hinders dispersion, and (b) we wish to analyse the size distribution of the mineral particles.

Use a <2 mm sieved air-dry soil (fine earth). Field sampling and sample preparation are described in Section 1.3. Place 10 g (±0.01) of soil in a 500 ml beaker. Add 10 ml of H_2O_2. Add also a few drops of octan-2-ol if available to reduce frothing. Allow to stand until frothing ceases and add another 10 ml of H_2O_2. When there appears to be no further reaction with fresh H_2O_2, gently heat over a low bunsen flame, stirring to break the froth. Add further H_2O_2 with gentle heating using in total about 100 ml of peroxide solution. Finally raise the temperature to boiling to complete the destruction of the organic matter. Allow to cool.

Dispersion of the soil

Prepare a dispersing agent by dissolving 50 g of sodium hexametaphosphate and 7 g of anhydrous sodium carbonate in water and making up to 1 litre. This contains 0.57 g of total reagent per 10 ml of solution, but this mass needs to be checked for the drying conditions used below because the reagents are present along with the particles in the determinations. To do this pipette 10 ml of solution into a dry weighed 100 ml beaker, and evaporate to dryness in an oven at 105 °C. Cool the beaker in a desiccator and reweigh.

Transfer the peroxide-treated soil quantitatively to a bottle of about 500 ml capacity or larger. This involves pouring the suspension through a funnel into the bottle, washing down the sides of the beaker with water from a wash bottle and if necessary dislodging material from the sides of the beaker with a rod, and transferring these washings into the bottle. The bottle should be about half full for effective shaking. If necessary make up to about 200 ml and then shake overnight on a mechanical shaker. If a shaker is not available shake vigorously by hand, and repeat at intervals over a few hours: dispersion may, however, not be complete.

The reagents disperse the soil by (a) adding sodium ions to increase the exchangeable sodium and cause a repulsion between particles, (b) adding hexametaphosphate which is adsorbed on to positive electrical charges on the sesquioxides and kaolinite clay, so preventing attraction to negatively charged clay, and (c) adding carbonate to raise the pH of the solution and so remove positive charge. Ch. 7 explains these properties.

Analysis by sedimentation and sieving

Sieves are available with apertures down to about 60 μm. Thus coarse sand can be separated on a 200 μm sieve, and fine sand on a 60 μm sieve. Sedimentation separates the silt and clay. If a 20 μm limit is used for fine sand, sieving is not practicable for this fraction and sedimentation is used. If no sieves are available sedimentation alone can be used.

Sedimentation principles

A spherical particle settles in a liquid with a sedimentation velocity which depends on the size of the particle, its density (Section 4.1) and the properties of the liquid. The settling velocity is derived from Stokes's Law for streamlined flow:

$$v = 2 g r^2 (\rho_s - \rho_l)/9\eta$$

where v is the sedimentation velocity ($m\,s^{-1}$), r the particle radius (m), g the gravitational force per unit mass ($9.81\,N\,kg^{-1}$), ρ_s the density of the particle (2600 kg m^{-3} is the average density for soil particles), ρ_l is the density of the liquid (998 kg m^{-3} at 20 °C for water) and η the viscosity of the liquid (1.002 × 10^{-3} N s m^{-2} at 20 °C for water).

Sand and silt particles are normally approximately spherical but clay particles are often plate shaped. The latter settle in water along an erratic path rather like a leaf falling in air and the term *equivalent spherical diameter* or *effective diameter* is used in this context. Thus a clay-sized particle is regarded as 2 μm in size if it settles at the same velocity as a 2 μm diameter sphere of the same density. If it is plate shaped it will probably have two dimensions larger than 2 μm and a third less than 2 μm.

Coarse sand particles settle so rapidly that the streamlined flow assumed in Stokes's Law no longer applies. However, even for these particles the sedimentation method is useful for some purposes.

Stokes's Law applies to the settling of spheres in stationary liquid. Temperature differences in the liquid will cause convection currents which have a serious effect on settling velocities of small particles. Hence experiments should be conducted in a constant temperature room. If this is not available choose a position away from direct sunlight or radiators, and in a room with minimum temperature fluctuations.

Tables 2.4 and 2.5 give sedimentation data. Both liquid density and viscosity vary with temperature. The method which follows involves the settling of particles through 10 cm. The settling times in Table 2.5 are used

Table 2.4 Sedimentation of spherical soil particles (ρ_s = 2.6 × 10^3 kg m^{-3}) in water at 20 °C

Particle diameter (μm)	Sedimentation velocity (m s^{-1})	Time to settle 10 cm (s)
200	3.47 × 10^{-2}	2.88
60	3.12 × 10^{-3}	32.02
20	3.47 × 10^{-4}	288 (4 min 48 s)
2	3.47 × 10^{-6}	28 800 (8 h)

Table 2.5 The effect of temperature on settling times

Temperature (°C)	Settling times through 10 cm				
	60 μm diam. (s)	20 μm diam. (min	s)	2 μm diam. (h	min)
16	35.4	5	19	8	51
18	33.7	5	3	8	24
20	32.0	4	48	8	0
22	30.4	4	34	7	37
24	29.1	4	22	7	16

in the following way. If sand, silt and clay are initially uniformly distributed through a column of liquid, and allowed to settle at 20 °C, then after 32 s sand particles larger than 60 μm initially at the liquid surface have settled beyond 10 cm, and 60 μm particles initially at the surface have reached 10 cm. The concentration of clay and silt at 10 cm depth has not changed and can be sampled with a pipette. The sample comes from a volume around the pipette tip, not only from the 10 cm depth, and so slight errors are involved.

The method – without sieves

After dispersing the soil, quantitatively transfer the contents of the shaking bottle to a 500 ml measuring cylinder and make up to 500 ml with water. Obtain a disc of cork or polystyrene and make a hole through the centre slightly smaller than the stem of a 20 ml pipette. Slide the pipette through the disc. Rest the disc on the cylinder and lower the pipette until the tip just touches the liquid surface. Measure the length from the tip to the disc (*d* centimetres). Now push the pipette through the disc until the length from the tip to the disc is d + 10 cm. The pipette now placed in the cylinder will pass 10 cm into the liquid (Fig. 2.7(a)).

Sampling the silt and clay Remove the pipette from the cylinder, and stir the suspension using a plunger (a rod with a perforated disc on the end) or by putting your hand over the end of the cylinder and shaking end over end. Thorough mixing is important (30 s). Allow to settle for 32 s (if at 20 °C) and draw off 20 ml of suspension by slowly introducing the pipette into the cylinder 15 s before the end of the period and filling the pipette so that the process takes about 5 s on either side of the required time (Note 1). The pipette-filling bulb should be checked before use and be ready for filling before the pipette is placed in the cylinder. Make sure it has adequate capacity to fill the pipette and do some trial runs on a cylinder filled with water before you attempt the actual experiment. Special pipettes are available (ELE International) with a tap which can be linked to a water vacuum pump for easier control. Transfer the 20 ml of suspension to a weighed dish or beaker (weigh to ±1 mg), dry at 105 °C, cool in a desiccator and reweigh. This gives the mass of silt + clay + a small residue of the dispersing agent.

Sampling the clay Start the sedimentation again after stirring and sample the clay at 10 cm depth 8 h later. Dry the 20 ml sample of suspension as above to give the mass of clay and a small residue of dispersing agent. Note that for the small mass of clay, errors can be large due to water absorption from the atmosphere after removal from the desiccator.

(a) **Sedimentation cylinder and sampling pipette**

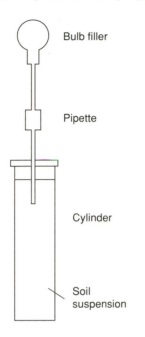

(b) **A nest of sieves**

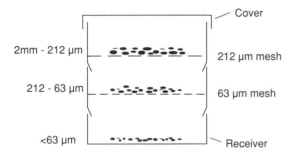

Figure 2.7 Sampling equipment for particle-size analysis by sedimentation and sieving. (a) Sedimentation cylinder and sampling pipette; (b) a nest of sieves.

Separating the sand After 8 h all of the sand (and some of the silt and clay) will be at the bottom of the cylinder. Gently pour away most of the supernatant liquid, and quantitatively transfer the sediment to a beaker which is more than 10 cm high. Mark the beaker using a felt-tip pen 10 cm above the base. Add water to the 10 cm mark, stir well, allow to settle for 32 s and carefully decant and reject the supernatant liquid retaining the sediment in the beaker. Repeat the stirring, settling and decanting until the supernatant is clear. All the silt and

clay has now been washed out of the sand. Transfer the sand to a weighed dish or beaker, dry at 105 °C, cool in a desiccator and reweigh (Note 2).

Determination of the air-dry water content of the soil Since the mass of sand, silt and clay is determined after drying at 105 °C, the initial mass of soil in this condition is also required. Weigh between 10 and 20 g (±0.01) of air-dry soil into a weighed dish or beaker and dry at 105 °C. Cool in a desiccator and reweigh. Calculate the mass of oven-dry soil in the 10 g sample of air-dry soil used in the analysis.

Calculations

The sand fraction in the soil sample has been determined directly:

percentage sand (*m/m*) = mass of sand × 100/mass of oven-dry soil

The silt and clay, and the clay fractions have been determined in 20 ml of suspension. Both these samples have some dispersing agent present (approximately 0.03 g but you have the weighings which give you the correct value). Subtract this mass to obtain the silt + clay and clay fractions in 20 ml of suspension. Determine the mass of silt by difference.

The total mass of silt in the soil sample = mass of silt in 20 ml × 500/20, and

percentage silt = total silt × 100/mass of oven-dry soil

The percentage clay can be similarly calculated.

The percentage sand + silt + clay should be less than 100 because of the mass of organic matter present in the oven-dry soil but destroyed by the peroxide pretreatment. For soils with less than 3 per cent organic matter, analytical errors may be larger than the error due to the organic matter. For more organic soils the shortfall will be more serious. The data for each should be re-expressed as a percentage of the mineral fraction, i.e. clay as a percentage of the sand + silt + clay. This can be done using the actual masses in the 10 g soil sample, or directly from the percentage values you have already calculated. For example,

percentage clay in the mineral fraction = percentage clay in whole soil × 100/percentage (sand + silt + clay) in whole soil.

Note 1 Errors are involved because of the short settling time relative to the time to draw the suspension into the pipette. For this reason the standard method involves the use of sieves. If the 20 μm limit between fine sand

and silt applies, the longer settling time of 4 min 48 s allows greater accuracy in the sedimentation method.

Note 2 The short settling time and disturbance of the sediment when decanting introduce errors. The coarse sand settles through 10 cm in about 3 s and so a very crude separation of coarse and fine sand can be made following the method described for sand.

The standard method – using sieves and sedimentation

The sand is separated from the silt and clay using a sieve, and can be subsequently fractionated by further sieving. The silt + clay and clay are determined by sedimentation.

Sieving After dispersing the soil place a funnel in the top of the 500 ml measuring cylinder, and place in the funnel a 63 μm sieve. Quantitatively transfer the contents of the shaking bottle through the sieve. Wash the residue on the sieve with about 200 ml of water. Do not make up to the mark.

Transfer the residue (sand) into a 250 ml beaker, dry at 105 °C and weigh. Separate into coarse and fine sand (or into other size fractions) by dry brushing the residue into a 212 μm sieve resting in a 63 μm sieve which is in its receiving cup. Cover and shake on a sieve-shaking machine for 15 min (Fig. 2.7(b)). Transfer the contents of each sieve into a dried and weighed beaker, dry at 105 °C and reweigh.

Any residue in the receiving cup after shaking is <63 μm and so should be washed back into the 500 ml measuring cylinder before making up to 500 ml (Note 3).

Sedimentation The suspension contains the silt + clay. After stirring the suspension immediately take a 20 ml sample from 15 to 20 cm depth. Transfer to a weighed dish or beaker, dry at 105 °C, cool in a desiccator and reweigh. This gives the mass of silt + clay in the 20 ml sample.

Sample the clay after 8 h settling following the method already described.

Note 3 Sieving is not practicable for separating fine sand if a 20 μm limit is being used. Pass the suspension through a 212 μm sieve to remove the coarse sand. Sample silt + clay and clay by pipette after sedimentation. Separate the fine sand by repeated sedimentation through 10 cm for 4 min 48 s and decanting as described under 'Separating the sand' in 'Method – without sieves'.

CUMULATIVE CURVES AND THE PRESENTATION OF DATA

Although particle-size analysis gives a percentage of sand (various fractions), silt and clay and can be used simply to allocate a textural class to the soil, information is often presented as a cumulative curve. This is particularly useful when more detailed analysis has been carried out using the hydrometer method or high-speed centrifugation, which provide information about particle-size distribution within the clay-sized fraction.

Cumulative curves are shown in Fig. 2.8(a). Because of the wide range of particle sizes a logarithmic axis is used on the abscissa. This means that there is no zero diameter on the axis. The clay percentage is plotted at 2 μm, clay + silt at 60 μm, clay + silt + fine sand at 200 μm, rising to 100 per cent at 2000 μm (= 2 mm), giving a distribution curve for the <2 mm fraction. The percentage in the silt and clay fractions can be found by difference as shown in the figure.

Cumulative curves show an important difference between soil material and parent material. Physically weathered rock material tends to have a narrow range of particle size and so the cumulative curve is steep. This is further emphasized if the material has been sorted by wind or water. In soils, weathering and the production of pedogenic mineral particles tends to give a wider range of particle size.

The use of logarithmic axes

Graph paper is available with one logarithmic axis and one linear axis known as semi-logarithmic graph paper. If you have data from an experiment which is to be plotted as a cumulative curve, arrange the paper as shown in Fig. 2.8(b). Note that the space between 1 and 2 is greater than between 2 and 3, and so on up to 10.

If this special graph paper is not available ordinary (linear axis) paper can be used. Mark the axis as shown with the numbers 1, 10, 100, etc., and also with the logs of the numbers (0, 1, 2, etc.). To position a number on the axis, e.g. 2, calculate its log = 0.30 and mark it on the axis. To read a number from the graph find the log of the number from the axis, e.g. log x = 1.78, and with a calculator find x = 60.

Section 2.4 Determination of the carbonate content of soils

The analysis involves the dissolution of carbonates in an excess of standard acid followed by back titration of the remaining acid. It determines calcium and magnesium carbonates together, but is often expressed as an equiv-

(a)

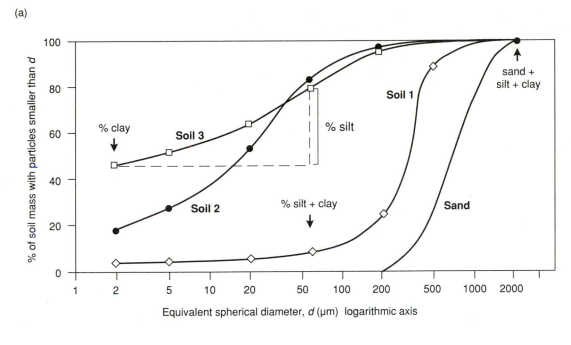

(b)

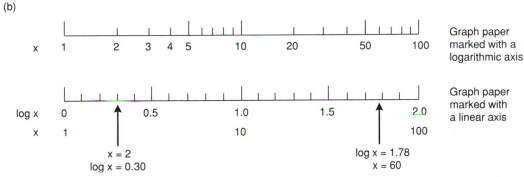

Figure 2.8 (a) Cumulative curves displaying particle-size distribution. Soil 1 is a sand-textured soil, soil 2 a silt loam, soil 3 a clay. The sand (particles between 2 and 0.2 mm) has been sorted by water; (b) linear and logarithmic axes.

alent amount of calcium carbonate ($CaCO_3$), i.e. the amount of pure $CaCO_3$ that would have reacted with the acid used in the analysis.

The method can be tested using $CaCO_3$ reagent (Note 1) and the determination can be carried out either on whole soil or on the particle-size fractions separated as in Section 2.3.

A summary of the principles involved in analytical work based on titrations is given in Section 2.5.

Principles

Soil carbonates are reacted with standard hydrochloric acid (HCl) and the excess acid back titrated with standard sodium hydroxide (NaOH).

$$CaCO_3 + 2HCl = CaCl_2 + H_2O + CO_2 \quad [2.1]$$

The molar mass of $CaCO_3$ is $100.1 \, \text{g mol}^{-1}$, and so 100.1 g of $CaCO_3$ reacts with 2 l of 1 M HCl (or 1 l of 2 M HCl). We have to choose a mass of soil and a volume and concentration of acid, so that all the carbonate reacts and leaves a suitable amount of acid for back titration:

$$HCl + NaOH = NaCl + H_2O \quad [2.2]$$

In this reaction 1 l of 1 M HCl reacts with 1 l of 1 M

NaOH, with phenolphthalein indicator giving an end point which is easy to observe.

Note that the acid reacts with the $CaCO_3$ and also reduces the pH of the soil from about 8 (the pH of calcareous soils) to a low value. Some acid is used in this second step, the amount being about $0.5 \, mmol \, H^+ \, g^{-1}$ soil per pH unit change (Section 8.4). The amount of acid used is small compared to the amount reacting with $CaCO_3$ unless the percentage of $CaCO_3$ is small, in which case significant errors may be involved. For these soils alternative methods measuring the volume of carbon dioxide (CO_2) released are used.

Suggested conditions for soil

If whole soil is being used, carry out a preliminary field test (Section 1.2) to estimate the amount of $CaCO_3$ present. In the analysis we require about 1 g of $CaCO_3$ to be present. A suitable mass of soil is therefore 100/approximate percentage $CaCO_3$. This can be reacted with 20 ml of 2 M HCl using about half the HCl in the reaction.

Reagents

Hydrochloric acid, 2 M.
Sodium hydroxide, 0.1 M.

Procedure

Grind about 25 g of <2 mm air-dry soil in a pestle and mortar. Weigh out 10 g (± 0.01) or the required amount of soil and transfer into a 250 ml conical flask. Pipette 20 ml of 2 M HCl into the flask and allow the reaction to proceed until effervescence ceases. Boil gently for 10 min to complete the reaction. Transfer the suspension quantitatively into a 100 ml volumetric flask through a funnel and filter paper (Whatman No. 1), washing with distilled water. Make up to the mark.

Pipette 10 ml of this solution into a 250 ml conical flask, add 50 ml of distilled water (approximately), add a few drops of phenolphthalein indicator and titrate with 0.1 M NaOH to give a pink colour. Repeat the titration on a second 10 ml sample of the solution.

Calculation

Method summary

Air-dry soil (?% $CaCO_3$)
$\downarrow$
10 g + 20 ml 2 M HCl $\rightarrow$ 100 ml
$\downarrow$
10 ml titrated against 0.1 M NaOH

Example The titration volume was 15.0 ml of 0.1 M NaOH. The number of mol of NaOH used in the titration is

$$0.1 \, mol \, l^{-1} \times 15.0/1000 \, l = 1.5 \times 10^{-3} \, mol$$

In the titration equal amounts (mol) of HCl and NaOH react together (Eq. 2.2). Therefore, the number of mol of HCl in 10 ml = 1.5×10^{-3} and the number of mol of HCl in the 100 ml flask is

$$1.5 \times 10^{-3} \times 100/10 = 0.015$$

This is the residue of HCl after reaction with the carbonate. The initial amount of HCl added to 10 g of soil was

$$2 \, mol \, l^{-1} \times 20/1000 \, l = 0.040 \, mol$$

Therefore, the acid which reacted with the carbonate is

$$0.040 - 0.015 = 0.025 \, mol \, HCl$$

Since 2 mol HCl react with 100.1 g $CaCO_3$ (Eq. 2.1), then the amount of $CaCO_3$ which reacted = $0.025 \times 100.1/2 = 1.25 \, g$.

Express the carbonate content of the soil as a percentage by mass. The percentage $CaCO_3$ in the air-dry soil = $1.25 \times 100/10 = 12.5$ per cent.

The result is normally given relative to the oven-dry mass of soil (Section 3.3). For example, if the air-dry soil contained 4 g H_2O per 100 g oven-dry soil, in 10 g of air-dry soil there would be 0.385 g of H_2O and 9.615 g of oven-dry soil. Therefore, the percentage $CaCO_3$ in the oven-dry soil is

$$1.25 \times 100/9.615 = 13.0 \text{ per cent}$$

Note 1 Using $CaCO_3$ reagent, weigh 1 g (± 1 mg) into the conical flask and follow the above procedure. The titration volume should be 20 ml.

Note 2 If your estimated $CaCO_3$ content using the field test was very inaccurate, you may find that you use either very little or all of the acid. If necessary repeat the experiment after adjusting the mass of soil so that between 20 and 80 per cent of the acid is used in the reaction.

Suggested conditions for particle-size fractions

In Section 2.3 the particle-size analysis used 10 g of soil. Carbonate determinations can be carried out on the sand fractions, and on the silt plus clay or clay fractions obtained from the 20 ml samples taken from the 500 ml of suspension. All these samples were oven dried. If they are to be used for carbonate determination they should have been dried in a beaker or conical flask. Again the amounts of $CaCO_3$ and acid have to be considered. Using your experience with measurements on the whole soil choose a suitable mass of sand and follow the procedure for the whole soil. For the silt and clay fractions, use 0.1 M HCl with each sample, adjusting the volume of acid so that between 20 and 80 per cent of the acid is used, and back titrate against 0.01 M NaOH. You can estimate the amount of acid needed by assuming that the percentage $CaCO_3$ in the whole soil is also the percentage in each size fraction, but as for the soil you may have to repeat the determination if this estimate is very inaccurate.

Example The clay content of the soil = 25 per cent. The $CaCO_3$ content of the soil (and assumed in the clay fraction) = 13.0 per cent. The mass of clay separated into 20 ml of suspension is

$$10 \text{ g} \times 0.25 \times 20/500 = 0.1 \text{ g}$$

The estimated mass of $CaCO_3$ in this clay = 0.1×0.13 = 0.013 g. From Eq. 2.1, 10.01 g $CaCO_3$ reacts with 2 l of 0.1 M HCl and so 0.013 g reacts with 2.6 ml. Thus if 10 ml of 0.1 M HCl is used, 7.4 ml remains and will be titrated against 7.4 ml of 0.1 M NaOH.

Section 2.5 Chemical equations, moles and titrations

Reacting masses

Chemical reaction can be described using a balanced chemical equation, i.e. for any element, the number of atoms in the reacting molecules (the reactants) is equal to the number in the products. For example, the reaction of calcium carbonate with hydrochloric acid to produce calcium chloride, water and carbon dioxide can be written as

$$CaCO_3 + 2HCl = CaCl_2 + H_2O + CO_2 \quad [2.3]$$

The equation is balanced and shows that one molecule of $CaCO_3$ reacts with two molecules of HCl. A knowledge of the mass of each molecule would give the ratio of reacting masses, thus giving a basis for analysis. However, the masses of atoms and molecules are very small and for analytical purposes a larger unit termed the *mole* (abbreviated *mol*) is used, which is defined as the amount of substance containing the same number of atoms or molecules as exactly 12 g of pure carbon-12. The number is $6.0220 \ldots \times 10^{23}$ and is known as the *Avogadro number*. The mass of 1 mol of an element or compound is called the *molar mass*, and has the units g mol^{-1}. Values are given in Appendix 2.

A chemical equation tells us the numbers of molecules of reactants and products involved in the reaction. Because 1 mol of any compound or element

contains the same number of molecules or atoms, then the equation also tells us the number of moles involved in the reaction. Thus, if we can write an equation for a reaction, and we know the molar masses we have a basis for calculating the relative masses of reactants and products: these relative masses in any one reaction are always the same.

Because in Eq. 2.3 one molecule of $CaCO_3$ reacts with two molecules of HCl, we also know that 1 mol of $CaCO_3$ reacts with 2 mol HCl. The molar mass of each compound can be obtained by summing the molar masses of the elements as follows:

Ca	C	O_3	+	2H	Cl	=
40.1 +	12 +	$(16)_3$	+	$2(1 +$	$35.5)$	=
	100.1				73	

Ca	Cl_2	+	H_2	O	+	C	O_2
40.1 +	$(35.5)_2$ +		$(1)_2 +$	16	+	12	$+ (16)_2$
	111.1			18			44

Thus 100.1 g of $CaCO_3$ reacts with 73 g of HCl, and they always react in the ratio 100.1:73.

Remember. For any element or compound

$$\text{mass} = \text{number of mol} \times \text{molar mass} \quad [2.4]$$

or, giving units,

$$\text{g} = \text{mol} \times \text{g mol}^{-1}$$

and a knowledge of any two of these values allows us to calculate the third.

Reagents in solutions

In analytical work solutions are often used as a convenient way of handling reagents. The concentration of the reagent in the solution can be expressed as g l^{-1}, or mol l^{-1}. The latter expresses the mass of the reagent dissolved in 1 litre in terms of a number of molar masses. Thus 36.5 g of HCl dissolved in 1 litre of solution would contain 1 mol of HCl per litre, is termed a 1 *molar* solution and is written as 1 mol l^{-1} or 1 M. The value 1 is the *molarity* of this solution.

Remember. For any solution

$$\text{molarity} = \text{number of mol/volume in litres} \quad [2.5]$$

In chemical calculations a knowledge of any two of these values allows the third to be calculated.

Returning to the reaction, 100.1 g of $CaCO_3$ will react with 2 l of 1 M HCl, since this volume and concentration contains 2 mol of HCl. If we wish to calculate the mass of $CaCO_3$ reacting with a known volume and concentration of acid, e.g. 5 ml of 0.1 M, then using Eq. 2.5

$$\text{number of mol} = 0.1 \text{ mol l}^{-1} \times 5/1000 \text{ l} = 5 \times 10^{-4} \text{ mol}$$

It is not necessary to calculate the mass of HCl in order to calculate the mass of $CaCO_3$ (this could be done using Eq. 2.4). Because 2 mol of HCl reacts with 100.1 g of $CaCO_3$ the mass of $CaCO_3$ reacting with this acid is

$$100.1 \times 5 \times 10^{-4}/2 = 0.025\,g$$

Titrations

In a titration two solutions are reacted together using an indicator to show when the reaction is complete. The solution with a known concentration is placed in a burette, and a known volume of a solution of unknown concentration is transferred by pipette into a conical flask. The indicator is added to the flask and solution added from the burette until the colour change gives the *end point* (Nuffield Advanced Chemistry II, Topic 12.4). The end point is the stage in the reaction when the reactants are balanced, i.e. the reaction is complete. For example, in the reaction

$$HCl + NaOH = NaCl + H_2O$$

1 mol of HCl reacts with 1 mol of NaOH. If 2.0 ml of 0.1 M NaOH was required to reach the end point then, using Eq. 2.5,

$$0.1\,mol\,l^{-1} \times 2/1000\,l = 2 \times 10^{-4}\,mol$$

of NaOH was used and must have reacted with 2×10^{-4} mol of HCl. This was present in the volume dispensed by the pipette. If this was 10 ml, then again using Eq. 2.5,

molarity $= 2 \times 10^{-4}\,mol/(10/1000)\,l = 2 \times 10^{-2}\,mol\,l^{-1}$

Using Eq. 2.4 the mass of HCl present is

$$2 \times 10^{-4}\,mol \times 36.5\,g\,mol^{-1} = 7.3 \times 10^{-3}\,g$$

The concentration of this solution is

$$7.3 \times 10^{-3}\,g \text{ in 10 ml}$$

or

$$7.3 \times 10^{-3} \times 1000/10 = 0.73\,g\,l^{-1}$$

The above calculation can be shortened as follows. We know that the number of mol of HCl must equal the number of mol of NaOH. From Eq. 2.5 we also know that the number of mol provided by a solution is equal to molarity $\times$ volume. Therefore

number of mol of HCl = number of mol of NaOH
$$x\,mol\,l^{-1} \times 10.0/1000\,l = 0.1\,mol\,l^{-1} \times 2.0/1000\,l$$

and so

$$x = 2 \times 10^{-2}\,M$$

In a similar way, assuming that the same experimental results apply in the reaction

$$2HCl + Ca(OH)_2 = CaCl_2 + H_2O$$

then the number of mol of HCl $= 2 \times$ the number of mol of $Ca(OH)_2$,

$$x\,mol\,l^{-1} \times 10.0/1000\,l = 2 \times 0.1\,mol\,l^{-1} \times 2.0/1000\,l$$

and so

$$x = 4 \times 10^{-2}\,M$$

Section 2.6 Projects

1. It is generally observed that across a river flood plain particles carried in flood water are distributed at right angles to the river according to size. Thus the sand is deposited in the levee bank, and the silt and clay further from the river. Take samples of surface and subsurface soil using an auger along a transect at right angles to a river and analyse the particle-size distribution to test the expected trend. Use the finger texture method initially as a field test.
2. Movement of soil particles on slopes tends to deposit the small particles at the foot of the slope, often causing deeper, heavier textured soils to develop. Take samples down a slope into a valley and analyse particle-size distribution to confirm this trend. Use the finger texture method initially as a field test.
3. In an area where soil is formed over limestone auger to determine depth of soil over rock along a transect from topslope to footslope. Sample the topsoil to determine $CaCO_3$ content. Is there evidence of soil movement on the slope?
4. It is generally found that particle size of soil material is much less uniform than in its parent material (Fig. 2.8). Compare the particle-size distribution of soil to that in the underlying material. Remember that soil may have formed on parent material moved by wind, water, ice or man.

Section 2.7 Calculations

1. A clay soil and a sandy soil have surface areas of 80 and 8 $m^2\,g^{-1}$ oven-dry soil respectively. If the air-dry soil has water adsorbed only on to the surfaces in a uniform layer 1 nm thick, calculate the water content of each soil. (*Ans.* 80 and 8 $mg\,g^{-1}$ oven-dry soil or 8 and 0.8%)
2. If the density of a smectite layer is 2.65 $g\,cm^{-3}$, calculate the surface area of the layers in 1 g of clay assuming that they are 1 nm thick. Neglect the area of the edges of the layers. (*Steps:* calculate the volume of clay per gramme $= 1/density$, imagine this volume as a pack of cards, calculate the number of cards in the pack, and their surface area). (*Ans.* 754 $m^2\,g^{-1}$)

3. A 10 g sample of soil was treated with 25.0 ml 2 M HCl. When the reaction was complete, the residual acid was titrated against standard NaOH solution. Calculation showed that 23.5 ml of the acid had been used in reaction with the soil. If the acid was reacting with $CaCO_3$ calculate the percentage of this by mass in the soil. (*Ans.* 23.5) If the acid was reacting with dolomite ($CaMg(CO_3)_2$) what would be the percentage of this in the soil? (*Ans.* 21.7)

4. Using the equation derived from Stokes's Law in Section 2.3 calculate the settling velocity and the time taken for iron oxide particles of 0.2, 0.02 and 0.002 mm diameter to settle through 10 cm assuming a particle density of 3 g cm^{-3}. (*Ans.* 4.36×10^{-2}, 4.36×10^{-4} and 4.36×10^{-6} m s^{-1}; 2.3 s, 3 min 49 s, 6 h 22 min)

5. A calcareous soil has a particle-size distribution of 40 per cent sand, 40 per cent silt and 20 per cent clay. If the $CaCO_3$ content of the soil was 5 per cent in the sand, 10 per cent in the silt and 20 per cent in the clay, calculate the particle-size distribution for the soil after removing the carbonate by reaction with acid, expressing your values as (a) percentages of the initial soil mass, and (b) percentages of the mass of soil after removal of the carbonate. (*Ans.* (a) 38, 36 and 16; (b) 42, 40 and 18)

6. From Fig. 2.8(a) determine the percentages of sand, silt and clay in the three soils and from Fig. 2.6 check that the correct textural classification has been given.

What errors would be introduced in the analysis if soil No. 3 contained an appreciable amount of iron oxide (density = 3000 kg m^{-3}) in the sand fraction? Similarly, what would be the effects be if soil No. 1 contained an appreciable amount of opaline silica (density = 2200 kg m^{-3}) in the silt fraction?

The clay fraction of soil No. 1 was further analysed giving the following: mass <0.5 μm = 35 per cent, mass <0.1 μm = 10 per cent. Replot the data from the figure extending the abscissa to display the extra results.

CHAPTER 3

Organic Materials - Alive and Dead

As soon as rock is exposed to the atmosphere organisms begin to colonize the surface. They are transported by wind and rain and find at the surface both an anchor in the rock particles and a source of nutrients as the rock minerals begin to dissolve in water. To be successful these organisms must be able to photosynthesize to produce carbon compounds for cell growth and multiplication. Once these cells are formed they can become a food source for other organisms which do not photosynthesize but rely on organic compounds as food. Thus an association of mineral particles with plant and animal life begins and on a microscopic scale a soil is formed. The organisms have to survive the extreme fluctuations of temperature and severe drying conditions at the rock surface. A striking example of the association of organisms with mineral particles is the growth of lichens on rock surfaces (Colour Plate 2). A lichen is a symbiotic association between a fungus and an alga which produces oxalic acid and causes biochemical weathering of the rock surface (Hawkesworth and Hill, 1984).

As the rock begins to be broken up by the effects of temperature fluctuations, frost and chemical weathering, roots can penetrate into the rock and the growth of higher plants begins. The environment beneath the surface is less harsh and is suitable for insects and other soil animals, with plant material providing their foods. Micro-organisms make use of plant and animal residues eventually forming a resistant residue which is a dark brown organic material called *humus*. This, together with living organisms, dead cells and mineral particles is now soil.

In temperate regions the earthworm is a major soil-forming organism. It ingests leaf and root material along with mineral particles, extracts components for its own use and leaves residues in the form of casts in which the organic material is in a suitable form for fungal and bacterial colonization. Much of the topsoil under grass in temperate regions has at some stage passed through the guts of earthworms. As a result, fresh litter is rapidly incorporated into the soil. Where earthworms are not present (acidic, dry or very wet conditions) litter accumulates and is only slowly broken down.

The term *soil organic matter* is used for all the organic material in soil including humus. In practice, the organic matter content is normally determined in the fine earth (Section 1.2), and living insects and root fragments will be included if they pass through a 2 mm sieve, as will all the micro-organisms. The mass of the living material is, however, small compared to the mass of dead material.

THE LIVING MATERIAL

Soil animals

Basic methods for extracting and identifying the large animals (macrofauna) are given in Jackson and Raw (1966) and Page (1982).

Plant roots

The mass of living roots in soil is known as the *root biomass*, and can be expressed relative to a mass or a volume of soil (Section 3.1). Examples of the root system of wheat are shown in Plate 3.1. Fig. 3.1 shows the amounts of root present in a winter wheat crop down to 1 m depth. The mass of the root system can be compared to the total mass of the plant shown in Fig. 11.4.

Nutrient uptake by plants is more closely related to root length than to mass since length governs the extent to which roots explore the soil volume. The measurement of root length is described in Section 3.1. Root extraction and measurement require much careful work. An annual crop is relatively simple to study because the distribution of roots is reasonably uniform over an area of the field compared to natural grassland or woodland. A ley (a grass or forage crop grown for 2—3 years) is also reasonably uniform, but is difficult to measure because of the amount of dead root material which has to be separated from living roots.

Micro-organisms

Basic methods for the culture and identification of soil micro-organisms are given in Jackson and Raw (1966) and Page (1982).

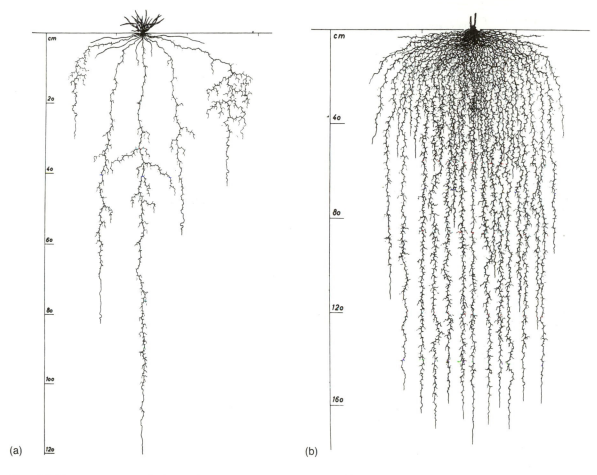

(a)   (b)

Plate 3.1 The root system of winter wheat grown in deep loam soil; (a) at the end of March and (b) at the end of June. From Kutschera (1960).

Table 3.1 Biomass and organic matter in soil

Location	Land use	Sample depth (cm)	Organic C (t ha^{-1})	Biomass-C (kg ha^{-1})	Biomass-C as % of organic C
England	Arable	0–23	29	660	2.2
England	Deciduous woodland	0–23	65	2180	3.4
England	Permanent grassland	0–23	70	2240	3.2
Nigeria	Secondary rain forest	0–15	19	760	4.0
Nigeria	Bush regrowth	0–17	27	700	2.6
Nigeria	Bush regrowth and cultivated for 2 years	0–16	22	370	1.7
Australia	Unimproved scrub	0–15	21	170	0.8
Australia	Improved pasture	0–15	35	430	1.2
Australia	Pasture	0–15	39	1170	3.0
Germany	Arable	0–10	14	300	2.2
Germany	Arable	0–10	28	910	3.2
Germany	Arable	0–10	32	620	1.9

From Jenkinson and Ladd (1981).

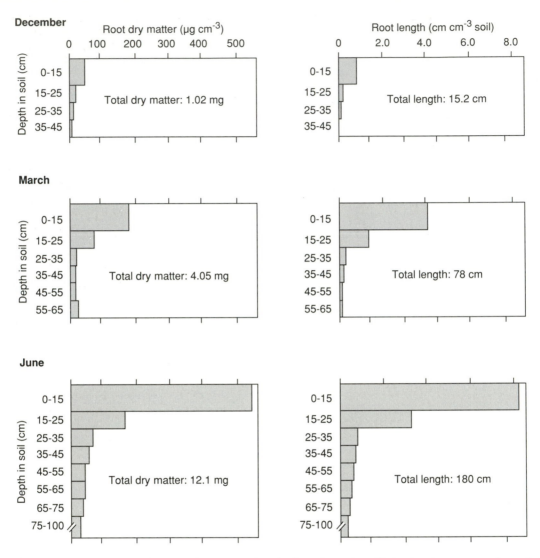

Figure 3.1 The root distribution of winter wheat grown in a clay loam in southern England at three stages of growth. The total dry matter and length are for the whole profile under 1 cm² of soil surface. From Russell (1977).

The mass of living micro-organisms in a soil is known as the *microbial biomass*. Its estimation is described in Section 3.2, and typical values are shown in Table 3.1. Much attention has been given to the biomass in recent years for two reasons:

1. During organic matter decomposition, nutrients pass through the microbial population and are released for plant use. This has been likened to the eye of a needle because the biomass is small compared to the total input of organic matter each year, and yet almost all the plant material has to pass through the biomass. The biomass and its activity are therefore indicators of soil fertility.

2. The microbial population may give an early warning of soil pollution by metals and pesticides (Section 6.9, Project 2 and Section 15.5). For example, metals added to soil in sewage wastes have been shown to inhibit the N-fixing activity of *Rhizobium* bacteria before any direct effects are seen on plant growth (Giller and McGrath, 1989). The effect of a pesticide on microbial respiration is a required test before it can be licensed for general use. The effects

of the activity of the biomass are considered in relation to soil aeration in Sections 6.2 and 6.3.

THE DEAD MATERIAL

Litter

Litter inputs to soils are difficult to measure under natural vegetation as is the mass of root material which dies and becomes part of the soil organic matter each year (Whittaker and Likens, 1975). For an annual crop, the root mass of the mature crop can be measured as can the mass of the crop residue. Inputs range from about 0.01 kg of dry material m^{-2} in desert and tundra regions to about 3.5 kg m^{-2} under tropical rain forest.

The composition of soil organic matter

The primary source of soil organic matter is plant material and its composition reflects that of the source. Decomposition through animals and micro-organisms causes changes in composition to produce humus in the form of highly complex and very large organic molecules. Soil organic matter is therefore a mixture of plant and animal material in various stages of decomposition together with humus.

The source of the elements in soil organic matter

Photosynthesis by plants combines CO_2 from the atmosphere with water from the soil to produce carbo-hydrates which are organic molecules containing carbon(C), hydrogen(H) and oxygen(O) with an elemental ratio CH_2O. In the process, O_2 is released into the atmosphere (Fig. 3.2). The net reaction can be represented as

$$CO_2 + H_2O \xrightarrow{\text{light}} O_2 + (CH_2O)$$

Plant roots take up from the soil nitrogen (N) (as NH_4^+ and NO_3^-), phosphorus (P) (as $H_2PO_4^-$) and sulphur (S) (as SO_4^{2-}), together with calcium, magnesium and potassium ions, all of which are macronutrients. Small amounts of the micronutrients iron, manganese, zinc, copper, boron, molybdenum, cobalt and chlorine are also taken up. These elements are incorporated into plant materials. When the plant dies the Ca^{2+}, Mg^{2+} and K^+ and the micronutrients are released rapidly into the soil; C, O, H, N, P and S are bound in the organic molecules but are slowly released as the plant material is decomposed through the action of animals and micro-organisms, which assimilate the elements they need, and release those not needed. Two processes are involved.

1. In *respiration*, O_2 is taken in by the organisms and CO_2 released:

$$(CH_2O) + O_2 \rightarrow CO_2 + H_2O$$

2. *Mineralization* is the release of N, P and S in the form of ammonium, phosphate and sulphate ions into the soil. The term 'mineralization' is used because the inorganic (mineral) forms of these elements are produced from the organic forms.

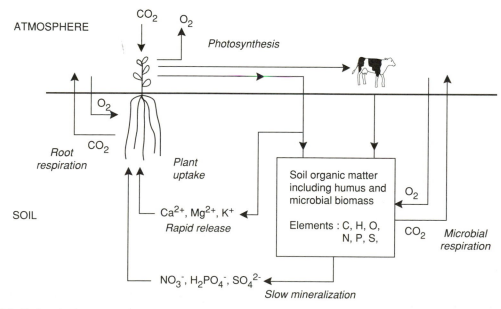

Figure 3.2 Carbon in the soil, plant and atmosphere.

Respiration is central to the understanding of soil aeration (Ch. 6). Mineralization of N, P and S is a key process in the supply of these elements to plants (Chs. 10 and 11). The distribution of C between soil organic matter, vegetation, atmosphere and sea is important because CO_2 is a 'greenhouse' gas. Burning of vegetation and losses of organic matter caused by cultivation both contribute to the 'greenhouse' effect (Wild, 1993).

Measurement of the organic matter content of soils

It is not possible to extract all the organic matter from soil and to weigh it: much of it is bound firmly on soil particles and extraction procedures change its composition. The simplest, although very approximate method of measurement, is to burn it off and determine the loss in mass, to give a value known as *loss on ignition*. This is a useful guide to differences between soils, but overestimates the actual content particularly in clay soils. It is more correctly a rough guide to the combined content of organic matter, clay and sesquioxides in soils. The method is given in Section 3.3.

Organic carbon

The standard method for determining organic matter is an indirect one in which the C in the organic material is oxidized to CO_2 and organic acids. The amount of an oxidizing agent used is measured and the mass of C is

determined. Assuming that soil organic matter contains on average 58 per cent C, a mass of organic matter is then calculated. This is called the dichromate method (Section 3.4). Inorganic C in soil, for example that in $CaCO_3$ is not determined in this method nor, for the most part, is C present in charcoal.

Amounts of organic matter in soils

The amounts of organic matter typically present in a range of soils are shown in Table 3.1. The amounts depend on both the annual input of organic matter and its rate of decomposition, the latter being highest in hot, humid climatic regions. The largest organic matter contents are found under temperate grassland where a high input is combined with a slow decomposition rate, and not under tropical rain forest. For a system at equilibrium, *turnover time* is the total amount of organic C divided by the input per year and the values indicate how long on average C atoms (added in plant material) remain in the soil before being respired as CO_2. Climate is a major factor influencing rate of decomposition and cultivation reduces turnover time. For example, at Rothamsted Experimental Station on similar clay loam soils, the turnover time under pasture is 25 years, but under continuous wheat only 16 years.

Studies of the kinetics of the decomposition process and the dating of soil organic matter using isotopic carbon (^{14}C) methods have shown that some of the added C is respired quickly, but some remains for very

Table 3.2 The concentration of soil organic matter in the topsoil (0–15 cm) after 17 years' cropping under different rotations at six UK experimental husbandry farms of the Agricultural Development and Advisory Service of the Ministry of Agriculture, Fisheries and Food

	Organic matter (%)		Mean annual rainfall (mm)	Clay content (%)
	Continuous arable	Rotation†		
Boxworth, Cambridgeshire	3.31	3.70	550	45
Bridgets, Hampshire	3.83	4.26	780	25
Gleadthorpe, Nottinghamshire	1.54	1.81	600	5
High Mowthorpe, Yorkshire	3.40	3.85	750	29
Rosemaund, Herefordshire	2.63	2.93	650	6
Trawscoed, Dyfed	5.64	5.98	1180	27

† 3 years grazed ley, 3 years arable.
From Batey (1988).

long periods of time. Mathematical analysis of the kinetics of the decay process (Section 3.5) has indicated that organic matter fractions with differing stabilities are present. Clay is considered to stabilize organic matter by protecting it both physically and chemically. Physical protection results from trapping in very small spaces between particles making the organic matter inaccessible to bacteria, and chemical protection results from adsorption on to clay surfaces such that bacteria cannot utilize it. Thus under similar climatic conditions and soil management, organic matter content is higher in soils with high clay content, as seen in Table 3.2. The sandy soil at Gleadthorpe maintains a much lower organic matter content than other sites. However, Trawscoed with its cool wet climate and slower decomposition rate maintains a higher organic matter content than High Mowthorpe with a similar clay content. Table 3.2 also shows that continuous arable plots have lower organic matter contents than rotation plots. Further discussion of the effects of management is in Section 13.3.

ORGANIC NITROGEN

Soil organic matter contains almost all the soil-N, much in the form of amino groups, $-NH_2$, originating from the proteins in the cells of plants, animals and micro-organisms.

Organic N is determined by decomposing organic matter in a boiling acid solution to form ammonium-N which is measured by steam distillation and titration (Section 3.6).

Amounts of organic nitrogen in soils

The amount of N in soils is closely related to the amount of organic matter. Although the chemistry of soil organic matter is complex its composition is broadly similar in all soils. Except for soils where largely undecomposed plant residues are accumulating (acid and flooded conditions) or where large amounts of fresh plant material have recently been incorporated, the ratios of elements in the organic matter are approximately constant: for agricultural soils in temperate regions the ratio of C:N is close to 10:1, that of C:P close to 50:1 and C:S close to 100:1. Since C is approximately 58 per cent of the organic matter, then N is approximately 0.058 times the organic matter content. Up to about 0.3 per cent N may be present in cultivated soils. These amounts of N are very large compared to plant uptake each year. A soil with 0.2 per cent N in the top 20 cm of soil contains about 5 t N ha^{-1} (Section 3.7) whereas a good wheat crop takes up about 150 kg N ha^{-1}. Thus all soils have large reserves of organic N which are only slowly mineralized and made available to plants.

Fresh plant material has C:N ratios varying from about 90:1 for cereal straw to 15:1 for legume residues. As decomposition proceeds C is respired and N assimilated by the microbial biomass. The biomass in soils receiving large amounts of straw may become starved of N and so for a time may be unable to make use of the carbon. Under such conditions much of the N mineralized by the soil population is immediately *immobilized* (taken up by the soil population) and so is not available to plants. Similar effects occur in pastures which are not well supplied with N: plant residues with a high C:N ratio accumulate, reducing the amount of mineral-N available for growth, and even after ploughing mineralization may be slow for a time (Section 11.4).

Table 3.3 The organic matter and microbial biomass in the unmanured soil under continuous wheat, Broadbalk, Rothamsted

Mass of soil to 23 cm	2610	t ha^{-1}
Organic matter	26	t C ha^{-1}
Nitrogen in the soil	2.7	t N ha^{-1}
Annual input of C	1.2	t ha^{-1} a^{-1}
Average turnover time of soil organic C	22	years
Radiocarbon age of soil organic C	1310	years
Number of fungal hyphae	7	million g^{-1}
Length of fungal hyphae	140	m g^{-1}
Volume of fungal hyphae	0.97	mm^3 g^{-1}
Number of bacteria and actinomycetes	1600	million g^{-1}
Fraction of the pore space occupied by organisms	0.35	%
Microbial biomass	220	μg C g^{-1}
Nitrogen in the biomass	95	kg N ha^{-1}
Flux of N through the biomass	38	kg N ha^{-1} a^{-1}
Phosphorus in the biomass	11	kg P ha^{-1}
Flux of P through the biomass	5	kg P ha^{-1} a^{-1}

From Jenkinson and Ladd (1981).

Many topsoils have C:N ratios between 9:1 and 12:1 and here microbial growth is limited by the C supply.

There is probably no soil in the world that has received more detailed study than that from the Broadbalk site at Rothamsted Experimental Station. Table 3.3 brings together the information relating to soil organic matter and the biomass (Jenkinson and Ladd, 1981).

USE OF LABORATORY DATA FOR FIELD STUDIES

Laboratory measurements are normally made using a known mass of dry soil. In the field we often require values relating to a hectare of land. The conversion method is given in Section 3.7. Approximate values commonly used for the mass of dry soil per hectare in a cultivated soil are 2000 t for a layer of soil down to 15 cm, and 2500 t to 20 cm.

The reliability of a measurement often has to be questioned in relation to the purpose for which it was measured. For example, the questions may be asked 'How reliable is my measurement of organic nitrogen as an indication of the organic nitrogen in the field from which the sample was taken' or 'From my measurements of organic carbon on samples from two field plots managed in different ways can I be sure that there are real differences?' These questions are discussed in Section 3.8.

FURTHER STUDIES

Ideas for projects are given in Section 3.9 and calculations in Section 3.10.

Section 3.1 Root sampling and measurement

Many methods have been devised for examining, sampling and measuring root systems (Böhm, 1979). A simple hand auger method is described here for taking known volumes of soil from which roots can be extracted. Stony soils, or those under woodland containing large roots cannot be sampled in this way.

Sampling procedure

A suitable auger (Plate 1.3) supplied by Eijkelkamp Equipment (Model 0501) consists of a cylindrical tube 15 cm long with an inside diameter of 7 cm and its lower end serrated for cutting into soil. Above this sampling tube a hollow shaft about 1 m long is attached. On the outside of the tube and shaft are marks at 10 cm intervals. The shaft is attached to a T-handle. Inside the hollow shaft is a rod with a disc at the bottom which works as a plunger inside the tube to force out the soil cores.

To take the cores, rotate and press the auger into the soil until the first 10 cm mark is reached, rotate several times and then pull out. To remove the core, turn the auger upside down, and press on the T-handle with your foot. Sampling is repeated for the 10–20 cm layer and so on to the required depth. Sandy soils should be moist to ensure that the core stays in the tube. Augering clay soils is difficult: drilling is easier if the auger is dipped in water between each sample, but a percussion auger is advised (Eijkelkamp Model 0502).

At least five replications are needed per plot to obtain reliable measurements.

Storing core samples

Ideally cores should be used within 1 or 2 days. Storage below 0 °C is probably the best way of keeping cores for later examination, and in heavy-textured soils freezing and subsequent thawing make it easier to disperse the soil during root washing. One day before washing, the cores are placed in water to thaw. Alternatively cores can be air dried, or oven dried at 70 °C. The roots become shrivelled but swell again when placed in water. Drying at 105 °C makes the roots too fragile.

Washing out the roots

A common method is as follows. Place the core in 1–2 l of water in a pail. After several hours or the next day, stir the soil–water mixture by hand to give a smooth suspension, breaking any remaining lumps. Allow the large soil particles to settle for a few minutes and the roots will tend to float. Pour the supernatant suspension and the roots on to a 0.5 mm sieve and wash with a spray of water (a hose attached to a sprinkler rose). Add water to the pail and repeat the process. Do this until no further roots are decanted.

Transfer the roots to a clean pail and stir in water. Decant on to the sieve and wash again. Finally place the roots in water ready for separating from organic debris.

The separation of roots from clay-rich soils is aided by carrying out the initial soaking in sodium pyrophosphate solution $(3 \, \text{g} \, \text{l}^{-1})$. If the soil is also calcareous, oxalic acid $(10 \, \text{g} \, \text{l}^{-1})$ or HCl (100 ml of 36 per cent *m/m* l^{-1}) are suitable soaking solutions.

Cleaning the roots

It is essential to clean the root sample before measurement. This is a tedious process. Place the roots in a flat dish of water on a black surface. Using tweezers, pick out the organic debris and dead roots. The main feature distinguishing living from dead roots is colour. Usually living roots are white when young, changing through yellow to brown as they age. When they die they often become grey and fragile. It is therefore easier to work with young plants, and to use field sites which do not contain large amounts of debris and dead roots. The ploughing of grass leaves a particularly difficult site for root measurements in a subsequent crop.

Storing the roots

If measurements are not made immediately after washing, store roots in 5 per cent (*v/v*) formaldehyde or 20 per cent (*v/v*) ethanol or at −20 °C.

Measurement

Determination of the mass of fresh roots

Place the roots on blotting paper to remove excess water, and weigh. This gives a mass in grams which is normally termed the *fresh weight*.

Determination of root dry matter

Dry at 100 °C overnight and weigh. Contamination of roots by soil particles may lead to errors. This can be checked by placing the weighed roots in a muffle furnace at 650 °C overnight and reweighing. The ash contains the plant ash but this is small compared to any soil residue. Subtract the mass of ash to give the corrected mass of dry roots, normally called *dry matter*.

Determination of root length

For very small root samples or single roots direct measurements can be made by placing the roots on graph paper, straightening them with tweezers so that they do not overlap, and estimating their length.

Normally the *intersection method* is used (Tennant, 1975). If roots are spread over a square grid (Fig. 3.3), their length is related to the number of intersections between the roots and the grid lines by the expression $R = \pi AN/2H$ where R is the total length (cm) of roots in an area A (cm^2) and N is the number of intersections between roots and the grid lines of total length H (cm). This relationship simplifies to $R = N \times$ a length factor where the length factor is 0.393, 0.786, 1.57 or 3.93 cm per intersection for grids with line spacings of 0.5, 1, 2 and 5 cm respectively. The relationship can be tested using cotton thread of known length spread on a paper on which a grid has been drawn.

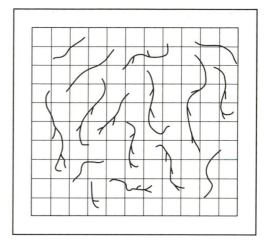

Example :   Grid size = 1 cm
Number of intersections = 64
Root length = 0.786 × 64
= 50.3 cm

Figure 3.3 The measurement of root length using the grid intersection method.

A shallow flat-bottomed dish of transparent plastic or glass measuring 30 × 40 cm is suitable. Alternatively a glass plate can be used. The grid can be drawn on paper which is placed under the dish. The spacing of the lines in the grid depends on the length of roots to be measured. For samples below 1 m length use a 1 cm grid, for samples up to 5 m use a 2 cm grid and for total root lengths up to 15 m a 5 cm grid is recommended. Alternatively for large root samples, measure the length of a subsample after determining root fresh weight.

The wet roots are placed in a little water in the dish and positioned randomly with tweezers so that they do not overlap. Intersections are then counted with both the vertical and horizontal lines. A hand tally counter facilitates the procedure.

Expressing the results

The core samples (7 cm diameter × 10 cm) have a volume of 385 cm^3. Results are normally expressed as mg fresh roots, μg dry roots or cm length cm^{-3} of soil. Fresh wheat roots contain about 0.1 g dry matter g^{-1} fresh weight, root length is between 50 and 250 m g^{-1} dry matter depending on age, and root diameter is about 0.3 mm. Similar values apply to most arable crops.

Section 3.2 The measurement of microbial biomass

Direct measurement of the mass of soil micro-organisms is a time-consuming and highly specialized research task. Dispersion of soil in water and examination of thin films of this dispersion by microscopy is the basis of the technique, which gives an estimate of volume of organisms. The densities of the main groups of organisms are needed to convert volume to mass (between 1.1 and $1.5\,\mathrm{g\,cm^{-3}}$).

More convenient indirect methods have been developed but only recently have standard methods been established and much research is still being done to improve them. The fumigation–extraction method given here is the simplest reliable procedure, (Note 1), (Amato and Ladd, 1988, modified by Joergensen and Brookes, 1990 and Ocio and Brookes, 1990).

The principle

When moist soil is placed in an atmosphere containing chloroform vapour micro-organisms are killed. A fraction of the cell constituents becomes soluble and can be extracted from the soil in potassium chloride solution. The N thus solubilized as amino acids and ammonium is estimated by reaction with ninhydrin (Nuffield Advanced Chemistry II, Topic 13.4) and measured as a purple complex using a spectrophotometer. The amount of N measured is directly proportional to the biomass initially in the soil: only about one-quarter of the biomass-N is released, but the fraction is approximately constant for different soils provided standard conditions are used. The proportionality constant has been established by comparing the amount of ninhydrin-reactive N with other measurements of biomass, primarily the fumigation–incubation method (Section 6.3).

Reagents and equipment

Ninhydrin reagent. Dissolve 0.8 g of ninhydrin and 0.12 g of hydrindantin in 30 ml of dimethyl sulphoxide. Add 10 ml of lithium acetate buffer. Prepare on the day of use. Alternatively the solution can be stored for several days if it is flushed with oxygen-free nitrogen gas for 30 min and kept airtight.

Lithium acetate buffer. Add 168 g of lithium hydroxide, LiOH.H$_2$O, to about 500 ml of water. Stir until about half dissolved. Add 293 ml of glacial acetic acid, CH$_3$COOH, and make up nearly to 1 l. Dilute 5 ml with 10 ml of water and measure the pH (Section 8.1). If the pH is not 5.2 ±0.05 adjust the stock solution with acetic acid or lithium hydroxide: 1 ml or 1 g respectively will change the pH by about 0.01 unit. Cover the solution and allow to cool overnight. Make up to 1 l and filter if the solution is not clear.

Ethanol–water. Dilute ethanol (95% *v/v*) with an equal volume of water.

Nitrogen standards. Dissolve 0.469 g of leucine in water and make up to 1 l. This contains $50\,\mu\mathrm{g\,N\,ml^{-1}}$. Into 100 ml volumetric flasks pipette 0, 5, 10, 15, 20 and 30 ml of this solution, add 50 ml of 4 M KCl to each flask and make up to the mark. These contain 0, 2.5, 5, 7.5, 10 and $15\,\mu\mathrm{g\,N\,ml^{-1}}$.

Potassium chloride solution. Dissolve 298 g of KCl in water and make up to 1 l. This is 4 M KCl. Dilute with an equal volume of water to obtain the extracting solution (2 M KCl).

Chloroform.

Glass bottles, 150 ml capacity.

Spectrophotometer and 1 cm cells.

Water bath, 100 °C.

The method

1. Preparation of the soil

This should be sampled moist from the field and passed through a 2 mm sieve. Partial drying makes it easier to crush and sieve. Air-dry soils can also be used although a substantial part of the microbial biomass is killed when soils are air-dried, stored and rewetted. Air-dry soils should be sieved and moistened to 40 per cent of their water-holding capacities (Note 2). Place four 25 g subsamples in unstoppered glass bottles and incubate at 25 °C for 2 weeks in a large closed container (a desiccator is suitable). Into the container place moistened filter paper to prevent the soils from drying. Open the container each day to maintain adequate aeration. This treatment allows the biomass to attain its stable level.

2. Fumigation

Place two subsamples in a vacuum desiccator along with a beaker containing 25 ml of chloroform. Evacuate the desiccator until the chloroform has boiled vigorously for 2 min. This causes chloroform vapour to penetrate the soil and kill the micro-organisms. Leave the soils in the chloroform vapour for 24 h and then remove the beaker of chloroform from the desiccator. (CARE: chloroform vapour is harmful and the liquid should only be handled and the vapour released in a fume cupboard.) Leave the desiccator open for a few minutes and then close and re-evacuate to aid removal of vapour from the soil. Open the desiccator again. Repeat the re-evacuation.

3. Extraction

Add to the four bottles of soil (two fumigated and two

unfumigated) 100 ml of 2 M KCl solution, cap and shake for 30 min. Filter through a Whatman No. 42 paper into a test tube. For determination of ninhydrin-reactive N, 2 ml of this filtrate is required.

4. Determination of ninhydrin-reactive N

Calibration

Into 50 ml test tubes, pipette 2 ml of each leucine standard. Add 1 ml of ninhydrin reagent slowly and mix thoroughly. Place the tubes in a boiling-water bath for 25 min. Cool to room temperature. Add 20 ml of ethanol–water, mix thoroughly and measure the absorbance values in a spectrophotometer at a wavelength of 570 nm using 1 cm cells with water in the blank cell. Section 10.1 explains the use of a spectrophotometer. Plot a calibration curve of absorbance against N concentration (Fig. 3.4).

Extracts. Develop the colour following the above method using 2 ml of each extract, and determine the N concentration from the calibration curve.

Method summary

Moist soil (? mg ninhydrin-N g^{-1} oven-dry soil)
↓
25 g + 100 ml
↓
Concentration measured (μg N ml^{-1})

Example. The non-fumigated and fumigated soils gave extracts containing 0.6 and 10.1 μg N ml^{-1} respectively. The difference between the two measurements

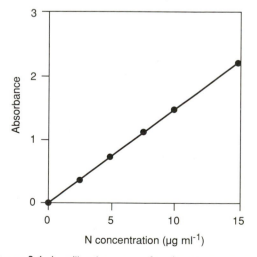

Figure 3.4 A calibration curve for the measurement of ninhydrin-reactive N.

(9.5 μg ml^{-1}) is the amount of ninhydrin-reactive N which was released by fumigation. The volume of the extract must have been 100 ml plus the volume of water in the moist soil. If the moist soil contained 0.24 g H_2O g^{-1} oven-dry soil, then 1.24 g of moist soil contains 0.24 g of H_2O, and 25 g contains

$$25 \times 0.24/1.24 = 4.84 \, \text{g} \, H_2O.$$

Thus the extract volume was 104.84 ml. The amount of ninhydrin-N in this extract was 9.5 × 104.84 = 996 μg. This came from 25 g of moist soil or 25 − 4.84 = 20.16 g of oven-dry soil. The ninhydrin-N is therefore

$$996/20.16 = 49.4 \, \mu\text{g} \, \text{N} \, \text{g}^{-1} \text{ oven-dry soil.}$$

The results

There is uncertainty about the factor used to calculate biomass from the amount of ninhydrin-reactive N, because of the difficulty, if not impossibility, of obtaining an absolute measure of biomass to allow calibration of the method. Following the Rothamsted methods (Ocio and Brookes, 1990) the following factors can be used. All values are μg g^{-1} oven-dry soil.

biomass-C = 31 × ninhydrin-N
biomass-N = 4.6 × ninhydrin-N

On average 50 per cent of the dry biomass is C and so

biomass (dry matter) = 62 × ninhydrin-N

Within a given experiment, treatment effects can be compared with confidence, but the method does not give absolute values and should be interpreted with caution. Data from different research groups may have been obtained using different experimental conditions and different conversion factors, e.g. Amato and Ladd (1988) and Carter (1991).

Note 1 Section 6.3 gives in outline the fumigation–incubation method for biomass determination. It requires the use of ethanol-free chloroform because fumigation would otherwise leave ethanol in the soil which would then be a C source used by micro-organisms to increase respiration. Other methods involve the determination of total C (Section 3.4) or total N (Section 3.6) in the KCl extract after fumigation. However the amounts present are low, causing difficulties in analysis.

Note 2 To determine the water-holding capacity, place a plug of cotton wool in a funnel, almost fill the funnel with moist soil, flood with water and allow to drain for about 1 h. Measure the water content as g H_2O g^{-1} oven-dry soil (Section 3.3). Water contents at field capacity are given in Table 5.2.

Section 3.3 The determination of water content and loss on ignition

Soils sampled from the field contain water, the amount depending on soil properties and the preceding weather conditions. Even when *air-dried* (Fig. 5.6) some water remains, the amount depending on soil texture and the humidity of the air in the drying room. The water content of soils is determined by drying at 105 °C which gives *oven-dry soil*. Results of soil analysis are normally expressed relative to a mass of oven-dry soil.

When an oven-dry soil is heated to 500 °C organic matter is burnt off and there is further loss of water. The mass loss between 105 and 500 °C is the *loss on ignition*.

Water content

Weigh about 10 g ($\pm$ 0.001) of soil into a weighed porcelain basin or crucible and place in an oven at 105 °C overnight. Cool in a desiccator and reweigh.

Example

Mass of air-dry soil	10.203 g
Mass of oven-dry soil	9.137 g
Mass of water lost	1.066 g

Water content = 1.066/9.137 = 0.117 g H_2O g^{-1} oven-dry soil. The value is often given as a percentage: 11.7 per cent which is 11.7 g H_2O per 100 g oven-dry soil. Water content can also be expressed relative to the mass of air-dry soil = 1.066/10.203 = 0.1045 g H_2O g^{-1} air-dry soil or 10.5 per cent. Note that because these two different percentage values can be obtained the full units should be stated. Where they are not given in the literature, it has to be assumed that a percentage is expressed relative to the oven-dry mass. Note also that data for plant material are normally given as *percentage dry matter* which is g oven-dry plant material per 100 g fresh plant material, and a water content if required would be g H_2O per 100 g fresh plant material (Section 9.3). Table 5.2 gives the water contents of air-dry soils of varying texture.

Loss on ignition

Place the sample in a furnace at 500 °C overnight. Cool in a desiccator and reweigh to give a mass of ignited soil. The mass lost by ignition is expressed relative to the mass of oven dry soil, i.e.

loss on ignition = 100 × (mass of oven-dry soil − mass of ignited soil)/mass of oven-dry soil

= g per 100 g oven-dry soil

If a laboratory furnace is not available, a pottery kiln adjusted to approximately 500 °C can be used, or the crucible can be placed in a sand bath and heated with a bunsen until the soil has lost its dark colour. Digital thermometers are available with metal probes which facilitate temperature control in the sand bath.

Loss on ignition as a measure of organic matter content

Soils which contain appreciable quantities of clay and sesquioxides lose 'structural' water between 105 and 500 °C. For example goethite, $FeOOH$, is dehydrated to haematite, Fe_2O_3, at between 280 and 400 °C. Most of the organic matter is burnt off at about 325 °C, but losses may continue at higher temperatures. Calcium carbonate loses CO_2 to form calcium oxide from about 770 °C. Thus loss on ignition is an approximate measure of organic matter content in sandy soils but may be up to twice the organic matter content in heavy textured soils.

Observations

There is a colour change during ignition associated with the removal of the black or dark-brown colour of the organic matter and the development of the stronger red–brown colours of the dehydrated iron oxides. Many surface soils change from a dark brown to a lighter red-brown as the colour of the iron oxides is seen without the masking of the organic materials. On the Munsell colour chart, (Section 1.2) there is a change of one or two steps in both value and chroma.

Section 3.4 The dichromate method for the determination of oxidizable carbon and soil organic matter

Soil organic C is almost completely oxidized by gently boiling for 2 h with an acid dichromate solution. The excess dichromate is determined by titration with ferrous sulphate. A simpler but less accurate modification involves boiling for 2 min over which time approximately 75 per cent of the organic C is oxidized.

Reagents and equipment

Ferrous sulphate, approximately 0.4 M. Add 5 ml of sulphuric acid (approx. 98% *m/m* H_2SO_4) to 1.5 l water (CARE). Dissolve in this approximately 320 g of ammonium ferrous sulphate, $(NH_4)_2SO_4FeSO_4.6H_2O$, and dilute to 2 l.

Potassium dichromate, 66.7 mM. Grind about 45 g potassium dichromate, $K_2Cr_2O_7$, to a powder, dry at 105 °C for 1 h or overnight and cool in a desiccator. Dissolve 39.23 g of the dried salt in about 700 ml water, add slowly with stirring 800 ml of H_2SO_4 (approx. 98 per cent *m/m* H_2SO_4) and cool. Add 400 ml of orthophosphoric acid (approx. 85% *m/m* H_3PO_4), stir until all the solids have dissolved, cool and dilute to 2 l.

Barium diphenylamine sulphonate indicator (poison). Dissolve 0.2 g of barium diphenylamine sulphonate in approximately 150 ml of warm water, add 20 g of barium chloride, $BaCl_2.2H_2O$, warm to dissolve, cool and dilute to 200 ml. Leave overnight and filter if necessary.

Digestion flasks: 500 ml conical or flat-bottomed flasks with cold finger or Liebig condensers.

Electric hot plate to maintain a temperature of 130–135 °C.

Oxidation of organic carbon

Grind about 40 g of <2 mm air-dry soil in a pestle and mortar and mix well. Transfer 1 g (±0.005) into a digestion flask and add 40 ml of dichromate solution. *Place on the hot plate and fit a condenser. Boil at 130–135 °C for 2 h. Remove the flask, cool and add about 100 ml of water. Add 2 ml of the indicator and titrate with the ferrous sulphate solution. The solution is a dirty brown colour initially, and then becomes purple just before the end point. At this stage titrate drop by drop until the colour changes to bright green. Record the volume of ferrous sulphate used, x (ml): in the example below this is taken as 18.5 ml. If this volume does not exceed 15 ml, repeat the determination using a smaller mass of soil (Note 1).

Standardization of the ferrous sulphate

This solution may oxidize on standing and so it should be standardized against the potassium dichromate solution along with each batch of soil determinations. Pipette 40 ml of the dichromate solution into a digestion flask. Continue as from the asterisk in the above paragraph. Record the volume of ferrous sulphate used in the titration, y (ml) : in the example below this is taken as 36.7 ml.

The principles

The reaction of organic C with dichromate can be represented (Note 2) as

$$2K_2Cr_2O_7 + 3C_{organic} + 8H_2SO_4 = 3CO_2 + 2Cr_2(SO_4)_3 + 2K_2SO_4 + 8H_2O. \qquad [3.1]$$

The organic C is oxidized by the chromate, $Cr_2O_7^{2-}$, which is reduced to chromic, Cr^{3+}. Oxidation and reduction are explained in Section 6.6. The equation tells

us that 2 mol of $K_2Cr_2O_7$ reacts with 3 mol of C. Section 2.5 explains the term mol.

The amount of dichromate used in the experiment is determined by titration against ferrous sulphate. The reaction is

$$K_2Cr_2O_7 + 6FeSO_4 + 7H_2SO_4 = Cr_2(SO_4)_3 + K_2SO_4 + 3Fe_2(SO_4)_3 + 7H_2O$$

In this reaction ferrous, Fe^{2+}, is oxidized to ferric, Fe^{3+}, by chromate which is reduced to chromic. The equation tells us that 1 mol of $K_2Cr_2O_7$ reacts with 6 mol of $FeSO_4$. This ratio governs the choice of 66.7 mM as the concentration of the dichromate solution which is 1/6 of the concentration of the ferrous sulphate, and gives approximately equal reacting volumes in the titration. Note that the ammonium sulphate in the reagent plays no part in the reaction. It is simply a convenient ferrous salt for preparation of the solution.

The calculation

1. The dichromate is the standard solution. Calculate the number of mol of $K_2Cr_2O_7$ present in 40 ml as follows:

   number of mol = concentration (mol l^{-1}) × volume (1) = 0.0667 × 40/1000 = 2.668 × 10^{-3} mol or 2.668 mmol.

2. Your standardization titration showed that 36.7 ml of ferrous sulphate reacted with 40 ml of dichromate which contained 2.668 mmol. After reaction with organic C, 18.5 ml reacted with the residual dichromate. Therefore the amount of residual dichromate = 2.668 × 18.5/36.7 = 1.345 mmol (Note 3). The amount of dichromate used to oxidize the organic C is (2.668 − 1.345) = 1.323 mmol.

3. Three mol of C react with 2 mol of K_2Cr_2O (Eq. 3.1). Therefore the number of mol of C oxidized = 1.323 × 3/2 = 1.985 mmol. The molar mass of C is 12 g mol^{-1}. Therefore the mass of C oxidized is

   1.985 × 10^{-3} mol × 12 g mol^{-1} = 23.82 × 10^{-3} g or 23.8 mg.

   If this is in 1 g of soil (but see Note 1) the C content is 23.8 mg C g^{-1} air-dry soil.

4. The above calculations can be reduced to a simple formula to give the C content of the soil:
   C content (mg C g^{-1} air-dry soil) = 48 (1 - *x/y*)/soil mass.

5. The C content should be expressed relative to the mass of oven-dry soil. If the air-dry soil contained 5 g H_2O per 100 g oven-dry soil (Section 3.3), then 105 g air-dry soil contains 100 g oven-dry soil. Thus 23.8 mg C is present in 100/105 = 0.952 g oven-dry soil. The C content is

$23.8 \times 1/0.952 = 25.0 \, \text{mg} \, \text{C} \, \text{g}^{-1}$ oven-dry soil.

6. If a value for organic matter is required, the assumption is made that this contains $0.58 \, \text{g} \, \text{C} \, \text{g}^{-1}$ organic matter or 58 per cent (Note 4). Therefore the organic matter content is

$25.0 \times 1/0.58 = 43.1 \, \text{mg}$ organic matter g^{-1} oven-dry soil or 4.31 per cent

Note 1 The mass of soil used may have to be decreased if the soil contains a large amount of organic matter. To ensure that all the C has been oxidized the final titration volume, x, should be at least 15 ml, i.e. a maximum of 25 ml of dichromate should react with the soil. The amount of organic C that can be oxidized by 25 ml of dichromate can be calculated. It contains 1.668×10^{-3} mol of dichromate and will react with $1.668 \times 10^{-3} \times 12 \times 3/2 = 0.03 \, \text{g}$ C. Thus 1 g of soil containing $0.03 \, \text{g}$ C (3% C or 5% organic matter) will react with this amount of dichromate. A 0.25 g sample of soil containing 12 per cent C or 20 per cent organic matter would use this dichromate.

Note 2 In Eq. 3.1 C_{organic} represents the C in the organic matter which is combined in complex molecules. The products of the oxidation are represented by CO_2, but are a mixture of CO_2 and small amounts of organic acids. However, for calculation it is assumed that the reaction proceeds such that 3 mol of C reacts with 2 mol of dichromate. The resultant errors are very small as shown by comparison with combustion methods (Page, 1982).

Note 3 You could at this stage calculate the molarity of the ferrous sulphate and from this calculate the amount of dichromate remaining after oxidizing the organic matter. This would be an unnecessarily complicated procedure.

Note 4 Soil organic matter, unless it includes large amounts of undecomposed plant residues, has a C content of about 58 per cent. Plant dry matter has between 40 and 45 per cent C. There is no satisfactory method for the determination of organic matter other than through an organic C measurement. Increasingly, data are recorded in terms of C rather than organic matter.

A modified method

If a temperature-controlled hot plate and suitable condensers are not available the oxidation can be carried out as follows. After adding the dichromate to the soil in a 500 ml conical flask, place it on a tripod and gauze, quickly bring to the boil and remove the bunsen. Allow to stand on the tripod for 2 min, applying the bunsen as necessary to maintain steady boiling. Remove and cool under running cold water (CARE). Under these conditions about 75 per cent of the organic C is oxidized. The calculations have to be adjusted (multiply by 4/3) to obtain the total soil C. Although less accurate than a complete oxidation, the method gives useful relative values provided the boiling conditions are standardized.

Section 3.5 The kinetics of organic matter decomposition

The rates at which processes cause changes in soil properties depend on many factors and mathematical analysis of the kinetics of these processes is complicated. A common procedure is to study a process in isolation from the soil as a means of gaining understanding of its role. For example, the weathering of limestone by acidic soil solution depends on the concentration of the acid, the reacting surface area of the limestone particles, the temperature and the rate at which the acid can move to the limestone surfaces and wash away the reaction products, i.e. the rate of leaching. This weathering process can be isolated from the soil, and the rate of reaction of limestone with acid can be studied at fixed temperature in a stirred solution, so examining the effect of acid concentration only (Nuffield Advanced Chemistry II, Topic 14.2).

Despite their complexities, chemical changes in soils are relatively simple to study in comparison to biological changes, where the activity of micro-organisms controls the rates, often through the production of an enzyme. The rates depend on the factors controlling microbial activity: temperature, water supply, nutrient supply and aeration along with the availability of the materials being broken down called the substrates. Again, parts of this complex system can be isolated and the kinetics of the process studied, e.g. the hydrolysis of urea by the enzyme urease produced by bacteria (Nuffield Advanced Chemistry II, Topic 13.5).

The rate of decomposition of organic matter depends on many biological processes. Although the loss of C can be followed by measuring the soil's respiration rate, or the decrease in organic C can be measured directly, the overall process is complex. Mathematical analysis has, however, proved useful in extending our understanding of the factors governing decomposition rates. It is based on the concept that soil organic matter is composed of various fractions which break down at different rates with standard kinetic analysis applied to each fraction.

The rate laws

Reactions are known that proceed at a constant rate, others at rates which depend on the concentration of one or more reactants. The number of reactants controlling the rate gives *the order of reaction*.

Zero-order or constant rate kinetics

Here the decomposition of A (the reactant) to give products proceeds at the same rate until all of A has decomposed and so

$$\text{rate} = -d[A]/dt = k$$

where k is the rate constant, $[A]$ is the concentration of A at time t, and the negative sign implies that $[A]$ decreases as t increases. Therefore

$$d[A] = -kdt$$

Integration gives

$$[A] = -kt + c$$

where c is a constant.

The value of $[A]$ when $t = 0$ is $[A]_0$. Substituting gives

$$[A] = [A]_0 - kt$$

This is a straight-line relationship with a negative slope when $[A]$ is plotted against t (Fig. 3.5(a)).

First-order kinetics

Here the rate of reaction is proportional to the concentration of A remaining. As the reaction proceeds $[A]$ decreases and the rate of reaction decreases (Fig. 3.5(b)).

$$\text{rate} = -d[A]/dt = k_1[A]$$
$$-(1/[A])d[A] = k_1 dt$$

Integration gives

$$-\ln[A] = k_1 t + c \qquad [3.2]$$

When $t = 0$, $c = -\ln[A]_0$ and so

$$-\ln[A] = k_1 t - \ln[A]_0$$
$$\ln[A]_0 - \ln[A] = k_1 t$$
$$\ln([A]_0/[A]) = k_1 t \qquad [3.3]$$

Thus if a reaction is first order then a graph of $\ln([A_0]/[A])$ against t will be a straight line with slope $= k_1$ (Fig. 3.5(c)).

The rate constant can be visualized through the graph, but an easier concept for general use is the *half-life* of the reaction. This is the time taken for the concentration to fall to half its initial value, $t_{1/2}$. Substituting in Eq. 3.3 gives

(a)

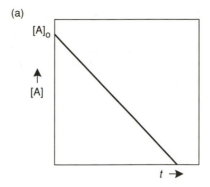

(b)

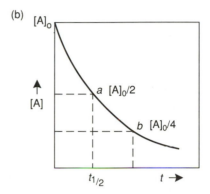

(c)

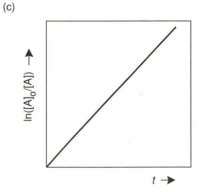

Figure 3.5 (a) Zero and (b) & (c) first-order kinetics.

$$\ln \frac{[A]_0}{[A]_0/2} = k_1 t_{1/2}$$

$$\ln 2 = k_1 t_{1/2}$$

and

$$t_{1/2} = 0.693/k_1 \qquad [3.4]$$

This shows that the half-life is constant and independent of the concentration chosen as the starting-point. Thus in Fig. 3.5(b) the time taken to reach a is the same as the time from a to b.

The decomposition of plant material in soils

The decomposition of plant material in soils can be followed if it is uniformly labelled with the isotope ^{14}C. The labelled C can be distinguished from the very much greater quantities of unlabelled C usually found in soils. This material is produced by growing plants in an atmosphere in which the CO_2 is labelled with ^{14}C which by photosynthesis becomes part of the plant organic carbon. When added to soil the normal decomposition processes release some ^{14}C by respiration as $^{14}CO_2$, and that remaining in the soil can be measured. Data are available for both Rothamsted and Nigeria and are shown in Fig. 3.6. In Nigeria the material decomposes at almost exactly four times the rate in England, reflecting the higher temperature in the humid tropical environment.

The decay curve is obviously not zero order: on first sight it looks like a first-order curve. This can be checked using the half-life principle. In England half the labelled C has been respired in 0.5 years, a further half after another 1 year and a further half after another 6.5 years. The data do not fit a first-order reaction. It looks as though some of the plant material is decomposing very quickly, but some is much more resistant.

There appear to be fractions of organic matter with differing stabilities.

Based on this conclusion, simple models can be devised and tested. For example, imagine that two fractions A and B decay following first order kinetics but at different rates. For example, A is initially 75 units decaying with a half-life of 0.1 years and B is 25 units with a half-life of 5 years. The decay curves for these fractions are shown in Fig. 3.7. The curve for $A + B$ which is the model's simulation of the decay of added organic C in the soil is similar to that in Fig. 3.6. The fit, however, is not perfect especially if longer times are considered.

Further analysis of the data is possible by computer modelling. Good simulation of soil organic C dynamics is obtained using five fractions decaying at different first-order rates, with C from one fraction being partly respired and partly passed on to a more resistant fraction. This is similar to the decay of a radioactive isotope, where products of the decay may themselves

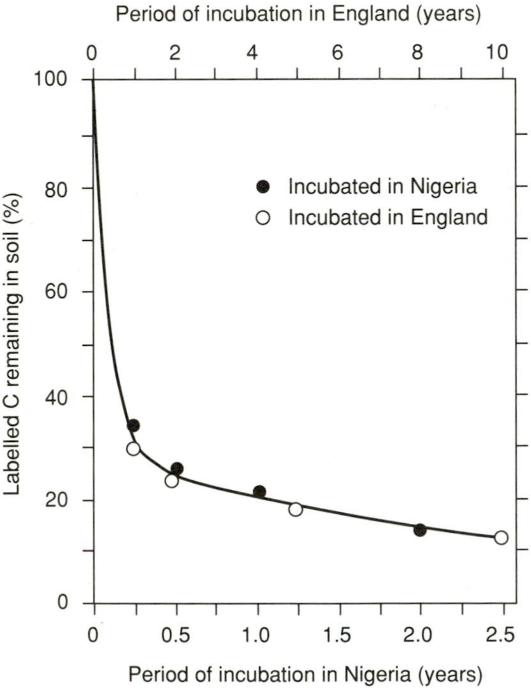

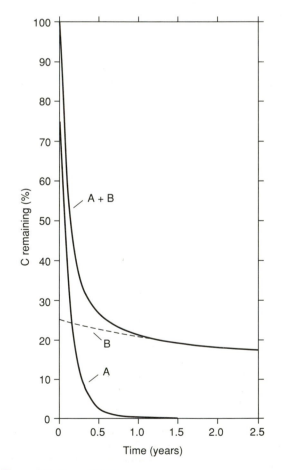

Figure 3.6 Decomposition of labelled ryegrass in soils of similar clay content in England (Rothamsted) and southern Nigeria. From Wild (1988).

Figure 3.7 Curve fitting.

Table 3.4 Predicted organic matter fractions and turnover times in the unmanured plot of the Broadbalk experiment, Rothamsted. Annual input of $C = 1.2\,t\,ha^{-1}$

Fraction	Amount of C ($t\,ha^{-1}$)	Turnover time (years)
Readily decomposed plant material	0.1	0.2
Resistant plant material	0.6	3.3
Microbial biomass	0.3	2.4
Physically protected organic matter	13.6	71
Chemically stabilized organic matter	14.6	2900

From Jenkinson (1981).

be radioactive with different half-lives to the parent element. An example of predicted fractions is given in Table 3.4. Rates of decay are not given as rate constants because C is moving through one fraction to another as well as being respired and a turnover time is a more useful concept. The turnover time (years) is the average time that a C atom resides in a given fraction and is equal to the amount of C in the fraction (assumed constant) divided by the amount of C entering (or leaving) the fraction each year. This approach suggests that about half the organic matter has a very long life. This is confirmed by radiocarbon dating which shows that the average age of all the organic matter in this soil is about 1500 years.

A computer assisted learning package dealing with the turnover of carbon in soils is available from the address given in Appendix 3.

Section 3.6 The determination of organic nitrogen

Organic N is converted to ammonium-N by sulphuric acid with sodium sulphate added to raise the boiling point to about 380 °C. Copper sulphate or selenium metal powder are added as catalysts. The conversion process is called a *digestion* and the solution which is produced, a *digest*. This process is often called a *Kjeldahl digestion* after its originator.

When the acid digest containing ammonium is made alkaline with NaOH the ammonium ions, NH_4^+, are converted to ammonia, NH_3, which is then released from the solution by steam distillation and condensed as ammonium hydroxide, NH_4OH. This is titrated against standard HCl. The procedure is summarized as follows:

$$\text{organic N} + H_2SO_4 \rightarrow$$
$$(NH_4)_2SO_4 + CO_2 + SO_2 + H_2O \quad \text{digestion}$$
$$(NH_4)_2SO_4 + \\ 2NaOH \rightarrow 2NH_3 + Na_2SO_4$$
$$2NH_3 + 2H_2O \rightarrow 2NH_4OH \quad\text{steam distillation}$$
$$NH_4OH + HCl \rightarrow NH_4Cl + H_2O \quad \text{titration [3.5]}$$

Reagents and equipment

Copper sulphate, $CuSO_4.5H_2O$.
Sodium sulphate (anhydrous), Na_2SO_4. } *or Kjeldahl tablets (Note 1)*
Sulphuric acid, approx. 98% m/m H_2SO_4.
Boric acid solution. Dissolve approximately 20 g of boric acid, H_3BO_3, in 1 l of water. Prepare fresh each week.
Indicator. Dissolve 0.1 g of methyl red and 0.2 g of bromocresol green in 250 ml of ethanol.
Sodium hydroxide solution. Dissolve 100 g of NaOH, in 200 ml of water.
Hydrochloric acid, 0.01 M.
100 ml digestion flask or tube (heat resistant, round bottom).
Heating rack to support the flask or tube over a bunsen, or an electrical heating block.
Fume cupboard.
Steam distillation unit.

Digestion

Weigh about 2 g (±0.01) of <2 mm air-dry soil into the digestion flask. Add 2.5 g of sodium sulphate and 0.5 g of copper sulphate or two Kjeldahl tablets (Note 1). Add about 4 ml of water and swirl the flask to wet the soil thoroughly. Add 6 ml of concentrated sulphuric acid. (CARE). Heat on a digestion rack in a fume cupboard gently at first until vigorous effervescence subsides and then gradually increase to full heat if using a bunsen, or raise to 380 °C if using a heating block. Continue boiling for 1 h after the digest is white with no charred organic matter remaining.

Cool, add 20 ml water and allow to stand for 30 s so that the sand particles settle, and decant the supernatant solution into a 100 ml volumetric flask. Repeat this process washing the sand and quantitatively transferring all the ammonium to the flask. Make up to the mark. Alternatively if a digestion tube has been used and it has a 100 ml calibration line, simply make the digested sample up to the mark (Note 2).

Distillation

Various designs of distillation equipment are available. Essentially a steam generator passes steam into a distillation flask which is connected via a splash trap to a

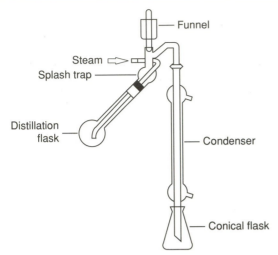

Figure 3.8 A steam distillation unit for the determination of ammonium-nitrogen.

condenser. A convenient apparatus is shown in Fig. 3.8 but other systems can be arranged from commonly available equipment. Set up the equipment and pass steam for 30 min. Take off the distillation flask and rinse with distilled water.

Place 10 ml of boric acid solution (Note 3) and a few drops of mixed indicator in a 250 ml conical flask and position it under the condenser. Pipette 50 ml of the digest into the funnel and release the digest into the distillation flask. Rinse the funnel with distilled water and release into the flask. In a similar way add 10 ml of NaOH solution. Pass steam and collect about 10 ml of distillate after the boric acid solution has changed from pink to green. Remove the conical flask, washing the tip of the condenser into the distillate and retain for titration. Remove the distillation flask and turn off the steam supply. (CAUTION. If you do this in the reverse order solution from the distillation flask will be sucked back into the steam generator). Rinse out the distillation flask with distilled water.

Titration

Titrate with 0.01 M HCl from the green colour through colourless to a pale pink colour. Record the volume, y (ml), (Note 4).

Calculation

Method summary
<2 mm air-dry soil (? mg N g^{-1} oven-dry soil)
↓
2 g → 100 ml digest
↓
50 ml titrated against 0.01 M HCl

Example In the titration 8.5 ml of acid was used. Equation 3.5 gives the titration reaction: 1 mol of HCl reacts with 1 mol of NH_4OH. Therefore the amount of acid used in the titration is

$$0.01 \, mol \, l^{-1} \times 8.5/1000 \, l = 8.5 \times 10^{-5} \, mol.$$

Because 1 mol of HCl reacts with 1 mol of NH_4OH, then the amount of NH_4OH in the conical flask is also 8.5×10^{-5} mol.

We now need to know the mass of N in this NH_4OH. One mol of NH_4OH contains 14 g N. Thus, the mass of N is

$$8.5 \times 10^{-5} \, mol \times 14 \, g \, mol^{-1} = 1.19 \times 10^{-3} \, g \, or$$
$$1.19 \, mg$$

This is the mass of N in 50 ml of the distillate. The mass in 100 ml = 2.38 mg N, which came from 2 g of soil. Therefore the N content of the soil is 1.19 mg g^{-1} air-dry soil (Note 5).

These calculations can be simplified for routine purposes to

$$mg \, N \, g^{-1} \, \text{air-dry soil} = 0.28y/\text{soil mass}$$

Oven-dry soil Express the answer as mg N g^{-1} oven-dry soil (Section 3.3). If the air-dry soil contained 5 g H_2O per 100 g oven-dry soil (105 g air-dry soil) then 1 g air-dry soil contains 100/105 = 0.952 g oven-dry soil. The N content is

$$1.19 \times 1/0.952 = 1.25 \, mg \, N \, g^{-1} \, \text{oven-dry soil}.$$

Per cent N The N content is often expressed as a percentage (m/m) of oven-dry soil. In this case $1.25 \times 10^{-3} \, g \, g^{-1}$ is 0.125 g per 100 g or 0.125 per cent.

Note 1 Suitable Kjeldahl tablets containing 1 g of sodium sulphate and 0.1 g of copper sulphate are supplied by BDH-Merck Ltd.

Note 2 This procedure introduces error because some of the 100 ml volume is occupied by fine particles decanted with the solution into the volumetric flask or remaining in the digestion tube. The digest could be filtered into the volumetric flask, but this is a slow procedure. For routine purposes the error (perhaps 0.2 ml of particle volume) is acceptable.

By making the digest up to 100 ml and distilling only 50 ml some solution remains if a second distillation is required: 25 ml could be used as it will not be possible to take up a further 50 ml into a pipette. For routine purposes it is more useful to run the whole experiment twice to give duplicate values rather than obtaining duplicate titrations on the distillate. Thus the whole digest can be quantitatively transferred to the distillation flask without making up to volume. The calculation would have to be adjusted.

Note 3 Boric acid is used to receive and trap the NH_4OH entering the conical flask. Because ammonia is unstable in NH_4OH (just as it is in the distillation flask), some would be lost if the distillate was collected alone. Boric acid is a weak acid. This means that it is primarily in the form of H_3BO_3, with only small amounts of H^+ dissociating from the boric acidic molecules. However, with NH_4OH it forms ammonium borate, $(NH_4)_3BO_3$. The mixture of H_3BO_4 and $(NH_4)_3BO_3$ is a buffer solution (Section 8.4) and holds the pH at a slightly acid value. However, when titrated with HCl (a strong acid which dissociates to release all its H^+), $(NH_4)_3BO_3$ reacts with HCl as though it were NH_4OH.

$$(NH_4)_3BO_3 + 3HCl = 3NH_4Cl + H_3BO_3$$

Viewed in terms of the calculation, the boric acid plays no part in the analysis; it simply ensures that ammonia is not lost.

Note 4

(a) A blank determination can be made as follows. Carry out the whole procedure but omit the soil. Subtract the blank titration value from all your soil titration values.

(b) The distillation procedure can be checked using a standard solution of known ammonium concentration. Dry ammonium sulphate, $(NH_4)_2SO_4$, at $105\,°C$ for 1 h and cool in a desiccator. Dissolve 1.321 g in water and make up to 1 l. Pipette 5 ml of this into the distillation flask and proceed as for the soil digest. This solution contains $0.28\,mg\,N\,ml^{-1}$, and 5 ml should require a titration of 10 ml of standard acid. A lower value will indicate that you are losing some ammonia in the distillation, and a higher value probably means that some NaOH has splashed through the trap and entered the condenser.

(c) The whole procedure can be checked using an organic N standard. Dry ethylenediaminetetra-acetic acid disodium salt, $[CH_2N(CH_2COOH)CH_2COONa]_2.2H_2O$. Dissolve 1.860 g in water and make up to 250 ml. This contains $0.56\,mg\,N\,ml^{-1}$. Pipette 5 ml into the digestion flask, add sodium sulphate, copper sulphate and the sulphuric acid (omit the water) and proceed as before. The distillate should contain 1.40 mg N and should require a titration of 10 ml of standard acid.

Note 5 Soil contains organic N, and mineral-N in the form of ammonium and nitrate. The measurement gives the amount of organic N plus ammonium-N, but excludes most of the nitrate-N which is lost during the digestion. Thus to refer to the measurement as organic N is not strictly correct. The value obtained is also occasionally called total N but this is not strictly correct because most of the nitrate is not included. Both errors are negligible for most purposes because the amounts of mineral-N are small. For example there may be between 5 and $50\,\mu g$ mineral-N g^{-1} soil compared to 0.2 per cent or $2000\,\mu g$ organic N g^{-1} in arable soils of temperate regions.

Section 3.7 Conversion of soil analysis values to field values

Example 1

A representative soil sample from the 0–20 cm layer in a field contains 1 per cent organic C. How much C is present per hectare (ha) in this layer of soil?

One per cent organic carbon is 1 g C per 100 g oven-dry soil or $0.01\,g\,C\,g^{-1}$ oven-dry soil. This can be restated as $0.01\,t\,C\,t^{-1}$ oven-dry soil.

One hectare of land is $10^4\,m^2$ and the volume of soil to 20 cm depth is $10^4\,m^2 \times 0.2\,m = 2000\,m^3$.

To bring these two values together we need to know the mass of soil in the calculated volume. The mass : volume ratio is the bulk density (Section 4.2). For agricultural topsoils, its value is often taken as 1.3 g oven-dry soil cm^{-3} which is also $1.3\,t\,m^{-3}$. Therefore 1 ha to 20 cm depth contains $1.3\,t\,m^{-3} \times 2000\,m^3$ which is 2600 t oven-dry soil. Similarly in 1 ha to 15 cm depth there is 1950 t. Thus the total C present (1 ha, 20 cm depth, $1.3\,t\,m^3$) is

$$0.01\,t\,C\,t^{-1}\ \text{soil} \times 2600\,t\ \text{soil ha}^{-1} = 26\,t\,C\,ha^{-1}$$

A useful way of expressing this result is $26\,t\,C\,ha_{2600\,t}^{-1}$. The commonly used value of 2500 t soil ha^{-1} for a 20 cm depth relates to a bulk density of $1.25\,t\,m^{-3}$.

Air-dry soil

For some purposes a mass of air-dry soil ha^{-1} may be needed. If an air-dry soil contains 5 per cent water (5 g H_2O per 100 g oven-dry soil), then 2600 t oven-dry soil is

$$2600 \times 105/100 = 2730\,t\ \text{air-dry soil}$$

Similarly 1950 t oven-dry soil is 2048 t air-dry soil. Table 5.2 gives typical water contents of air-dry soils of varying texture.

Example 2

Often comparisons need to be made between plots treated in different ways. For example, there is renewed interest in the changes in the amounts of organic C and N resulting from land use for crop production. For this purpose, soil analysis values need careful interpretation. For example, cultivating a virgin soil will mix organic matter initially concentrated near the surface through the cultivated layer. This will decrease the organic C content near the surface, and increase it at depth. However cultivation loosens the soil, decreasing its bulk density and a given depth of soil contains a smaller mass of soil after cultivation. With time, organic C will be lost from the soil, and the bulk density is likely to increase. Measurement of the C percentage in soil sampled from a standard depth will thus be of limited value. A valid comparison between sites requires samples to be taken from layers down through the profile. Thickness, bulk density and C percentage should be measured for each layer, and the C content of the layers then summed to give the total C in the profile. Where erosion occurs after cultivation, the comparison of sites becomes even more difficult.

Section 3.8 The reliability of data

Soils are not uniform in the field. Properties vary with depth in a profile, so giving horizons. Even within a horizon properties vary from one point to another. Across a plot or field there can be marked variations in the properties of the topsoil or of lower horizons. The methods used to obtain a representative sample from a profile, and to obtain a representative sample from a plot or field are described in Section 1.3. These are known as bulked samples.

A bulked sample from a field is normally air dried, passed through a 2 mm sieve and thoroughly mixed. A subsample can then be taken for chemical analysis, which hopefully represents the bulked sample, with the bulked sample representing the field site. If several subsamples from the bulked sample were analysed a set of values would be obtained all being slightly different. We obviously cannot have complete confidence in any single value and a mean value is normally calculated. However, if another set of subsamples were analysed a slightly different mean would be obtained, and so we cannot even have complete confidence in our mean value. To add to these problems, another bulked sample from the field would give a different set of mean values, and we cannot have complete confidence in our sampling procedure.

Let us assume that we do have a bulked sample which represents a horizon, plot or field and deal with the problems of variability between the analysis values for different subsamples. A measure of variability provides information about the uniformity of the mixed bulked sample and the reliability of the analytical method. It also allows comparison of measurements on bulk samples from different sites.

How reliable is a single analysis value?

Imagine that the C contents (x) of nine subsamples of soil A were measured. The values of x were 2.31, 2.17, 2.20, 2.12, 2.01, 2.05, 2.03, 2.21 and 2.13 per cent. The range of values is 0.30 (2.01–2.31), with a mean value $\bar{x}$ equal to 2.137 ($\bar{x} = \sum(x)/n$ where $\sum(x) =$ the sum of the values of x, and n is the number of values). The measured values are just a few of the total population of values that would be obtained if all of the bulked sample was put through the analysis using a large number of subsamples. If we did this the mean for the whole bulked sample (μ) would be obtained. This is known as the *true value* or *population mean* which we would like to measure, but cannot because of practical limitations. The data would also give a measure of the reliability of the analysis and the uniformity of the subsamples through the spread of values around this mean.

Statistical analysis allows us to quantify the variability between the nine subsamples and to estimate the properties of the total population of subsamples from the data for the set of values. You may have to learn or revise some statistics at this stage: Garvin (1986) and Rowntree (1981) give useful introductions. Note that statisticians normally talk of a *total population of values* and a *sample from a population*. Because in soil analysis, sample refers to a soil sample, we use here total population of values and a *set of values*.

The *standard deviation* is the commonly used measure of variability. For a set of values, it is termed s, and can be determined using a scientific calculator. For the nine values above, $s = 0.0973$. As the number of measurements increases, so the standard deviation of the set of values approaches that for the total population of subsamples. The standard deviation of the total population is termed σ. It can be estimated from a set of values, and again using a calculator the nine values above give $\sigma = 0.0918$. Note that many calculators use the terms σ_{n-1} ($= s$) and σ_n ($= \sigma$) for the standard deviations of the set of values and the total population respectively.

The standard deviation can be used to give a measure of variability which can be easily visualized. If the population of values is normally distributed, then it is known that 95 per cent of the values fall within the probability limits $\mu \pm 1.96\sigma$. Using the above set of values we do not know μ which is the mean of the whole population and so the limits cannot be calculated. However, we do have an estimate of σ and so the interval between the limits can be calculated. This is $2 \times 1.960\sigma$ and has a value of 0.36. Thus if we measured the organic C content of only one subsample we could be 95 per cent confident (there would be a probability of 0.95) that the result would fall within this interval around the unknown population mean value. As the number of measurements increases, the more closely does $\bar{x}$ approach μ, and the more closely can the limits be obtained.

Note that the interval (0.36) is larger than the measured range of values (0.30) because if we made more measurements a few would probably be even further from the mean than those in our set of values. In other cases the interval may be smaller than the range.

How reliable is a mean value?

To improve the reliability of data, soil analyses are normally replicated, and a mean value is given. The reliability of this mean can be judged if we look at the variability between a number of mean values. However it would be a tedious procedure to have to measure many subsamples to obtain several mean values in order to determine their variability. Statistics gives a way of estimating the *standard deviation of the mean values* known as the *standard error*, $s_{\bar{x}}$, from only one set of measurements: $s_{\bar{x}} = s/\sqrt{n}$ where s is the standard deviation of the measured values and n is the number of values. Using the standard error, probability limits can be found for a mean value in a similar way to that described above for a single measurement. However in this case the variability of the measurements is used to determine the limits within which the 'true' value lies with a probability of 95 per cent, known as the *95 per cent confidence limits*.

Imagine that measurements had only been made on the first three subsamples of soil A (2.31, 2.17, 2.20). For these $\bar{x} = 2.227$, $s = 0.0737$ and $s_{\bar{x}} = 0.0426$. It is

known that the 95 per cent confidence limits are $x \pm ts_{\bar{x}}$. This expression is similar to that for the probability limits of all the individual values ($\mu \pm 1.96\sigma$) and the value of t (Table 3.5) decreases towards the limiting value of 1.96 as the number of values measured to give the mean increases. The reason for this is that as n increases so $_{\bar{x}}$ approaches μ and $s_{\bar{x}}$ approaches σ. The value of t decreases as n increases because the greater the value of n, the smaller will be the variability between the means and the greater the confidence we can have in our mean value as an estimate of the true value.

For the three measurements above, $n = 3$ and $t = 4.30$. The 95 per cent confidence limits are $2.227 \pm (4.30 \times 0.0426)$ or 2.044–2.410. Thus by measuring three replicate values we find a mean value of 2.227 and we can be 95 per cent confident (there is a probability of 0.95) that the true value lies between 2.044 and 2.410.

Presentation of results

When mean values are shown in tables of results standard errors or confidence limits are often listed also. If mean values are plotted on graphs, the confidence limits can usefully be added to form a bar (Fig. 3.9) which is the *confidence interval*.

Is the carbon content of soil A greater than soil B?

If we take bulked samples from two sites and analyse their C contents, replicate determinations provide confidence limits for each set of measurements. These can be displayed on a graph. If the confidence intervals do

Table 3.5 Table of t distribution for determining 95% confidence limits and for the Student's t test

Degrees of freedom†	t	Degrees of freedom	t
1	12.71	10	2.23
2	4.30	12	2.18
3	3.18	15	2.13
4	2.78	20	2.09
5	2.57	30	2.04
6	2.45	40	2.02
7	2.37	60	2.00
8	2.31	120	1.98
9	2.26	∞	1.96

† Degrees of freedom: $= n - 1$ for 95% confidence limits;
$= n_A + n_B - 2$ for Student's t test.

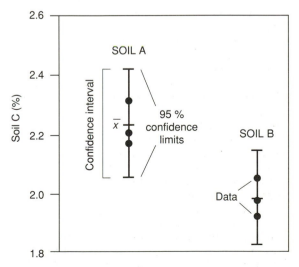

Figure 3.9 The display of confidence limits on graphs.

not overlap then it is likely that the means are significantly different. With increasing overlap it becomes less likely that significant differences exist. It is not correct, however, to draw firm conclusions from overlap of confidence intervals.

Example

Compare the soil data above (soil A) with another soil (B) which gave the following values: 1.98, 2.05, 1.92 for which $n_B = 3$, $\bar{x}_B = 1.983$, $s_B = 0.0651$ and $s_{\bar{x}B} = 0.0376$. The 95 per cent confidence limits of the mean $= 1.983 \pm (4.30 \times 0.0376)$ or 1.821–2.145.

The limits for soil A are 2.044–2.410. It would seem that the means might not be significantly different because their confidence intervals overlap (Fig. 3.9).

Student's t test: a simple test of significance

Calculate the *standard deviation of the difference between the means*. This is given by $s_{(\bar{x}A - \bar{x}B)} = (s^2_{\bar{x}A} + s^2_{\bar{x}B})^{1/2}$. Its value is $(0.0426^2 + 0.0376^2)^{1/2} = 0.0568$. Calculate the actual difference between the means, $\bar{x}_A - \bar{x}_B$, which is $2.227 - 1.983 = 0.224$. Calculate the ratio $(\bar{x}_A - \bar{x}_B)/s_{(xA - xB)}$. Its value is 3.94 and tells us that the difference between the means is almost four times the standard deviation of the difference between the means. This ratio, which is t, indicates the probability of the true values of A and B being significantly different. However, the smaller the number of measurements made to determine each mean, the greater t must be to indicate a significant difference. Table 3.5 gives the values of t which indicate a significant difference depending on the number of degrees of freedom which is $n_A + n_B - 2$ (= 4 in the above example). The value of t indicating a significant difference with a 95 per cent probability for 4 degrees of freedom is 2.78. Our calculated value of t is 3.94, indicating a greater than 95 per cent probability of a significant difference between the soil samples. Values of t for larger numbers of degrees of freedom and for other probabilities can be found in Garvin (1986) and Mead and Curnow (1983), both books explaining Student's t test in more detail.

Section 3.9 Projects

1. Collect earthworm casts and compare their loss on ignition and C and N contents with soil from the horizons in the profile below the collection site (Edwards and Lofty, 1977).
2. Compare the loss on ignition or C contents of (a) a field which permanently grows grass, and an adjacent field which is cultivated, (b) a lawn and an adjacent shrubbery from which fallen leaves are removed every autumn, (c) the horizons of a soil under coniferous woodland and deciduous woodland, or (d) a marshy area and an adjacent well-drained area.
3. Make plant 'pots' out of 50 cm lengths of plastic drain pipe, 15 cm diameter, by gluing (Araldite) a plastic plant pot saucer on to one end of the pipe and drilling drainage holes. Fill these pots with sieved soil (a garden sieve, approx. 10 mm aperture) and plant with wheat or some other plant with roots which can be easily handled (maize and other cereals, brassicas). Further information on pot techniques is in Section 9.2. Water every few days such that a little water drains from the base of the pot. After emergence sample fortnightly, separate the roots and determine dry matter and root length. Compare to data in Fig. 3.1. Calculate root mass per hectare if this crop were growing in a field.
4. Determine the biomass of soils from the sites in (2) above. Comment on the relationships between biomass and the measured soil properties.
5. Obtain a soil which is likely to have a low organic matter content (cultivated field, shrubbery, subsoil). Air dry and pass through a 2 mm sieve. Mix into the soil various treatments: sucrose as a C source (10 g sucrose kg^{-1} soil), N (100 mg N kg^{-1} = 286 mg NH_4NO_3 kg^{-1}), sucrose and N, compost, powdered leaves or straw (10 g oven-dry material kg^{-1}). Moisten the soils (including an untreated soil) to 40 per cent of the water-holding capacity (Section 3.2, Note 2) and determine the effects of the treatments on the biomass by sampling at convenient intervals.
6. Using the information in Section 3.5 simulate the decay of the C in plant residues (Figs. 3.6 and 3.7) using the two fraction model. Vary the amounts of A and B and their half-lives in order to obtain the best fit to the experimental data. Use the same approach with the data in Fig. 13.8 for the site at Rothamsted which was taken from old grassland into arable crops.

Section 3.10 Calculations

1. A soil contains 3 per cent organic matter. If the organic matter contains 58 per cent C, and its C:N ratio is 10:1, calculate the percentage of C and N in the soil. (*Ans.* 1.74 C, 0.17 N)
2. A soil has an organic matter content of 5.2 g per 100 g oven-dry soil. If this soil when air dry contained 2.3 g H_2O per 100 g oven-dry soil, calculate the organic matter content as g per 100 g air-dry soil. (*Ans.* 5.08).
3. It has been estimated that under pastures at Wye College in Kent there are 6 million earthworms

ha^{-1}. If their total mass is 1.7 t, calculate the average mass per worm. (*Ans.* 0.28 g)

Estimates suggest that 2 per cent of the soil in the top 10 cm layer passes through the worms each year. What is the mass of soil per hectare passing through the worms, and how much passes through each worm per year? (Assume 1300 t soil ha^{-1} in the 10 cm layer.) (*Ans.* 26 t, 4.3 g)

Express as a ratio the mass of soil passing through the earthworms per year : mass of earthworms. (*Ans.* 15:1)

Estimates show that the number of earthworm burrows which come to the surface are between 0.5 and 10 million per hectare. Calculate the number per square metre. (*Ans.* 50–1000)

If the average diameter of the burrows is 6 mm, calculate the burrow area as a fraction of the soil surface. (*Ans.* 0.001–0.02)

If these are acting as drainage holes, calculate the diameter of a cylindrical pipe which would have the same area as the burrows per square metre. (*Ans.* 4.2–19.0 cm)

4. Typical elemental concentrations in maize straw, bacteria and soil organic matter are given in Table 3.6. Calculate the C:N ratio for each material. (*Ans.* 86:1, 5.6:1, 12.9:1)

If soil bacteria are feeding on the straw and assimilating 60 per cent and respiring 40 per cent of the C, what percentage of their N can be derived from the straw? (*Suggestion*: start with 1 g of straw; 60 per cent of its C is taken up by the bacteria. Assume that all the N in the straw is available for the bacteria and

compare this to the amount of N required by the bacteria to maintain their C:N ratio). (*Ans.* 11)

Using the same principles calculate the percentage of the P and S in the bacteria which can be derived from the straw. (*Ans.* 7, 63). From what sources might the extra N, P and S be obtained?

5. Fig. 11.4 shows the distribution of dry matter between the various parts of a wheat crop. Calculate the inputs of carbon to the soil at harvest as t ha^{-1} if the straw is (a) ploughed into the soil or (b) burnt before ploughing. Assume that the dry matter contains 43 per cent C. (*Ans.* (a) 3, (b) 0.4).

Assume that these inputs are occurring each year in a soil containing 26 t C ha^{-1} (Table 3.3). Calculate an average turnover time in each case if the C content of the soil is not changing from year to year. (*Ans.* 9 and 65 years). *Note*: for similar soil and climatic conditions the larger input would lead to a larger amount of C in the soil once the soil C had come to equilibrium with the cropping system and climate.

6. In Fig. 3.1 the total mass of wheat roots in June under 1 cm^2 of soil surface to a depth of 1 m occurs in a soil volume of 100 cm^3. If the mass of soil in this volume is 130 g (it has a bulk density of 1.3 g cm^{-3}, Section 4.2), calculate the percentage of roots (*m/m*) in the whole profile. Also calculate the percentage in the top 15 cm of soil. (*Ans.* 0.009, 0.04).

If the organic matter content of the top 15 cm of soil is 3 per cent (*m/m*), express the mass of roots as a percentage of the organic matter in this layer. (*Ans.* 1.3)

Calculate the average distance between roots in the soil in June. *Suggestion*: calculate the average root density L (cm root cm^3 soil). Imagine the root growing through the centre of a block of soil L centimetres long and 1 cm^3 in volume. Calculate the dimensions of the block. The nearest root is in the centre of an adjacent similar block of soil. (*Ans.* 0.75 cm)

If the dry matter content of roots is 6 per cent of their fresh weight and fresh roots have a density of 1 g cm^{-3}, calculate the total volume of roots in June, as a percentage of the soil volume. (*Ans.* 0.2)

Table 3.6 Typical values for the elemental composition of dry material (mg g^{-1}).

Element	Maize straw	Bacteria	Soil organic matter
C	430	430	580
N	5	77	45
P	1	24	12
S	3	8	5

CHAPTER 4

The Arrangement of Particles and Pores: Soil Structure

SOILS ARE POROUS

Particles in soil are in contact with one another. Between the particles spaces or *pores* are present which contain air and water. During soil formation large spaces are formed by physical and biological action, and so soils are even more porous than would be expected from a random packing of particles. This is an important soil characteristic; soils are described as being unconsolidated, which implies that they can be consolidated, i.e. pores may be reduced in volume when a force is applied. Soils are characteristically less consolidated than the parent materials from which they are formed.

Soil *porosity* is part of the property known as *soil structure* which includes the arrangement of particles into *aggregates* (groups of particles, also called *peds*), and the size, shape and distribution of pores both within and between the aggregates. It is of vital importance in the ability of soils to support plant, animal and microbial life. The spaces hold water, allow drainage, allow entry of O_2 and removal of CO_2 from the soil, allow roots to penetrate, and are indirectly responsible for modifying the mechanical properties of soils so that cultivation can be carried out successfully.

The porosity of unstructured material

All materials composed of particles contain pores. In Fig. 4.1 spherical sand particles are shown in open and close packing. Their calculated porosities are $0.48 \, cm^3$ of pores cm^{-3} of material for open packed spheres and between 0.26 and 0.3 for close packing depending on the arrangement. The porosity of the materials composed of particles of mixed size is less than that for uniform particles and for soil materials the range of particle size would result in very small porosity values (about $0.2 \, cm^3 \, cm^{-3}$) when fully consolidated. Also, because clay-sized particles tend to have a platey shape they can fit together in parallel arrangement (Fig. 2.2) and give low porosities. For example, a clay parent material typically has a porosity when dry of $0.25 \, cm^3 \, cm^{-3}$. Thus the fact that surface horizons of soils have porosities between 0.4 and $0.6 \, cm^3 \, cm^{-3}$ shows clearly

that the development of soil structure involves the formation of a large volumes of pores.

The processes involved in the development of structure are:

1. *Physical:*
   (a) drying and wetting which cause shrinkage and swelling with the development of cracks and channels;
   (b) freezing and thawing which create spaces as ice is formed.
2. *Biological:*
   (a) the action of roots, which remove water resulting in the formation of spaces by shrinkage, release organic materials, and leave behind organic residues and root channels when they die;
   (b) the action of soil animals which move material, create burrows and bring mineral and organic residues into close association;
   (c) the action of micro-organisms which break down plant and animal residues, leaving humus as an important material binding particles together.

The formation of soil structure thus requires both physical rearrangement of particles and the stabilization of the new arrangement. Stability is particularly associated with organic materials linking mineral particles together and with the clay minerals and sesquioxides. It is here that the chemical properties of the system become important, the binding between clay particles depending on the ions associated with their surfaces and in soil solution (Section 14.4).

The measurement of soil porosity

As soils become more porous they also become less dense, and porosity can be determined from a measurement of the dry density of soil sampled so as to preserve its natural structure. This density depends both on the proportion of the total volume of soil occupied by air since the air in the pores contributes no weight when the soil is in air, and also on the density of the particles themselves. Therefore both *particle density* (Section 4.1)

(a)

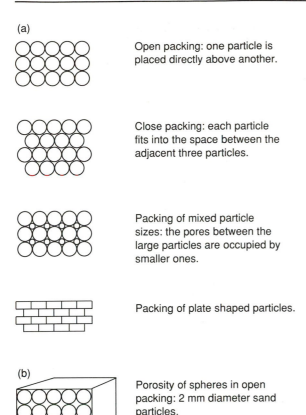

Open packing: one particle is placed directly above another.

Close packing: each particle fits into the space between the adjacent three particles.

Packing of mixed particle sizes: the pores between the large particles are occupied by smaller ones.

Packing of plate shaped particles.

(b)

Porosity of spheres in open packing: 2 mm diameter sand particles.
Number of sand grains per cm^3 = $5 \times 5 \times 5 = 125$
The volume of the sand grains = $125 \times 4\pi r^3/3$
= 0.52 cm^3 with 0.48 cm^3 of pores.
The porosity is 0.48 cm^3 per cm^3.

Figure 4.1 (a) Particles packed in various ways; (b) porosity of spheres in open packing.

and the density of the whole soil, known as *dry bulk density* (Section 4.2), have to be determined in order to calculate porosity. *Void ratio* is an alternative measure of pore space commonly used in soil mechanics (Section 4.2).

Values for the densities of particles commonly present in soils are given in Table 4.4. For soils with low organic matter contents (<3%) which includes many cultivated soils, an average value of 2.6 g cm^{-3} can be used. These are often termed *mineral soils*. For soils with higher organic matter contents average particle density is lower and can be measured or calculated (Section 4.1). Soils with organic matter contents >15 per cent are often called *organic soils*.

Soil porosity depends on the structure of the soil. It therefore varies depending on:

- texture and organic matter content, since these both control to some extent the type of structure which develops;
- depth in the profile, because structure-forming processes are most active near the soil surface and consolidation increases with depth;
- management, because this causes changes in organic matter content over time and applies forces to soils which may either loosen or compact them (Colour Plates 5 and 7). The terms consolidation and compaction are defined below.

Organic matter and the associated biological activity in soils are of major importance in maintaining soil porosity. Typical values are given in Table 4.1. High porosities are also found in soils of the humid tropics containing appreciable contents of iron oxide or allophane because of the presence of pores within the mineral particles. Sandy soils, particularly fine sands, tend to contain smaller amounts of organic matter than heavier-textured soils, structure is less stable and, even though they are easily loosened by cultivation, natural consolidation rapidly follows when they are rewetted.

Table 4.1 Typical densities and porosities of soils

	Particle density (g cm^{-3})	Dry bulk density (g cm^{-3})	Porosity (cm^3 cm^{-3})
Cultivated mineral soils, plough horizons			
medium–heavy texture	2.60	0.8–1.4	0.69–0.46
light texture	2.60	1.4–1.7	0.46–0.35
Subsoils and parent materials	2.65	1.5–1.8	0.43–0.32
Grassland and woodland, A horizons	2.4	0.8–1.2	0.67–0.50
Peats	1.4	0.1–0.3	0.93–0.79

The porosities are calculated from the bulk densities and particle densities. The bulk densities are for soils sampled at field capacity.

Pores have varying shape and size

Observation of soil structure in the field is confined to the appearance of aggregates and to the pores which are visible without magnification or with $\times 10$ magnification using a hand lens. This is termed *macrostructure* and the pores are termed *macropores*, being larger than about $50\,\mu\text{m}$ (0.05 mm). An example is shown in Colour Plate 4(a) and in diagrammatic form in Fig. 4.2. Field methods are available for the description of macrostructure and for estimating the percentage of macropores (Section 1.2). A method for measuring macroporosity is described in Section 4.2. By subtracting the macroporosity from the measured total porosity, a value is found for the volume of smaller, non-visible pores in the soil. These are *micropores*, less than about $50\,\mu\text{m}$ in size. Large thin sections of soil (Colour Plate 5(a) and (b)), although difficult to make, allow

more critical observation of macropores and can be combined with point-counting methods to give quantitative estimates of macroporosity.

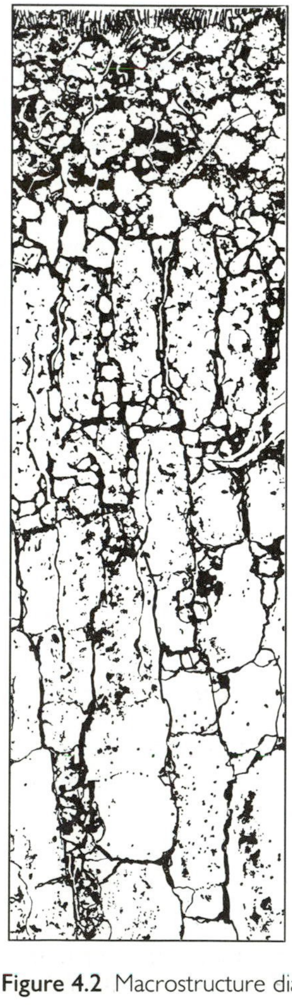

Classification of pores according to function

It may seem arbitrary to classify pores as visible or non-visible and even more arbitrary to choose $50\,\mu\text{m}$ as the size limit. Apart from being a convenient classification, it is related to the function of pores: the macropores are those which allow rapid drainage of water after heavy rainfall or irrigation, and once these pores are emptied drainage becomes very slow (Ch. 5). The soil is then said to be at *field capacity*. Further removal of water by evaporation from the soil surface or transpiration from leaves removes water from micropores, and a second critical state in relation to soil water is reached when plants permanently wilt, termed the *permanent wilting point* (Ch. 12). Here water is held only in pores less than about $0.2\,\mu\text{m}$ in size. The amount of water held between these two states is termed the *available water capacity*. This leads to three classes of soil pores as shown in Table 4.2.

The relationship between pore size and function means that the volume of macropores (and thus the volume of micropores) can be determined from the total

Fine angular blocky and granular

10 cm

Blocky and prismatic

30 cm

Prismatic breaking to blocky

Figure 4.2 Macrostructure diagram.

Table 4.2 Classification of pores by size and function

Pore class	Pore size (μm)	Pore function
Transmission = macropore	>50	Drainage after saturation
		Aeration (movement of O_2 and CO_2) when the soil is at field capacity. Both aeration and drainage require pores to be continuous in the vertical direction
		Root penetration: for many arable crops pores >0.2 mm are required
Storage = micropore†	50–0.2	Store water available for plant use
Residual = micropore†	<0.2	Hold water so strongly that it is not available to plants. This water is mostly associated with clay-sized particles and controls to a large extent the mechanical strength of the soil

† Storage and residual pores are together classified as micropores.

Table 4.3 Typical volumes of pores ($cm^3\,cm^{-3}$) as indicated by water contents at field capacity and the permanent wilting point in soils of varying texture. A more detailed set of values is in Table 12.1

	Texture			
	Light	Medium	Heavy†	Peats‡
Transmission pores	0.2–0.3	0.10–0.15	0.05–0.15	v
Storage pores	0.05–0.15	0.20–0.25	0.15–0.2	v
Residual pores	0.05–0.1	0.15–0.2	0.25–0.35	v
Total porosity	0.35–0.45	0.45–0.55	0.50–0.70	about 0.8

† Heavy textured soils shrink and crack during drying, increasing the volume of transmission pores. Much of the available water is released by shrinkage of very small pores (Section 4.3). In such soils changes in pore size during drying complicate the interpretation.

‡ v: Values are variable and difficult to measure because of extensive shrinkage and slow reswelling. A saturated peat typically loses little water by drainage, so storage + residual pore volume is about 0.8. At the wilting point about 0.7 g water g^{-1} oven-dry soil may remain. Relative to the volume of wet peat, this gives a residual pore volume of about 0.1 and thus a storage pore volume of 0.7.

porosity and the water content of the soil at field capacity (Section 4.2). Similarly the water content at the wilting point (Ch. 12 and Table 12.1) gives the volume of residual pores, which subtracted from the micropore volume gives the storage pore volume. Typical values for these three pore classes are shown in Table 4.3. There are three critical conditions:

1. If transmission pore volume is below about $0.1\,cm^3\,cm^{-3}$ then drainage problems may occur.
2. If storage pore volume is below about $0.15\,cm^3\,cm^{-3}$ then water availability is likely to be restricted.
3. Residual pore volumes above about $0.2\,cm^3\,cm^{-3}$ suggest that the soil may have difficult mechanical properties, being plastic and sticky when wet and hard when dry.

The term *microstructure* is used for the arrangement of particles and pores which can be seen using a microscope (Plates 2.2 and 2.4 and Fig. 4.3). Micropores and the arrangement of clay-sized particles are important aspects of microstructure. Residual pores dominate the pore system in domains of clay-sized material. Storage and residual pores are developed by structure-forming processes. A sample of clay parent material taken from below the depth of drying in a soil profile may contain only fine storage and residual pores. On drying, shrinkage occurs decreasing further the size of these pores, but with the introduction of cracks which are large transmission pores. In contrast, the surface horizon

Soil crumb thin section, light microscope (x 70 magnification)

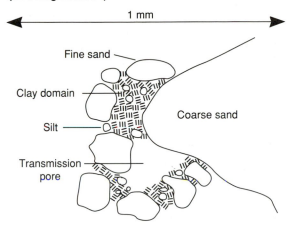

Electron microscope (x 3000 magnification)

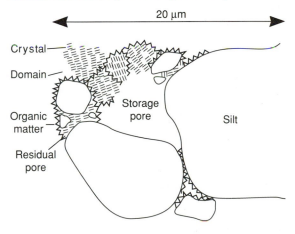

Figure 4.3 Microstructure diagram.

may have a distribution of pore size which gives excellent properties in terms of physical conditions and biological activity, showing the effectiveness of the structure-forming processes. Section 4.3 describes a method for studying swelling and shrinking properties of clays and soils, and Section 14.4 deals with the behaviour of clays in saline and sodic soils.

THE STABILITY OF SOIL STRUCTURE

Over a period of time loose, freshly cultivated soils consolidate. *Consolidation* is natural settling due to the force of gravity. A soil under natural vegetation does not consolidate: it is in equilibrium with its environment and its structural stability enables it to withstand natural forces. *Compaction*, like consolidation, causes a reduction in porosity but the term is used for the effect of external forces applied to soils. The most common examples are compaction under the wheels of farm machinery (Plate 4.1, and Section 4.4) or under the feet of animals and man (Bullock and Gregory, 1991, Ch. 6). A ploughshare, in lifting and turning soil, transmits a force to the soil at the base of the furrow via the wheels of the plough or tractor. The share may also smear the soil surface so rearranging particles into a more compact skin. These combined effects result in the formation of a *plough pan* (Fig. 4.4 and Plate 4.2) which may restrict the drainage of water and the penetration of air and roots. Any cultivation tool may compact and smear soil, and the damage can have important effects on plant growth. Raindrop impact can be particularly damaging to exposed soil surfaces, with the force of impact disrupting aggregates and allowing particles to be washed into closer packing. This can form a cap on the soil surface which can reduce seedling emergence. It will also reduce the rate at which water can enter the soil and may cause runoff and erosion (Ch. 12).

The resistance of soil structure to damage

The ability of soil to withstand these effects depends both on structural stability and on the water content of the soil. The structural stability of soils varies under

Plate 4.1 The effects of compaction by tractor wheels on the infiltration of water into soil. Photograph by D. Campbell, Scottish Centre of Agricultural Engineering.

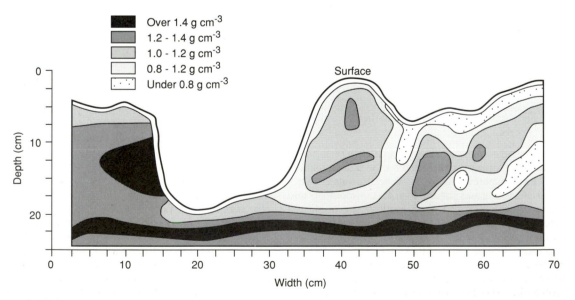

Figure 4.4 Bulk density distribution in a soil before ploughing (left) and after ploughing (right). Compaction of the furrow wall by the tractor wheel and the plough, and a clear ploughpan can be seen. By D. Campbell, Scottish Centre of Agricultural Engineering.

Plate 4.2 Ploughing a medium textured soil, southern England. Soil conditions are good and there is little smearing on the top of the turned soil and in the bottom of the furrow. Compaction due to the tractor wheel in the furrow is likely to be the only deleterious effect on soil structure.

natural conditions, with the fine sands and silts being weak as a result of their low content of clay minerals, sesquioxides and humus. Soils rich in kaolinite clays and iron oxide in the humid tropics may also be weak due to the association of these materials into very small aggregates which behave like silt and fine sand. Soils influenced by sodium salts (Section 14.4) are also naturally unstable.

Cultivation of land for crop production leads to reduced structural stability as the organic matter content of soil decreases. All arable cropping has this effect unless very large inputs of organic manure are used each year. The extent to which stability can be reduced before crop yields are affected or problems such as erosion begin will vary depending on the type of soil. Successful soil management requires that the organic matter is maintained above a critical content depending on soil type, the site, climate, and the use of the soil. Ley–arable systems (rotations of grass and arable crops) maintain larger organic matter contents than do continuous arable crops (Section 13.3).

The mechanical strength of a soil decreases as it be-

comes wetter, and compaction and smearing are problems associated with wet soils. Successful soil management involves not taking machinery over land or cultivating when it is too wet. When soils are too dry, cultivation may produce large lumps of soil rather than the tilth required for the establishment of a crop. Thus there is a range of water contents for any soil over which it can be cultivated successfully. This is known as the *cultivation window*; it is the number of days the soil is in a suitable state for cultivation when soils are rewetting at the end of a dry season or are drying at the end of a wet season. Stable structure and large organic matter contents extend this period and so increase the flexibility of soil use by the farmer. The same principles apply to the cultivation of a vegetable garden. Here maximum flexibility of management may be important because of interest in year-round cropping and because many people cannot choose when they have free time for cultivations. A large organic matter content is more easily maintained on this small area through the use of composts where the economics of the operation are not a major consideration.

Summary

Deterioration of soil structure occurs in two ways. Firstly, by compaction, and secondly when stability decreases which is normally the result of a decrease in organic matter content. The latter makes the soil more sensitive to compacting forces.

The measurement of structural stability

Structural stability is commonly measured by wet sieving in which soil is shaken under water in a nest of sieves. As the aggregates break down they pass through the sieves and the masses of stable aggregates of given sizes can be measured (Page, 1982). Similarly, the Emerson (1967) test gives a stability index for 3–5 mm diameter aggregates when immersed in water. Both methods give a quantitative measure of the stability of aggregates under standard but arbitrary conditions, and as such only give an empirical assessment of structural stability in the field.

Soil structure can be protected against damage

A knowledge of the factors controlling structural stability and the processes involved in damaging soil structure has led to significant changes in soil management in temperate climatic regions of the world. More powerful machinery, although often applying higher stresses to soils, allows a greater area of land to be cultivated in a given time, and may allow cultivations to be carried out only within the cultivation window. Larger tyres with low inflation pressures reduce stress on soil for a given load (Section 4.4 and Colour Plate 3). The trend towards production of autumn-sown cereal crops instead of spring-sown crops in Britain has led to cultivations being carried out in the autumn when soils are drier than in the spring thus avoiding in most years the problems of cultivating wet soils. Cereal crops are now produced using the *tramline* system in which all machinery uses the same wheel tracks through the field once the crop has been drilled (Colour Plate 6). Compaction is thus restricted to a very small part of the field, and if required deep tine cultivation along the tramlines can be carried out in the autumn to loosen the soil.

Horticultural crops are also produced using a 'bed' system with tractor wheelings restricted to the same tracks throughout the growth period, and on a garden scale, raised beds serve a similar purpose. More difficult in many agricultural and intensive horticultural systems is the maintenance of structural stability as this requires large organic matter inputs. Mixed farming systems using ley-arable rotations and livestock enterprises are the traditional effective way of maintaining organic matter contents. For limited areas of land large amounts of organic manures can be brought in from farms with dairy or beef units or from riding stables. Maintenance of structural stability by these methods is part of the broader problem of maintaining soil fertility, which together with organic farming systems, is discussed in Ch. 13.

Damage to soil structure is reversible

Because soil structure is developed by natural processes, suitable management may allow structure to be reformed over a period of time. The use of grass leys following arable crops is the most common way of regenerating structure. Not only is the organic matter slowly increased, but the crop itself develops a dense root system and encourages greater activity of earthworms and other organisms. The incorporation of organic manures has a similar effect.

The effects of compaction can be counteracted to some extent by cultivation. Although the soil is loosened and a system of large transmission pores produced, the aggregates themselves may still be compact. The loosened soil will consolidate, with structural stability governing the long-term effectiveness of the treatment. It can, however, be very effective in removing problems associated with plough pans and compaction in tramlines. The effects of various cultivation treatments are examined in Section 4.4.

FURTHER STUDIES

Ideas for projects are given in Section 4.5 and calculations are in Section 4.6.

Section 4.1 The determination of the density of soil particles

Table 4.4 gives the definition of density, commonly used units and values for soil minerals and soil organic matter. Determination involves the measurement of the volume of a known mass of particles. The soil is dispersed (the particles separated) in water and all the air is expelled from the suspension. In a known volume of suspension the volume occupied by the particles is then found.

Method

Weigh a 250 ml beaker (± 0.01 g). Add about 25 cm^3 of <2 mm oven-dry soil and reweigh. Add about 50 cm^3 of water and boil gently for 30 min. This disperses the soil and expels the air (Note 1). Cool the suspension by

Table 4.4 The determination of soil particle density

Definition

$$Density = mass/volume$$

Units

$kg\,m^{-3}$
$g\,cm^{-3}$
$t\,m^{-3}$ (numerically equal to $g\,cm^{-3}$)
For convenience in the laboratory, $g\,cm^{-3}$ is used.

Values for soil materials

	$(g\,cm^{-3})$
Water	1.0
Quartz and clay minerals	2.65
Iron oxide	>3.0
Calcium carbonate	2.71
Opaline silica	2.2
Organic matter	1.2–1.5

For other materials see Nuffield Advanced Science *Book of Data* (1984) or Kaye and Laby (1973).

standing the beaker in running water. Weigh a 250 ml volumetric flask ($\pm 0.01\,g$). Pour the suspension through a funnel into the flask using a wash bottle to ensure that all the soil particles have been transferred. Make up to the mark and weigh the flask plus suspension.

Calculation (using typical data)

	(g)
Mass of beaker	92.44
Mass of beaker + dry soil	117.76
Mass of dry soil (*a*)	25.32
Mass of flask	92.36
Mass of flask + suspension	358.01
Mass of suspension (*b*)	265.65

The mass of water in the flask can now be found as $b - a = 240.33\,g$. Since the density of water is $1\,g\,cm^{-3}$ (Note 2), then the volume of water in the flask = $240.33\,cm^3$.

The volume of particles in the flask =
volume of the flask – volume of water = $250 - 240.33$
$$= 9.67\,cm^3$$

and thus the density of the particles = $25.32/9.67 = 2.62\,g\,cm^{-3}$.

Depending on the use to be made of the measurement, a number of replicate determinations should be

made. For a sandy loam, the following were obtained: 2.45, 2.42, 2.46, mean 2.443; standard error = 0.0120; 95 per cent confidence limits = 2.391–2.495 (Section 3.8).

Note 1 It is difficult to disperse the soil completely and remove all the air. Even after boiling for 30 min trapped air may give errors (low particle densities) particularly in heavy-textured soils.

Note 2 The density of water at 24 °C is $1.00\,g\,cm^{-3}$ and varies slightly with temperature. For practical purposes here, a density of $1\,g\,cm^{-3}$ can be assumed.

Particle density of organic soils

The average particle density of a mixture of organic and mineral particles can be calculated if the organic matter content is known, and if it is assumed that the mineral particles have a density of 2.65 and the organic matter $1.4\,g\,cm^{-3}$.

Calculation

For a soil with 15 per cent organic matter, there is 15 g of organic matter in 100 g of oven-dry soil, with 85 g of mineral particles. Since volume = mass/density, then for each fraction the volume of the particles can be found:

$$volume\ of\ organic\ matter = 15/1.4 = 10.71\,cm^3$$
$$volume\ of\ mineral\ matter = 85/2.65 = 32.08\,cm^3$$
$$total\ volume = 42.79\,cm^3$$
$$average\ particle\ density = 100/42.79 = 2.34\,g\,cm^{-3}$$

Other values are 3 per cent, $2.58\,g\,cm^{-3}$; 5, 2.54; 10, 2.43; 20, 2.25; 30, 2.09 and 50, 1.83. Thus for cultivated soils with organic matter contents up to about 4 per cent, a value of $2.6\,g\,cm^{-3}$ can normally be assumed.

Section 4.2 The determination of dry bulk density and porosity

Dry bulk density

The mass of oven-dry soil present in a given volume of naturally structured soil has to be found. The water content at the time of sampling has to be considered because of the swelling and shrinking of soils.

Plate 4.3 Equipment used to measure the bulk density of soils. The guide plate and cylinder (right) is placed on the surface of the soil. The sampling cylinder (centre) fits into the driving tool (left), being held in place by a rubber O-ring. The driving tool and cylinder are then inserted into the guide cylinder and hammered into the soil. The driving tool allows soil to extend beyond the end of the sampling cylinder, so avoiding compaction.

Equipment (Plate 4.3)

Open ended metal cylinder: 5 cm internal diameter × 5.1 cm long × 3 mm thick together with caps.
Driving tool, hammer and knife.

Method

Samples may be taken vertically into the soil or horizontally from the face of a soil pit. Prepare the soil surface with a trowel or knife. Place the cylinder against the soil with the driving tool in place. Gently hammer the cylinder into the soil until the soil projects 3 mm out of the cylinder. Excavate the cylinder plus soil leaving extra soil extending from each end of the cylinder. Trim the soil flush to the ends of the cylinder and put on the caps. The excavation tool aids removal of the cylinder and trimming of the soil. If the samples are to be used for water content determinations (Section 5.1) place the sample in a polythene bag. Label and close the bag.

In the laboratory, weigh the sample moist if required, dry at 105 °C to give oven-dry soil and re-weigh. The mass of the empty cylinder plus caps is also required. If many samples are required and cylinders and caps are reasonably uniform, then the average mass of one cylinder plus caps can be used for all the

samples, since errors arising from mass differences are small compared to sample variability.

Calculation of dry bulk density (using typical data)

	(g)
Mass of cylinder + caps + dry soil	224.28
Mass of cylinder + caps	77.02
Mass of oven-dry soil	147.26

Volume of cylinder = $\pi r^2 L = \pi \times 2.5^2 \times 5.1 = 100\,cm^3$. Therefore oven-dry bulk density is

$$147.26/100 = 1.47\,g\,cm^{-3}$$

Sample Variability

Because soils are heterogeneous in the field (spatially variable in their properties), bulk density varies appreciably from place to place even within the same horizon. Particularly where large pores occur there will be much variability between samples. Replicates are therefore needed. For a cultivated sandy loam, 0–25 cm horizon, the following values were obtained: 1.44, 1.50, 1.58, mean 1.507, standard error = 0.0406, 95 per cent confidence limits = 1.333–1.681.

The use of small cylinders tends to overestimate field bulk density because the operator is inclined to avoid cracks and channels. The method given below under 'Problem soils' samples a larger volume of soil and gives a better estimate of bulk density.

Swelling and shrinking

Soils containing appreciable amounts of organic matter or clay shrink when dried (Section 4.3). The soil is consolidated (the bulk density increases) by a lowering of the soil surface and cracks open up, so causing difficulties in choosing a representative sample. Soil between cracks will have a greater bulk density than that of the whole soil. For these reasons, and because sampling is easier, the soil should be sampled moist, preferably at field capacity, and the moisture content measured and recorded.

Problem soils

Soils containing large roots, large pores or many stones are difficult to sample using bulk density cylinders. The method described in Section 1.2 to measure the stone content of soils should be used, modified as follows.

Sampling Excavate the soil from a volume 30 × 30 × 20 cm. Weigh the moist soil. Mix thoroughly and place a subsample in a polythene bag. Seal and label. In the laboratory, weigh the moist subsample, dry at 105 °C and reweigh. Calculate the oven-dry mass of the whole sample in the field.

Measuring the soil volume With care, soil can be excavated from the above volume ($18\,000\,cm^3$). However, it is preferable to measure the volume after excavation as follows. Pour into the hole 2 cm diameter 'Allplas' plastic balls (Capricorn Chemicals Ltd, Lisle Lane, Ely, Cambridgeshire CB7 4AS) until the balls are level with the top of the hole. Weigh the balls required to fill the hole. Determine the ratio of ball mass to volume by filling a similar shaped container of known volume (a biscuit tin) with balls and weigh.

Example

	(g)
Mass of moist soil in the field	30 370
Mass of moist subsample	105.20
Mass of oven-dry subsample	84.16
Therefore, mass of oven dry soil in the field	24 296
Mass of balls required to fill a tin $20 \times 20 \times 10\,cm^3$	528
Mass of balls required to fill the sample hole	2 425

Therefore, volume of the sample hole = $4000 \times 2425/528 = 18\,371\,cm^3$ and bulk density = $24\,296/18\,371 = 1.32\,g\,cm^{-3}$

The bulk density of <10 mm sieved soil and the fine earth bulk density

Following the method in Section 1.2 the oven-dry mass of soil (<10 mm or <2 mm) in the field sample can be found. When divided by the volume of the hole, the bulk density of the sieved soil is found.

Aggregate bulk density

The method given in Section 1.2 for determining the density of porous stones should be used.

Conversion of laboratory to field data

The conversion of laboratory bulk density values ($g\,cm^{-3}$) into field values ($t\,ha^{-1}$) is given in Section 3.7.

Calculation of porosity

The volume of the particles in a known volume of soil has to be calculated, the remainder of the volume being pore space.

Calculation (using typical data)

Particle density = $2.65\,g\,cm^{-3}$ particles
Dry bulk density = $1.50\,g\,cm^{-3}$ oven-dry soil

If $1\,cm^3$ soil contains 1.50 g particles, the volume of these particles can be calculated from the particle density:

2.65 g particles occupy $1\,cm^3$
1.50 g particles occupy $1 \times 1.50/2.65 = 0.57\,cm^3$

Therefore in $1\,cm^3$ of soil, $0.57\,cm^3$ is occupied by particles and the volume occupied by pores is $1 - 0.57 = 0.43\,cm^3$. The porosity (often given the symbol θ) = $0.43\,cm^3$ pores cm^{-3} soil or 43 per cent (*v/v*). Because porosity is expressed as a volume ratio it can be given without units, i.e. $\theta = 0.43$.

The calculation is summarized by the equation

$$\text{porosity} = 1 - (\text{bulk density/particle density}) \quad [4.1]$$

Bulk density and porosity of peats and organic soils

The bulk density is low compared to mineral soils, typically between 0.2 and $0.3\,g\,cm^{-3}$ for organic soils and even lower for peats, reflecting both the small particle density and the large porosity. For a peat soil with a bulk density of $0.25\,g\,cm^{-3}$ and a particle density of $1.4\,g\,cm^{-3}$, the porosity is $0.8\,cm^3\,cm^{-3}$ (Table 4.3). For this bulk density the masses of soil in the field are 375 and $500\,t\,ha^{-1}$ for the 0–15 and 0–20 cm layers respectively (Section 3.7). However, much shrinkage occurs during drying, and the values obtained depend on the moisture content when sampled.

Void ratio

In a soil which undergoes a change in volume by compaction, swelling or shrinking, porosity is altered because of changes in both pore space and soil volume. There are advantages in using a measure in which pore space is related to particle volume or mass which does not change. The void ratio is cm^3 pores cm^{-3} particles. It can be determined from bulk density and particle density as follows. If we consider 1 g of oven-dry soil, then the volume of soil = 1/bulk density, particle volume = 1/particle density and the difference is the volume of pores. Thus

$$\text{void ratio} = (1/\text{bulk density} - 1/\text{particle density}) / (1/\text{particle density})$$

which simplifies to (particle density/bulk density) − 1. In the example above its value is $(2.65/1.50) - 1 = 0.77\,cm^3$ pores cm^{-3} particles.

The determination of macro- and microporosity

Observation and estimation

Because macropores (<0.05 mm) are visible without magnification, estimates of macroporosity can be made in the field (Section 1.2). Thin sections show the distribution of pores more clearly, especially if magnified. The difference between measured total porosity (above) and estimated macroporosity gives an estimate of microporosity.

Colour Plate 5(a) and (b) shows the soil fabric from the surface of a permanent grassland site in southern Scotland. There is clearly a significant volume of visible pores, amounting to perhaps 30 per cent of the area. Provided this also represents the volume distribution of pores, the macroporosity is about $0.3 \, cm^3 \, cm^{-3}$. This estimate can be quantified using *point counting* methods (Fitzpatrick, 1980).

Measurement in the field

Macropores (transmission pores) are by definition drained at field capacity and micropores (storage + residual pores) are water-filled. The water content at field capacity is therefore a measurement of microporosity and the difference between total porosity and microporosity is a measure of macroporosity.

Soil is brought to field capacity by watering the soil in the field until drainage occurs. As a guide, soils at the end of the dry summer period require about 200 l of water per m^{-2} of soil surface to bring a 1 m depth of soil to field capacity (Table 12.1). If the soil is already close to field capacity, smaller amounts of water will be needed. After watering, the soil should be covered with a polythene sheet for 2 days to prevent evaporation and allow drainage.

Sample the soil using bulk density cylinders, and determine the water content of the soil by weighing moist, drying at 105 °C and weighing again.

Example

	(g)
Mass of moist soil + container (a)	256.68
Mass of dry soil + container (b)	224.28
Mass of water ($a-b$)	32.40
Mass of container (c)	74.02
Mass of oven-dry soil ($b-c$)	150.26

Calculation

Calculate the dry bulk density as shown in Section 4.2. For the data above, this is $1.50 \, g \, cm^{-3}$.

Calculate the water content as g H_2O g^{-1} oven-dry soil = $0.22 \, g \, g^{-1}$ or $0.22 \, cm^3 \, H_2O \, g^{-1}$ oven-dry soil.

The volume of 1 g of oven-dry soil (the specific volume, Section 4.3 Note 1) can be calculated from the bulk density: $1 \, g / 1.50 \, g \, cm^{-3} = 0.667 \, cm^3$.

Thus at field capacity there is $0.22 \, cm^3$ of H_2O in $0.667 \, cm^3$ of soil or $0.22/0.667 = 0.33 \, cm^3 \, H_2O \, cm^{-3}$ soil which is the microporosity.

The total porosity of this soil having a bulk density of $1.50 \, g \, cm^{-3}$ is $0.43 \, cm^3 \, cm^{-3}$. Therefore the macroporosity = $0.43 - 0.33 = 0.10 \, cm^3 \, cm^{-3}$.

The microporosity can be divided into residual and storage pore volume by measuring the water content at the wilting point (Section 5.2).

Table 4.3 gives the range of porosities for soils of varying texture.

Section 4.3 The swelling and shrinking of soils

The effects of shrinkage are clearly seen in the surface of heavy-textured soils at the end of a dry period. Cracks form in a roughly hexagonal pattern and can be several cm wide and up to 50 cm deep (Colour Plate 21(a)). Similar effects occur when a trench or road cutting allows exposed subsoils to dry.

Shrinkage occurs as a result of water evaporating from soil in which the pores are predominantly so small that instead of air entering a rigid framework of particles and pores as water is removed, the clay particles move together to reduce the size of the pores. Thus the whole body of clay material has to shrink. Vertical shrinkage can be accommodated by a lowering of the soil surface, but horizontal shrinkage can only occur with the formation of vertical cracks. Drying from the faces of the cracks at depth in the soil can then begin. The stresses resulting from this localized drying cause horizontal cracks to form which penetrate from the face of the vertical ones (Fig. 4.5) and in a freshly exposed clay material the development of soil structure is initiated. Rewetting causes swelling, but the crack faces do not become exactly realigned, and on them a skin of compressed material begins to form (Colour Plate 4(b)). The next drying and wetting cycle will tend to open up the same cracks which have now become planes of weakness, with further development of the skins which are called *slickensides*.

An example of vertical shrinkage in a clay soil is shown in Fig. 4.6. The year 1976 was very dry, and the

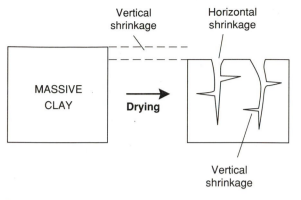

Figure 4.5 Cracking of a clay by drying.

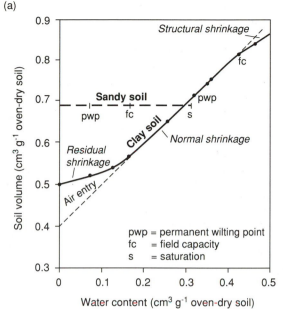

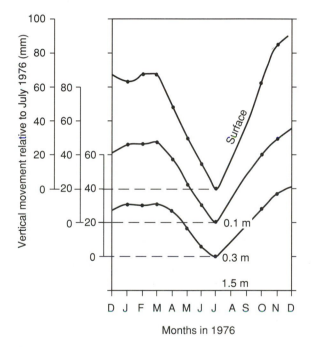

Figure 4.6 Vertical movement at various depths in a soil on Gault Clay at Compton Beauchamp, near Oxford, UK. The soil contained 42–55 per cent clay, predominantly smectite. Data by D. Payne.

surface fell by 67 mm in the first 6 months of the year and then rose by 90 mm by the end of the year.

The shrinkage of aggregates of clay soil is shown in Fig. 4.7(a). They were sampled when wet, and their volume and water content determined during drying. Three stages are observed:

1. *Structural shrinkage* occurs initially when clay begins to shrink, but at the same time air enters a few large pores.

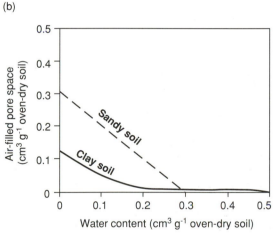

Figure 4.7 (a) The shrinkage of aggregates of clay soil sampled from the Drayton Experimental Husbandry Farm, Stratford-upon-Avon. (0–40 cm depth; 57% clay, predominantly smectite; particle density $2.65\,\mathrm{g\,cm^{-3}}$). (b) The dashed line shows the behaviour of non-shrinking sandy soil with a bulk density of $1.45\,\mathrm{g\,cm^{-3}}$. From Lawrence *et al.* (1979).

2. *Normal shrinkage* occurs when the change in soil volume is equal to the change in water content. No air enters, giving a 45° line on the figure. This covers the range of soil water contents from field capacity to well below the wilting point for this soil.

3. *Residual shrinkage* occurs at small water contents

when cracks begin to form and air enters these and very small pores.

In contrast, a sandy soil shows almost no shrinkage, all the water loss occurring as a result of the emptying of pores which do not change in size during drying (Fig. 4.7(a)). The amounts of air-filled pore space in the two soils are shown in Fig. 4.7(b). Soils with intermediate textures and clay soils with well-developed structure (surface horizons) have properties between those shown in the figure.

Changes in the water content of clay soils cause changes in consistency. Commonly used indices of consistency are the *plastic and liquid limits*. Smith and Mullins (1991) summarize the methods and Archer (1975) gives data relating limits to texture and organic matter content.

The measurement of shrinkage during the drying of soils

Wet remoulded soil is packed into a shallow circular container, and allowed to dry slowly. Its change in water content is measured by weighing, and shrinkage by measuring the diameter of the soil disc.

Equipment

Metal or glass dishes, approximately 1 cm deep × 10 cm diameter.
Soils. About 500 g each of a clay subsoil, a sand and a sandy loam soil.
Vaseline.

Method

Add water to the clay soil and work it with a rod until a smooth dough is produced. It should be just at the point where it is changing from being plastic to flowing as a liquid and all the air bubbles should have been released. The clay will continue to absorb water over many hours and it should be left overnight and further water added as required. For the sand and sandy loam, add water and work with a rod until they are approaching saturation, but have not become liquids.

Measure the internal dimensions of the dishes and coat their interiors with Vaseline to prevent the clay from sticking to the dish. Weigh each dish.

Transfer the dough into a dish, working the soil until the dish is completely filled. Prepare duplicates of each soil. With a metal ruler, trim off excess soil so that it is flush with the top of the dish. Weigh the dish + soil. Allow to dry slowly in the air, weighing each day and measuring the mean diameter of the disc of soil in the dish. The latter requires several measurements at different angles across the disc. When the weight has

become almost constant, dry at 105 °C overnight and determine the oven-dry mass of soil (Section 3.3).

Calculation

Calculate the specific water content (Note 1) for each time of weighing as $cm^3 H_2O g^{-1}$ oven-dry soil.

Calculate the volume of the soil disc at each time as follows. Initially the disc diameter is d_1 centimetres and the thickness is l_1 centimetres. The volume is $l_1 \times \pi d_1^2/4 cm^3$. When the diameter has decreased to d_2, it can be assumed that the thickness has also been reduced in proportion to the change in diameter, i.e. that the disc is shrinking homogeneously. Thus the new thickness $l_2 = l_1 d_2/d_1$ and the new volume is $l_1 d_2/d_1 \times \pi d_2^2/4$. Express as specific volumes, cm^3 soil g^{-1} oven-dry soil.

Data handling

An example of the data for a clay is shown in Fig. 4.8. The remoulded clay is saturated initially and normal shrinkage occurs down to a water content of about $0.3 cm^3 g^{-1}$. Note that soil volume = particle volume + water volume + air volume. If no air entered the clay, the normal shrinkage curve would continue to the y-

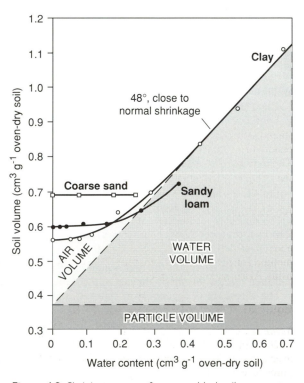

Figure 4.8 Shrinkage curves for remoulded soils.

axis with the intercept being the point where, in an imaginary soil with no air or water,

$$\text{soil volume} = \text{particle volume} = 0.38 \, \text{cm}^3 \, \text{g}^{-1}$$

The actual intercept is the point where, with no water,

$$\text{soil volume} = \text{particle volume} + \text{air volume} = 0.56 \, \text{cm}^3 \, \text{g}^{-1}$$

and so the air volume in the dry soil is $0.18 \, \text{cm}^3 \, \text{g}^{-1}$. The divergence of the measured line from the normal shrinkage curve gives the air volume at any water content.

Specific soil volume is the reciprocal of density. Thus

$$\text{particle density} = 1/\text{specific particle volume} = 1/0.38 = 2.63 \, \text{g cm}^{-3}$$

and

$$\text{(dry) bulk density} = 1/\text{specific (dry) soil volume} = 1/0.56 = 1.79 \, \text{g cm}^{-3}$$

Compared to soil values in Table 4.1, the remoulded dry clay is very dense, hard and strong. The remoulded and dried sandy loam has a bulk density of $1.67 \, \text{g cm}^{-3}$, again showing the effects of damaged soil structure.

The Drayton data Analysis of the data in Fig. 4.7(a) is not so straightforward as Fig. 4.8 because some air enters the soil during structural shrinkage. However, the particle density was measured ($2.65 \, \text{g cm}^{-3}$, Section 4.1) giving a specific particle volume $= 1/2.65 = 0.377 \, \text{cm}^3 \, \text{g}^{-1}$. At any point on the graph,

$$\text{soil volume} = \text{air volume} + \text{water volume} + 0.377$$

and because water volumes are known, air volumes can be calculated as shown in Fig. 4.7(b). Only $0.01 \, \text{cm}^3$ air g^{-1} enters during structural shrinkage, remaining constant during normal shrinkage and increasing during residual shrinkage to $0.12 \, \text{cm}^3 \, \text{g}^{-1}$ in the oven-dry soil. The data show that the angle of the straight line part of the curve is 44°, confirming normal shrinkage.

An alternative method

The measurement of soil volume lacks precision in the above method. Sections 1.2 and 4.2 give an alternative method for measuring the volume of coated aggregates using Archimedes' Principle.

Equipment

Saran resin. To 1 l of butanone add 200 g of Saran resin F220 (Aldrich Chemical Company). Stir occasionally over a day or two to dissolve. For aggregates with large pores, use $250 \, \text{g} \, \text{l}^{-1}$.

Obtain a few aggregates of wet clay soil. Tie a thread around each and weigh. Coat with Saran resin by dipping slowly and allow the solvent to evaporate. Weigh again. Determine aggregate volume by the method of Section 1.2. Hang in the air and allow to dry. The resin coating (a) is permeable to water vapour, but is not permeable to liquid water, (b) maintains a slow rate of drying which is needed to prevent premature formation of surface cracks, and (c) shrinks with the aggregate. Weigh and measure volume occasionally. When the mass is almost constant, dry at 105 °C overnight and determine the oven-dry mass of the aggregate.

Note 1 'Specific' is used to denote physical quantities relative to the mass involved. Thus a specific volume = volume/mass. In Fig. 4.7, the term is used for soil volume, air volume and water content. It is also applied to the surface areas of clay minerals: in Table 2.3, $\text{m}^2 \, \text{g}^{-1}$ is a specific surface area.

Section 4.4 The effects of vehicles and implements on porosity

Vehicle wheels

Information obtained from experiments carried out at the Scottish Centre of Agricultural Engineering (Smith, 1987) can be used to analyse the effects of machinery wheels on soil porosity. In these experiments a sandy loam soil had been ploughed in the autumn and harrowed and levelled by light rolling in the spring. A tractor and trailer were driven across the plot (Colour Plate 7) and bulk density measured both under and away from the wheelings. Table 4.5 gives information regarding the treatment, Fig. 4.9 gives the bulk density values and Fig. 4.10 shows the equipment and its effects.

The wheels created a depression of 11 cm, and the soil must have increased in strength until it could withstand a stress of 98 kPa under the trailer wheel. The increase in strength was associated with an increase in bulk density and a decrease in porosity. These changes are associated primarily with a decrease in transmission pore volume.

Data analysis

A hypothesis can be stated as follows: *the wheels cause a compaction of the soil which is accommodated only by vertical movement of soil.* If this is correct, the volume of the rut equals the decrease in soil volume in the column under the rut which must also equal the decrease in

Table 4.5 Loads and stresses imposed by machinery wheels of a tractor and wheeled trailer.

Wheel	Wheel load (kN)	Tyre–soil contact area (m²)	Tyre–soil contact pressure (kPa)
Tractor-front	6.9	0.08	86
Tractor-rear	19.1	0.25	76
Trailer	31.2	0.32	98

The load is the force resulting when gravity acts on the mass of the machinery. A mass of 0.1 kg produces a load of 1 N (newton). The pressure applied to the soil (termed a stress) results from the force being spread over the contact area.

Pressure in pascals (Pa) = force/area ($\mathrm{N\,m^{-2}}$).

The trailer wheel (31.2 kN = 31.2 × 100 kg = about 3 t) spreads its load over 0.32 m² of area producing a pressure of 98 kPa. Using more common units, the pressure on the soil is about $10\,\mathrm{t\,m^{-2}}$ or about $15\,\mathrm{lb\,in^{-2}}$ and this is related to the internal pressure of the tyre. Big wheels are used to spread the load over a larger area to reduce the applied pressure. For this purpose they are inflated to a low pressure. If the above trailer wheel was large enough, the inflation pressure could be halved, approximately doubling the contact area to 0.64 m² and reducing the pressure applied to the soil to about $5\,\mathrm{t\,m^{-2}}$, the actual values depending on the contribution of tyre strength and other factors. Thus the damage caused to soil structure by the wheel would be reduced (Colour Plate 7).

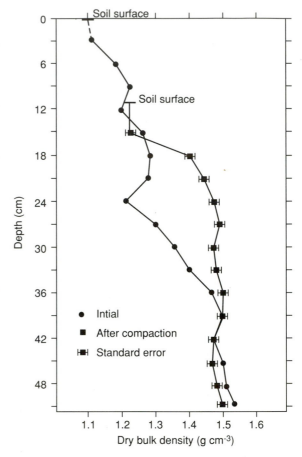

Figure 4.9 Dry bulk densities before and after passage of vehicle wheels (Smith, 1987).

pore volume in the column. This hypothesis can be tested using the bulk density data to calculate changes in pore volume. It is assumed that the site was initially uniform, so that differences in bulk density are caused by the treatments.

Consider a 1 cm² column of soil under the wheel down to 51 cm depth, this being the limit of the measurements. Bulk density was determined at 3 cm intervals, and so each point in Fig. 4.9 can be considered to represent the density within a segment of soil on either side of the depth of measurement. For example in the untreated soil, the first measurement is at 3 cm depth, is about 1.11 g cm³ and represents the soil from 0 to 4.5 cm. The second measurement represents the segment from 4.5 to 7.5 cm and so on in 3 cm increments. In the compacted soil the 0–11 cm segment is the rut, and the first density measurement represents the 11–16.5 cm segment, again followed by 3 cm segments. Table 4.6 shows the data before and after compaction. The bulk density values have been read from Fig. 4.9. A particle density of 2.6 g cm⁻³ is assumed (a cultivated mineral soil, Section 4.1). The particle volume = bulk density/2.6 cm³ cm⁻³ (Section 4.2) and pore volume = 1 − particle volume (cm³ cm⁻³) (Eq. 4.1)

The pore volume per segment is equal to pore volume,

($\mathrm{cm^3\,cm^{-3}}$) × segment volume (cm³). The total porosity in the soil column has then been calculated, and is 25.46 cm³ initially and 18.43 cm³ after compaction, a decrease of 7.03 cm³. Thus, assuming there are no changes below 50 cm, a rut of 11 cm³ has been produced with only 7 cm³ decrease in pore space in the soil column.

The hypothesis is clearly not correct because 4 cm³ of the rut volume must have been accommodated by soil movement other than vertical compression. Smith (1987) has analysed these changes in more detail. Soil is squeezed horizontally out of the column under the wheel and may have caused some heaving of the soil on either side of the rut. Fig. 4.10 shows the changes in the 17 segments in diagrammatic form together with the effects of other machinery. The density of shading represents the bulk density, the widening of the segments indicates horizontal soil movement and the different widths of the columns indicate the widths of the

Table 4.6 The calculation of pore volumes before and after compaction

Depth of segment (cm)	Bulk density (g cm^{-3})	Particle volume (cm^3 cm^{-3})	Pore volume (cm^3 cm^{-3})	Pore volume per segment (cm^3)
Before compaction				
0–4.5	1.11	0.43	0.57	2.57
4.5–7.5	1.19	0.46	0.54	1.62
and subsequent 3 cm segments to				
49.5–52.5	1.52	0.58	0.42	1.26
			Total porosity	25.46
After compaction				
0–11	No soil			
11–16.5	1.23	0.47	0.53	2.92
16.5–19.5	1.41	0.54	0.46	1.38
and subsequent 3 cm segments to				
49.5–52.5	1.49	0.57	0.43	1.29
			Total porosity	18.43

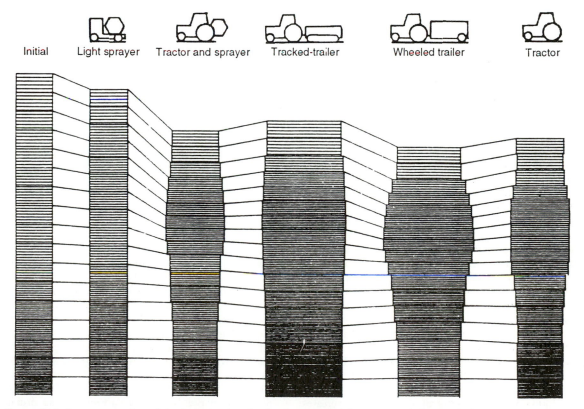

Figure 4.10 A representation of the bulk density of soil columns and their segments before and after the passage of five vehicles (Smith, 1987).

wheels. The light sprayer has the least effect, and the loaded wheeled-trailer the greatest effect. The tracked-trailer (an unusual machine) has less effect as a result of spreading the same load over a larger soil area, the soil contact pressure being reduced from 98 to 25 kPa.

Cultivations

Cultivations are carried out for a variety of reasons: to bury crop residues, to bury or kill weeds, to prepare for the following crop a seedbed with a suitable tilth (soil structure) and occasionally to break up a compacted soil

Table 4.7 Cultivation treatments

Treatment	Primary cultivation (autumn)	Secondary cultivation (spring)
Deep ploughing	Mouldboard ploughing once to 30–35 cm	Harrowing (twice)
Shallow ploughing	Mouldboard ploughing once to 15–20 cm	Harrowing (twice)
Chisel ploughing	Chisel ploughing to the maximum practical depth on three occasions during the autumn and winter	Harrowing (twice)
Zero ploughing	None	None

Notes:
1. A mouldboard plough inverts the surface soil (see Plate 4.2).
2. A chisel plough breaks up the soil without inversion.
3. Harrowing loosens the top few cm of soil like a rake.
Further information on cultivation methods is given in Davies *et al.* (1972).

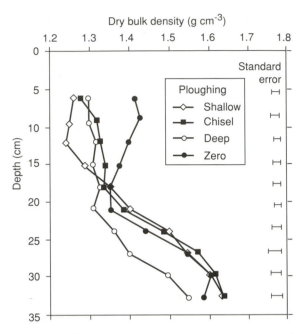

Figure 4.11 The effects of cultivation treatments on soil bulk density (Soane, 1975).

layer at depth in the profile (a pan). The changes in bulk density caused by cultivations can be measured as an indication of structural changes.

In experiments at the Scottish Centre of Agricultural Engineering (Soane, 1975) four cultivation treatments were imposed on a soil which had been used for arable crops. Following a cereal crop, autumn cultivations were carried out, followed by secondary cultivations in the spring (Table 4.7). Barley was then sown, and in the zero ploughing treatment *direct drilling* was used where the seed is placed in a slit cut in the uncultivated soil surface. Dry bulk densities shown in Fig. 4.11 were measured during the growing season.

The results can be used to answer the following questions:
To what depth might this soil have been cultivated in previous years?
- The zero ploughing treatment is the best indicator of cultivation in previous years: although the soil surface is affected by the machinery there are unlikely to be significant effects at depth. The increases below 20 cm are typical of an uncultivated soil or of a cultivated soil with some compaction (a plough pan). The discontinuity in the curve at about 20 cm suggests that previous cultivations reached this depth. In an uncultivated soil, the bulk density would continue to decrease to the surface.
What are the effects of the direct drilling machinery on the zero plough treatment?

- If the soil had been ploughed to about 20 cm in previous years bulk densities similar to those for the shallow ploughing treatment would have been produced. Consolidation would subsequently occur. Harvesting and other machinery would compact the surface layer. Drilling would produce further compaction at the surface, and the increase in bulk density above 20 cm reflects these changes.
To what depth have the shallow ploughing and chisel ploughing had an effect?
- Comparing these treatments with zero ploughing shows that both have loosened the soil above 17.5 cm, the mouldboard plough having the greatest effect. Below this depth the two treatments cause compaction giving a (further) development of a plough pan.
What changes have occurred in the deep ploughing treatment both as a result of ploughing and subsequently?
- The soil to 30–35 cm depth is inverted. Before ploughing it presumably had a bulk density distribution similar to the zero ploughing treatment. Soil initially below 20 cm has been much loosened and left with a low bulk density at the surface. Surface soil must also have been loosened but subsequent consolidation occurs at depth. The 0–15 cm layer produced by deep ploughing has a higher bulk density than the shallow ploughed soil, reflecting its initial high bulk density at depth.

Section 4.5 Projects

1. Using bulk density measurements determine porosity in the following situations:
   (a) Under wheelings in a cultivated area. This can be in an agricultural setting, or in a school garden.
   (b) A comparison of a playing field area where games in wet conditions have puddled and compacted soil, with an area of similar soil away from the playing area (see Bullock and Gregory, 1991, Ch. 6).
2. On a cultivated area measure the effects of stresses applied to soil. This can be done by increasing the load on a given area. Use a plastic dustbin supported on a wooden frame on a 'foot' which is a length of wood with a 10×10 cm cross-section. Add water to the bin in known volumes (1 l weighs 1 kg) and measure the penetration of the foot. Similarly a fixed load can be applied to feet having different areas.

   Plot the relationship between penetration and stress. Do you get the same relationship for feet of different sizes? Measure the size of a cow's hoofprint. If the mass of the cow is 500 kg, and it has a minimum of two hooves on the ground, calculate the stress and predict the hoof penetration for your soil.

   Carry out these experiments immediately after saturating the soil. Cover the soil and measure again 2 days later when the soil has drained to field capacity (Section 4.2) or after a longer period of drying.

   Carry out similar experiments in a grass field.
3. Using golf balls, marbles and ball bearings determine their particle density, and bulk density when packed in a container of known volume. Compare your results to calculated values following Calculation 5 in Section 4.6.

   Using sieves, or the method of settling in water given in Section 2.3, separate builders' sand into particles with known size ranges. Determine particle density, and dry bulk density when present in a container of suitable volume. Again, compare to calculated values. Use the whole mixed sand in a similar experiment to show that mixed particle sizes give much higher bulk densities.
4. Using bulk density cylinders take samples from several sites with different types of soils (e.g. light versus heavy texture, arable versus grassland). Determine bulk density and porosity.
5. Use the infomation in Smith's (1987) paper as the basis for data analysis following the approach in Section 4.4.
   (a) Complete the data in Table 4.6 to check the total porosity values given.

(b) Fig. 4.6 seems to show that no horizontal soil movement occurred under the light sprayer. Check this conclusion using Smith's Fig. 2 to calculate the total soil porosity after the passage of the machine.
(c) The conclusions of (a) and (b) can be checked by calculating the mass of soil under the wheeling in each treatment. If there has been no horizontal movement of soil under the wheeling, then the mass soil is unchanged. You will have to assume that there are no changes below 51 cm. For each segment multiply soil volume by bulk density to give mass per segment. Sum the masses.
(d) Using Smith's Table 1 and Fig. 2, plot compression (rut depth) against ground pressure. Smith's Fig. 10 helps to explain why the data do not fit to a simple relationship. Calculate the wheel contact areas for the two treatments in Fig. 10.
(e) Using the data in Fig. 4.11 calculate the increase in height of the soil surface caused by the three ploughing treatments compared to zero ploughing. The depths given are relative to the new soil surfaces.
   Suggestion: calculate the mass of soil per segment, and sum the masses. The differences in total mass relate to soil in the segment not sampled below 35 cm (Fig. 4.12). Calculate the depth of this segment assuming that it has a bulk density equal to that at 33 cm.
6. Section 4.3 describes a project to measure shrinkage of soils.

Section 4.6 Calculations

1. A cylinder of soil, 5 cm diameter and 5 cm long, was sampled in the field when the soil was at field capacity. The moist soil weighed 156.5 g and after drying at 105 °C the soil weighed 111.0 g. Calculate:
   (a) the water content of the soil ($g\,g^{-1}$ oven-dry soil). (*Ans.* 0.41)
   (b) the dry bulk density of the soil ($g\,cm^{-3}$). (*Ans.* 1.13)
   (c) the volumetric water content ($cm^3\,cm^{-3}$). (*Ans.* 0.46)
   (d) the volume of solids in 1 cm³ soil assuming that the particle density is $2.6\,g\,cm^{-3}$. (*Ans.* 0.43 cm³)
   (e) the volume of pores in 1 cm³ dry soil. (*Ans.* 0.57 cm³)
   (f) the volume of transmission pores in 1 cm³ soil. (*Ans.* 0.11 cm³)

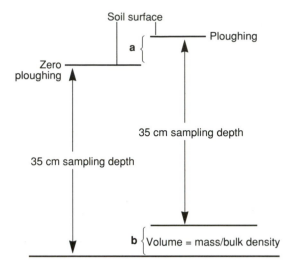

a = increase in the height of the soil surface.

b = equal to the increase in the height
of the soil surface

Figure 4.12 The change in height of the soil surface after cultivation.

At the wilting point this soil contained $0.08\,\text{cm}^3$ water cm^{-3} soil. Calculate the volume of storage and residual pores in $1\,\text{cm}^3$ soil. (*Ans.* 0.38, $0.08\,\text{cm}^3$)

2. At field capacity the water content of a sandy loam was $20\,\text{g}$ water per $100\,\text{g}$ oven-dry soil. If the dry bulk density was $1.3\,\text{g}\,\text{cm}^{-3}$ calculate the water content as $\text{cm}^3\,\text{H}_2\text{O}\;\text{cm}^{-3}$ soil. (*Ans.* 0.26) If this soil was compacted to $1.5\,\text{g}\,\text{cm}^3$ without any water being squeezed out, calculate the new water content values both as g per $100\,\text{g}$ and $\text{cm}^3\,\text{cm}^{-3}$. (*Ans.* 20, 0.3) If compaction continued, a stage would be reached when water would have to be squeezed out (saturation). Calculate the bulk density at which this occurs. Assume that the particle density is $2.6\,\text{g}\,\text{cm}^{-3}$. (*Ans.* 1.71) Note that water would come out before this stage because of the presence of trapped air.

3. The following particle densities were obtained for samples from A and C horizons of a soil profile: A, $2.3\,\text{g}\,\text{cm}^{-3}$; C, $2.6\,\text{g}\,\text{cm}^{-3}$. Calculate the percentage organic matter in each sample if the density of mineral particles is $2.65\,\text{g}\,\text{cm}^{-3}$ and that of organic matter is $1.4\,\text{g}\,\text{cm}^{-3}$.
Suggestion: let the mass of organic particles be x, and mineral particles $(1 - x)\,\text{g}\,\text{g}^{-1}$ soil. Express the volumes of the particles in each fraction in terms of x. Sum to give the total volume of particles in $1\,\text{g}$ of soil. Also calculate the total volume of the particles as $\text{cm}^3\,\text{g}^{-1}$ from the particle density of the sample. Equate the two total volumes and solve for x. (*Ans.* A 17%, C 2.2%).

4. A plough horizon contains 0.17, 0.2 and $0.12\,\text{cm}^3$ of transmission pores, storage pores and residual pores respectively cm^{-3} of soil. A tractor runs over this soil and the surface is depressed by $2.0\,\text{cm}$. If this compaction were uniformly accommodated throughout the top $20\,\text{cm}$ of soil by a reduction in the volume of transmission pores, what fraction of these pores would be lost? (*Ans.* 0.59) What was the average bulk density in the plough horizon before and after compaction? Assume a particle density of $2.6\,\text{g}\,\text{cm}^{-3}$. (*Ans.* 1.33 and $1.59\,\text{g}\,\text{cm}^{-3}$).

5. Following the approach in Fig. 4.1, calculate the porosity of a bed of spherical sand particles in open packing for several particle sizes to show that porosity is independent of particle size when packed in this way. Using algebra, prove this independence.
   It is more difficult to determine the porosity of spheres in close packing. Observe the arrangement using marbles, and attempt to calculate porosity. (Note that there are various possible spatial arrangements.)

CHAPTER 5

Water in Soils

Soil pores are occupied by either water or air. Fluctuations in water content occur with climate controlling the inputs of water and drainage and evaporation from the soil and leaf surfaces controlling losses. The ability of soil to allow drainage, to hold water for plant use and to hold some water so firmly that plants cannot use it depends on the sizes, shapes and continuity of the soil pores. Table 4.2 gives a classification of pores in terms of size and function.

The presence of water in soils has a marked effect on their physical properties. For example cohesion (resistance to cultivation) and strength to withstand loads (resistance to compaction) often increase as soils become drier. These properties are also influenced by texture, organic matter content and structure through their effects on (a) the distribution of pores of various sizes (known as the *pore size distribution*) and thus on the distribution of water in these pores, and (b) the stability of the structural arrangement which depends on the strength of the bonds between particles. The total porosity of soils and typical values for the volumes of transmission, storage and residual pores in soils of different texture are given in Tables 4.1 and 4.3.

Soil water is also of central importance in the following respects:

- It supplies water requirements of plants
- It is the environment in which microbial activity occurs.
- It is the soil solution which contains dissolved ions and molecules including nutrients for plants and microbial life.
- It is the medium in which chemical reactions occur, particularly at the interface with particle surfaces.
- It transports clay-sized particles and solutes, thus having a major role in soil formation, acidification, salinization and leaching of pollutants into groundwater.
- It displaces air which may lead to poor aeration and the associated changes in biochemical processes.

WATER CONTENT OF SOILS

The maximum amount of water that a soil can hold is its *saturated water content*, which depends on the total porosity (Section 4.2). In practice, this rarely occurs in the field because even when soils are temporarily flooded, or where soil is below the water table, trapped bubbles of air remain. Table 4.3 shows that the total porosity and thus the saturated water content is typically between 40 and 60 per cent of the soil volume. After soils have drained to *field capacity* (Ch. 4 and Section 12.2) water typically occupies between 10 and 55 per cent of the soil volume. Transpiration through plants can reduce the water content to a limit which is between 5 and 35 per cent which is called the *permanent wilting point*. Further drying occurs by evaporation until the *air-dry* state is reached, the water content then depending on the prevailing relative humidity of the air and on soil texture and organic matter content. Air-dry water contents range from almost zero in sandy soils to about 8 per cent *m/m* in heavy clays at relative humidities of between 50 and 60 per cent.

Water contents can be measured and expressed in various ways depending on the purpose of the measurement. Methods are given in Section 5.1.

WATER RETENTION BY SOILS

Two observations illustrate the ability of soils to hold on to water:

1. When drainage ceases in a wet soil (the state of field capacity) gravitational force is still acting on the water, and must be balanced by a force holding the water in the soil.
2. When soils have dried to the wilting point, water remaining in the soil cannot be extracted by the suction exerted by the root surface.

Water retention depends on forces acting between water molecules and hydrophillic (water-loving) particle surfaces; these forces include hydrogen bonds, van der Waals' forces and electrical attraction to exchangeable cations. There are two related results. Firstly, water is held in small pores and in the necks between larger air-filled pores, as shown for a drained soil in Fig. 5.1(a) and secondly, thin films of water are held on to particle surfaces enclosing air-filled pores. Because of the attraction between water molecules and the walls of

(a)   (b)

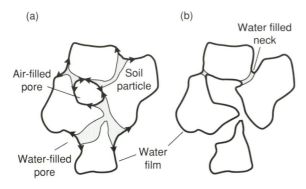

Figure 5.1 The distribution of water in soil.

the pores, the water is pulled in the directions indicated by the arrows, the liquid–air interfaces are curved and the water is in a state of tension, leading to the term *soil water tension.*

An alternative way to view this concept is that if a drop of water at atmospheric pressure is brought into contact with moist soil, it will be sucked into the soil by the water tension. This demonstrates that the pressure in the soil water is less than the pressure in the added water, and so water moves into the soil as a result of the pressure difference. The amount by which the pressure is less than atmospheric is termed the *soil water suction* or soil water tension.

Some water can be removed from the soil in Fig. 5.1(a) by applying a suction greater than the soil water suction: this is, in effect, what plant roots do. Fig. 5.1(b) shows the distribution of the remaining water. The films of liquid have a reduced thickness, some large pores have been emptied as a result of air–water interfaces retreating into smaller pores and necks and the remaining water is bounded by water–air interfaces with smaller radii of curvature. The soil water suction is now greater than before (the water pressure has been reduced further below atmospheric pressure. The relationships between soil water suction, curvature and the size of water-filled pores are developed in Section 5.2, and have been used to calculate the suctions shown in Table 5.1 at which pores of given sizes empty, known as the *critical suctions.* Suction has the unit of pressure, which is the pascal, abbreviated Pa, which is equivalent to a force of 1 newton acting over an area of 1 square metre ($1 \, \mathrm{N \, m^{-2}}$). Note that because of the irregular shapes of soil pores and necks, the quoted sizes are effective pore diameters, i.e. the diameter of a cylindrical pore or circular neck which would empty at a given suction.

The table shows the wide range of pore size and indicates why the soil is such a suitable environment for

Table 5.1 The relationships between pore size and the soil water suction required to empty them during drying

Pore size, diameter (μm)	Critical soil water suction (kPa)	Equivalent hydraulic head, h_w (m H$_2$O)	Comments
20 000	0.015	0.002 (2 mm)	A 2 cm crack
4 000	0.075	0.008 (8 mm)	An earthworm channel
300	1.0	0.10 (10 cm)	The diameter of a cereal root
60–30	5–10	0.5–1.0	Soil water suction at field capacity. Transmission pores are >50 μm
2	150	15	Size of a bacterial cell. Limit of 'readily available' water. Upper size limit of a clay-sized particle. Storage pores are 50–0.2 μm
0.2	1500 (1.5 MPa)	150	Water suction at the wilting point. Residual pores are <0.2 μm
0.003	100 000 (100 MPa)	10 000	Water suction in air-dry soil. The pore size is approximately 10 × the size of a water molecule

the many scales of biological activity. Examples are given below.

- Earthworm channels are freely drained, allowing easy supply of oxygen required for aerobic respiration and very rapid movement of water.
- Roots cannot normally penetrate into pores smaller than themselves: thus the pores into which they grow are also drained at field capacity allowing good access of oxygen.
- Available water is held in storage pores where the close association of particle surfaces and pore water facilitates rapid replenishment of nutrient ions when uptake by roots occurs.
- Water movement to roots occurs primarily through water-filled storage pores. Their size and often their close proximity to roots and to each other, together with water films and water-filled residual pores, mean that movement can continue over a wide range of water contents.
- Bacteria need to be bathed in a water film to function normally, and most also need a supply of oxygen; these requirements are met at the water contents allowing plant growth.

A well-structured soil not only has a distribution of pore size which allows the soil to function in the above ways but the pores are also stable to applied forces (Ch. 4).

The concepts of field capacity, wilting point and available water will be discussed in Ch. 12. Note that a range of values is given in Table 5.1 for field capacity indicating that differences occur between soils of different textures; the range covers the minimum diameter of transmission pores given in Table 4.2 (50 μm). For the wilting point a suction of 1.5 MPa (0.2 μm pore diameter) is generally accepted as the suction at the *permanent wilting point* (Section 12.2), and a pore diameter of 0.2 μm is used to classify residual pores.

The measurement of soil water suction

Methods are given in Section 5.3. Tensiometers are commonly used and the principles of their operation illustrate the way in which the *equivalent hydraulic head* in Table 5.1 is a measure of soil water suction.

Fig. 5.2 shows a tensiometer consisting of a reservoir of water in a manometer (a U-tube) connected to the soil water via a porous ceramic cup. The meniscus is initially at level A, the same height as the cup. Depending on the soil water suction, water will be drawn into the soil through the porous wall of the cup lowering the meniscus to B. The pressure in the cup is reduced until at equilibrium it equals that in the soil water, and has a value

$$P = -h_w g \rho_w \qquad [5.1]$$

(a)

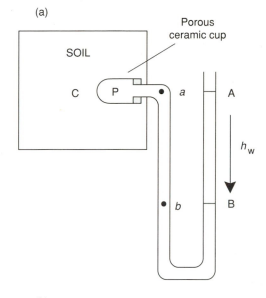

(b)

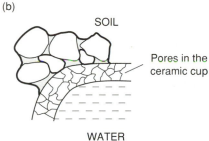

Figure 5.2 Tensiometers and soil water suction. An enlarged view of part of the ceramic cup is shown in (b).

relative to atmospheric pressure, where h_w is the difference in height (m), g is the gravitational force per unit mass $(9.81 \, \text{N kg}^{-1})$ and ρ_w is the density of water $(1000 \, \text{kg m}^{-3})$. The derivation of this equation is given in Section 5.2, Note 1. Substituting values gives $P = -9810 \, h_w$ pascals and so the soil water suction $(= -P)$ is $+9810 \, h_w$ pascals. The value of h_w is the equivalent (negative) hydraulic head and it is useful to remember that an equivalent hydraulic head of 1 m is equivalent to a suction of approximately 10 kPa. In effect the soil water suction is balanced by pressure due to the weight of the column of water between a and b which is hanging in the left-hand side arm of the manometer: the water at b is at atmospheric pressure.

Table 5.1 shows that the equivalent hydraulic head of water in soil varies from values that can be easily measured using this technique (up to about 8 m) to

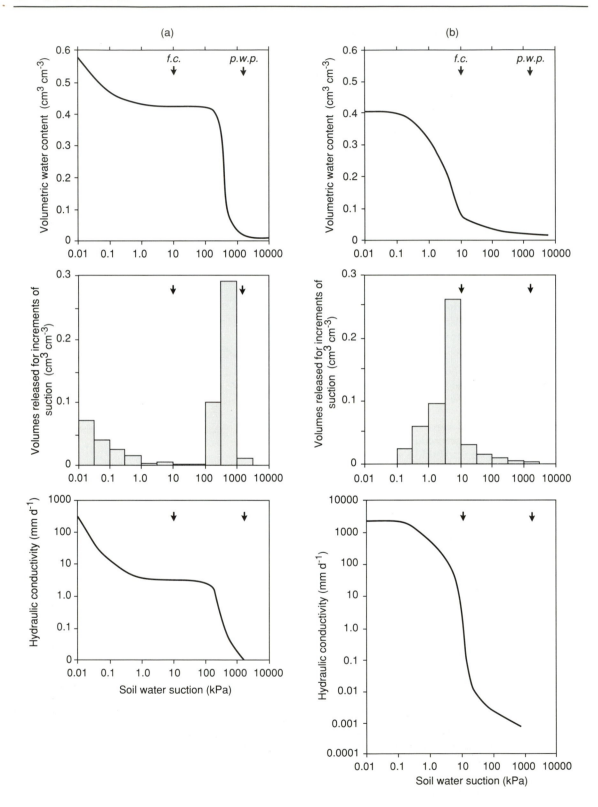

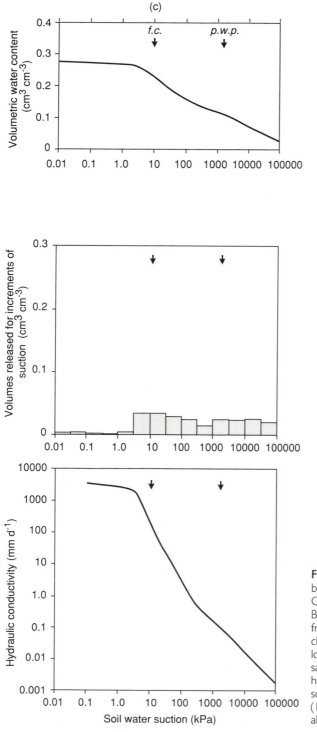

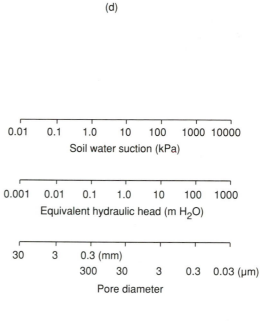

f.c.   Field Capacity

p.w.p.   Permanent Wilting Point

Figure 5.3 Water release characteristics, pore size distributions and hydraulic conductivities. (a) A chalk rock (Upper Chalk) at West Ilsley, Berkshire, taken from 2 m depth. Based on Gardner *et al.* (1990). (b) A sand-textured soil from Sadoré in Niger (0–20 cm, 91% sand, 5% silt, 4% clay). Data from Hoogmoed and Klaij (1990). (c) A sandy loam from Wellesbourne, Warwickshire (0–45 cm, 70% sand, 12% silt, 18% clay, 1.2% organic carbon). The hydraulic conductivity was measured by the method described by Rowse (1975). Data from Rowse and Stone (1978). (d) A comparison of soil water suction values, equivalent hydraulic heads and pore diameters.

values well beyond the practical range. However, the concept is useful even when measurements can no longer be made in this way.

WATER RELEASE CHARACTERISTICS OF SOILS

The relationship between soil water content and soil water suction is known as the *water release curve* or *water release characteristic*. Examples are shown in Fig. 5.3 for a chalk rock and two soils. The materials release their water at different suctions depending on their pore size distribution. For example, the chalk contains a few large fissures which drain at small suctions and occupy about 15 per cent of the volume of chalk. The remainder of the pores drain at suctions greater than 30 kPa, and occupy about 40 per cent of the volume.

Because the relationship between the size of pores and the suction at which they drain is known (Section 5.2), the water release curve can be transformed into a graph showing pore size distribution. In its simplest form this is a histogram showing the volumes of water released between given limits of suction. The x-axis is labelled both in terms of the suction and in terms of the maximum diameter of a pore which can remain water-filled at that suction. The pores in the chalk are primarily either greater than about 1 mm or between 3 and 0.3 μm, the latter being shown in Plate 2.1: the chalk has a simple bimodal distribution. In contrast, the sand (Fig. 5.3(b)) has lost much of its water when it has drained to field capacity, with pores primarily in the range 30–1000 μm which is a unimodal distribution. It drains freely and has a small available water capacity, the value depending on the actual suction at field capacity (between 5 and 10 kPa). The sandy loam (Fig. 5.3(c)) has been used for continuous vegetable production leading to poor structure with only about 0.05 cm³ cm⁻³ of transmission pores. There is a uniform distribution of pores between 10 kPa and 100 MPa.

The properties of soils with other textures can be summarized as follows:

- Medium-textured soils, if managed well, have a distribution of pore size which gives good drainage and available water characteristics and have a small content of residual pores associated with good mechanical properties (Table 4.2).
- Clays have pores which vary in size as the soil wets and dries (Section 4.3). Unless they are well managed they can be nearly saturated at field capacity, shrinking to produce large cracks as they dry and having difficult mechanical properties. Grass crops tend to maintain good structure in the topsoil.

The air-entry suction and the capillary fringe

There is a characteristic soil water suction at which air begins to enter the pore space known as the *air-entry suction*. This is the suction required to empty water from (and therefore draw air into) the largest pore in the soil. The air-entry suction has special significance where a water table is present, the depth of which can be determined using a *dipwell* which is a vertical hole down which a sensor can be lowered until it touches the water. There will be a layer of saturated soil above the water table which is called the *capillary fringe* (Fig. 5.4). Provided that there is no upward movement of water to roots or to the soil surface, the height of the fringe will be equal to the hydraulic head required to cause air entry into the largest pores. This can be illustrated by Fig. 5.2(a). Imagine that the soil water suction is the air-entry value which is equivalent to the hydraulic head, h_w. The water level B can now be considered as the level of the water table which, if present, would maintain the soil water suction at C equal to the air-entry value, and so h_w would be the thickness of the capillary fringe.

In a clay soil the capillary fringe may extend to about 40 cm above the water table and even at 1 m there may only be about 10 per cent of the soil volume occupied by air. This value is often taken as a rough indication of the air content of the soil below which aeration problems may occur. Ch. 6 develops this topic further. In the sand textured soil illustrated in Fig. 5.3(b) the fringe only extends to about 1 cm above the water table and will probably be well aerated 10 cm above the water table.

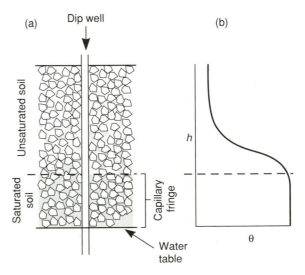

Figure 5.4 A capillary fringe. The soil particles are greatly magnified in (a). The change in water content, θ, with height above the water table, h, is shown in (b).

A further example of a capillary fringe is in a pot of soil placed in a dish of water which maintains a water table 1 or 2 cm from the base of the pot. Large pores are required in the potting compost or soil for there to be air-filled pores almost down to the base. For example, pores of 3 mm diameter are needed if air-filled pores are to be present 1 cm above the water level. Composts contain peat, coco fibre or other materials to maintain these large pores. Lack of adequate porosity is the principal reason why overwatering is a common source of problems with house plants and in pot experiments.

WATER MOVEMENT

Water in soil is rarely still, and an understanding of water movement has many useful applications. For example, how can we establish the depth from which water moves upward in response to uptake by plant roots? This has important implications in determining the amount of water available to crops. A related question concerns the depth to which solutes such as nitrate or pesticides have to move before they are out of reach of the root system and so must then move irrevocably towards the groundwater.

Water moves if it is subject to unbalanced forces. Three forces commonly act on the water:

1. *The earth's gravitation force* acts to move water downwards.
2. *Differences in soil water suction.* Water tends to move from where the suction is small (wet soil, relatively large pressure) to where it is larger (dry soil, relatively low pressure). Thus, as in the case of water distribution in pipes, movement is in the direction of decreasing pressure.
3. *Water moves by osmosis* from regions of low solute concentration to regions of higher concentration separated by a semi-permeable membrane (Nuffield Advanced Biology I, Chs. 8 and 9). However, osmosis is rarely involved in large-scale water movement in soils because of the lack of semi-permeable membranes: a difference in salt concentration is quickly lost by diffusion of dissolved salts. The only exception is when water moves as a vapour in dry soils because the air–water interface acts as a membrane allowing the passage of water but not of solutes. Osmosis does, however, result in localized movement of water into interlayer spaces of clay crystals causing swelling (Section 14.4) and is important when considering water uptake by plants, where water has to traverse plant membranes (Section 14.3).

These three forces are brought together when considering water movement or availability of water to plants by the use of the concept of *soil water potential.*

This concept is based on the principle that water moves in the direction of decreasing energy status, i.e. towards lower elevation, smaller pressure or larger solute concentration, and that soil water potential is a measure of the potential energy content of the soil water. Section 5.4 develops this concept: the principles are used to examine the direction of movement in a profile subject to rain and evapotranspiration, leading to conclusions about the amount of rainfall which moves downward to replenish groundwater resources and the amount taken up by plants or lost by evaporation.

The rate of water movement

The rate of flow of water through soil is important in the context of the hydrological cycle and plant–soil relationships as illustrated by the following questions. To what extent is water uptake by plants limited by the rate of water movement to roots? During drought, to what extent can the water requirements of vegetation be sustained by upward movement from a water table? How intense does rainfall have to be to cause surface ponding and runoff? How do soil properties influence the rate of loss of water by evaporation from the soil surface?

Section 5.5 explains the principles governing the rate of water movement in soils, and shows that the rate of flow depends on both the driving force measured by the water potential gradient (the change of potential with distance) and the *hydraulic conductivity* of the soil which is a measure of the ease with which water moves through the soil. The hydraulic conductivity depends on a number of factors:

- *Water content.* The greater the volumetric water content the greater the cross-sectional area of water-filled pores through which water can flow.
- *Tortuosity of the pore system.* The more convoluted the system of water-filled pores, the further the water has to move to travel the same distance. The tortuosity increases as water content decreases because water has to move around air-filled pores.
- *The size of water-filled pores.* Water is virtually stationary where it is in contact with a particle surface. The flow rate increases to the middle of pore just as the current is strongest in the middle of a river. Thus, for a given potential gradient, the flow is more rapid through a large pore, being directly proportional to the fourth power of the pore radius. This has far-reaching consequences. For example, Table 5.1 shows that there might be a 100 times difference in pore radius between an earthworm channel and the largest water-filled pore at field capacity. For the same driving force, the rate of flow would be $100^4 = 10^8$ times faster in the earthworm

channel. The largest pores holding water at the wilting point are about 1000 times smaller still, with rate of movement being reduced further by a factor of 10^{12}. This means that hydraulic conductivity decreases very rapidly as soils dry because the larger pores can no longer conduct water, and that at a given water content there is a very wide range of water velocities in the soil matrix, causing dispersion of solutes as water percolates through the soil (Sections 11.5 and 15.4).

When a soil is saturated, all the pores can conduct water and conductivity is at its maximum, termed the *saturated hydraulic conductivity*. This, and the hydraulic conductivity in unsaturated soils, depends on the size, number, orientation, distribution and continuity of water-filled pores. In Section 5.5 the quantitative relationships between conductivity and water content are discussed further and methods for measuring hydraulic conductivity are described. Fig. 5.3 gives examples of the relationships. Note that the chalk has a high conductivity when the large fissures are filled with water, decreasing very sharply when these drain. The sand loses its ability to conduct water at around field capacity because most of its pores are emptied and the necks of water which remain are poorly connected. In the sandy loam the conductivity does not decrease so sharply due to the larger amounts of retained water. A poorly structured clay has a low saturated hydraulic conductivity but is able to maintain a modest conductivity at large suctions because of the continuous network of small pores which retain water.

There are practical implications to the above differences:

- The chalk when saturated can allow a downward flow of about 15 mm rain h^{-1}. A typical storm event in the UK has a rain intensity of about 2 mm h^{-1}. Thus, there is no likelihood of soils over chalk becoming waterlogged unless the soil itself has a low conductivity. This is one of the reasons why chalk downland is a favoured area for racehorse training.
- A large clod of clay soil will dry in air with slow movement of water to the surface until the whole clod reaches the air-dry state (Section 4.3). Similarly a clay profile can continue to lose water steadily by evaporation for very long periods causing deep drying.
- Evaporation of water from the surface of a sandy soil soon forms a thin layer of dry soil at the surface with a very low hydraulic conductivity, which is an effective barrier to further evaporation. Thus the soil below remains moist unless water is taken up by roots. This property is the basis of *fallowing* to preserve stored soil water in areas where rainfall is erratic (Ch. 12).

THE COMBINED EFFECTS OF SUCTION AND MOVEMENT ON WATER SUPPLY TO PLANTS

The rate at which seeds germinate depends to a large extent on the rate at which they can absorb water. This is controlled by the seed characteristics which determine the suction they can develop at the seed surface, by the soil water suction and by the hydraulic conductivity of the soil. Whether they finally germinate depends primarily on the soil water suction. These principles are the foundation for considering water availability to plants in Ch. 12, and are introduced in Section 5.6, Project 3 as an illustration of the use of the data in Fig. 5.3.

FURTHER STUDIES

Ideas for projects are in Section 5.6 and calculations in Section 5.7.

Section 5.1 Methods for measuring and expressing soil water contents

Water contents are expressed in three ways:

1. *Gravimetric* water content is the mass of water per unit mass of oven-dry soil. The method of measurement is given in Section 3.3. Typically the units are $g\,g^{-1}$, $g\,kg^{-1}$ or as a percentage ($g\,H_2O$ per 100 g oven-dry soil).
2. *Volumetric* water content (normally termed θ) is the volume of water per unit volume of soil, typically $cm^3\,H_2O\,cm^{-3}$ soil and is often expressed as a dimensionless fraction or as a percentage by volume. Values are directly comparable to porosities expressed in the same units. The principles have been introduced in Section 4.2 where methods are described for determining macro- and microporosities by measuring the volumetric water content at field capacity and the total soil porosity.
3. *A depth of water* per unit depth of soil, typically mm H_2O, which allows direct comparison with amounts of rain or irrigation water.

The determination of volumetric water content

Equipment

As for the determination of bulk density (Section 4.2).

Method

Obtain a bulk density sample of the moist soil in its weighed cylinder, put on the end caps and place in a polythene bag. Tie the bag to prevent evaporation. In the laboratory weigh the moist soil + cylinder. Dry in an oven at 105 °C overnight (to constant weight) and reweigh. Record the length and internal diameter of the cylinder.

Calculation

Using the data in Section 4.2, the mass of oven-dry soil is 147.26 g, the volume of the cylinder is 100 cm^3 and the (dry) bulk density is 1.47 g cm^{-3}. If the mass of moist soil is 188.53 g then the sample contains 41.27 g of water. The gravimetric water content of the soil is

$$41.27/147.26 = 0.28 \, \text{g g}^{-1}$$

The volumetric water content is calculated assuming that 1 g H$_2$O occupies 1 cm^3 H$_2$O, and so

$$\theta = 41.27 \, \text{cm}^3 \, \text{H}_2\text{O divided by } 100 \, \text{cm}^3 = 0.41$$

If the data are available for gravimetric water content and bulk density, then

volumetric water content (θ) = gravimetric water
content × bulk density   [5.2]

Giving units, cm^3 H$_2$O cm^{-3} soil = g or cm^3 H$_2$O g^{-1} soil × g soil cm^{-3} soil. In the above case, $\theta = 0.28 \times 1.47 = 0.41$. Compared to values in Table 4.3, the soil might be a medium-textured soil at field capacity (storage and residual pores filled with water).

Water content expressed as a depth of water

Imagine a 1 cm cube of soil at a volumetric water content of 0.41 cm^3 H$_2$O cm^{-3} soil (Fig. 5.5). If no soil particles were present, the water would occupy a layer $1 \times 1 \times 0.41$ cm. If the soil were initially completely dry

and 0.41 cm of rain entered the surface of the cube and distributed itself through the soil, the volumetric water content would be 0.41. Thus the value of θ is also the water content expressed as cm H$_2$O cm^{-1} depth of soil.

Example Ten mm of rain falls and distributes through the top 20 cm of soil. The volumetric water content increases by 1/20 = 0.05. If the bulk density of this layer is 1.3 g cm^{-3}, then the increase in the gravimetric water content (Eq. 5.2) is 0.05/1.3 = 0.038 g H$_2$O g^{-1} soil.

Air-dry water contents

The water content of air-dry soil is normally expressed gravimetrically (Section 3.3). Values depend on the relative humidity of the atmosphere where the soil has been allowed to dry and on the specific surface area of the soil (Section 4.3, Note 1), since this determines the extent of the water films retained by the soil particles. In this condition much of the water is held by the exchangeable cations which reside on the surfaces of clay particles and humus, and so clay mineralogy and organic matter content also influence the water content.

Specific surface area is closely related to particle size, and so a knowledge of texture gives a good indication of the likely air-dry water content of a soil. Fig. 5.6 gives

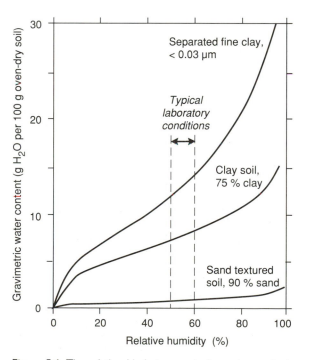

Figure 5.6 The relationship between air-dry water content and relative humidity of the air. Data from Thomas (1924).

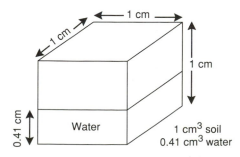

Figure 5.5 Water content of soil expressed as a depth of water.

data for soils of varying texture. Relative humidities in temperate regions are commonly between 50 and 60 per cent, and so air-dry water contents up to about 10 per cent are likely. The figure includes a fine clay separated from a clay soil by sedimentation. Large amounts of water are held on its very extensive surface. Section 5.7, Calculation 5 develops these ideas further.

Typical water content values

Table 4.3 gives typical values of volumetric water content for soils of different textures. Gravimetric values are given in Table 5.2; the air-dry water contents are taken from Fig. 5.6 and Thomas (1924) and the remaining values are calculated as follows.

Permanent wilting point and field capacity Based on Table 4.3, typical total porosities of the light-, medium- and heavy-textured soils were taken to be 0.4, 0.5 and 0.6 cm^3 cm^{-3} respectively. Using Eq. 4.1, these soils have bulk densities of 1.56, 1.30 and 1.04 g cm^{-3}. The ranges of residual porosity and residual + storage porosity were then used to calculate the gravimetric water contents. For example, a heavy-textured soil with a residual pore volume of 0.35 cm^3 cm^{-3} has a gravimetric water content of

$$0.35/1.04 = 0.34 \, \text{g H}_2\text{O g}^{-1} \text{ oven-dry soil or 34 per cent}$$
$$\text{(Eq. 5.2)}$$

Saturation The ranges of total porosity and thus the volumetric water contents when saturated are given in Table 4.3. Taking an example, the medium-textured soil has a total porosity of 0.55 cm^3 cm^{-3} and a bulk density of 1.17 g cm^{-3} (Eq. 4.1) and so the gravimetric water content is

$$0.55 \, \text{g H}_2\text{O cm}^{-3} \text{ soil}/1.17 \, \text{g soil cm}^{-3} = 0.47 \, \text{g g}^{-1} \text{ or}$$
$$\text{47 per cent}$$

Section 5.2 Soil water suction and its measurement

THE PRINCIPLES

When water drains out of soil under the force of gravity or is removed by evaporation or transpiration it is held at a suction (a reduced or negative pressure) due to the adhesion of water molecules to particle surfaces and the cohesion forces acting between water molecules (Hillel, 1982). Consider water in the cylindrical pore in Fig. 5.7, and imagine that a root is trying to extract water at

Table 5.2 Typical values for the gravimetric water contents of soils based on Fig. 5.6 and Table 4.3

Texture	Gravimetric water content (g H$_2$O per 100 g oven-dry soil)			
	Air dry	Wilting point	Field capacity	Saturation
Light	1–2	3–6	6–16	21–31
Medium	2–5	12–15	27–35	31–47
Heavy	5–10	24–34	38–53	38–90

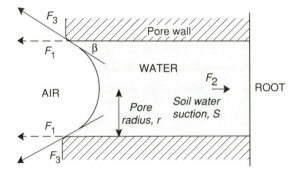

Figure 5.7 The relationship between suction and the radius of curvature of a meniscus at an air–water interface.

the right-hand end of the pore. The angle (β) is due to the resultant of the forces pulling water in the direction indicated by force, F_3. Force F_1 is the component of F_3 which acts in direct opposition to F_2 produced by the root suction. Thus at equilibrium, $F_1 = F_2$. Each force will be considered in turn.

The force due to root suction By definition, pressure = force/area. Giving units, $1 \, \text{Pa} = 1 \, \text{N m}^{-2}$. The suction applied by the roots on the pore water is S pascals, and operates over the cross-sectional area of the pore, πr^2 square metres. Thus $F_2 = S\pi r^2$ newtons.

The force due to adhesion and cohesion At the line of contact between the meniscus and the pore wall (the perimeter of the meniscus), the force is expressed as a force per unit length of contact, γ newtons per metre and is termed the *surface tension*. It acts along the water surface at the point of contact, i.e. at an angle β. Because it acts around the whole perimeter which has a length $2\pi r$ metres, $F_3 = 2\pi r\gamma$ newtons. F_1 is the component of F_3 which acts to oppose F_2 and is found by resolving using trigonometry: $\cos \beta = F_1/F_3$ and so $F_1 = 2\pi r\gamma \cos \beta$.

At equilibrium the two forces are equal Because $F_1 = F_2$, $S\pi r^2 = 2\pi r \gamma \cos \beta$. Rearranging gives $S = 2\gamma \cos \beta/r$.

If initially the soil was saturated and the soil water was at atmospheric pressure, $S = 0$, $\cos \beta = 0$ and so $\beta = 90°$, i.e. the meniscus is flat. As S increases, the meniscus retreats into the pore, and β decreases until eventually a maximum value of S is attained when $\cos \beta = 1$ and $\beta = 0°$. Any further attempt by the root to increase the suction simply draws the water back into the pore and empties it. Thus the critical suction required to empty a pore occurs when $\cos \beta = 1$ and $S = 2\gamma/r$. At 20 °C the value of γ for water in contact with soil particles is 0.075 N m^{-1} and so

$$S = 0.15/r \qquad [5.3]$$

where S is in pascals and r in metres. The values of critical soil water suction in Table 5.1 have been calculated using this equation.

THE MEASUREMENT OF SOIL WATER SUCTION

Tensiometers

A simple tensiometer is shown in Fig. 5.2. Soil water suction is determined from Eq. 5.1 by measuring h_w (m H$_2$O) which is the equivalent hydraulic head. With water in the manometer, its use is limited to soil water suctions near to field capacity. Using a mercury manometer extends its range to about 80 kPa. These can be made relatively easily and are illustrated in Fig. 5.8.

Making a tensiometer

Equipment
Porous ceramic cup, 22 cm diameter available from Soil Moisture Equipment Co., P.O. Box 30025, Santa Barbara, CA 93105 USA.
Polythene tubing with a 3 mm internal diameter (approx.).
Nylon or PVC tubing with an internal diameter approximately equal to the external diameter of the porous cup.
Nylon T-piece as illustrated in Fig. 5.8(b).
Araldite.
Wooden stand and mm scale.
Mercury and small plastic vials. CARE. Mercury vapour is harmful: handle in a fume cupboard or in an open space.

Preparation Glue the ceramic cup into the end of the nylon tube with Araldite as shown in Fig. 5.8(b). The

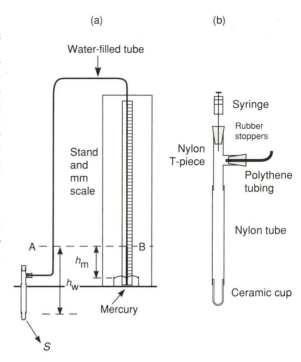

Figure 5.8 (a) A simple tensiometer with mercury manometer, and (b) an enlarged view of the tensiometer.

length of the tube is chosen depending on the maximum depth to which the tensiometer will be inserted in the soil. Similarly glue the nylon T-piece to the other end of the tube. Insert the capillary tube through a rubber bung and press into the side arm of the T-piece. Attach the capillary tube to the stand with its end dipping into a vial containing a few ml of mercury. Using a hypodermic syringe and boiled, cooled water, inject water into the top of the T-piece until the tube is completely filled. Now insert the needle through a rubber bung as shown and press the bung into the T-piece. Continue injecting water forcing it through the capillary tube; when completely filled, water will bubble out through the mercury. Remove the syringe and bung, and press another (unpierced) bung into the T-piece ensuring that no air bubbles are trapped in the tubing. Store the tensiometer with the cup in a beaker of water until required.

Insertion into soil

The tensiometer is inserted at the required depth in the soil as follows. Using a screw auger (Section 1.3), drill a hole of the same diameter as the tensiometer to within about 1 cm of the required depth. Insert the tensio-meter, finally pressing firmly so that good contact is made between the ceramic cup and the soil. Initial addition of a little water or of a slurry of kaolin and water will help the insertion. The slurry is particularly useful in stony soils where the ceramic cup may otherwise not make good contact with the soil. The addition of water has little effect on the soil water suction once it has been drawn into the surrounding soil.

The operation of the tensiometer and measurement of soil water suction

The suction in the soil water draws water out of the ceramic cup and mercury rises in the tube until equilibrium is attained (Fig. 5.8(a)). The pressure in the water in the capillary tube at A and B is the same. At A it is less than atmospheric pressure due to the soil water suction, S, and the column of water, h_w metres (Note 1). At B it is less than atmospheric pressure due to the column of mercury, h_m metres. Equating these two effects in terms of suctions gives $S + h_w g \rho_w = h_m g \rho_m$ where ρ_w and ρ_m are the densities of water and mercury (1000 and 13 590 kg m^{-3}) respectively and g is the gravitational force per unit mass (9.81 N kg^{-1}). Inserting values gives

$$S = 1.33 \times 10^5 h_m - 9810 h_w \text{ pascals} \quad [5.4]$$

The value of h_m When a capillary tube is inserted into mercury, the level of mercury in the tube is depressed below the surface in the vial. Thus the effective value of h_m is from the position of the depressed surface to the top of the column of mercury. The correction is only a few mm, and can be found by filling a short piece of capillary tube with water and inserting it into the mercury. If the tube is moved to the side of the vial (glass or clear plastic), the position of the depressed surface can be seen and the depression measured accurately enough for the purposes of the experiment.

The limits of use The maximum suction which can be measured with this equipment is about 80 kPa, equivalent to about 600 mm of mercury, at which stage air is drawn through the ceramic cup and the mercury column falls. However, this covers much of the range of suction between field capacity and the limit of readily available water (Table 5.1).

Note 1 In Fig. 5.9 the difference in water pressure between a and b is due to the force of gravity acting on the water. In effect the weight of the water is increasing the pressure at b. If the tube has a radius r (m), the volume of water between a and b is $\pi r^2 h_w$, and its mass is $\pi r^2 h_w \rho_w$ (kg). The gravitational force g (N kg^{-1}) acts

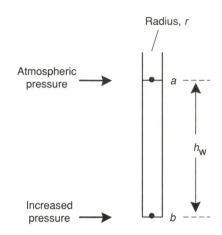

Figure 5.9 The pressure differences in a tube of water.

on this mass of water to give a force of $\pi r^2 h_w \rho_w g$ (N). This acts on the water at the base of the tube with an area πr^2 and so the increased pressure at b (force/area) is $h_w \rho_w g$ (Pa). In Fig. 5.2(a) the column of water between a and b is hanging below a so causing the pressure at a to be $h_w \rho_w g$ (Pa) less than at b. The pressure at b is the same as the pressure at the open water surface B which must be at atmospheric pressure. In Fig. 5.8(a) a column of water is hanging below A and a column of mercury is hanging below B. In each case the above expression can be used with h_m and ρ_m substituted for the mercury column.

Commercially available tensiometers

These consist of a porous ceramic cup connected to a pressure gauge, for example the Quickdraw Soil Moisture Probe available from ELE International. The gauge is calibrated in bars (0–100 centibars), an old unit of pressure. The conversion factor is 1 bar ≈ 100 kPa, and so 1 centibar ≈ 1 kPa. The limit of measurement is again about 80 centibars.

The filter paper method

Filter paper is a porous material and, like soil, has a water release characteristic. This characteristic has been determined for Whatman No. 42 papers by equilibrating papers to known suctions using a range of methods, all of which are beyond the scope of this book (Hamblin, 1981; Smith and Mullins, 1991). The results are shown in Fig. 5.10. Three sets of data are included obtained with different batches of papers over a number of years, and are reliably consistent. The measurement of soil water suction involves equilib-

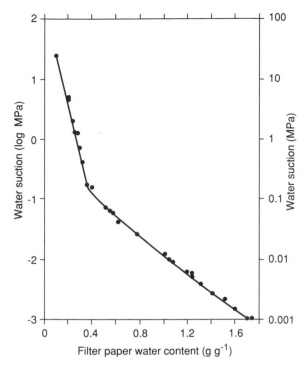

Figure 5.10 The water content–water suction relationship for Whatman No. 42 filter papers. For help with the use of a log scale, see Section 2.3 and Figure 2.8. From Hamblin (1981).

ration of a paper with a moist soil until their water suctions are equal. The water content of the paper is determined and from the figure the suction is found. C. E. Mullins provided information on this method.

Equipment

Whatman No. 42 filter papers, 55 mm diameter. Note that although alternative papers can be used for most other purposes in this book, Fig. 5.10 only applies to these Whatman papers.
Plastic bottles with caps, 65 mm diameter × 50 mm deep or similar.
Electronic balance weighing to 1 mg.
Cardboard box, open fronted and lined with moist paper.
Insulated picnic box.
Lightweight plastic weighing bottles with caps.
Plastic adhesive tape.
Artist's brush.

Method

Disturbed samples Half fill a plastic bottle (replicated)

with soil and place a filter paper (labelled) centrally on top of the soil. Add just sufficient soil to fill the container when the cap is screwed down (about 180 g in total). Tape the cap with adhesive tape to prevent moisture loss and place in an insulated picnic box standing on an expanded polystyrene block in a constant temperature room or in a room where there is little temperature fluctuation. Temperature fluctuations would cause water to evaporate from the soil and condense on the walls of the container. Equilibration time depends on the soil water content: 6 d is recommended as a standard, but equilibration is nearly complete after 2 d.

Tare a weighing bottle on the balance, and place the open bottle in the cardboard box. Remove the plastic bottle from the insulated box and in the cardboard box pour the soil and paper on to a dish, pick up the paper with tweezers, brush off loose soil, place the paper in the weighing bottle, close the cap and weigh immediately. Moist papers lose water by evaporation if waved about in the air in the laboratory at a rate of about $1 \, \text{mg s}^{-1}$. The filter papers weigh about 300 mg (oven dry) and contain between 60 and 550 mg H_2O after equilibration.

Dry the paper at 105 °C and reweigh.

Calculate the water content of the paper as $g \, H_2O \, g^{-1}$ oven-dry paper, and from Fig. 5.10 determine the water suction in the paper and thus the soil water suction.

The main error in this technique results from poor temperature control which allows water to condense on the walls of the bottle. This reduces the water content of the soil and may wet the paper if it comes into contact with the walls during equilibration or when tipping the soil out of the bottle. Contamination of the paper by soil only introduces minor errors.

Normally the gravimetric water content of the soil would also be measured (Section 3.3).

Undisturbed cores Cores sampled for bulk density determination (Section 4.2) or for root measurements (Section 3.1) can be used. After trimming the face of the core, place it on a filter paper (cut if necessary so that it is smaller than the diameter of the core) in a closed container. Equilibrate and measure as above. Alternatively the core can be cut and the paper inserted between adjacent faces.

Direct measurement in pots or in the field Insert the paper in a slit cut in the soil with a spatula and close the soil above the paper. The paper should be weighed immediately it is removed from the soil. It is not necessary to dry each paper since standard papers have almost equal masses: oven-dry a few papers beforehand and determine the average mass.

Resistance blocks

Resistance block soil moisture meters can be used to measure suctions between 20 and 1500 kPa. In principle the method is similar to the filter paper method. A ceramic block (or other suitable material) is placed in the soil and absorbs water until it is in equilibrium with the soil water. Its electrical resistance is measured, and is related to its water content. With calibration a measure of suction is obtained. Equipment is available from ELE International.

Other methods

A simple method to bring soil samples to known suctions is given by Pritchard (1969). Standard methods are discussed by Smith and Mullins (1991).

Section 5.3 The measurement of the water release characteristics of soils

Two methods are described: the Haines method is used to measure the water release characteristic between saturation and about 15 kPa suction (slightly drier than field capacity), and the filter paper method (Section 5.2) is used between 1 kPa and 50 MPa (wetter than field capacity to air dry). More advanced methods are discussed by Smith and Mullins (1991).

The Haines method

The method is based on the principles of the tensiometer shown in Fig. 5.2. The soil is placed in a Buchner funnel made with a porous sintered glass plate in the base which is connected via a water-filled polythene tube to a graduated glass tube (Fig. 5.11). The soil is initially saturated. By lowering the graduated tube a suction can be applied to the soil water which is drawn through the sintered plate until equilibrium is re-established. Soil water suction is measured by the equivalent hydraulic head, and changes in soil water content by changes in water level in the graduated tube. The final water content of the soil is determined gravimetrically. The procedure will be described initially using sand, and then modified for measurement with soil.

Equipment

Glass Buchner funnel with sintered glass plate. Obtainable from Fison's Scientific Equipment with a Grade 4 plate which has a pore size of 10–16 μm and so can withstand

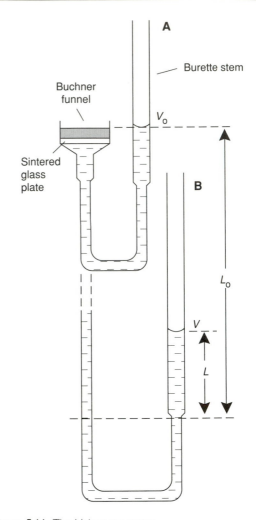

Figure 5.11 The Haines apparatus.

an applied suction of 9–15 kPa (0.9–1.5 m H_2O) before air is drawn through (Eqs 5.1 and 5.2).
Polythene tubing (1.5 m) and a calibrated glass tube. A 50 ml burette stem without the tap is ideal.
Retort stand, clamps and clips.
Sand.

Method

Set up the equipment shown in Fig. 5.11. Place the funnel upside down in a bucket of boiled and cooled water and with a vacuum pump draw water through the funnel into the tube until filled. Ensure that there are no air bubbles in the system. Clamp the funnel and the tube at the top of the stand and adjust to position A in the figure. Add boiled water, or pour water from the tube as required.

Weigh about 50 g (±0.01) of air-dry sand and pour it into the water in the funnel. Tap the funnel, adding water if necessary to ensure that the sand is saturated. Lower the tube to draw water out of the sand (pouring water away if necessary) until the sand is just saturated (no free water above the sand) and the meniscus in the tube is level with the surface of the sand. The initial setting should leave the water level near to the bottom of the tube. Cover the funnel with Clingfilm (pierced in a few places) to prevent evaporation. Proceed as follows:

- Measure the height of the meniscus, L_o (cm), above a fixed reference point (the retort stand or bench top), and record the volume, V_o (cm^3), in the calibrated tube.
- Lower the tube by about 3 cm; a little water will be drawn out of the sand into the tube. When the system has equilibrated (about 3 min for sand), record the new values of L and V (Fig. 5.11, position B).
- Repeat the above step, recording equilibrium values of L and V, until the sand appears to have lost most of its water and V is changing very little. Lower the tube now in steps of 10 cm or more until the meniscus is about 100 cm below the sand ($L_o - L \approx$ 100 cm). Note that for sand there is a well-defined stage (a critical range of suction) at which air 'breaks through' into the pores and water is released.
- Finally, weigh a 100 ml weighing bottle. Without changing the position of the funnel or calibrated tube, clamp the polythene tube just below the

funnel. Transfer most of the sand to the weighing bottle and determine its gravimetric water content, g H$_2$O g^{-1} oven-dry sand (Section 3.3). Determine also the water content of a sample of the air-dry sand used in the experiment.

Data handling

An example of results for a sand is shown in Table 5.3. The hydraulic head (column b) is calculated from the difference in height of the sand and the meniscus (column a). Volumes of water withdrawn from the sand are calculated using column c. The footnote to the table gives calculations needed to determine the mass of water present in the sand at the end of the experiment (1.18 g). Assuming that 1 g of H$_2$O occupies 1 cm^3 the final value in column e is 1.18 g from which the other values in the column are calculated. Column f can then be calculated (values in column e/50.32).

Conversion to volumetric water contents The particle density of sand is 2.65 g cm^{-3} and thus its specific volume (Section 4.3, Note 1) is 1/2.65 = 0.377 cm^3 g^{-1}. Thus saturated sand with a gravimetric water content of 0.235 g H$_2$O g^{-1} sand (column f) has

$$0.235 \text{ cm}^3 \text{ H}_2\text{O} + 0.377 \text{ cm}^3 \text{ sand} = 0.612 \text{ cm}^3 \text{ total volume}$$

and its volumetric water content = 0.235/0.612 = 0.384 cm^3 cm^{-3}. Assuming that the volume of the bed of sand does not change, the values in column g are calculated by dividing column f values by 0.612.

Table 5.3 Data for the Haines experiment using sand, sieved to give 0.5–2 mm particles

a Height of the meniscus above the bench (cm)	b Hydraulic head $h_w = L_0 - L$ (cm)	c Volume reading (cm^3)	d Volume withdrawn, $V_0 - V$ (cm^3)	e Volume in the sand (cm^3)	f Gravimetric water content (g g^{-1} oven-dry sand)	g Volumetric water content, θ (cm^3 cm^{-3})
$L_0 = 84.10$	0	$V_0 = 33.05$	0	11.83	0.235	0.384
$L = 78.70$	5.40	$V = 32.30$	0.75	11.08	0.220	0.359
74.10	10.00	32.09	0.96	10.87	0.216	0.353
71.50	12.60	31.40	1.65	10.18	0.202	0.330
69.80	14.30	29.80	3.25	8.58	0.170	0.278
66.70	17.40	26.90	6.15	5.68	0.113	0.185
60.35	23.75	23.80	9.25	2.58	0.051	0.083
47.20	36.90	22.90	10.15	1.68	0.033	0.054
22.50	61.60	22.65	10.40	1.43	0.028	0.046
3.30	80.80	22.40	10.65	1.18	0.023	0.038

Mass of air-dry sand used = 50.34 g
Water content of air-dry sand = 0.4 g kg^{-1} oven-dry sand;
Mass of oven-dry sand used = 50.32 g;
Final water content of the sand = 23.5 g kg^{-1} oven-dry sand;
Mass of water finally present in 50.32 g oven-dry sand = 1.18 g.

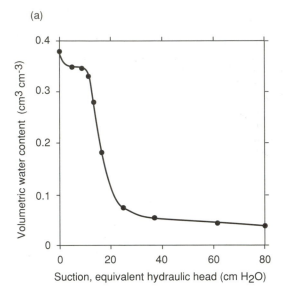

(a)

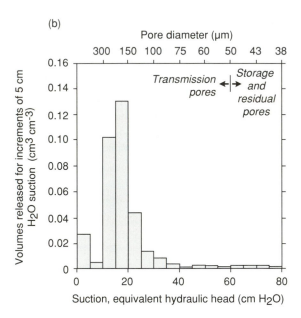

(b)

Figure 5.12 (a) The water release characteristic, and (b) the pore-size distribution of 0.5–2 mm sand.

Pore size distribution The volumes of water released for each 5 cm increase of suction are plotted in Fig. 5.12(b). The x-axis is also labelled as pore size calculated by combining Eqs 5.1 and 5.3 to give $r = 0.15/9810h$ where r and h are in metres, or diameter, $d \approx 3000/h$ where d is in μm and h is in cm. The pores which empty between 10 and 30 cm suction have diameters between 300 and 100 μm, and occupy $0.288 \, \text{cm}^3 \, \text{cm}^{-3}$, which is 75 per cent of the total pore volume, and 88 per cent of the pore volume is in transmission pores ($>50 \, \mu$m). A sand of this particle size in the field would drain to a water content at field capacity of only about $0.05 \, \text{cm}^3 \, \text{cm}^{-3}$ or $0.03 \, \text{g g}^{-1}$, indicating the small amounts of available water held by coarse sandy soils (Tables 12.1 and 5.2 and Fig. 5.3b). For materials of this sort with a narrow range of particle and pore size and no structural pores, the predominant pore size is about one-fifth of the particle size (about 200 μm and 1 mm respectively in this case).

A modified method for soils

It is much easier to learn how to handle the equipment and the data by determining the water characteristic of a sand before moving to soil measurements. The experiment takes little time because of the rapid equilibration of the sand, and errors can be corrected.

The range of suctions used in the Haines experiment empties mainly the transmission pores ($>50 \, \mu$m, Table 4.2). In soils these are structural pores, and so there is little value in carrying out the experiment with air-dry sieved soil, other than to obtain an approximate value for the water content at field capacity (50–100 cm H_2O). Storage and residual pores are not altered much by drying and sieving. Thus, to determine a meaningful water release characteristic the experiment should be run using an undisturbed core sampled and trimmed as for bulk density determination (Section 4.2), but using a cylinder which fits into the Buchner funnel and is about 2 cm thick. The bottom of this disc of soil must be carefully trimmed so that it sits in contact with the sintered plate. The soil is best left in its metal cylinder during the experiment.

After sampling and trimming the moist soil, place it in the funnel initially with only a few mm of water present. Allow the soil to wet, maintaining the water level to saturate the soil over a period of a few hours. Alternatively, stand the soil on wet tissue paper and add water to the paper occasionally to wet the soil slowly to saturation.

Proceed as for the sand but allow much longer for equilibration at each step: check that the meniscus has finished moving by measuring at intervals. The time required for equilibration depends on soil texture:

The water release characteristic is plotted in Fig. 5.12(a). A small amount of water is withdrawn initially which is 'free' on the surface of the bed of sand. Most of the water is released between 10 and 30 cm H_2O suction.

sandy soils may need about an hour at each step but clay soils need several weeks. Evaporation must be prevented. Finally remove the soil + ring, weigh moist, oven-dry and weigh again. Weigh and measure the dimensions of the empty cylinder after washing and drying.

Data handling

Calculate the oven-dry mass of soil, the final water content and the volume of the core (Section 4.2). With data recorded as in Table 5.3, calculate the values for columns e and g. Column f is probably not needed. Examples of water release characteristics and pore size distributions are given in Fig. 5.3.

When used with soil the sintered plate may become partially blocked hindering equilibration. Clean by placing in detergent solution (Decon) and boiling.

The filter paper method

Soil samples are prepared with a range of water contents between air-dry and field capacity. The filter paper method of Section 5.2 is used to measure soil water suction and gravimetric water content is determined.

Equipment

The equipment for the filter paper method.
Soil. 6 kg of air-dry <2 mm soil.

Method

Calculate the amounts of water which will bring 750 g samples of soil to water contents between air-dry and field capacity as follows. Table 5.2 gives typical values of gravimetric water contents of soils. Work with about eight samples. Place each sample in a polythene bag and add by pipette the required volume of water with mixing and shaking. Store the bags for 2 weeks with the necks folded in a container (an insulated picnic box) which will prevent evaporation and allow each sample to become uniformly moistened.

Measure the soil water suction by the filter paper method (three replicates) and the gravimetric water content of each sample (Section 3.3).

Plot the data as in Fig. 5.3. The experiment can usefully be combined with measurements of the effects of soil water on the germination of seeds (Section 5.6, Project 3).

Section 5.4 Soil water potentials and the direction of water movement

Water moves in soil as a result of gravitational forces, in response to differences in soil water suction and by osmosis. Osmosis is normally only of importance when considering movement into roots. A common situation in the field is that soil near the surface is drier than soil at depth. Hence there is a gradient of soil water suction producing an upward force on the soil water. This is opposed by the downward gravitational force. The two must be considered together if we wish to know in which direction the water will move.

An example is shown in Fig. 5.13(a) where simple tensiometers with water manometers have been inserted at A and B in a soil profile. The tensiometers indicate that the soil at A is relatively dry (has a greater suction) compared with the soil at B. This is shown by h_a being greater than h_b. The direction of water flow can be judged by comparing the water levels in the two open limbs of the manometers (as measured by L_a and L_b). Because water is free to flow through the soil between A and B, it would be expected that water would flow between the tensiometers until an equilibrium is reached where the two water levels are the same, i.e. $L_a = L_b$. Hence, water would move down through the soil, drying soil A and wetting soil B. However, if the soil at A is much drier (Fig. 5.13(b)), then the upward force due to the suction gradient exceeds the gravitational force and water moves upwards from B to A. In order to consider these two forces together we make use of the principle that water will move from where its potential energy is large (because of high elevation or small suction) to where the potential energy is small (because of low elevation or large suction). We have therefore to consider how elevation affects the *gravitational potential* and how adhesion to soil particles (the soil matrix) affects the *matric potential*. These terms will now be explained.

Potential energy

In moving a mass of material vertically there is a change in potential energy. Work has to be done to raise the body against the force of gravity which can be released again as the material falls: as the body is raised, energy is stored and the potential energy (ψ) increases. By definition, the work done on the body is equal to the change in its potential energy and is measured by the product of force applied (newtons, N) × distance (metres, m) moved. The unit of energy (and of work) is the joule (J) and giving units $1 J = 1 Nm$. For a body being moved against gravity, the gravitational force

(a)   (b)

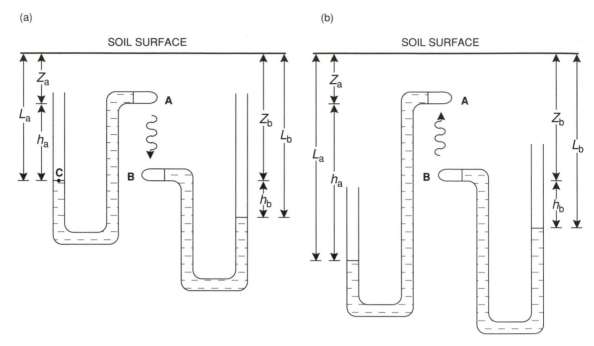

Figure 5.13 Soil water suction gradients and gravitational forces in soils.

g ($\mathrm{N\,kg^{-1}}$) acting on a mass M (kg) has to be overcome, which requires a force of $M \times g$ newtons. Thus if the body is raised through a distance L metres, the increase in potential energy, ψ, is MgL (units, $\mathrm{kg} \times \mathrm{N\,kg^{-1}} \times \mathrm{m} = \mathrm{N\,m}$ or J) and must be equal to the energy used in raising the body.

Using Fig. 5.13(a) the concept of potential energy can be applied directly to the difference between water at the soil surface (an open dish of water at that elevation often termed *free water*) and the water in the manometer tube at C. Both are at atmospheric pressure, but because of the difference in elevation, M (kg) of water at the surface has a potential energy MgL_a (J) greater than at C. Thus we say that the water at C has a potential energy of $-MgL_a$ (J) relative to the standard reference condition (the free water at the surface). Expressed per kilogram of water, its potential energy is $-gL_a$ ($\mathrm{J\,kg^{-1}}$).

Soil water potential

We now have to consider the water in the soil at A; because water in the tensiometer has come to equilibrium with the water at A, the two must have the same energy status (otherwise further movement would occur). Thus the potential energy of the water at A is also $-gL_a$ relative to the standard reference condition, and is termed the *soil hydraulic potential*. If, as assumed

from here on the osmotic effect (Section 14.3) is negligible, it can be regarded as the *soil water potential*, ψ. It may at first sight seem strange that the potential energy of the water at A is related to depth L_a when it is only actually at depth Z_a. This difference shows that the soil water potential is less than in the reference condition for two reasons:

1. The soil water is below the soil surface, and has a *gravitational potential*, ψ_g, depending on depth Z_a.
2. The water is held by the soil matrix and has a *matric potential*, ψ_m, which is related to the soil water suction since both depend on the adhesion of water to soil particles.

In order to handle conveniently these two terms the water potential can be expressed as energy per unit volume of water, i.e. as $\mathrm{J\,m^{-3}}$. Since $\psi = -gL_a$ ($\mathrm{J\,kg^{-1}}$) and ρ_w has the units $\mathrm{kg\,m^{-3}}$, then $\psi = -\rho_w gL_a$ ($\mathrm{J\,m^{-3}}$). Note that this equation is similar to Eq. 5.1 and that the units of ψ (strictly $\mathrm{J\,m^{-3}}$) are dimensionally equivalent to the units of pressure (Pa) since $1\,\mathrm{J\,m^{-3}} = 1\,\mathrm{N\,m\,m^{-3}} = 1\,\mathrm{N\,m^{-2}} = 1\,\mathrm{Pa}$.

The gravitational potential at A can now be expressed as $-\rho_w gZ_a$ since it is simply the potential energy of free water at elevation A compared to the reference condition. Since $\psi = \psi_g + \psi_m$, then the matric potential must be $-\rho_w gh_a$ and is numerically equal to the negative value of soil water suction (Eq. 5.1). Thus

$$\psi = -\rho_w g Z_a - \rho_w g h_a \qquad [5.5]$$

with the units $\mathrm{J\,m^{-3}}$, or numerically

$$\psi = -9810 Z_a - S \qquad [5.6]$$

where ψ has the units $\mathrm{J\,m^{-3}}$, Z_a is in m and S in Pa. Because the units $\mathrm{J\,m^{-3}}$ are equivalent to Pa, the three terms in the above equations are sometimes expressed for convenience in Pa. Similarly, because Eq. 5.5 can be written

$$\psi = -\rho_w g (Z_a + h_a) = -\rho_w g L_a$$

soil water potential is sometimes expressed as an equivalent hydraulic head, L_a. Note that the term equivalent hydraulic head can be used both for soil water suction (Table 5.1) and soil water potential.

The direction of water movement

Water moves in the direction of decreasing soil water potential, i.e. down a gradient of potential. Rate of movement is discussed in Section 5.5. Where there is no change in potential between two points, there is no movement of water, a condition termed *zero flux*. Field capacity is a special case of zero flux: if ψ_g is expressed as an equivalent hydraulic head, it changes by $1\,\mathrm{mm\,m^{-1}}$ depth. If in the soil the change in ψ_m is equal and opposite to ψ_g, i.e. the soil is slightly drier towards the surface, then the gradient of ψ is zero and no movement occurs. Conditions in a column of soil allowed to drain in the laboratory are discussed in Section 5.5, Note 1. However, field capacity under most circumstances is the water content at which hydraulic conductivity has fallen to such a value that downward water movement becomes negligible even though a potential gradient may be present.

Conclusion

The determination of soil water potential requires only the measurement of soil water suction and depth in the profile. The former can be found using tensiometers or the filter paper method. Gradients of water potential indicate the direction of water movement. A case study from Niger follows to illustrate how measurements can be used to determine the direction of water movement in a profile and Section 5.6, Project 4 suggests how the methods could be applied in a temperate region.

A CASE STUDY: SOIL WATER DYNAMICS IN SUB-SAHARAN SAVANNA

A project carried out at the ICRISAT (International Crops Research Institute for Semi-Arid Tropics) Sahelian Centre in Niger, West Africa by the Institute of Hydrology at Wallingford and the University of Reading aimed to determine the extent to which water losses from the soil were by drainage or by upward movement resulting from evaporation from the soil surface and from vegetation. In savanna regions the problem is complex because the vegetation consists of occasional clumps of bushes with deep roots interspersed with relatively shallow rooted annual grasses (Colour Plates 8(a) and (b)). Hence, the pattern of water movement is likely to depend strongly on proximity to bushes which can extract water from deep in the soil profile.

The site chosen was close to Niamey, the capital of the Republic of Niger. There is a distinct wet season (June to October) during which virtually all the rain falls (Fig. 12.8). The average annual rainfall is similar to London (650 mm) but the evaporative demand of the hot, dry air (Ch. 12) is large, and there is insufficient rainfall to keep pace with demand. Even so, because the rain falls in erratic, intense storms and the soil has a small storage capacity, some water penetrates to depth and ultimately replenishes the groundwater resources which are exploited for domestic water supplies using boreholes. Hence there is considerable interest in understanding how changing the vegetation as land is cleared for agriculture will change the patterns of deep percolation and groundwater recharge. The soil is very

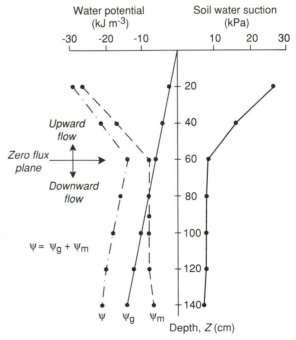

Figure 5.14 The changes in soil water potential and soil water suction with depth under grass at Sadoré in Niger. Data by V. L. Grime, Reading University and the Institute of Hydrology.

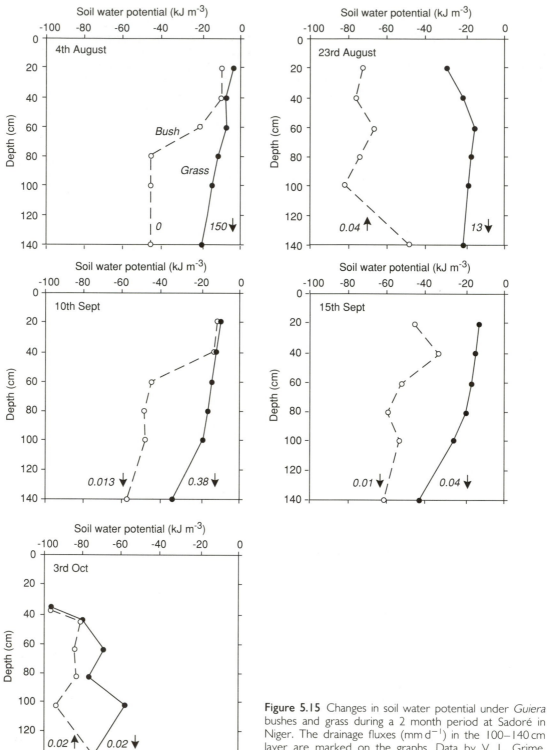

Figure 5.15 Changes in soil water potential under *Guiera* bushes and grass during a 2 month period at Sadoré in Niger. The drainage fluxes (mm d^{-1}) in the 100–140 cm layer are marked on the graphs. Data by V. L. Grime, Reading University.

sandy (West et al., 1984), and its properties are shown in Fig. 5.3(b).

Core samples were taken at various times from under the bushes and under grass; depth of sample was recorded and soil water suction measured using the filter paper method (Section 5.2). It was expected that water below the reach of roots would move downwards in response to gravity, whereas water within the root zone would be sufficiently depleted to generate an upward flow from the wet soil below.

Fig. 5.14 shows the suction profile (the changes in suction with depth) beneath grass on 23 August 1990. The soil at 20 and 40 cm depth was much drier (larger suction) than below. The suctions at 60 cm and below were 7–8 kPa, values typical of field capacity (Table 5.1). In the upper part of the soil profile there was a large suction gradient and in the lower part a small suction gradient, both tending to move water upwards.

To determine whether these suction gradients were sufficient to overcome gravity, soil water potential, ψ, was calculated using Eq. 5.5, and is plotted together with values of gravitational, ψ_g, and matric potential, ψ_m, in Fig. 5.14. Note that the profile of ψ_m is a mirror image of the suction profile, and that the ψ_g profile is a straight line since it depends simply on depth. The direction of water movement can now be gauged from the slope of the water potential profile ψ. Above 40 cm the water must be moving upwards towards a more negative water potential, whereas below 80 cm the flow must be downwards. Somewhere in between (around 60 cm), at a point where the graph is vertical (no change of potential with depth), there is apparently no flow.

This critical depth is known as the *zero flux plane* and is marked on the figure. During dry periods the depth of the zero flux plane can be a good indicator of the depth from which roots have taken up water.

Figure 5.15 shows a sequence of soil water potential profiles through the rainy season which began on 24 July. The dashed line shows measurements close to a clump of *Guiera senegalensis* bushes (multi-stemmed trees growing several metres tall) and the solid line is for a grass clearing (9 m from the nearest tree). The date of 4 August was sufficiently into the rainy season for the soils to be at their wettest. The 'grass' profile was close to field capacity, but the bush profile was relatively dry below 60 cm. It seems that the bushes produced a good canopy of leaves soon after the rains began which provided a large demand for water; successive rainfall events were mainly replenishing water that had been used by the bushes rather than wetting the soil profile to increasing depth. Alternatively, the leaf canopy may have acted as an umbrella, reducing the amount of water that entered the soil. At both sites, the slope of the water potential profile suggests that water was moving downwards.

Between 4 and 23 August there was little rain, and soil water was depleted mainly through uptake by vegetation. By 23 August, the zero flux plane had reached 60 cm under grass. This is the example discussed above (Fig. 5.14). It seems that grass roots had grown down to 60 cm. This would have required a growth rate of $2 \, \text{cm} \, \text{d}^{-1}$ since the beginning of the rains, which is reasonable for annual grasses. In contrast, the zero flux plane beneath the bushes is below the maximum depth of measurement. Overall it seems that water is generally moving upwards in the soil profile which reflects the greater rooting depth and water demand of the bushes. Certainly, little water would have been lost as drainage under the bushes, whereas some drainage would have occurred in the grass clearings. Rates of drainage are calculated in Section 5.5.

Between 23 August and 10 September there was 100 mm of rainfall which recharged the surface layers with water, and on 10 September the zero flux plane in both sites was close to the surface, indicating that water was draining through both soil profiles. From mid-September there was little more rain, and by 3 October both profiles were very dry with water moving upwards.

Section 5.5 Hydraulic conductivity and rate of water movement

Hydraulic conductivity: definition and units

The rate, F, at which water moves through a column of soil or sand, known as the water flux, is expressed by Darcy's Law:

$$F = -K \, \mathrm{d}\psi/\mathrm{d}x \qquad [5.7]$$

where K is the hydraulic conductivity and $\mathrm{d}\psi/\mathrm{d}x$ is the gradient of water potential (Section 5.4). Equations of this type are also used to express the relationship between movement of gases and their concentration gradients (Section 6.5), the movement of nutrients (Chs 9 and 10) and the movement of heat. Hydraulic conductivity is a measure of the ease with which water can move through the soil, and the potential gradient is a measure of the force driving water through the soil. Rearranging the equation gives $K = F/(-\mathrm{d}\psi/\mathrm{d}x)$, and so K is equal to the rate of water flow that is achieved by unit driving force. The negative sign indicates that the force operates towards smaller potentials.

Various units can be used for the terms in Eq. 5.7. Perhaps the most widely used and most convenient are as follows.

- *Water flux* is expressed as the volume of water passing through unit cross-sectional area of soil per unit time, $m^3 m^{-2} s^{-1}$, which is equivalent to a water velocity, $m s^{-1}$. This velocity can be thought about as the rate of change of the depth of a pool of water as it infiltrates the underlying soil. The velocity of water in the soil pores is faster than F, because the flow of water is not distributed across the whole soil volume. For example, imagine a puddle of water on the surface of a sand with a porosity of $0.5 cm^3 cm^{-3}$. If the level of the puddle decreases by $10 mm h^{-1}$, the water flux in the soil is also $10 mm h^{-1}$, but the average pore water velocity is $20 mm h^{-1}$ because half of the soil volume is non-conducting (Section 11.5).

- *Water potential* is conveniently expressed as an equivalent hydraulic head, metres, because the units of water potential gradient are then $m m^{-1}$, i.e. a dimensionless parameter.

- *Hydraulic conductivity* has the same units as water flux ($m s^{-1}$) if $d\psi/dx$ is dimensionless, and can be thought about as the flux achieved when $d\psi/dx = -1$, since substitution in Eq. 5.7 gives $F = K$. Since the gradient of water potential due to gravity alone is -1 when expressed as an equivalent hydraulic head (ψ_g decreases by $1 m$ for each $1 m$ increase in soil depth), the hydraulic conductivity can be thought of as the rate of water flow achieved by gravity alone. For example, the hydraulic conductivity of a saturated soil (no suction gradient) covered with a very shallow puddle of water is equal to the flux through the soil and is therefore measured by the rate at which the water level in the puddle decreases.

The relationships between water-filled pore characteristics and hydraulic conductivity

The size and continuity of water-filled pores determine the hydraulic conductivity of a soil (pages 85–6). Relationships can be derived mathematically by using the water release curve to determine the pore size distribution, considering the pores as capillary tubes and applying Poiseuille's equation which relates the rate of flow through a capillary tube to the radius of the tube and the pressure gradient in the tube. Marshall and Holmes (1988) describe this approach, and an example of its use for a sand is given by Marshall (1958).

THE MEASUREMENT OF HYDRAULIC CONDUCTIVITY

There is no standard method for measuring hydraulic conductivity of soils. Four techniques are described here.

1. Saturated hydraulic conductivity is measured in repacked columns of soil in the laboratory.
2. Infiltration of ponded water into soils provides the basis for measuring saturated hydraulic conductivity in the field using a double ring infiltrometer.
3. The hydraulic conductivity of soils at water contents between saturation and field capacity is determined by measuring the water content of an unsaturated column of soil through which water is flowing.
4. Hydraulic conductivity of drier soils is determined by measuring the rate of evaporation from the top of a column of soil supplied with water at its base.

Although (1) and (2) are standard methods, measurements of unsaturated hydraulic conductivity are normally made using more sophisticated techniques than (3) and (4) which are beyond the scope of this book (Smith and Mullins, 1991).

The measurement of saturated hydraulic conductivity in the laboratory

The method is most easily used with sand as a means of learning how to use the equipment and handle the data (Green and Ampt, 1911). A modification for a core of soil is also described.

Equipment

Leaching tube as shown in Fig. 5.16(a) which can be made from a 15 cm length of 5 cm diameter glass tubing. Into one end insert a bung drilled to take a short length of 5 mm diameter tubing. Connect to the base of the leaching tube a glass outflow tube bent as shown in the figure.
Absorbent cotton wool.
Measuring cylinder.
Sand.
Stopclock.

Method

Set up the equipment on a stand as shown in figure. Insert a small plug of cotton wool in the base of the tube, clip the outflow tube and half fill the leaching tube with water. Now allow water to flow out of the equipment to completely fill the outflow tube, making certain no air bubbles are trapped. Pour a layer of sand into the leaching tube (about 5 cm deep) tapping and swirling to allow air bubbles to escape. Insert another cotton wool plug above the sand.

Attach to a 500 ml volumetric flask a length of rubber tubing and a clip as shown in the figure. Fill the flask with water, close the clip, insert in the leaching tube and open the clip. Open the clip on the outlet tube.

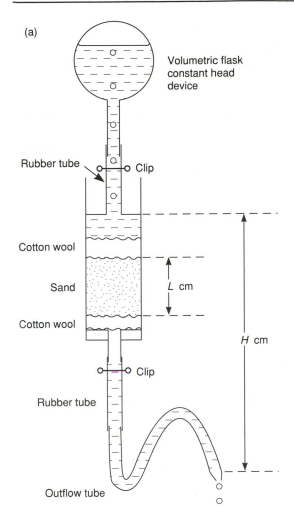

(a)

Volumetric flask
constant head
device

Rubber tube — Clip

Cotton wool

Sand   L cm

Cotton wool

H cm

Clip

Rubber tube

Outflow tube

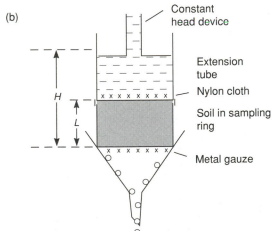

(b)

Constant
head device

Extension
tube

Nylon cloth

Soil in sampling
ring

Metal gauze

Figure 5.16 Equipment for measuring the saturated hydraulic conductivity of (a) sand and (b) soil.

Adjust the position of the volumetric flask and the outflow tube so that the head, H, is between 2 and 15 cm to give a flow of about 1 drop s^{-1}. Fast flow rates lead to turbulent flow through the sand and Darcy's Law no longer applies.

When flow is steady measure its rate, Q ($cm^3 s^{-1}$), using a measuring cylinder and stopclock. Measure several times and take a mean value. Measure the length of the sand layer, L_1 (cm); measure at several positions around the tube and take a mean value. Measure the head, H (cm). Because sand has a large conductivity the heads are small; use a transparent ruler and a small mirror to avoid parallax error. Repeat the determination with several different heads (five values are adequate).

Remove the volumetric flask and the top plug of cotton wool, add sand to give a 10 cm layer, L_2, and replace the cotton wool plug. Repeat the above measurements.

Finally measure the internal diameter of the leaching tube, and calculate its cross-sectional area, A (cm^2).

Calculations

The resistance to flow of water through the system depends on both the sand and the equipment (the cotton wool and outlet tube primarily). Calculation of the hydraulic conductivity of the sand requires the equipment resistance to be taken into account as follows.

Darcy's Law is analogous to Ohm's Law which describes the conduction of electricity through a resistor which can be written:

current = potential difference/resistance

For the above experiment this can be written

$$Q/A = H/R \qquad [5.8]$$

where R is the resistance of the sand plus the equipment. Applying the units of measurement, $cm^3 s^{-1}/cm^2$ = cm/R and so R has the units s.

For each stage of the experiment (the 5 and 10 cm sand layers), plot Q/A on the y-axis against H on the x-axis. Fit a straight line to the points and determine its slope (Section 9.6). The value of R is the reciprocal of the slope, R_1 and R_2 being for the 5 and 10 cm layers respectively.

Again using an electrical analogy, the resistance of the sand, R_{sand}, and of the equipment, $R_{equipment}$, can be considered to be 'in series'. The data for the two depths of sand are now used to separate the two resistances. With the 5 cm layer of sand $R_1 = R_{sand 1} + R_{equipment}$ and with the 10 cm layer of sand $R_2 = R_{sand 2} + R_{equipment}$. Subtracting the two equations gives $R_{sand 2} - R_{sand 1} = R_2 - R_1$. Thus $R_{sand 2} - R_{sand 1}$ (and

$R_2 - R_1$ also) is the resistance of the added layer of sand which has a length $L_2 - L_1$. Calculate the *resistivity* of the sand (its resistance per unit length) as $(R_2 - R_1)/(L_2 - L_1)$ which, as measured, has the units $s\,cm^{-1}$, and multiplied by 100 has the units $s\,m^{-1}$. Resistivity is the reciprocal of conductivity and so

$$K\,(m\,s^{-1}) = 1/\text{resistivity}\,(s\,m^{-1})$$

Note that the term *resistance* applies to the whole 'resistor', in this case the whole leaching system or the whole column of sand. The *resistivity* of the sand is its resistance per unit length and unit area of column.

Alternative equipment

A more sophisticated constant head device is described on page 104 and in Section 14.5 using a Marriotte bottle, and a leaching tube with the sand or soil retained between discs of nylon cloth.

Modification of the method for soil

Although the above method can be used for sieved soil repacked into the leaching tube, the measured hydraulic conductivity bears little relationship to that in the field except perhaps for sandy soils. Saturated hydraulic conductivity is very dependent on the size and continuity of transmission pores and so must be measured using soils with their natural structure; a core of soil sampled as for bulk density determination (Section 4.2) can be used. After trimming the core faces place it on a pad of wet tissue paper and cover to prevent evaporation. Over a few days occasionally add water to the tissue paper so that the core slowly wets and comes close to saturation. Connect an extension tube to the bulk density cylinder using water proof tape. Place a disc of metal gauze on the base of the cylinder and stand the equipment in a funnel as shown in Fig. 5.16(b). Cover the soil surface with a disk of nylon cloth. Conduct the experiment as before. The equipment resistance is negligible, and so only the single soil thickness is required. Note that in this case Eq. 5.8 can be written $Q/A = KH/L$ and applying units, $cm^3\,s^{-1}/cm^2 = K \times cm/cm$ and so K has the units $cm\,s^{-1}$. If Q/A is plotted against H, the slope is K/L.

Using Eq. 5.8 determine the resistance of the soil. Hence calculate the resistivity and the saturated hydraulic conductivity. Measurements may have limited validity for the following reasons:

- The conductivity of a small core of soil may not be representative of soil in the field because of the distribution of large cracks and channels.
- Trimming the surface of the soil in the cylinder may smear openings to the pores so reducing conductivity.

- There may be air trapped in the core even after careful wetting.
- There may be spaces between the cylinder and the soil sample down which water can flow.

The measurement of saturated hydraulic conductivity in the field

When soil is flooded, the infiltration rate is initially rapid, but falls to a constant rate which is a measure of the hydraulic conductivity of the saturated soil (Fig. 5.17). Flooding a small area necessarily involves water moving both laterally and vertically and a double ring infiltrometer is used, the outer ring creating a guard zone of saturated soil around the central ring where water is assumed to be moving vertically and where measurements are made. Saturated hydraulic conductivity is possibly the most variable of the commonly measured properties of soils in the field with a single large crack or earthworm channel allowing very rapid flow. Thus, many measurements are needed to characterize a plot or field, and data should be treated with caution.

Equipment

Double ring infiltrometer, consisting of inner and outer metal cylinders 25 cm deep and 30 and 80 cm diameter respectively.
A 50 cm ruler.
Large water containers and an adequate supply of water (150 l perhaps).

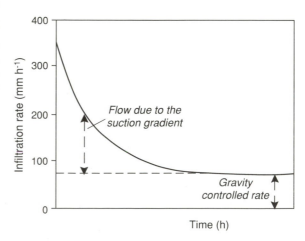

Figure 5.17 The changes in infiltration rate with time after flooding.

Method

Hammer or press the outer ring a few cm into the ground to make a good seal between the soil and the ring. Firmly press the inner ring into the ground avoiding disturbance of the soil surface. Place a disc of thin cloth over the soil in the inner ring to protect the soil when water is poured in. The technique involves maintaining the water level between 5 and 10 cm above the soil in both the inner and outer rings, and measuring the rate of infiltration in the inner one. Large volumes of water are needed. For example, the sandy soil in Fig. 5.3(a) has a saturated hydraulic conductivity of $2240 \, \text{mm} \, \text{d}^{-1}$ or $1.5 \, \text{mm} \, \text{min}^{-1}$ which in a 80 cm diameter ring is $750 \, \text{cm}^3 \, \text{min}^{-1}$. For a 30 cm ring the rate is $106 \, \text{cm}^3 \, \text{min}^{-1}$. Initially infiltration is even more rapid.

A convenient way to measure the infiltration rate is shown in Fig. 5.18. A ruler is placed in the inner cylinder and firmed into the soil. Variations in water level can be recorded as a reading on the sloping mm scale. Prepare five containers each with 2 l of water and one containing 5 l. It is helpful to have two persons to carry out the initial stages of measurement.

Pour 5 l of water into the inner ring and at the same time add water to the outer ring to give the same level. Immediately record the position of the water surface on the ruler and the volume of water added to the inner ring; 5 l of water in the inner ring gives a depth of about 7 cm and a reading of about 13 cm on the ruler for the arrangement shown in the figure. At suitable intervals depending on infiltration rate, record the water position on the ruler. Add 2 l volumes of water as required to maintain a water depth of between 5 and 10 cm, recording the additions. Keep the water level in the outer ring within 1 or 2 cm of that in the inner ring. Continue recording until the infiltration rate is constant.

Data

For the arrangement shown in Fig. 5.18, a 1 cm change of water position on the ruler is equivalent to a depth

change of 0.556 cm which, for a ring area of $706.8 \, \text{cm}^2$ is a volume change of $393 \, \text{cm}^3$. Calculate the cumulative amount of infiltration (cm^3) and plot against time (min). For various times determine the slope of the curve which is the infiltration rate ($\text{cm}^3 \, \text{min}^{-1}$). Convert your values to $\text{mm} \, \text{h}^{-1}$, and plot as shown in Fig. 5.17. The saturated hydraulic conductivity is the final gravity-controlled rate.

Note that there is a positive hydrostatic pressure resulting from the water in the ring. This has a negligible effect on the final infiltration rate.

An example of infiltration into a dry soil at the beginning of the rainy season in Niger is shown in Fig. 12.5.

The determination of unsaturated hydraulic conductivity by infiltration

The principle of this method can be understood using Fig. 5.19. A column of soil is saturated and then allowed to drain, with water entering the soil surface at a constant rate. When the rate at which water leaves the column has decreased to equal the input rate, the water content in the soil is constant throughout the column except for the very bottom where the water content increases sharply to saturation in order to enable water to drip from the base at atmospheric pressure. With no gradient of matric potential, the flow rate is equal to the hydraulic conductivity because the gradient of gravitational potential has a value of 1 m equivalent hydraulic head m^{-1} (Eq. 5.7). The operator therefore chooses a hydraulic conductivity (the flow rate) and determines the water content required to maintain this value.

Soil samples

Because the experiment determines hydraulic conductivity in soils wetter than field capacity, the natural system of transmission pores should be maintained during sampling. Two approaches are possible.

1. *Undisturbed cores.* A 6 cm diameter PVC pipe (a drainpipe), about 30 cm long, drilled at 5 cm intervals with 10 mm diameter holes and having a sharpened rim can be hammered down into the soil profile provided the soil is light or medium-textured and free from stones. Place a block of wood on the top of the pipe to avoid damage. When the pipe is a few cm into the soil excavate around it, and then hammer in further. Continue until a 30 cm core has been obtained. Excavate under the pipe and trim the bottom of the soil column. Place on the base of the soil a disk of nylon cloth. Glue to the base of the pipe a plastic disc with outflow tube (Fig. 5.19).

    The soil may be compacted during sampling so altering its conductivity. Special sampling tools are

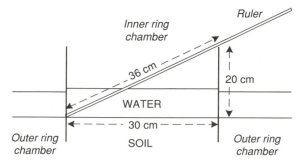

Figure 5.18 A double ring infiltrometer.

Marriotte bottle

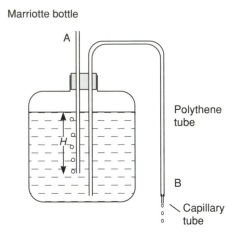

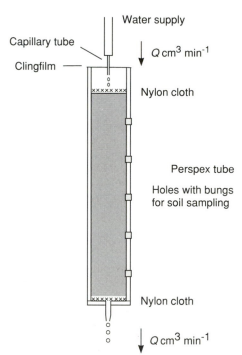

Figure 5.19 Steady-state infiltration in a soil column.

available from Eijkelkamp Agrisearch Equipment giving less compaction and deeper cores. Also there may be spaces between the core and the wall of the tube, increasing conductivity at water contents close to saturation.

2. *Repacked cores*. Light-textured soils from arable sites (<2 mm air dry) can be repacked into a leaching column prepared from a plastic drainpipe or Perspex tube as in Fig. 5.19. Add the soil in ap-

proximately 5 cm layers, tapping the column on the bench a few times between each addition to pack the soil. Once rewetted the soil may settle further. Its bulk density is, however, normally less than in the field, increasing its hydraulic conductivity when at water contents close to saturation, but giving useful measurements near to field capacity.

Equipment

Leaching column. As above, containing an undisturbed core or repacked soil.
Marriotte bottle with capillary tube outlet (Fig. 5.19).
Sampling tube and extrusion rod. A cork borer (10 mm external diameter) is suitable.
Weighing bottles.

Method

Arrange the leaching column and Marriotte bottle on suitable stands. Before starting the experiment fill the bottle and blow through tube A (Fig. 5.19) to fill tube B. Water will drip from the capillary tube. Adjust the depth H and the elevation of B so that the outflow is about $1\,\mathrm{ml\,min}^{-1}$ which for this column diameter will give an application rate of about $500\,\mathrm{mm\,d}^{-1}$. Measure the rate by collecting several samples over known periods and measuring their volume or weight. The Marriotte bottle maintains a constant flow through the capillary tube even though the water level is falling in the bottle. Clip the outlet tube while the soil column is being prepared.

Attach an 80 cm length of polythene tube to the column outflow as a U-tube, and connect it to a large funnel. With the funnel below the base of the column pour water into the funnel to fill the tube (but not to wet the soil). Ensure that air bubbles are released from the tube. Raise the funnel slowly adding water as required so that the soil column is wetted from the base upwards to give eventually a saturated column. Clip the polythene tube close to the column outflow. Place a disc of nylon cloth on the top of the soil column, and begin to apply water from the Marriotte bottle. Immediately remove the polythene tube from the base of the column and allow water to drain until the outflow rate equals the inflow. For slow flow rates, prevent evaporation using Clingfilm both on the top of the column and on the collecting vessel.

With the water still flowing, take soil samples from the column by inserting the cork borer through the holes and pushing out the soil with a rod into a weighing bottle. Determine the water content of the samples (Section 3.3) and calculate a mean value.

Repeat the experiment with different flow rates.

If the relationship between hydraulic conductivity and suction is required, determine the water release characteristic using the Haines equipment in Section 5.3.

Data

An example of the results which might be obtained for a sandy soil is shown in Fig. 5.3(b). The hydraulic conductivities at saturation and field capacity are 2240 and $3 \, mm \, d^{-1}$ respectively. Thus for a 6 cm diameter tube having a cross-sectional area of $28.3 \, cm^2$ and a potential gradient of $1 \, m \, m^{-1}$, flow rates for this soil of between 260 and $0.4 \, cm^3 \, h^{-1}$ would be expected.

Note 1 If the water input is stopped, the column will continue to drain until a matric potential gradient develops which is equal and opposite to the gravitational potential gradient, i.e. the soil is drier at the top of the column. In a soil profile, this would be the field capacity state but in a drained column, its wetness at the base depends on the characteristics of the nylon cloth, the shape of the tube and the outflow conditions. Thus although we cannot predict the suction at the base of the column, we do know that the suction at the top, expressed as an equivalent hydraulic head, is greater than that at the base by an amount equal to the length of the column in metres.

The determination of unsaturated hydraulic conductivity by evaporation

The principles of this method (Moore, 1939) can be understood using Fig. 5.20. A column of moist soil has a water table maintained at its base. Evaporation dries the surface soil, producing a matric potential gradient through the column. When a steady state has been established, which may take many days, the rate of loss by evaporation equals the rate of uptake from the water table, which is a measure of the flow rate through the column. Either the water content of soil samples taken from the sampling holes is determined and related to matric potential by measuring the water release characteristic, or tensiometers are inserted in the tube to measure suction directly. Using Eq. 5.7, the flow at all points in the column is the same and can be found from the rate of loss from the water-table cup. Thus hydraulic conductivity values can be calculated from water potential gradients.

A simple alternative technique is described by Rowse (1975): the mathematical analysis is, however, more complex.

Equipment

Soil column, as in the infiltration method above, with a water-table cup, about 2.5 cm diameter.

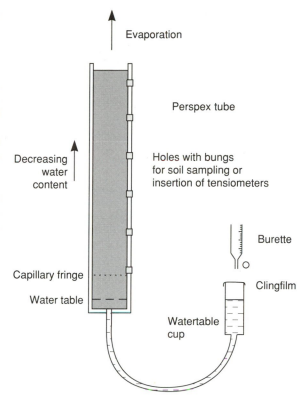

Figure 5.20 Steady-state evaporation from a soil column with a water table.

Method

Either an undisturbed core, or a repacked column can be used (see the infiltration method). The column length should be about 30 cm for light-textured soils (Note 2). Saturate the soil from the base as in the infiltration method, and then lower the funnel to allow the column to drain (maintain the rim of the funnel at the required water-table position and allow to overflow). When flow ceases, connect the water-table cup to the outflow tube, and as evaporation from the soil surface proceeds, maintain the water at a marked level by adding water daily from a burette and recording the volumes used. Cover the cup with Clingfilm to prevent evaporation.

The rate of water loss decreases to become steady after a period which may be days or weeks depending on soil characteristics. Ideally the column should be maintained both at constant temperature and with a constant evaporation rate. However, if installed in a laboratory with a reasonably constant environment, useful data can be obtained. The evaporation rate can be increased using a fan to maintain a flow of air over the soil surface.

When a steady rate of evaporation has been established sample soil from the column and determine the distribution of water content. Determine the water release characteristic of the soil using the filter paper method of Sections 5.2 and 5.3. Alternatively, tensiometers can be inserted through the sampling holes when the column is set up (Section 5.2) to give a direct measure of suction.

Data

Express the soil water potential, ψ, as an equivalent hydraulic head at each sampling point using Eq. 5.6. Plot ψ against depth in the column. From the slope of the graph determine the potential gradient $d\psi/dx$ (mm^{-1}) at each sampling depth. The steady rate of water loss is known ($cm^3 d^{-1}$). Measure the internal diameter of the soil column and calculate its cross-sectional area, so giving the rate of flow of water through the column, Q ($cm^3 d^{-1} cm^{-2} = cm d^{-1}$ re-expressed as $mm d^{-1}$). Using Eq. 5.7 calculate the hydraulic conductivity at each sampling depth, and plot the values against suction or water content.

Note 2 It is difficult to chose the 'correct' length for the soil column. However, the column must be long enough for drainage alone to lower appreciably the water content at the surface and thus reduce the hydraulic conductivity also.

A CASE STUDY CONTINUED: SOIL WATER DYNAMICS IN SUB-SAHARAN SAVANNA

The direction of water movement in soils under savanna in Niger has been analysed in Section 5.4. Darcy's Law is now used to calculate drainage fluxes using the water potential gradients shown in Fig. 5.15 and the hydraulic conductivity data in Fig. 5.3(b).

Example

Using the data for 10 September in Fig. 5.15 the calculations are as follows, and the results are shown in Table 5.4.

- The average water potential gradient between 100 and 140 cm depth is
$$d\psi/dx \simeq (\psi_{140} - \psi_{100})/(Z_{140} - Z_{100})$$
where the subscripts refer to the depths. The gradients are $(-58 + 48)/40 = -0.25$ for the bush site and $(-34 + 19)/40 = -0.375$ for the grass site, and the units are $kJ m^{-3} cm^{-1}$. As shown above, more con-

Table 5.4 Data showing the rate of drainage from the Sadoré site in Niger

	Bush site	Grass site
Water potential ($kJ m^{-3}$)		
At 100 cm	−48	−19
At 140 cm	−58	−34
Water suction (kPa)		
At 100 cm	38	9
At 140 cm	44	20
Average	41.0	14.5
Hydraulic conductivity ($mm d^{-1}$)	0.005	0.1
Water potential gradient ($m m^{-1}$)	−2.55	−3.82
Drainage flux ($mm d^{-1}$)	0.013	0.38

venient units are $m m^{-1}$ where water potential is expressed as an equivalent hydraulic head. Since $1 kJ m^{-3}$ is equivalent to $1 kPa$ (Section 5.4), and $9.81 kPa$ is equivalent to $1 m$ (Eq. 5.4), the above gradients are -2.55 and $-3.82 m m^{-1}$ respectively. The negative sign indicates that the potential is becoming more negative (smaller) as depth increases.

- Water suction at the two depths is calculated using the water potential values and Eq. 5.6. The mean hydraulic conductivity for the 100–140 cm layer is determined from Eq. 5.3 using the average suction in the layer.
- Equation 5.7 is used to calculate the drainage flux through the layer.

The following conclusions can be drawn:

- The potential gradients show that water is draining out of the profile at both sites.
- The gradient under grass is a little larger than at the bush site, but because of the wetter soil, the hydraulic conductivity is about 20 times larger and consequently, the drainage flux is about 30 times larger. This illustrates a general principle, that although the potential gradient determines the direction of flow, the hydraulic conductivity has an overriding effect on its magnitude.
- A drainage flux of $0.4 mm d^{-1}$ under grass is small but will be an important component of the water budget of the profile. If maintained over a period of 4 weeks for example, 11 mm of water would move down to contribute to the recharge of groundwater. Losses by evapotranspiration might be 5 mm per d^{-1} from vegetation well supplied with water, but could be of similar magnitude to the drainage flux as the profile begins to dry.

Drainage fluxes have been calculated for the other sampling dates and are marked on Fig. 5.15. On 4 August following intense rainfall, sampling coincided with a flush of water moving out of the profile under grass which could not have been maintained for more than a few hours: $153\,\mathrm{mm\,d^{-1}} = 6\,\mathrm{mm\,h^{-1}}$, compared to a storm of perhaps 20 mm. On 23 August the grass site is losing $13\,\mathrm{mm\,d^{-1}}$ by drainage and presumably also losing about $5\,\mathrm{mm\,d^{-1}}$ by evaporation which is drying the 0–60 cm layer. Under the bushes a large potential gradient between 100 and 140 cm is only slowly pulling water up into the profile because of the low hydraulic conductivity. By 3 October both sites are relatively dry and again despite large gradients between 100 and 140 cm, the fluxes are small.

This case study is continued in Section 12.2 where water losses by evaporation are considered.

Section 5.6 Projects

1. Obtain samples of sand from a builders' merchant or a garden centre and determine their water release characteristics using Haines equipment (Section 5.3). Make use of the data as suggested in Section 5.7, Calculation 1.
2. The Haines equipment allows transmission pore characteristics to be determined, and so is an indication of how well drained a soil is likely to be. Obtain samples of similar texture (light or medium) from sites under varying management and compare their water release characteristics over the range from saturation to field capacity. The filter paper method (Section 5.3) allows the characteristics to be extended to the air-dry condition.
3. Following the methods of Section 5.3 (filter paper method) and using a sand, a loam and a clay, prepare samples of each soil containing a range of water contents between air-dry and field capacity. When the soils have equilibrated, use some of the sand to determine the gravimetric water contents and soil water suctions. Place samples of the soil in petri dishes and into each sow 20 seeds (barley, lettuce) covering the seeds with soil. Place the covered dishes in a closed propagator to prevent evaporation. Open the dishes daily and observe and record the numbers of seeds which germinate, i.e. seeds which have a radical more than 2 mm long.

   Plot for each soil the final germination percentage (a) against water content and (b) against soil water suction. You should find that final germination depends on suction rather than water content.

   Plot for each petri dish the germination percentage against time. From each graph determine the time required for 50 per cent of the seeds to germi-

nate. Plot this time (a) against water content and (b) against soil water suction. Are the results similar to those for final germination? You should find that modest suctions in the sand restrict the rate of germination because of low hydraulic conductivity and poor liquid contact with the seeds. The experiment can be extended by moistening the soils with salt solutions rather than water, so reducing the soil water potential (Sections 5.4 and 14.3). In this case you should find that germination depends on soil water potential (matric plus osmotic) rather than soil water suction (matric potential) or water content.

4. With a Jarret auger take soil samples to obtain profiles of water suction similar to those shown in Figs. 5.14 and 5.15 using the filter paper method (Section 5.3). Sample at intervals as soils are remoistened after a dry period, and interpret your data as in the Niger case study.
5. Use the infiltration method (Section 5.5) to measure the saturated hydraulic conductivity of the following sites: an uncultivated woodland or grassland, an adjacent cultivated area. In the cultivated area compare soil which has not been affected by wheelings with an area which has been compacted (Section 4.4). Observe the distribution of pores in the top 25 cm layer of soil (Section 1.2). Sample and measure the bulk density (Section 4.2). Draw conclusions from your data and observations in terms of the distribution of large continuous pores.

Section 5.7 Calculations

1. The data in Table 5.5 were measured using the Haines equipment (Section 5.3) with sand sieved to give known particle sizes. Calculate as in Table 5.3, and plot the results including the data from Fig. 5.12. Calculate the effective pore diameter at the air-entry suction and plot the results to show that pore diameter is approximately one-fifth of the median particle size.
2. Check that the values given in Table 5.2 have been correctly derived from Table 4.3 and Fig. 5.6.
3. Following the methods given in Section 5.5, check that the fluxes of water through the 100–140 cm layer given in Fig. 5.15 are correct.
4. Replot Fig. 5.12(a) using the same x-axis as Fig. 5.3, and compare the results to the sand and sandy loam soils.
5. Calculate the approximate water film thickness over the surface of the separated fine clay in Fig. 5.6 if the clay is predominantly smectite with a surface area of $500\,\mathrm{m^2\,g^{-1}}$ and is in an atmosphere at 60 per cent relative humidity. (*Ans.* 0.28 nm) Imagine that the clay is like a pack of cards (Figs. 2.2 and 14.8).

Table 5.5 Water content – water suction data for graded sands

| Sand particles size (mm) | | | | | |
| 2.8–2.0 | | 0.5–0.2 | | 0.2–0.1 | |
Suction (cm H_2O)	θ (cm^3 cm^{-3})	Suction (cm H_2O)	θ (cm^3 cm^{-3})	Suction (cm H_2O)	θ (cm^3 cm^{-3})
0	0.454	0	0.402	0	0.455
1.8	0.428	2.3	0.386	4.6	0.420
3.1	0.364	9.0	0.380	15.8	0.411
4.0	0.286	23.0	0.370	28.0	0.406
5.6	0.168	30.0	0.348	40.9	0.400
7.7	0.087	35.9	0.304	50.4	0.391
15.2	0.069	42.6	0.231	59.6	0.385
24.5	0.055	49.0	0.172	69.6	0.316
44.4	0.043	74.9	0.109	82.2	0.226
89.4	0.020	85.3	0.090	96.0	0.180

What is the film thickness between adjacent clay layers? (*Ans.* 0.56 nm) Compare this to the values shown in Fig. 14.9.

6. A sandy loam sampled from the field had a gravimetric water content of 15 per cent *m/m* and a bulk density of 1.4 g cm^3 (Section 4.2). Calculate its volumetric water content. (*Ans.* 0.21 cm^3 cm^{-3})

7. A wet clay soil sampled as in Calculation 6 had a water content of 40 per cent *m/m* and a bulk density of 1.25 g oven-dry soil cm^{-3} of wet soil volume. After oven drying the soil core was removed from the bulk density cylinder (5.1 cm long × 5 cm diameter) and was found to be 4.5 cm long and 4.4 cm in diameter. Calculate the total porosity and the volume of air-filled pore space in the wet core. (*Ans.* 0.52 and 0.02 cm^3 cm^{-3}) Calculate the porosity of the core after drying expressed as cm^3 pores cm^{-3} of dry soil volume. (*Ans.* 0.3) What is the bulk density of the dry core? (*Ans.* 1.83 g cm^{-3} of dry soil volume)

8. A column of sand-textured soil 10 cm long and 5 cm diameter is saturated with water, and under a head (*H*) of 20 cm of water (Fig. 5.16) the flow rate is 6.8 cm^3 min^{-1}. Assuming that the only resistance to flow is in the sand, calculate its saturated hydraulic conductivity. (*Ans.* 1.7 mm min^{-1}) Compare this to the value in Fig. 5.3(b).

CHAPTER 6

Air in Soils – Supply and Demand

Oxygen is an essential requirement for life in soils. Soil animals, plant roots and the majority of micro-organisms use O_2 and release CO_2 in the process of respiration by which they obtain energy. Thus soils can be considered to 'breathe' to maintain biological activity, with O_2 moving into the soil and CO_2 moving out into the atmosphere. The breathing process is, however, primarily by diffusion of gases in contrast to our breathing which is by mass movement of air into and out of our lungs.

Oxygen is used by soil organisms to oxidize carbohydrates (Table 6.1). Plants are the primary source of carbohydrates: energy from the sun is stored by photosynthesis in these compounds. Within the plant carbohydrates are translocated to the roots where they are oxidized, releasing the energy required for plant growth. Oxygen is taken in and CO_2 is released at the root surface. Soil animals use plant materials or other animals as food, again oxidizing carbohydrates to obtain energy. The O_2 demand by soil animals is large individually but overall is small compared to plant roots and microbial populations. The latter have a demand similar in magnitude to roots. They use carbohydrates in plant and animal residues as part of the process of breakdown of soil organic matter.

The demand for O_2 in soils is therefore greatest in the surface horizons where there are the maximum numbers of roots, micro-organisms and animals. A soil's *respiration rate* (the rate of O_2 use) depends on the activity of these organisms which is controlled by temperature, soil organic matter content, soil water content and nutrient supply. If the supply of O_2 does not meet the demand, respiration rate will have to decrease. Under these conditions the soil may become *anaerobic*. This term indicates that parts of the soil have no O_2 or that the O_2 supply is so poor that biological processes requiring O_2 can no longer continue. Soil *aeration* is strictly the process by which O_2 is replenished and carbon dioxide removed, but is a term often used in a more general way to describe the condition of the soil, i.e. the concentrations of O_2 and CO_2 present.

The process of respiration can be described by the following equation in which glucose is oxidized to carbon dioxide.

$$C_6H_{12}O_6 + 6O_2 = 6CO_2 + 6H_2O + Energy \quad [6.1]$$

Further information on this process can be found in Nuffield Advanced Biology I, Ch. 5 and Bryant (1971).

THE MEASUREMENT OF RESPIRATION

Field measurements

The measurement of respiration in the field is beyond the scope of this book. It involves isolating a volume of soil in a container, with little disturbance of its natural organization of horizons and structure. This is known as a soil *respirometer*. Early work involved excavating soil, inserting a metal tank, and replacing the soil in the tank, keeping the horizons in natural order. A sandy soil repacks relatively easily but a heavy-textured soil is very different after repacking. Modern techniques use *undisturbed cores*, obtained by forcing a cylinder (normally fibreglass) into the soil, excavating, placing a steel plate underneath, lifting out with a crane and resiting in a field laboratory (Plates 6.1 and 6.2).

With the tops open these cores are known as *lysimeters*: to be used as a respirometer the top of the cylinder has to be enclosed with a material that allows light to the plants beneath. Air is passed through the

Table 6.1 Oxygen consumption by soils ($g\,m^{-2}\,d^{-1}$). To convert to volumes, $32\,g\,O_2 = 22.4\,l$ at s.t.p.†

Temperate regions, moist soil, summer	Bare ground	1–12		
	Forest	10–20		
	Arable crops	4–20		
Humid tropics	Forest	8–50		
		January		July
		Cropped	Bare	Cropped	Bare
Rothamsted, brassica crop		2	0.7	24	12
Soil temperature at 10 cm depth (°C)		3		17	

Rothamsted data from Currie (1970).
† Standard temperature and pressure.

Plate 6.1 Removal of an undisturbed core of soil to form a lysimeter. Photograph by AFRC Letcombe Laboratory.

Plate 6.2 Installation of lysimeters in a feld laboratory. Once installed the surface of the lysimeter is level with the surrounding soil, a crop can be grown over the whole area and growth in the lysimeter monitored. There is underground access to the lysimeter so that measurements can be made on the soil and drainage water. Photograph by AFRC Letcombe Laboratory.

Plate 6.3 A respirometer consisting of a metal cylinder forced into the soil covered by a plastic dome with gas entry, exit and sampling ports. Photograph by AFRC Letcombe Laboratory.

enclosure, and changes in O_2 or CO_2 content measured. The cover inevitably alters the environment in which the plants are growing and changes soil temperature, so the results obtained are not entirely representative of field conditions.

A simpler field system involves forcing a metal cylinder down into the soil to isolate an area of soil surface, and then covering and measuring as for conventional respirometers (Plate 6.3). Interpretation of results may be complicated by horizontal movement of gases through the soil underneath the cylinder.

Data obtained using these and other methods are given in Table 6.1. Although reported values are very variable it is clear that temperature has a major effect on respiration, and that cropped soil respires at a rate approximately double that of a bare soil (Section 6.1).

Laboratory measurements

The respiration rate of the microbial biomass can be measured relatively simply in the laboratory (Section 6.2), showing the effects of temperature and organic matter (Section 6.3). The measurements can be used for predictions regarding soil aeration (Section 6.5), and as an indicator of harmful effects of pollutants in soils (Section 3.2).

SOIL AERATION: THE SUPPLY OF OXYGEN AND RELEASE OF CARBON DIOXIDE

If respiration were not occurring, air in soil would have the same composition as the atmosphere. This has about 21 per cent by volume of O_2, 0.03 per cent CO_2 and 79 per cent N_2 and inert gases. The amount of O_2 in this imaginary soil, is related to the volume of air in the soil, normally expressed as *air-filled porosity*, cm^3 air cm^{-3} soil, and measured as described in Section 6.4. It is dependent on the total porosity of the soil and therefore on structure, and on the water content because as this increases the volume of air-filled pores has to decrease. Even in a soil containing a large amount of air-filled pore space, the O_2 may only maintain microbial respiration for 1 or 2 days depending on the respiration rate if the surface of the soil were sealed to prevent access of O_2 (Section 6.4).

Mechanisms of gas movement

The uptake of O_2 by respiration reduces its concentration in the soil air, and that of CO_2 is increased. To maintain the concentration of O_2 and to remove the CO_2 these gases move in and out of the soil by two processes. Bulk movement normally termed *mass flow* occurs as a result of changes in pressure or temperature whereby expansion or contraction forces the gas in and out as though the soil were a pair of bellows. Rain water moving down through the soil can force gas ahead of it, and can carry dissolved gases. These processes are generally much less important than *diffusion*, which is the movement of gas molecules by random thermal motion in response to the concentration gradients formed in the soil.

Diffusion occurs primarily through the air-filled spaces. The gases do dissolve and diffuse slowly in water, but the rate of diffusion in air is about 10 000 times faster than in water. Thus water forms a very effective barrier to gas movement and as water content increases so the rate of movement of gas decreases. For easy movement, air-filled spaces have to be continuous, particularly in the vertical direction through the soil, because a film of water between two air-filled pores will block the diffusion path.

The distribution of oxygen and carbon dioxide

Active sites

The main sites of O_2 uptake and CO_2 production in the soil are the microbial cells and plant roots. Fig. 6.1 shows the complexity of the system in diagrammatic form based on electron microscope photographs such as those in Plates 2.2 and 6.4. The gases have to diffuse

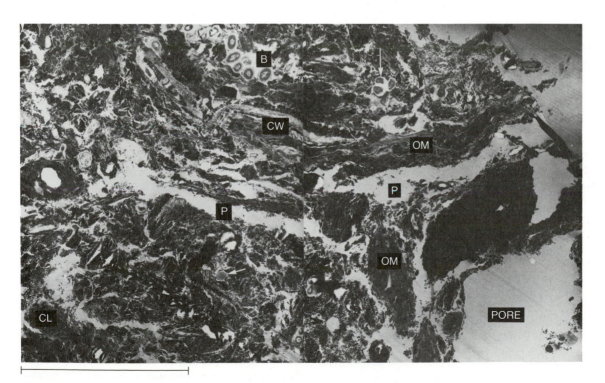

Plate 6.4 Transmission electron micrograph of soil in the rhizosphere adjacent to a clover root. The soil is dominated by clay (CL). Pores are marked P. Old cell wall remnants (CW) support a small colony of bacteria (B) and isolated bacteria are scattered throughout the soil fabric (arrows). OM is soil organic matter. Photograph by R. Foster. The bar represents 10 μm.

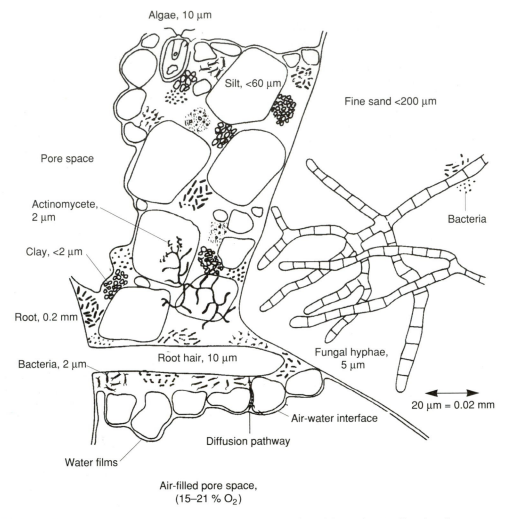

Figure 6.1 The relationship between active sites of respiration, mineral particles and water films in soil.

through films of water surrounding the organisms and the lowest concentrations of O_2 and the highest concentrations of CO_2 are at their surfaces. They can apparently continue to respire even when the O_2 concentration at their surface has fallen to very low values: the presence of O_2 at the active site indicates that diffusive supply is meeting the organisms' demand. Provided that the concentration of O_2 in air-filled pores is near to 21 per cent, O_2 supply to organisms can be maintained through an adjacent water-filled space about 1 cm thick (Section 6.5). This is known as a *critical distance*. It is inversely related to respiration rate.

Distribution in a uniform sandy soil

The concentration of O_2 and CO_2 depends on the balance between demand and supply (Section 6.5). Even in a relatively uniform, freely draining soil these factors vary with both depth and time. The demand is greatest in moist surface horizons because of the higher organic matter content, biomass activity and root density. Climatic conditions affect demand through temperature and water supply, the demand being highest in the summer after rain. The diffusion of gases normally becomes more restricted with depth because subsoils are usually wetter and more compact, but after heavy rain aeration may be more restricted in the surface. Thus the composition of soil air is very variable both within a profile and from day to day. Oxygen concentrations are at their lowest during a wet period in the summer and not in the winter when despite the wetter conditions O_2 demand is low.

Measurement of the composition of soil air is beyond the scope of this book. It requires air to be withdrawn from a known depth using a narrow tube and a suitable syringe, followed by analysis using gas chromatography. Alternatively O_2 diffusion electrodes can be used, but have to be inserted at the depth of measurement, which may disturb the soil and change its aeration. For this reason they are more useful in respirometers where they can be inserted from the side rather than from the soil surface. Using these methods the concentration of O_2 in sandy soils is found to remain above about 19 per cent even at 60 cm depth and the CO_2 does not rise above about 1 per cent. Drainage following heavy rain is rapid and a good system of transmission pores (Table 4.2) ensures that active sites are not more than the critical distance from an air-filled pore. For these soils aeration is normally good with no likelihood of plant growth being affected by lack of O_2.

Distribution in heavy-textured soils

Poor aeration commonly occurs in heavy-textured soils. Respiration rates are similar to those in sandy soils, but O_2 supply can be very restricted for three reasons:

1. These soils may be in low lying positions in the land surface and be subject to high *groundwater tables*. Near to the water table even large pores remain water filled (Section 5.3). Thus, although the soil may have a good system of transmission pores, large volumes of the soil may be saturated and anaerobic.
2. Even where heavy-textured soils are not affected by a groundwater table, many of them overlie clay subsoils which have low permeability. After heavy rain, water is held up above the subsoil, to form a *perched water table*. This can again cause anaerobic zones to be present.
3. Some heavy-textured soils may not be influenced by either type of water table, but after heavy rain poor structure with a sparse distribution of transmission pores may lead to volumes of soil remaining saturated and anaerobic.

Note that discussion centres around the condition of soil after heavy rain. Drier soils have a better system of air-filled pores, and aeration is rarely a problem once a soil begins to dry out. Soil at field capacity is holding the maximum amount of water after drainage, only transmission pores are air filled and it may remain in this condition for some time. Thus it can be a critical condition in terms of aeration.

The distribution of anaerobic zones in these soils is shown in Fig. 6.2. In a flooded soil the thin surface layer is aerobic, its depth being the critical distance. However, during temporary flooding pockets of air will be trapped so maintaining aerobic zones for some time.

The special case of marsh and paddy soils is discussed below.

The effects of anaerobic conditions

The direct effect of a lack of O_2 is that organisms which rely on this gas for their metabolism can no longer survive. Thus soil animals move away from a flooded area, the nature of the plant community changes in favour of those which are tolerant to anaerobic conditions, and the microbial population changes towards those species which can survive or which prefer these conditions. The changed microbial population is less able to decompose organic matter and a build-up of plant residues on the soil surface begins, eventually forming a layer of peat. Thus plant, animal and microbial ecology are linked with soil aeration.

Crop growth

Apart from rice, crops suffer from anaerobic conditions. In severe cases metabolism of roots is affected so that toxins are produced which damage or kill the plant. Similarly toxins can be produced by soil micro-organisms. Under less severe conditions N is lost from soil when nitrate in soil solution is *denitrified* (reduced to nitrogen and nitrous oxide gases, Section 6.6 and Ch. 11). Yield may be reduced by shortage of N, but effects are highly variable due to the range of factors influencing both plant growth and denitrification. Soil temperature is critical, with temporary flooding in the summer being much more damaging than a permanent high water table, which may have a positive effect on yield because of good water supply. In the winter a high water table or temporary flooding has little effect on growth because soil temperatures and respiration rates are low.

Soil properties

The microbial populations which become active in anaerobic soil have to respire without using O_2, a process known as *anaerobic respiration*. Both organic and inorganic compounds are reduced by micro-organisms as a means of obtaining energy, and they make use of either organic compounds or CO_2 as a source of C.

The biochemistry of these processes is summarized in Section 6.6. Central to these processes is the transfer of electrons and protons from a compound which is being oxidized to one that is being reduced. In aerobic respiration the electrons and protons are transferred from carbohydrate to O_2 to form water, whereas in anaerobic respiration they are transferred to either organic or inorganic compounds producing a wide range of reduced materials. Oxygen is the compound

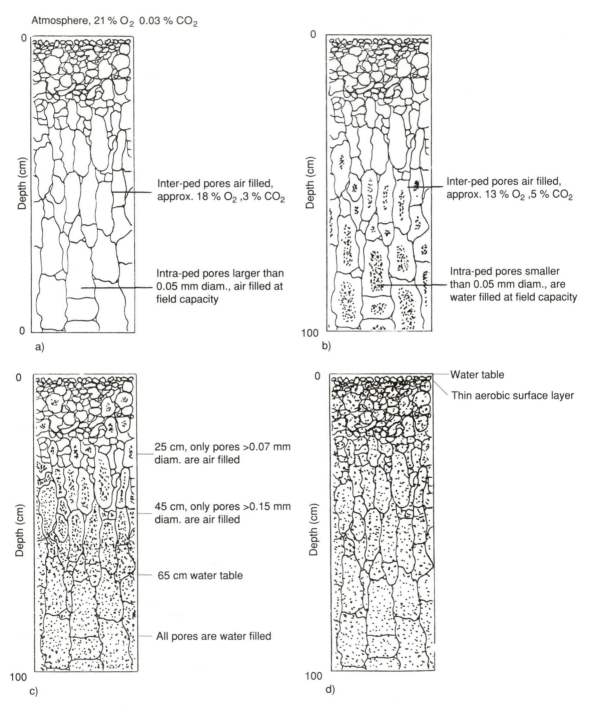

Figure 6.2 The distribution of aerobic and anaerobic zones in soils under different drainage regimes. (a) Well drained at field capacity and fully aerobic; (b) imperfectly drained at field capacity with anaerobic micro-sites; (c) poorly drained at field capacity due to a high water table or impermeable subsoil; (d) waterlogged soil. The dotted regions are anaerobic.

most easily reduced, and in its presence anaerobic respiration does not occur. When O_2 is not present, compounds are reduced in sequence depending on their ease of reduction with more severely reduced conditions developing as time proceeds. The severity of reduction is measured by the *redox potential* of the soil (Section 6.7). Aerobic soils have redox potentials of about $0.6\,V$, whereas anaerobic soils have potentials between 0.4 and $-0.2\,V$.

Reduction causes important changes in soil properties. Almost immediately after becoming anaerobic denitrification occurs with serious implications for crop production. At about the same time various organic compounds are produced including ethene (ethylene) gas and ethanoic (acetic) acid which can interfere with plant growth. As the soil becomes more severely reduced, iron and manganese oxides and hydroxides are reduced (Section 6.6) producing the characteristic patterns of grey- and brown-coloured soil known as *mottles* or *gley* features (Colour Plate 10). Their degree of development and the depth in the soil at which they occur are a good guide to soil aeration and drainage (Fig. 6.3). An alternative approach is the use of soil wetness classes (Robson and Thomasson, 1977). Reduced iron can be

transported in soil water and reprecipitated as an iron-cemented layer or *iron pan* often associated with a fluctuating water table. Eventually in very severely reduced conditions sulphate is reduced to sulphide, some of which is lost as hydrogen sulphide gas, H_2S, and some precipitated as iron sulphide. The bad smell of a freshly exposed reduced soil is due to H_2S and other reduced organic compounds. The final stage is a build-up of peat as a result of permanent flooding (Colour Plate 10).

Drainage of land for agricultural use

Most heavy-textured soil in Britain is artificially drained by underground pipes or by mole drains. The principles are discussed by Davies *et al.* (1972). Drainage of marshes and peatlands for agricultural use has been part of agricultural development for the last 200 years, and the fate of the small areas which remain in developed countries is a matter of concern (Burton and Hodgson, 1987).

PLANT GROWTH IN FLOODED SOILS

Plants which grow naturally in flooded soils do not

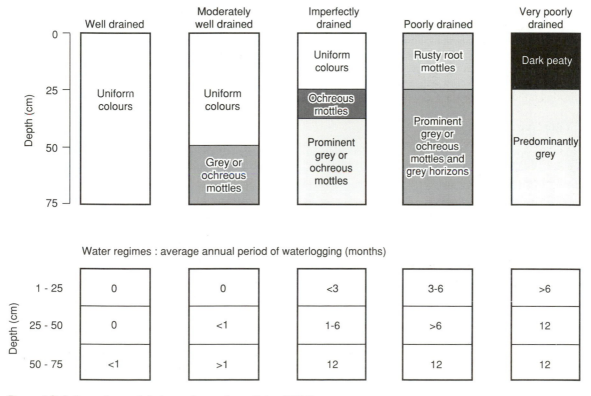

Figure 6.3 Soil aeration and drainage classes. From Batey (1971).

respire anaerobically as in the case of some micro-organisms. Their structure allows O_2 to be transported from the leaves down through the plant to the roots. Thus the O_2 demand of the roots is met via the leaves and stems rather than by the soil. Rice grown in wet-lands or paddy soils is an economically important plant which grows in this way (IRRI, 1988 and de Datta, 1981). It provides 21 per cent of the energy supply in the world's food. Some varieties grow in normal drained soils (dryland rice), but most rice (about 87 per cent of the area) is grown in anaerobic soils (paddy rice). The total production is 464 million t (145 million ha) compared to 430 million t of wheat (239 million ha). Thus the average yield of rice per hectare is almost twice that of wheat.

Paddy soil management is of much interest because it appears to contradict so many of the principles that are normally involved in maintaining soil physical con-ditions for crop production. Rice is germinated in drained aerobic conditions because of its high O_2 re-quirement at this stage. The paddy fields are levelled areas which are surrounded by a bank. They are flooded to give about 6 cm of water above the soil sur-face and puddled by animals or machinery (Colour Plate 9). This destroys soil structure and makes the soil almost impermeable to water so retaining the flood water. The soil quickly becomes anaerobic. The young rice is transplanted into this soil and grows on to matur-ity, at which stage the land is allowed to dry to facilitate harvesting. This crop production system raises many questions such as 'Why grow rice in this labour inten-sive way if some varieties will grow in normal soils?' Section 6.8 deals with some of these questions.

FURTHER STUDIES

Ideas for projects are in Section 6.9 and calculations are in Section 6.10.

Section 6.1 Respiration rates, carbon loss and temperature effects

Soil respiration rates are normally expressed as a mass of O_2 or CO_2 m^{-2} of soil surface d^{-1}. However, it is easier to imagine and measure an amount of gas in terms of volume. We may also wish to calculate amounts of CO_2 from amounts of O_2 and to determine volumes of gases at different temperatures and pres-sures.

Conversions
Gas volumes

Equation 6.1 tells us that 1 mol O_2 is used to produce 1 mol CO_2 (Section 2.5). From the gas laws it is known that 1 mol of any gas at standard temperature and pres-sure (s.t.p., 0 °C and 1 atmosphere pressure) occupies 22.4 l (Nuffield Advanced Chemistry I, Topic 3). Thus the volumes of O_2 used and CO_2 produced are the same in the reaction represented by Eq. 6.1. The ratio, volume of CO_2 produced/volume O_2 used is called the *respiratory quotient*. Measurements of soil respiration confirm that the respiratory quotient is approximately 1. Only if a soil becomes partially anaerobic does the CO_2 produced exceed the O_2 used.

Masses to volumes

Using the above relationship we can state that 1 mol of O_2 (32 g) or 1 mol of CO_2 (44 g) will occupy 22.4 l at s.t.p. Thus the respiration data for bare ground (Table 6.1) given as 1–12 g O_2 m^{-2} d^{-1} can be converted into volumes as follows:

$$32\,g\ O_2\ \text{occupies}\ 22.4\,l$$
$$12\,g\ O_2\ \text{occupies}\ 22.4 \times 12/32 = 8.4\,l\ \text{at s.t.p.}$$

The range of values is 0.7–8.4 l.

Assuming a respiratory quotient of 1 the range of values for CO_2 production must also be 0.7 to 8.4 l. However the mass of CO_2 is different from the mass of O_2 and can be calculated as follows:

$$32\,g\ O_2\ \text{will produce}\ 44\,g\ CO_2$$
$$12\,g\ O_2\ \text{will produce}\ 44 \times 12/32 = 16.5\,g\ CO_2$$

and the range of values is 1.4–16.5 g.

Volumes at different temperatures and pressures

The volume of a given mass of gas increases as tempera-ture rises or pressure falls. The relationship is expressed through the ideal gas equation:

$$pV = nRT$$

where p is pressure in atmospheres (atm), V is volume (l), n is number of moles of gas, R is the gas constant = 0.082 atm l K^{-1} mol^{-1} and T is temperature in degrees K = 273 + °C. Again taking the above example, the volume of gas at any temperature or pressure can be calculated.

Example We calculated that 12 g O_2 occupied 8.4 l at s.t.p. This could have been calculated using the ideal gas equation,

$$V = nRT/p = (12/32) \times 0.082 \times 273/1 = 8.4\,l$$

The value of n is 12/32 because the number of mol = mass/molar mass (Section 2.5).

The volume at any other temperature can similarly be calculated. Thus at 20 °C which is 293 K and keeping the pressure constant

$$V = (12/32) \times 0.082 \times 293/1 = 9.0 \, l$$

The approach can be simplified by writing $V_1/V_2 = T_1/T_2$ at constant pressure and $V_1/V_2 = p_2/p_1$ at constant temperature, where subscripts 1 and 2 indicate the two conditions.

The dependence of respiration rate on temperature

Biological activity increases as temperature rises, and so the mass of CO_2 released from a soil depends on soil temperature. This varies with depth and on a diurnal and annual basis. However, laboratory measurements show that temperature dependence can be described by the equation $R = R_0 Q^{T/10}$ where R is the respiration at T °C, R_0 the respiration at 0 °C and Q is the Q_{10} factor. The value of Q is close to 3 for soils.

Fig. 6.4 shows changes in respiration rate for 1 year in a soil at Rothamsted. The curves were considered to represent the data for February to August and for September to January. They are the theoretical changes in respiration with temperature assuming that $Q = 3$, and $R_0 = 1.2$ or $0.9 \, \mathrm{g \, m^{-2} \, d^{-1}}$ (see also Calculation 4, Section 6.10).

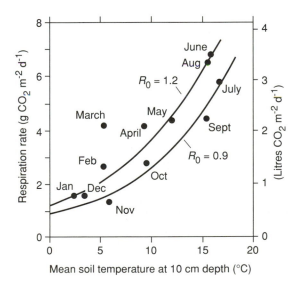

Figure 6.4 The relationship between daily soil respiration rate and mean soil temperature at Rothamsted for a bare soil (October 1960–September 1961). From Monteith *et al.* (1964).

The differences between the two periods may relate to the availability of plant material for decomposition early in the year and to the drier soil conditions later in the year. Microbial respiration is often assumed to be negligible below 5 °C: the measured rates in December and January may result from higher temperatures below 10 cm.

Section 6.2 The laboratory determination of the respiration rate of the microbial biomass

The CO_2 produced by respiration in a moist soil is trapped by absorption in NaOH solution. The amount of NaOH remaining after a known time period is determined by titration with standard acid. No specialized equipment is required.

Soil

Drying and rewetting soil results in a flush of biological activity. Ideally to obtain values which represent those of the soil in the field, obtain a field-moist sample, dry in the air only so far as is needed to push the soil through a 2 mm sieve (or a garden sieve if a 2 mm sieve is not available). Alternatively using an air-dry <2 mm soil, moisten with about 10 g water per 100 g air-dry soil and store in a polythene bag for 1 week. Do not seal the bag: loosely fold to allow access of air. Open the bag and shake to aerate the soil each day.

Reagents and equipment

Respiration flasks. A 250 ml conical flask with a rubber bung forms a simple respirometer (Fig. 6.5). Into the bung screw a small hook and suspend a vial (2 cm diameter, 8 cm long) from the hook with thread. Adjust so that the vial hangs freely in the flask when the bung is in place.

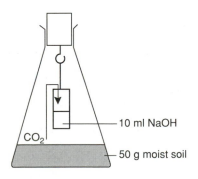

Figure 6.5 A simple laboratory respirometer.

Sodium hydroxide, approximately 0.3 M. Dissolve 12 g of NaOH in distilled water in a 250 ml beaker. Cool, transfer to a 1 l volumetric flask and make up to the mark.
Hydrochloric acid, 0.1 M.
Barium chloride, 1 M. Dissolve 61 g of $BaCl_2.2H_2O$ in water and make up to 250 ml.
Phenolphthalein indicator. Dissolve 1 g of phenolphthalein in 100 ml of ethanol.

Measurement

In duplicate, weigh 50 g (± 0.1) of moist soil into respiration flasks. Pipette into the vials 10 ml of approximately 0.3 M NaOH solution, place the vial and bung in the flask and firmly press into place. Avoid spilling from the vial by carrying out this step with the flask held on the bench. Ensure a gas tight seal with adhesive tape. Set up two control flasks in the same way but use 50 g of sand instead of soil. Note the time each bung was placed in the flask. Store these flasks in the dark for 1 week at room temperature or in an incubator at the required temperature.

Determine the water content of the moist soil at the time the flasks are set up. Into duplicate 100 ml beakers weigh about 10 g (± 0.01) of moist soil. Record the mass of the beaker and the mass of soil. Dry overnight at 105 °C and reweigh. Calculate g water per 100 g oven-dry soil (Section 3.3).

Titration

To familiarize yourself with the titration, pipette 10 ml of the 0.3 M NaOH solution into a 250 ml conical flask, add about 10 ml of water and add by pipette 10 ml of 1M $BaCl_2$ solution. Add 6 drops of phenolphthalein solution and titrate with 0.1 M HCl until the colour changes from red to colourless (approx. 30 ml, but often less because the NaOH solution may absorb CO_2 during storage). This titration value is not used in the calculations.

For each of the respiration flasks in turn, remove the adhesive tape and then the bung and vial taking care not to spill solution. As before, hold the flask on the bench as you remove the bung. Quantitatively transfer the NaOH solution to a 250 ml conical flask (rinse out the vial with distilled water into the flask), and follow the titration procedure above. Record the volume of HCl used and the time the vials were removed from flasks.

The principles

Sodium hydroxide solution has an affinity for CO_2. It is absorbed and forms sodium carbonate in solution.

$$2NaOH + CO_2 = Na_2CO_3 + H_2O \quad [6.2]$$

All the CO_2 is taken up from the flask. In the controls, this is simply from the air in the flask, but with soil, respired CO_2 is also taken up.

The mixed solution of Na_2CO_3 and NaOH cannot be titrated directly with HCl because both would react. Barium chloride is added to precipitate the carbonate.

$$Na_2CO_3 + BaCl_2 = BaCO_3 + 2NaCl$$

Because the solution in the flask remains alkaline until the end point is reached, the barium carbonate does not react with the acid. Thus the acid only reacts with the residual NaOH.

$$NaOH + HCl = NaCl + H_2O \quad [6.3]$$

Calculation

Draw up a table of results. Typical data are given in Table 6.2 in order to explain the calculation. It relates to a soil containing about 3 per cent organic matter and an incubation temperature of 25 °C.

Table 6.2 An example of the data from the respiration experiment

Column i Flask contents	ii Titration (ml)	iii Mean titration (ml)	iv HCl used = NaOH remaining (mol)	v NaOH reacted with CO_2 (mol)
Soil	14.75	14.70	1.47×10^{-3} (*a*)	1.015×10^{-3} (*b-a*)
Soil	14.65			
Control	24.80	24.85	2.485×10^{-3} (*b*)	
Control	24.90			

Column ii The titration of the controls (Note 1) is larger than the soil flasks because in the latter the respired CO_2 has reduced the amount of NaOH remaining (Note 2). Eq. 6.3 tells us that 1 mol of acid reacts with 1 mol of base. (Section 2.5 gives the principles.)

Column iv Calculate the amount of acid used for the soil titration (use the mean titration value, column iii). For the soil, this is

$$0.1 \, mol \, l^{-1} \times 14.70/1000 \, l = 1.47 \times 10^{-3} \, mol$$

and is also the number of mol of NaOH which has reacted with the acid (column iv). Thus *a* and *b* are the amounts of NaOH in the vials at the end of the experiment, and the difference is the amount of NaOH which has reacted with the respired CO_2.

Column v Equation 6.2 tells us that 2 mol NaOH react with 1 mol CO_2. The molar mass of CO_2 is 44 g mol^{-1}. Therefore the amount of CO_2 which has reacted with NaOH in the flasks is

$$1.015 \times 10^{-3} \, \text{mol} \times 44 \, \text{g mol}^{-1} = 0.0447 \, \text{g } CO_2$$

The respiration rate is therefore 0.045 g CO_2 per 50 g moist soil per week (or the time of storage). To bring this into a standard form, you need to know the mass of oven-dry soil in 50 g of moist soil (for example 45.2 g) and the respiration time (for example 6 d, 23 h and 32 min which is 603 120 s). Therefore the respiration rate, expressed as g CO_2 g^{-1} air-dry soil s^{-1} is

$$0.0447/(45.2 \times 603\,120) = 1.64 \times 10^{-9} \, \text{g g}^{-1}\text{s}^{-1}$$
(Note 3)

Note 1 The amount of CO_2 initially in the flasks is very small, and its effect on the titration is automatically corrected by using controls. The controls are needed primarily because NaOH is unstable in air, absorbing CO_2. Thus its concentration may change if it is left in a vial on the bench. Also, the concentration of the stock solution may change. It would not be satisfactory to titrate a 10 ml sample at the end of the respiration period and to assume that this gives the concentration when it was placed in the vials.

The effect of the CO_2 initially in the flask can be calculated as follows. A 250 ml flask contains air with 0.03 per cent by volume of CO_2. Therefore the volume of CO_2 is

$$250 \times 0.03/100 = 0.075 \, \text{cm}^3 \quad \text{or} \quad 7.5 \times 10^{-5}\,\text{l}$$

From the gas laws (Section 6.1), 1 mol of gas at s.t.p. occupies 22.4 l. Therefore the amount of CO_2 in the flask is

$$7.5 \times 10^{-5}\,\text{l}/22.4 \, \text{l mol}^{-1} = 3.4 \times 10^{-6}\,\text{mol}$$

This would react with $2 \times 3.4 \times 10^{-6}$ mol NaOH (Eq. 6.2). The concentration of NaOH remaining for titration would therefore be reduced by 6.8×10^{-6} mol. The amount of HCl used would also be reduced by this number of mol. The reduction in the titration volume would be

$$6.8 \times 10^{-6}\,\text{mol}/0.1 \, \text{mol l}^{-1} = 6.8 \times 10^{-5}\,\text{l} \quad \text{or} \quad 0.07 \, \text{ml}$$

Note 2 The soil titration value should be between ¼ and ¾ of the control. If its value is too large, errors are also large when a difference is determined in column v. If its value is small, most of the NaOH has been used and the respired CO_2 may not all have been absorbed.

The experiment may have to be repeated with the respiration time suitably adjusted.

Note 3 This value can be compared to field respiration rates as follows. Assume a field bulk density of 1.3 g cm^{-3}, with the 0–20 cm layer of soil respiring at the calculated rate. Below this depth, the organic matter content is small and respiration is small. Therefore the volume of respiring soil per square metre of soil surface is

$$100 \times 100 \times 20 = 2 \times 10^5 \, \text{cm}^3$$

The mass of respiring soil is

$$2 \times 10^5 \, \text{cm}^3\,\text{m}^{-2} \times 1.3 \, \text{g cm}^{-3} = 2.6 \times 10^5 \, \text{g m}^{-2} \text{ soil surface}$$

The respiration rate is

$$2.6 \times 10^5 \, \text{g m}^{-2} \times 1.64 \times 10^{-9} \, \text{g g}^{-1}\text{s}^{-1} \times 86\,400 \, \text{s d}^{-1}$$
$$= 37 \, \text{g } CO_2 \, \text{m}^{-2}\text{d}^{-1}$$

This is larger than the values shown in Fig. 6.4, but relates to a temperature of 25 °C.

Section 6.3 The activity of the microbial biomass in soils

Respiration rate measured by the method in Section 6.2 is an indication of the activity of the microbial biomass. Temperature has already been shown to have an important effect on this activity in the field (Section 6.1) and similar effects are found in laboratory incubations. Organic matter content also has a major effect on biomass activity as shown by soils from the Classical Experiments at Rothamsted Experimental Station (Table 6.3). These long-term experiments have produced a range of organic matter contents on otherwise similar soils.

Many natural processes have a marked effect on biomass activity. Drying kills a large part of the microbial population, but on rewetting, populations multiply very rapidly and have available to them C and N in the dead microbial cells. This results in a flush of activity with high initial respiration rates. This effect is shown in Fig. 6.6 where the biomass has been killed by chloroform fumigation, all traces of chloroform removed and the moist soil then inoculated with soil micro-organisms and incubated at 25 °C for 10 d. Bacterial and fungal cells which survive these treatments give to soil resilience in terms of biological activity. Some treatments have a long-term effect, however, and are more serious in relation to the way we manage our soils. Heavy metals added to soil in sewage sludge persist indefinitely, and field plots at Woburn Experimental Farm which received sludges 20 years ago still show the

Table 6.3 The effects of soil organic matter on biomass activity

Soil	Fertilizer or manure	Crop	Organic C (%)	Respiration in 10 d (μgC g^{-1} soil)
Broadbalk	None	Wheat	1.00	102
Broadbalk	None	Fallow	0.84	86
Broadbalk	Manure	Wheat	2.47	174
Broadbalk	Manure	Fallow	2.33	132
Broadbalk	Fertilizer	Wheat	1.09	71
Broadbalk	Fertilizer	Fallow	1.09	79
Broadbalk	None	Woodland	3.49	424
Broadbalk	None	Grass†	3.23	465
Park Grass	None	Grass	3.20	322

Manure: 37 t ha^{-1} a^{-1} since 1844.
Fertilizer: N, P, K$^+$, Na$^+$ and Mg^{2+} annually.
† This is part of the Broadbalk Wilderness which has woody plants cut each year. Herbaceous plants therefore dominate.
From Jenkinson and Powlson (1976).

Table 6.4 The effects of sludge containing heavy metals on biomass activity in soils from Woburn Experimental Farm

Treatment	Microbial biomass (μg g^{-1} soil)	Soil respiration in 10 d (μgC g^{-1} soil)	Biomass respiration in 10 d† (mgC g^{-1} biomass)
Fertilizer	293	56	197
Manure (rate 1)	340	63	195
Manure (rate 2)	411	62	152
Sludge (rate 1)	212	58	279
Sludge (rate 2)	207	65	347
Sludge compost (rate 1)	233	58	268
Sludge compost (rate 2)	237	66	284

† Biomass respiration (mgC g^{-1} biomass) = 1000 (soil respiration μgC g^{-1} soil)/μg biomass g^{-1} soil. Means of four plots.
Fertilizer: applied annually between 1942 and 1967.
Manure: 5.2 or 10.4 t dry matter ha^{-1} a^{-1}.
Sewage sludge: 8.2 or 16.4 t ha^{-1} a^{-1}.
Sewage sludge/straw compost: 5.9 or 11.8 t ha^{-1} a^{-1}.
From Brookes and McGrath (1984)

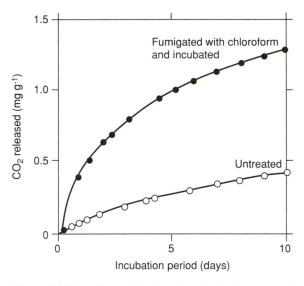

Figure 6.6 The effects of chloroform fumigation on soil respiration. From Jenkinson and Ladd (1981).

The use of respiration as a measure of biomass: the fumigation-incubation method.

The extra CO_2 respired over a 10 d period at 25 °C by fumigated soils is closely correlated with the biomass before fumigation, and is used as a method for its measurement (Jenkinson and Powlson, 1976). It has been found in neutral soils that biomass C (mgC g^{-1} soil) = 2.2 × F where F is the C respired by the fumigated soil during a 10 d incubation period at 25 °C minus that respired by the unfumigated control both in mgC g^{-1} soil. The factor 2.2 shows that the fraction of the biomass-C mineralized to CO_2 during the incubation is 0.45 (= 1/2.2): it has been obtained by incorporating known amounts of micro-organisms into soils and measuring the effect on respiration, and by direct counting techniques (Jenkinson and Ladd, 1981). Measurements using the fumigation–incubation method are the basis for calibrating fumigation-extraction methods (Section 3.2).

Method

Prepare the soil and fumigate following the method in Section 3.2 but use 50 g soil samples. Ideally, ethanol-free chloroform should be used: otherwise the ethanol becomes a substrate for subsequent microbial growth and respiration is slightly increased. After fumigation measure respiration during 10 d at 25 °C following the methods of Section 6.2; the volume of the conical flask should be increased to 500 ml to ensure that adequate

effects on biomass and its activity (Table 6.4). Although the biomass is still reduced by these old treatments (predominantly copper and nickel), soil respiration is little affected, indicating that the remaining microbial population is more active per gram of biomass than in the uncontaminated soils. Further information is given by Chander and Brookes (1991a, b).

O_2 is available. Subtract the respiration of the unfumigated soil from that of the fumigated soil to obtain the flush of respiration. Multiply by 2.2 to obtain the biomass-C.

Microbial spores survive the chloroform treatment in neutral soils and recolonize the soil to give the flush of respiration. In acid soils, inoculation of the fumigated soil with a small amount of unfumigated soil is necessary: to both fumigated and unfumigated samples add 50 mg of unfumigated soil and mix thoroughly before incubation.

Section 6.4 The measurement of air-filled porosity

The determination of the total porosity of a soil is described in Section 4.2, and is normally expressed as a volume ratio, cm^3 pores cm^{-3} soil. In a moist soil part of this space is occupied by water. If the water content is measured as $cm^3 H_2O$ cm^{-3} soil (Section 5.1), then by difference the air-filled porosity can be determined. Typical values can be obtained from Table 4.3 which shows that air-filled porosities at field capacity (the transmission pore volume) range from about 0.05 to $0.3\ cm^3\ cm^{-3}$. The air-filled porosity when the soil is at the wilting point is total porosity minus residual pore volume with a range from 0.25 to $0.35\ cm^3\ cm^{-3}$.

How long can respiration be maintained if no gas exchange occurs?

The respiration rate and air-filled porosity are required for this calculation. Use the respiration rate found at 25 °C in Section 6.2, i.e. $1.6 \times 10^{-9}\ g\ CO_2\ g^{-1}\ s^{-1}$, and an air-filled porosity of $0.2\ cm^3\ cm^{-3}$ initially containing 21 per cent O_2.

Mass of O_2 in the pores

Oxygen in the air-filled pores is 21 per cent of $0.2\ cm^3\ cm^{-3} = 0.042\ cm^3$. The amount of O_2 in this volume can be calculated using the gas equation (Section 6.1), $pV = nRT$. The number of mol, n is

$$pV/RT = (1 \times 0.042/1000)/(0.082 \times 298) = 1.72 \times 10^{-6}\ mol.$$

One mol of O_2 is 32 g. Therefore the mass of O_2 is

$$1.72 \times 10^{-6}\ mol \times 32\ g\ mol^{-1} = 5.5 \times 10^{-5}\ g\ O_2\ cm^{-3}\ soil.$$

If the bulk density is $1.3\ g$ oven-dry soil cm^{-3}, then the soil contains

$5.5 \times 10^{-5}/1.3 = 4.2 \times 10^{-5}\ g\ O_2\ g^{-1}$ soil.

Respiration rate

The respiration rate is in terms of CO_2 production ($1.6 \times 10^{-9}\ g\ CO_2\ g^{-1}\ s^{-1}$). A value for O_2 use is needed. Because $1\ mol\ O_2$ (32 g) produces $1\ mol\ CO_2$ (44 g) during respiration, then the rate of O_2 use is

$$1.6 \times 10^{-9} \times 32/44 = 1.16 \times 10^{-9}\ g\ O_2\ g^{-1}\ s^{-1}$$

Duration of aerobic conditions

The mass of $O_2\ g^{-1}$ soil and the respiration rate are now in the same units. Therefore the amount of O_2 divided by rate of use $(g\ g^{-1}/g\ g^{-1}\ s^{-1})$ gives the time required (s) for all the O_2 to be used:

$$4.2 \times 10^{-5}/1.16 \times 10^{-9} = 36\ 207\ s = 10.0\ h$$

Calculation

Rework the above data for a field soil at 15 °C and then at 5 °C assuming a Q_{10} factor equal to 3 (Section 6.1).

Section 6.5 Predictions regarding soil aeration

It has been shown in Section 6.4 that the O_2 present in the air-filled pores of a moist soil can, without replenishment, only supply the needs of the microbial biomass for 1 or 2 days depending on temperature. If plant roots are also respiring, the soil will remain aerobic for only about half as long because respiration is approximately doubled. This situation only occurs if access of O_2 is prevented by water saturating the soil surface.

A more normal situation is that O_2 concentration decreases with depth, and if the respiration rate and water content are large enough, the concentration will fall to zero at a certain depth with anaerobic soil below. This situation can be analysed mathematically if certain assumptions are made regarding the system.

Assumptions

Consider a soil with the same water content, porosity and respiration rate at all depths. This will only ever apply (and then only approxmately) to a plough horizon of a sandy soil. Imagine a distribution of O_2 as shown in Fig. 6.7. The O_2 concentration in the soil air at the surface is 21 per cent since it is in contact with the atmosphere. Below L no O_2 is present and the soil is anaerobic.

Simple conclusions

All the O_2 needed by a profile has to pass through the

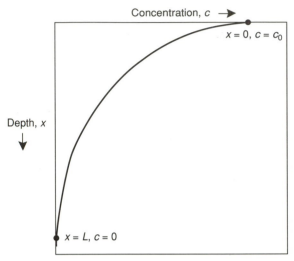

Figure 6.7 The change in oxygen concentration with depth in a uniform soil profile.

soil surface. If all the soil to depth L is respiring at the same rate independent of the O_2 concentration, then half-way down the profile ($x = L/2$) only half the O_2 has to pass and at $x = 3L/4$ only $\frac{1}{4}$ of the total O_2 has to be moving down. At $x = L$, no O_2 is moving. The shape of the curve can be explained on this basis, because the rate of movement of gas by diffusion at any depth depends on its concentration gradient at that depth. This principle is expressed by Fick's law of diffusion.

In a soil segment (Fig. 6.8) of thickness dx and area $1\,cm^2$ with a concentration difference dc across the faces, the rate of diffusion $F = -D\,dc/dx$, where rate of diffusion for our purposes is cm^3 O_2 passing through $1\,cm^2$ soil s^{-1}, c is in cm^3 $O_2\,cm^{-3}$ soil, x is in cm and D is the diffusion coefficient in $cm^2\,s^{-1}$. The coefficient is defined by the above relationship: its dependence on soil properties is discussed below.

Mathematical analysis

If the concentration decreased linearly with depth (dc/dx = constant) then Fick's Law could be used to determine the depth of aerobic soil. However, because the gradient and flow are both changing with depth, a derivation from Fick's Law sometimes called Fick's second law has to be used. This states that where a concentration gradient exists and flow results, then the rate of change of concentration at a given point depends on the rate of change of concentration gradient with distance as follows: $dc/dt = D\,d^2c/dx^2$. There is no respiration term here. The equation tells us what would happen if a non-linear concentration gradient existed and movement occurred by diffusion, so tending to make the concentration uniform through the soil.

In the real system diffusion occurs in response to the gradient, and the gradient is being maintained by respiration. Again taking a small segment of soil (Fig. 6.9), O_2 is being used in the segment at a rate of M (cm^3 $O_2\,cm^{-3}$ soil s^{-1}). Thus F_2 is less than F_1 (the flow in and out of the segment) and the tendency for the concentration to increase is counteracted by the removal of O_2 by respiration and so $dc/dt = D\,d^2c/dx^2 - M$. Note that the units of c/t and M are the same and M is acting in opposition to dc/dt. Thus M has a negative sign.

If we assume a steady state, the diffusion of O_2 into the segment just balances the O_2 use, there is no change in the concentration or its gradient with time and so

$$dc/dt = 0 = D\,d^2c/dx^2 - M$$

and

$$d^2c/dx^2 = M/D \qquad [6.4]$$

We can now use the boundary conditions shown in Fig. 6.7, i.e. at $x = 0$,

$$c = c_0$$

and at $x = L$,

$$c = 0 \quad \text{and} \quad dc/dx = 0$$

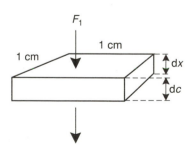

Figure 6.8 Conditions in a segment of soil through which oxygen is diffusing.

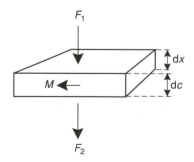

Figure 6.9 Diffusion through a soil segment in which respiration is occurring.

Integrating Eq. 6.4 is straightforward because neither M nor D vary with x. So

$$dc/dx = Mx/D + \text{constant}$$

Substituting $dc/dx = 0$ when $x = L$ gives

$$0 = ML/D + \text{constant}$$

and so the constant $= -ML/D$. Therefore

$$dc/dx = Mx/D - ML/D$$

Integrating again gives

$$c = Mx^2/2D - MLx/D + \text{constant}$$

Substituting $c = c_0$ when $x = 0$,
$$c_0 = \text{constant}$$
and

$$c = Mx^2/2D - MLx/D + c_0$$

or rearranging

$$c - c_0 = Mx(x - 2L)/2D \qquad [6.5]$$

This gives a relationship between O_2 concentration and depth. To determine L, the depth of the aerobic layer, we can substitute using the third boundary condition, that $c = 0$ when $x = L$. This gives

$$0 - c_0 = ML(L - 2L)/2D$$

and

$$L^2 = 2Dc_0/M \qquad [6.6]$$

The parameters

The theoretical analysis has been based on a uniform sandy soil. Thus the following parameters have been chosen for calculation: bulk density $= 1.5\,\text{g cm}^{-3}$, total porosity $= 0.42\,\text{cm}^3\,\text{cm}^{-3}$, gravimetric water content at field capacity $= 0.13\,\text{g g}^{-1}$ so that the air-filled porosity $= 0.2\,\text{cm}^3\,\text{cm}^{-3}$ (Tables 4.1 and 4.3).

c_0 This is the concentration of O_2 in the soil surface, $\text{cm}^3\,O_2\,\text{cm}^{-3}$ soil, assuming 21 per cent O_2 in the gas phase. Taking an air-filled porosity (ϵ) of $0.2\,\text{cm}^3\,\text{cm}^{-3}$, $c_0 = 0.042\,\text{cm}^3\,\text{cm}^{-3}$.

M This is the rate of O_2 use in $\text{cm}^3\,O_2\,\text{cm}^{-3}$ soil s^{-1}. Taking a respiration rate at 25 °C of $1.6 \times 10^{-9}\,\text{g}$ $CO_2\,\text{g}^{-1}\,\text{s}^{-1}$ (Section 6.2) which gives $1.16 \times 10^{-9}\,\text{g}$ $O_2\,\text{g}^{-1}\,\text{s}^{-1}$, two changes are required:

1. Convert from g to mol and then to $\text{cm}^3\,O_2$ at 25 °C:

$$1.16 \times 10^{-9}\,\text{g}\,O_2\,\text{g}^{-1}\,\text{s}^{-1} = 1.16 \times 10^{-9}\,\text{g}/32\,\text{mol}\,\text{g}^{-1}$$
$$= 3.63 \times 10^{-11}\,\text{mol}\,O_2\,\text{g}^{-1}\,\text{s}^{-1}$$

Using the gas equation (Section 6.1) $pV = nRT$,

$$V = 3.63 \times 10^{-11} \times 0.082 \times 298/1$$
$$= 8.87 \times 10^{-10}\,\text{lg}^{-1}\,\text{s}^{-1}$$
$$= 8.87\,10^{-7}\,\text{cm}^3\,O_2\,\text{g}^{-1}\,\text{s}^{-1}$$

2. Taking a bulk density of $1.5\,\text{g}$ oven-dry soil cm^{-3}, the respiration rate $M = 8.87 \times 10^{-7}/1.5 = 5.9 \times 10^{-7}\,\text{cm}^3\,O_2\,\text{cm}^{-3}\,\text{s}^{-1}$.

D This is the diffusion coefficient of O_2 in the moist soil in $\text{cm}^2\,\text{s}^{-1}$. To obtain this value we need to know the diffusion coefficient of O_2 in air: its value D_A is $0.23\,\text{cm}^2\,\text{s}^{-1}$. The coefficient in soil is less than this value because particles and water occupy some of the space, and the molecules have to take a tortuous path through the soil; their progress is impeded by the soil. Oxygen diffusion through water is so much slower than through air that it can be neglected here. The two systems are shown in Fig. 6.10 together with the relationship between D and D_A which has been determined experimentally. F_A and F_S are diffusion rates in air and soil respectively. To obtain D both α and ϵ are needed.

The term α is an impedance factor. It depends on air-filled porosity ϵ ($\text{cm}^3\,\text{cm}^{-3}$). It has been determined experimentally for moist sandy materials (Fig. 6.11). This, and the equation $D = \alpha\epsilon D_A$, show that diffusion decreases as water content increases both because the volume of air-filled pore space available for diffusion

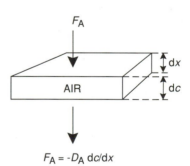

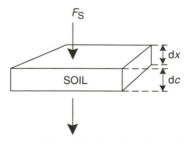

Figure 6.10 The effect of soil particles on the diffusion of gas through a soil segment.

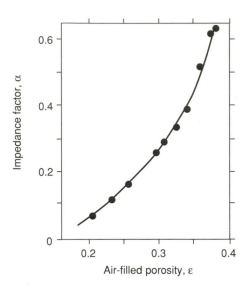

Figure 6.11 The relationship between the impedance factor, α, and the air-filled porosity, ϵ, for a moist sand. From Currie (1970).

decreases and also because the pathway becomes more tortuous.

For the soil conditions we have considered $\epsilon = 0.2\,cm^3\,cm^{-3}$ and so $\alpha = 0.06$. Thus $D = 0.06 \times 0.2 \times 0.23 = 2.76 \times 10^{-3}\,cm^2\,s^{-1}$.

The solution to Eq. 6.6, $L^2 = 2Dc_0/M$

Substituting values

$$L^2 = 2 \times 2.76 \times 10^{-3} \times 0.042/5.9 \times 10^{-7}$$

and

$$L = 19.8\,cm$$

Thus calculation predicts that for these conditions (25 °C, field capacity water content in a uniform sandy soil) the depth of the aerobic layer is about 20 cm. Using respiration rates more typical of soil temperatures in Britain (Section 6.2, Note 3), critical depths of 30–60 cm are predicted. The soil however is not uniform, having lower organic matter content and, in the summer, lower temperature with depth. It is therefore unlikely that anaerobic conditions will develop in this type of soil, except perhaps in humid tropical conditions.

Further calculations

Rework the above calculations for the following situations:

(a) Consider other soil temperatures (15 and 5 °C), cal-

culating changed respiration rates assuming a Q_{10} factor of 3 (Section 6.1). Values of D_A are 0.21 and 0.20 $cm^2\,s^{-1}$ respectively. (*Ans.* 32.8 and 55.3 cm)

(b) Calculate the depth of aerobic soil under drier conditions (air-filled porosity = 0.3) at the same bulk density. (*Ans.* 61.8 cm).

(c) Calculate the depth of aerobic soil for less compact conditions. Assume a bulk density of 1.3 $g\,cm^{-3}$ and the same gravimetric water content. Note that not only is ϵ increased by loosening the soil, but also M is reduced because there is now a smaller mass of soil respiring per cubic centimetre of soil. (*Ans.* 78.5 cm).

Aeration of a saturated soil

Equation 6.6 can also be used in this case to calculate the depth of aerobic soil when diffusion is occurring only through water-filled pores. This is the critical distance.

The parameters

c_0 This is the concentration of dissolved O_2 in cm^3 $O_2\,cm^{-3}$ soil. The volume of O_2 dissolved in water in equilibrium with the atmosphere at 25 °C and atmospheric pressure is $5.9 \times 10^{-3}\,cm^3\,O_2\,cm^{-3}$ water. For a soil with a bulk density of 1.5 $g\,cm^{-3}$, the total porosity is 0.42 $cm^3\,cm^{-3}$ (particle density = 2.6 $g\,cm^{-3}$), and this is then also the water-filled porosity. Therefore

$$c_0 = 0.42 \times 5.9 \times 10^{-3} =$$
$$2.48 \times 10^{-3}\,cm^3\ O_2\,cm^{-3}\ soil$$

M This has the same value as before if we assume that respiration rate is unaffected so long as O_2 is available. Therefore

$$M = 5.9 \times 10^{-7}\,cm^3\,O_2\,cm^{-3}\,s^{-1}$$

D The diffusion coefficient of O_2 in water D_w is $2.6 \times 10^{-5}\,cm^2\,s^{-1}$. Again $D = \alpha\epsilon_w D_w$ where ϵ_w is the water-filled porosity (0.42 $cm^3\,cm^{-3}$). The value of α is found to be about 0.66 for this condition, as for the diffusion of O_2 in a completely dry soil since the spaces involved are the same in both cases (Fig. 6.11). Therefore

$$D = 0.66 \times 0.42 \times 2.6 \times 10^{-5} = 7.2 \times 10^{-6}\,cm^2\,s^{-1}$$

The solution to Eq. 6.6

Substituting values

$$L^2 = 2 \times 7.2 \times 10^{-6} \times 2.48 \times 10^{-3}/5.9 \times 10^{-7}$$

and

$$L = 0.25\,cm\ or\ about\ 3\,mm$$

At a more normal temperature of 15 °C, M is about ⅓ of the above value and L is about 5 mm.

In a paddy soil the depth of the aerobic layer at the surface of the soil will be only a few mm. In a drained soil at field capacity transmission pores should not be more than about 1 cm apart if all the soil is to remain aerobic. These conditions are depicted in Fig. 6.2.

Section 6.6 The chemistry of aerobic and anaerobic respiration

AEROBIC RESPIRATION

The overall process can be summarized by restating Eq. 6.1:

$$C_6H_{12}O_6 + 6O_2 = 6CO_2 + 6H_2O + energy$$

The transfer of electrons, e^-, and protons, H^+, which occurs in the overall process can be shown in two stages:

$$C_6H_{12}O_6 + 6H_2O = 6CO_2 + 24H^+ + 24e^- \quad [6.7]$$

$$24H^+ + 24e^- + 6O_2 = 12H_2O + energy \quad [6.8]$$

These equations summarize a large area of biochemistry with many contributing reactions. The electrons are said to be donated by glucose and accepted by O_2.

ANAEROBIC RESPIRATION

Carbohydrate can only be partially oxidized in the absence of O_2. The process can be illustrated by the oxidation of glucose to pyruvic acid, which also occurs as the first step in aerobic respiration:

$$C_6H_{12}O_6 = 2CH_3COCOOH + 4H^+ + 4e^-$$

This cannot occur on its own and anaerobic micro-organisms link this oxidation to electron acceptors which are substances which can be reduced. The reductions are specific to a given species or group of species.

The following reductions are important in soils:

Denitrification
$$2NO_3^- + 12H^+ + 10e^- = N_2 + 6H_2O$$
nitrate nitrogen gas

Fermentation
$$CH_3CHO + 2H^+ + 2e^- = CH_3CH_2OH$$
acetaldehyde ethanol

Iron reduction
$$Fe(OH)_3 + 3H^+ + e^- = Fe^{2+} + 3H_2O$$
ferric hydroxide ferrous [6.9]

Methane production
$$CO_2 + 8H^+ + 8e^- = CH_4 + 2H_2O$$
carbon dioxide methane

These reactions occur in sequence as increasingly severe reducing conditions develop. Micro-organisms transfer electrons from the compound being oxidized into the compound being reduced and can be thought of as a pump building up a 'pressure' of electrons to push the reduction reactions forward.

REDOX POTENTIALS

Using the idea of a pressure makes the concept of the redox potential easier to understand. When platinum is placed in a solution in which oxidation and reduction reactions are taking place, electrons move on to the metal surface depending on their 'pressure' in solution. This gives to the metal an increasingly negative electrical potential as the 'pressure' of electrons increases.

A potential cannot be measured in isolation but has to be measured with reference to another potential (Nuffield Coordinated Sciences, Physics, Ch. P17 and Nuffield Advanced Chemistry II, Topic 15). A simple illustration of this is the way altitudes are given as heights above a standard height, normally sea level. This applies also in relation to water potential (Section 5.4). Redox potential (symbol E_h) is therefore defined as the potential of the platinum electrode relative to the potential of a hydrogen electrode. The details of the latter are not important here because it is a difficult electrode to use, and a calomel reference electrode is normally substituted. The potential of the reference electrode is not affected by the solution, and so the potential difference measured as a voltage between the two electrodes varies only with the potential of the platinum surface. When a calomel reference is used, the measured voltage is 0.248 V less than with a hydrogen electrode. Compared again to the measurement of height, it is as though the reference point had been set 248 m above sea level: altitudes relative to sea level would then all be 248 m more than those measured using this reference point (Fig. 6.12). Thus when a calomel electrode is used

$$E_h \text{ (volts)} = E_{measured} + 0.248 \text{ V}$$

Definition

The redox potential is a measure of electron 'pressure' and the severity of reducing conditions. In soils it indicates which compounds are likely to have been reduced.

Numerical relationships

The physical chemistry of oxidation–reduction systems leads to numerical relationships between redox

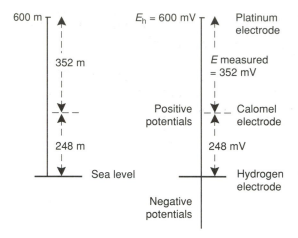

Figure 6.12 The relationship between redox potential and the potential measured using a platinum and a calomel reference electrode.

potentials and concentrations of oxidized and reduced compounds in soils (Nuffield Advanced Chemistry II, Topic 15). In aerobic soils, Eq. 6.8 shows that O_2 is being reduced. For this system

$$E_h = 1.23 + 0.015 \log P_{O_2} - 0.059\text{pH} \quad [6.10]$$

where P_{O_2} is the fractional volume of O_2 in the atmosphere with which the solution is in equilibrium. The equation shows that E_h depends on three terms:

1. 1.23 which is a standard measure of how easily O_2 is reduced;
2. P_{O_2} which is directly related to the O_2 concentration in the soil solution;
3. pH which is a measure of the effect of hydrogen ions on the reduction of O_2. Both electrons and hydrogen ions drive forward the reaction shown by Eq. 6.8.

For the normal atmosphere P_{O_2} is 0.21 (21% O_2), and in aerobic soils the value is often between 0.18 and 0.2. In a neutral (pH 7) aerobic soil therefore, E_h is theoretically close to 0.8 V. In practice, a freshly prepared soil paste often has a value of about 0.6 V, possibly due to the small O_2 concentration in the water or to locally small E_h values around colonies of micro-organisms.

A similar analysis shows that the reduction of nitrate occurs at about 0.4 V. When all the nitrate has been reduced redox potentials are normally stabilized between 0.1 and 0 V by the reduction of iron compounds which are often present in amounts large enough to *poise* (hold) the redox potential at these values for many weeks. The relationship derived for Eq. 6.9 is

$$E_h = 1.06 - 0.059 \log [\text{Fe}^{2+}] - 0.177\text{pH} \quad [6.11]$$

where $[\text{Fe}^{2+}]$ is the concentration (strictly the activity,

Section 7.1, Note 2) of ferrous ions in solution $(\text{mol}\,l^{-1})$. Using this equation, a neutral soil in which the redox potential has fallen to 0 V theoretically has a ferrous concentration of about 10^{-3} M in soil solution produced by the reduction of ferric hydroxide.

The role of pH in oxidation-reduction reactions

Both O_2 and ferric hydroxide are reduced more easily in acid conditions. In Eq. 6.9 the reduction of ferric hydroxide is driven forward by both protons and electrons as for the reduction of O_2 (Eq. 6.8). The equations suggest that the protons involved are simply those produced by the oxidation of glucose. However, in soil the concentration of protons depends on the natural pH of the soil modified by the reduction process.

Extended definition

Redox potential and pH together control the reduction of compounds, and together they indicate the severity of reduction in a soil.

It is therefore useful to think as follows. 'Oxidizing conditions are being maintained by the presence of O_2' or 'reducing conditions are severe enough to have produced a 10^{-3} M concentration of ferrous ions in solution.' In the former case substituting $P_{O_2} = 0.21$ into Eq. 6.10 gives

$$E_h = -0.059\text{pH} + 1.219 \quad [6.12]$$

and theoretically any combination of E_h and pH can be present in contact with this O_2 concentration provided the above equation is satisfied. Similarly for ferric hydroxide, $[\text{Fe}^{2+}] = 10^{-3}$ substituted in Eq. 6.11 gives

$$E_h = -0.177\text{pH} + 1.23 \quad [6.13]$$

and any combination of E_h and pH can be present when iron has been reduced to this extent provided this equation is satisfied. These relationships are plotted in Fig. 6.13 as *stability diagrams*. The slopes of the lines are -0.059 and -0.177 V per pH unit respectively. All points on the line indicate the same severity of reduction, despite changing E_h values. Because Fe(OH)_3 often poises reduced soils the Fe(OH)_3-Fe^{2+} line is an indication of the E_h-pH conditions present.

Conclusion

Both E_h and pH should be determined to measure the severity of reduction in soils. For comparison between soils with different pH values the E_h value 'corrected to pH 7' (symbol E_h^7) is often used. This is the E_h value which would occur at pH 7 for the same severity of reduction. For the Fe(OH)_3-Fe^{2+} system E_h^7 can be calculated as follows:

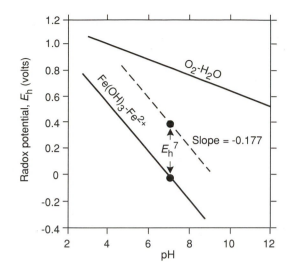

Figure 6.13 E_h–pH relationships in aerobic and anaerobic soils. O_2–H_2O stability line: Eq. 6.12, $P_{O_2} = 0.21$; $Fe(OH)_3$–Fe^{2+} stability line: Eq. 6.13, $[Fe^{2+}] = 10^{-3}$ M (strictly an activity). The broken line is explained in the text.

$$E_h{}^7 = E_h - 0.177 \,(\text{soil pH} - 7)$$

Graphically we move the point along the $Fe(OH)_3$-Fe^{2+} line in Fig. 6.13 to pH 7. In reduced soils we do not know what system is poising the soil. However, for convenience all corrections are made using a slope of -0.177 V per pH unit. In graphical terms, the measured E_h and pH point is plotted and a line parallel to the $Fe(OH)_3$-Fe^{2+} line is drawn (the dotted line on Fig. 6.13) through this point. The $E_h{}^7$ value can then be read from the graph. For aerobic systems a slope of -0.059 is used.

Changes in pH caused by reduction

If an acid soil is flooded and then poised by the reduction of iron compounds, pH normally rises to between 6 and 7. This can be explained by rewriting Eqs 6.7 and 6.9

$$C_6H_{12}O_6 + 6H_2O = 6CO_2 + 24H^+ + 24e^-$$
$$24Fe(OH)_3 + 72H^+ + 24e^- = 24Fe^2 + 72H_2O$$

To transfer $24e^-$ to $Fe(OH)_3$ an extra $48H^+$ are required above those produced by the oxidation of the glucose molecule, and are taken out of the soil solution. The pH value therefore rises. Other reactions occur in alkaline soils reducing pH to about the neutral point. Thus the pH of all flooded soils tends to move towards neutrality.

Thus the E_h measured in flooded soils is often close to $E_h{}^7$, and can be used as an approximate guide to the severity of reduction even if the pH is not known.

Section 6.7 The measurement of redox potential

A platinum electrode and a suitable reference electrode (Fig. 6.14) are inserted into wet soil, and the potential between them measured as a voltage difference using a millivoltmeter.

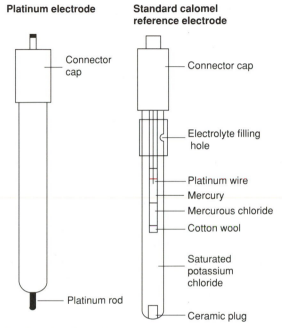

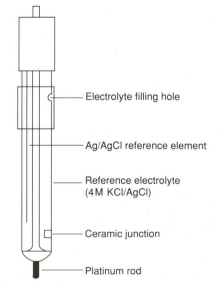

Figure 6.14 Platinum and calomel electrodes.

Equipment

Platinum electrode. These can be bought separately or combined with a reference electrode. A platinum electrode can be made from a platinum wire (2 mm diameter) soldered to an insulated copper wire. Fix the platinum wire into the end of a glass tube (Araldite) leaving about 1 cm of wire exposed taking care not to contaminate the exposed platinum surface. The lead running through the glass tube is connected to the voltmeter.
Reference electrode. A calomel electrode is commonly used.
Millivoltmeter. pH meters are millivoltmeters and most can be set to read millivolts.

Reagents

A standard redox solution. Dissolve 3.92 g of ferrous ammonium sulphate, $Fe(NH_4)_2(SO_4)_2.6H_2O$, and 4.82 g of ferric ammonium sulphate, $FeNH_4 (SO_4)_2.12H_2O$, in 50 ml of distilled water in a 250 ml beaker.

Add 5.6 ml of concentrated sulphuric acid (approx. 98% *m/m* H_2SO_4) (CARE), mix and allow to cool. Transfer to a 100 ml volumetric flask and make up to the mark. This contains 0.1 M Fe^{2+} and 0.1 M Fe^{3+} in approximately 1 M H_2SO_4 and has a potential of 430 mV (platinum–calomel) at 25 °C.
Sulphuric acid, approximately 1 M. To 50 ml of distilled water in a 250 ml beaker add 5.6 ml of concentrated H_2SO_4 (approx. 98% *m/m* H_2SO_4) (CARE), mix and allow to cool. Transfer to a 100 ml volumetric flask and make up to the mark.

Measurement

Connect the electrodes to the meter as instructed by the manufacturers. Support the electrodes in a beaker of distilled water. Before use, place the platinum electrode in 1 M H_2SO_4 to clean its surface. Wash the electrodes with distilled water, remove excess water with a tissue and insert both electrodes in the standard redox solution. Following the manufacturer's instructions, set to 0 mV, and then read the voltage between the electrodes (Note 1). If this is 430 ± 10 mV, the electrodes are functioning correctly.

Wash the electrodes and insert into wet soil (Note 2). Allow the reading to become steady (about 1 min) before recording the voltage. (Remember that $E_h = E_{measured} + 0.248$ V, Section 6.6.)

The platinum surface should be cleaned in acid and rinsed with distilled water between readings.

Note 1 The standard redox solution is not used to standardize the meter in contrast to the use of a buffer solution when measuring pH. The meter gives a direct reading in millivolts: the standard solution is used only to check that the electrodes and meter are operating correctly. Incorrect readings probably indicate that the platinum is not clean or that electrical contacts are not adequate.

Note 2 A liquid contact is required between the electrodes. In a soil suspension or paste the calomel electrode should simply make contact with the supernatant liquid. The platinum electrode should be lowered into the soil.

Measurement in 'undisturbed' soil Exposure of wet soil to air will allow oxidation to occur. Thus measurements in the field at the face of a soil pit or in lumps of soil brought back into the laboratory are made by pushing the platinum electrode into soil below the oxidized surface. Soil taken from the field should be placed in a polythene bag immediately after sampling, air excluded and the bag sealed. Measurements should be made as soon as possible. The soil must be wet enough for the calomel electrode to make liquid contact with the soil surface.

The platinum surface is not damaged by pushing the electrode into the soil although the wire may be bent if it touches a stone. If pH is also being measured (Section 8.1), more care is needed and a toughened glass electrode is recommended. Make a hole in the soil with a rod of the same diameter as the electrode before insertion. Coarse sand and stones can damage the glass electrode surface.

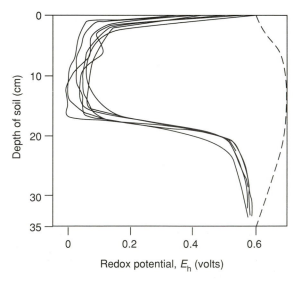

Figure 6.15 Redox potentials in a paddy soil (full lines) and in an adjacent well-drained soil (dotted line). From De Gee (1950).

Redox potentials in flooded soils

Redox potentials were measured down through a paddy soil at six positions between two rice plants 42 cm apart (Fig. 6.15). The graphs show clear changes with depth, but at any given depth the potentials are variable due to the effects of rice roots carrying O_2 into the soil.

Only the upper few cm of soil are aerobic, as predicted by the calculations in Section 6.5. Redox potentials are then close to 0 mV down to about 17 cm, which probably indicates the depth to which the soil has been puddled. The deeper soil is naturally drained with adequate air-filled pore space to maintain aerobic conditions.

Section 6.8 Questions about paddy rice production

Why grow rice in flooded soils?

- As well as being adapted to flooded conditions, rice is very sensitive to water stress. Only slight drying of the soil causes reduction in growth. Dryland varieties, although less sensitive, yield on average about half that of paddy rice primarily because of water stress.
- Flooding prevents the growth of a wide range of weeds which would otherwise compete with the rice plants.
- Nitrogen is supplied to the plants naturally through the fixation of N_2 into the cells of blue-green algae which thrive in the surface water, followed by mineralization to release ammonium ions when the cells decompose. Thus for thousands of years (until the development of fertilizers), production was maintained by this N at a level well above that of cereal crops in drained soils.
- Phosphorus (as phosphate) is released from iron compounds when they are reduced in anaerobic soil, increasing its availability.
  Similarly silicon (as silicate), manganese and iron, which are required in larger amounts by rice than by other crops, are made more available.

What happens to nitrogen compounds in anaerobic conditions?

Nitrate present in reduced soil is quickly lost by denitrification to N_2O and N_2. Mineralization of organic N releases ammonium which is stable in these conditions and can be taken up by rice. However, ammonium which diffuses up to the surface water will be oxidized (nitrified) to nitrate which may diffuse down again into the reduced layer where it will be denitrified. The sup-

ply of N by mineralization of organic matter occurs with only small losses, but fertilizer-N, usually applied as urea, $CO(NH_2)_2$, to the surface water, is very inefficiently used. It is hydrolysed to ammonium and nitrified to nitrate in the oxidized soil. Ammonia is lost by volatilization from the warm water. Some nitrate enters the reduced soil and before being taken up, is lost by denitrification. Only about 30 per cent of the applied nitrogen enters the crop and there are serious losses of NH_3, N_2O and N_2 to the atmosphere, with implications in relation to the greenhouse effect (Wild, 1993). In contrast, well-managed wheat takes up about 80 per cent of the applied N.

Why does organic matter not build up in paddy soils?

In permanently flooded soils peat develops, yet in paddy soils the amount of organic C tends to remain constant. Thus inputs of C in plant residues and animal manures must be balanced by losses. The inputs are lower than in a natural marsh, and the drying between cropping seasons allows oxidation to proceed for part of the year. However, fermentation in the flooded soil causes large losses of methane. This again has implications in relation to the greenhouse effect.

Section 6.9 Projects

1. For soils sampled from various sites having different vegetation and management, determine organic C (Section 3.4) and biomass activity using the respiration method (Section 6.2). Plot your data together with that in Table 6.3. Relate your results to those in Table 3.1 and Calculation 6 (Section 6.10).
2. Obtain a soil from a cultivated site (vegetable garden or arable field). Air dry and pass through a 2 mm sieve. To 100 g samples of soil add heavy metals as listed in Table 6.5 and remoisten with 20 ml water per 100 g.

Table 6.5 Amounts of reagents needed for the treatment of soils with heavy metals.

Metal	Soil concentrations: recommended upper limit (mg kg^{-1})	Reagent	Mass of reagent per 100 g of soil (mg)
Zinc	300	$ZnCl_2$	63
Copper	135	$CuCl_2$	29
Nickel	75	$NiCl_2.6H_2O$	30
Cadmium	3	$CdCl_2.2\frac{1}{2}H_2O$	0.6
Lead	250	$PbCl_2$	34

Also add 10 times and 100 times these treatments. To add small amounts of reagent make up a suitable solution and add the required volume to 100 g of soil. For example, 3 mg cadmium kg^{-1} could be added by dissolving 0.6 g of cadmium chloride in 1 l of water, adding 1 ml of this to 20 ml of water and using this to moisten the soil. Store the soils for 10 d and then measure respiration rates over 10 d. The basic treatment levels are equal to the recommended maximum soil concentrations, Section 15.5 (Giller and McGrath, 1989; ADAS, 1987).

3. Obtain samples of soil from an area of marsh. Sample from several depths if possible. Place the wet soil in polythene bags and mould the bag to work out any trapped air. Seal to exclude air. As soon as possible measure redox potential and pH. Plot your results as in Fig. 6.13 to determine the severity of reduction.

4. Almost fill a bucket with stone-free soil. Flood with water and work it into a smooth paste allowing the air to escape. Store in a warm place, maintaining a layer of water a few cm deep over the soil. Measure the distribution of E_h and pH at 1-week intervals after flooding. Plot your results as in Fig. 6.13.

Section 6.10 Calculations

1. If 1 m^2 of a cultivated soil has a total organic C content of 7 kg and is respiring at a rate of 10 g CO_2 m^{-2} d^{-1}, calculate the fraction of the soil organic C lost per day. (*Ans.* 0.04) If the mean respiration rate over a year was 3 g CO_2 m^{-2} d^{-1}, and the organic C content was maintained by inputs of plant residues, calculate the turnover time (Section 3.5) for the C. (*Ans.* 23.5 years)

2. The annual input of straw and roots to an arable soil is 0.7 kg dry matter m^{-2}. If this contains 43 per cent C and the soil organic matter content was not changing from year to year calculate the average respiration rate from the soil as g CO_2 m^{-2} d^{-1} assuming a constant rate throughout the year. Compare your value to those given in Table 6.1. (*Ans.* 3)

3. A soil has an air-filled porosity of 0.3 cm^3 cm^{-3}, and an O_2 content of 18 per cent (*v/v*) in the soil air. Calculate the amount of O_2 present as m^3 m^{-3} soil. (*Ans.* 0.054) Convert this into a mass of O_2 m^{-3} soil if the soil temperature was 15 °C. (*Ans.* 73 g)

4. In Fig. 6.4 the curves were drawn by Monteith *et al.* (1964) to represent the data for February to August, and for September to January. Imagine that a single curve is needed to represent all the data. Using the information in Section 6.1, choose values of R_0 and Q_{10} and plot calculated values of R to obtain a good fit. A better simulation of field conditions would have to take account of the effects of drying and availability of decomposable plant material. Can the data in Table 6.1 be fitted to a Q_{10} factor of 3?

5. The effects of chloroform fumigation on microbial activity in a soil are shown in Fig. 6.6. Using the information in Section 6.3, calculate the biomass C in this soil. (*Ans.* 537 µg g^{-1}) Using the data in Section 3.2, calculate the biomass-N and the biomass in this soil. (*Ans.* 130 µg g^{-1} and 1.07 mg g^{-1})

6. Plot the relationship between organic C (t ha^{-1}), (*x*-axis) and biomass-C (kg ha^{-1}), using the data from Table 3.1. Draw the line of best fit (a straight line) and give the equation for this line in the form $y = Ax$ (A has the units kg t^{-1}). The fitting of the line can be done by eye or using a linear regression analysis (Section 9.6). (*Ans.* $y = 26x$) Compare your results to those in Fig. 1 in the paper by Anderson and Domsch (1980) where the line of best fit has been drawn. Determine the slope of this line: the units are µg g^{-1}/g 100 g^{-1}. Re-express the slope value in kg t^{-1} to allow comparison.

7. Replot Fig. 6.13 to show the area on the graph and hence the E_h and pH values that could be found in aerobic soils if the O_2 concentration in the soil air can be between 0 and 21 per cent, and in anaerobic soils if the Fe^{2+} concentration can be between 10^{-5} and 10^{-2} M. You will have to take a real value for the 'zero' O_2 system: measurements have shown that aerobic respiration continues as long as the O_2 concentration at the surface of an organism is more than about one-hundredth of the atmospheric concentration, so use a lower limit of 0.2 per cent (P_{O_2} = 0.002).

8. In two soils E_h and pH were measured: soil A 0.40 V, pH 4.0; soil B 0.30 V, pH 7.0. Plot the values on Fig. 6.13. Which of these soils is more severely reduced?

9. The first stage of denitrification is the reduction of nitrate to nitrite. The nitrite has a transient existence, being reduced further to N_2 and N_2O. The equations for this first stage are

$$NO_3^- + 2H^+ + 2e^- = NO_2^- + H_2O$$
$$E_h = 0.83 - 0.03 \log([NO_2^-]/[NO_3^-]) - 0.059 pH$$

Calculate the relationship between E_h and pH when half the NO_3^- has been reduced and [NO_3^-] = [NO_2^-]. Plot this relationship on Fig. 6.13. What are the implications of your findings for very acid soils?

10. Using the data in Fig. 6.15, assume that the redox potentials in the reduced soil are being poised by the reduction of Fe(OH)$_3$ to Fe^{2+}. What range of Fe^{2+} concentrations would be present in the soil solution at 10 cm depth? Assume a pH of 6.5 in the solution. (*Ans.* 1.5–29 mM)

CHAPTER 7

Particle Surfaces and Soil Solutions

Many important chemical properties of soils are controlled by the reactions which occur between water in the soil and particle surfaces. Soil water is a solution containing a range of cations, anions and organic molecules normally in small concentrations. Particle surfaces react with this water, the main processes being

- *dissolution* and *precipitation* of salts and minerals and
- *adsorption* and *desorption* on the surfaces of clay minerals, sesquioxides and humus.

The amounts of cations, anions and organic molecules held on the particles are normally large compared to the amounts in soil solution.

As a result of these processes the composition of the solution in a soil is controlled within surprisingly narrow limits as soil conditions continually change due to wetting and drying, temperature fluctuations, plant uptake of nutrients and mineralization of organic matter. The control of soil solution composition in this way is known as *buffering*. It distinguishes a soil from a moist sand where there is very little interaction between the particles and water. Buffering prevents large leaching losses, and maintains a supply of nutrients in solution to plant roots.

DISSOLUTION AND PRECIPITATION

Salt or mineral particles when placed in water will dissolve and provided enough solid is present, dissolution will eventually cease when the concentration of dissolved ions has reached *saturation*. In this condition an equilibrium exists between the solid and the solution which is now saturated with respect to this solid. However, if water evaporates from this solution the concentration of dissolved ions increases and the solution becomes *supersaturated*. This condition is unstable, and normally precipitation reduces the concentration of ions to the saturation value again. These principles apply to the weathering of minerals in soils (Section 2.1). In terms of the buffering of soil solutions, the soil tends to maintain a saturated solution with respect to the main salts or minerals present. Thus during wetting and drying, following leaching or plant uptake of ions or after fertilizer applications, the soil solution will tend

to return to its equilibrium composition. The rate at which equilibrium is re-established may, however, be slow. Soil solutions therefore are not normally at equilibrium but are often close to this condition.

Some salts are very soluble. Common salt, NaCl, is an example and only under arid conditions is this present in soils (Colour Plate 21(b)). Gypsum has a lower solubility and is commonly present in soils of semi-arid regions, but is dissolved away in moist temperate regions (Colour Plate 11). Gypsum is followed by calcite, feldspars, micas, quartz, clay minerals and sesquioxides in decreasing order of solubility, such that under humid tropical conditions clay minerals and sesquioxides often predominate. The more soluble salts and minerals of arid regions lead to saline soil solutions, in contrast to dilute solutions in humid regions (Table 7.1).

ADSORPTION AND DESORPTION

These terms cover a variety of reactions of which ion exchange involving electrostatic interactions, and chemical adsorption to form surface compounds are the most important.

Surface charge and ion exchange

Soil clay minerals, sesquioxides and humus particles all have electrical charges associated with their surfaces. Some clay minerals have a permanent negative electrical charge which holds exchangeable cations close to their surfaces (Section 2.2). Other clay minerals and sesquioxides have a charge which varies with pH, being positive at low pH and negative at high pH, holding either anions or cations respectively (Section 2.2). Humus also has negatively charged sites, the number increasing as pH rises (Section 7.1).

These particles contribute to the total soil charge. As a result soils fall into three main groups:

1. *Group i.* Mineral soils dominated by permanent negative charge on their 2:1 clay minerals;
2. *Group ii.* Organic soils dominated by negative charge on humus which varies with pH;
3. *Group iii.* Mineral soils dominated by sesquioxides, kaolinite and allophane with variable charge.

131

Table 7.1 Exchangeable cations and soil solution composition.

	Soils		
	A	B	C
Soil			
pH	7.0	5.9	4.7
C (%)	nd	13.5	1.5
Clay (%)	nd	12.0	10.7
CEC $(cmol_c kg^{-1})$†	40.3	29.4	2.31
Exchangeable cations $(cmol_c kg^{-1})$			
Ca^{2+}	} 29.0	16.6	1.24
Mg^{2+}		2.11	0.35
K^+	0.8	1.01	0.23
Na^+	10.5	0.17	0.03
Al^{3+}	nd	nd	} 0.46
H^+	nd	nd	
Solution‡			
pH	nd	6.2	4.95
Ion concentrations $(mmol\, l^{-1})$			
Ca^{2+}	16.2	1.18	8.02
Mg^{2+}	19.2	0.29	3.97
K^+	0.51	0.85	3.45
Na^+	145.0	0.52	nd
Al^{3+}	nd	0.02	0.07
SO_4^{2-}	53.0	0.46	0.10
Cl^-	105.0	0.22	0.84
NO_3^-	nd	2.29	30.75§
HCO_3^-	3.3	− 0.12	nd

† CEC determined at pH 7, except for C where CEC is the effective cation exchange capacity.
‡ Measurements were made on a saturation extract for soil A, and on solution extracted from moist soils by centrifugation for soils B and C.
§ The high nitrate concentration is presumably the result of mineralization of organic matter or fertilizer inputs.
nd = Not determined.
A = Saline soil, California. From Richards (1954).
B = Yellow-brown loam, New Zealand. From Edmeades *et al.* (1985).
C = Ultisol, USA. From Elkhatib *et al.* (1987).

There are, however, no sharp boundaries between the groups and for any soil, charge will depend on the amounts of the three types of surfaces present and, for soils in Groups ii and iii, on the pH and solution concentration also. Soils of temperate regions tend to be in Group i. The increased weathering occurring in humid tropical regions tends to produce acidic Group iii soils which may have positive charge and only small amounts of negative charge. Group ii soils tend to occur in the cool moist regions of the world where the rate of turn-over of organic matter is slow, but local conditions are important with poor drainage reducing the rate of decomposition. Soils in desert regions often have small amounts of charge due to the lack of pedogenic minerals and humus.

Cation exchange capacity and exchangeable cations

There are always sufficient cations held by electrostatic forces on soil particle surfaces to balance the surface negative charge known as the *cation exchange capacity* or CEC. The cations commonly held in this way are calcium, magnesium, potassium and ammonium, with increasing amounts of sodium in saline soils, and hydrogen and aluminium in acid soils. Each of these is termed *exchangeable*, e.g. exchangeable calcium. The methods used to measure exchangeable cations and CEC are described in Sections 7.2 and 7.3 for neutral and acidic soils respectively. Using an ion exchange process the exchangeable cations are displaced and measured in solution. The results are expressed as $cmol_c kg^{-1}$, a unit which is explained in Section 7.2. Soil CEC values range between about 2 and $60\,cmol_c kg^{-1}$ depending on clay content, clay type and humus content.

- *Clays*. Table 2.3 gives the CEC values of the clay minerals. A soil containing 40 per cent smectite clay $(100\,cmol_c kg^{-1}$ clay) would have a CEC of about $40\,cmol_c kg^{-1}$ soil if the clay surfaces were uncontaminated: a large content of 2:1 clay is needed to give values at the top end of the range.
- *Humus* may contribute a significant part of the CEC even in a mineral soil because its effective charge can be up to about $100\,cmol_c kg^{-1}$ humus in neutral soils (Section 7.1). Thus 3 per cent humus in a soil may contribute up to $3\,cmol_c kg^{-1}$ to the CEC, depending on the characteristics of the humus, pH, soil solution concentration, interaction of humus with clays and sesquioxides, and on bonding of cations to humus by a process known as *chelation* (Section 7.1)
- *Sesquioxides* make little contribution to the CEC unless the pH is is above about 7 (Section 7.5), and even at pH 8.5 which is the maximum found in all except sodic soils the negative charge of uncontaminated surfaces is only about $2\,cmol_c kg^{-1}$ sesquioxide. Taking this value, a soil containing 5 per cent sesquioxide would have only about 0.1 cmol negative charge kg^{-1} soil associated with the 50 g of sesquioxide present. The actual value depends on pH, soil solution concentration, binding of cations and anions and interaction with humus and clays. The importance of sesquioxides is in Group iii soils where, with 1:1 clays and at low pH, they are the main source of anion exchange capacity.

Table 7.2 The range of CEC values commonly found for mineral soils of varying texture.

	CEC ($cmol_c\ kg^{-1}$)
Sand	2–4
Sandy loam	2–12
Loam	7–16
Silt loam	9–26
Clay and clay loam	4–60

- *Soils.* The range of CEC values for soils is shown in Table 7.2. The wide range of values for each textural group reflects differences in the amount and type of clay and organic matter content.

Table 7.1 gives the CEC of three soils together with their exchangeable cations. Soil A is in Group i. Soil B has both permanent charge clay and a large organic matter content: it has characteristics of both Group i and ii soils. Soil C is in Group iii.

The effective cation exchange capacity

Because the charge on humus and sesquioxides varies with both pH and salt concentration in solution, CEC is measured in two ways:

1. At pH 7 using the methods developed for soils of temperate regions (Section 7.2);
2. At the pH and soil solution concentration of the soil in the field. This is known as the *effective cation exchange capacity* or *ECEC* (Sections 7.3 and 7.6).

Factors controlling the amounts of exchangeable cations in soils

The amounts of the various exchangeable cations filling a soil's exchange capacity depend on inputs and losses as shown in Fig. 7.1. The processes are as follows:

- Dissolution of minerals by rain and groundwater and atmospheric deposition in coastal regions are the main natural inputs, along with fertilizers, lime and irrigation in agricultural systems.
- Leaching is the main loss, with those cations having least affinity for exchange sites being most easily leached. Affinity depends on electrostatic attraction between the cation and the surface which in turn depends on the charge and radius of the ion. The size of the (hydrated) ion determines how close it can come to the exchange sites. Thus Na^+ is (a) lost more easily than Ca^{2+} which is lost more easily than Al^{3+} and (b) lost more easily than K^+.
- Nutrient cations are cycled between soil and plant.

Table 7.1 illustrates some of these effects. Soil A has

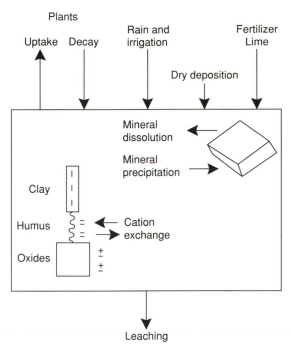

Figure 7.1 The processes which influence the amounts of exchangeable cations in soils.

a large input of soluble salts from groundwater and irrigation water, and Na^+, Mg^{2+} and Ca^{2+} dominate the CEC. Soil B has Ca^{2+} dominating its CEC, reflecting the presence of minerals dissolving to release Ca^{2+}, and probably the use of lime in agricultural management. Acidity in soil C has caused Al^{3+} to dissolve from soil minerals to occupy 20 per cent of its CEC.

Because of the range of factors influencing exchangeable cations it is only possible to give broad guidelines regarding their expected amounts.

- Ca^{2+} plus Mg^{2+} varies from less than 10 per cent of the CEC in very acid soils to almost 100 per cent in neutral soils. Linked with these changes H^+ plus Al^{3+} varies from zero in neutral soils to more than 90 per cent of the CEC in very acid conditions.
- $Ca^{2+}:Mg^{2+}$ ratios vary between 5:1 and 1:2.
- K^+ may be up to 5 per cent of the CEC.
- Na^+ is almost zero in leached soils but may be up to 50 per cent of the CEC in sodic soils.

Exchangeable cations and the composition of soil solutions

The ratios of cation concentrations in soil solution reflect the ratios of exchangeable cations. Table 7.1 gives examples. The cation charge in solution is always balanced by anion charge. In leached soils the amounts of cations in solution may be only 1/50 to 1/100 of the amounts held in exchangeable form, and the solution composition is buffered by the exchangeable cations. In saline solutions, the cations present reflect their input as soluble salts; in these soils there may be more cations in solution than on exchange sites and the solution may not be buffered to any extent by exchangeable cations.

The measurement of solution composition is described in Section 7.4, with the special case of saline soils developed further in Section 14.2.

Three examples of the buffering of soil solutions are discussed below, and the concept is developed further in Section 9.4.

1. Buffering and leaching

Rain enters a neutral soil and dilutes the soil solution. The concentrations of Ca^{2+} and Mg^{2+} ions in rain (Table 14.1) are very much smaller than in soil solution (Table 7.1), whereas the concentration of H^+ is generally larger. Thus H^+ moves on to cation exchange sites displacing Ca^{2+} and Mg^{2+} into solution, a process termed *cation exchange*. The system moves towards an equilibrium between exchangeable cations and solution cations depending on the affinity of the cations for the exchange sites. Using Table 7.1 soil B as an example, the amounts of exchangeable cations are so large compared to those in the equilibrium solution, that a displacement of exchangeable cations can occur to replenish the solution, with little overall change in the soil. Given time however, the exchange sites become decreasingly dominated by Ca^{2+} and Mg^{2+} with the development of soil acidity (see also Ch. 8 and Fig. 8.1).

The cation exchange process between H^+ and Ca^{2+} can be written

$$R^{2-}\,Ca^{2+} + 2H^+ \rightleftharpoons R^{2-}\,H_2^+ + Ca^{2+}$$

where R is a negatively charged surface. The solution cations have to be balanced by anions to maintain electroneutrality in soil solution.

2. Buffering and nutrient uptake by plants

A plant selectively takes up the nutrient ions it requires for growth. For example, the uptake of potassium from soil solution decreases its concentration and disturbs the equilibrium between exchangeable and solution cations. Cation exchange will tend to re-establish the equilibrium, with K^+ being released from exchange sites to be replaced by other cations from solution. For example,

$$R^{2-}\,K_2^+ + Ca^{2+} \rightleftharpoons R^{2-}\,Ca^{2+} + 2K^+$$

The concentration of K^+ (and other cations in solution)

is normally small, and without cation exchange would be quickly exhausted.

Because roots take up a range of cations, multi-ionic exchange occurs to re-establish equilibrium. The progressive depletion of exchangeable K^+, Ca^{2+} and Mg^{2+} ions requires that other cations take their place, and H^+ and eventually Al^{3+} are involved. Hydrogen ions are released during mineralization of organic matter and by the roots themselves.

3. Buffering and fertilizer applications

Fertilizers are normally applied to soils as soluble salts, e.g. potassium chloride which dissolve in solution and disturbs the equilibrium between exchangeable and solution cations. Equilibrium is re-established by K^+ displacing the range of cations from exchange sites. In neutral soils Ca^{2+} is the predominant cation and so the main process is

$$R^{2-}\,Ca^{2+} + 2K^+ \rightleftharpoons R^{2-}\,K_2^+ + Ca^{2+}$$

Thus the K^+ concentration is only slightly increased. The chloride remains in solution and is eventually leached along with cations (primarily Ca^{2+}) from the soil. This aspect of fertilizer use could be criticized by those favouring organic farming systems. The fertilizer increases the chloride concentration in soil solution and increases the loss of $CaCl_2$ (Section 9.4).

Positive charge and anion exchange capacity

Sesquioxides and allophane are the main source of positive charge in Group iii soils, together with smaller amounts on the edges of kaolinite crystals. Fig. 7.7 shows that the positive charge on goethite and gibbsite can be up to about $6\,cmol_c\,kg^{-1}$ at pH 4 and 5 mM solution concentration. In soils of the humid tropics these minerals contribute up to about $2\,cmol_c$ positive charge kg^{-1} soil (Table 7.3). Thus *anion exchange capacities* (abbreviated AEC) are small compared to CEC values of Group i and ii soils (up to $60\,cmol_c\,kg^{-1}$). Although positive charge is not called the *effective* anion exchange capacity, in practice it is normally determined at the pH and concentration of the soil solution.

An important distinction needs to be drawn between charge holding ions by electrostatic attraction which exchange freely (*non-specific adsorption*) and charge which is neutralized by ions held firmly by chemical bonds (*specific adsorption*). Although cations have different affinities for negative sites on clay surfaces, they all exchange freely. Part of the negative charge on humus also holds cations in this way and part holds specifically adsorbed cations (Section 7.1). Similarly, the affinity of anions for positive sites varies greatly. Chloride and nitrate exchange freely, but sulphate (in acid soils) and

Table 7.3 Ion exchange properties of soils from the humid tropics

Soil order†	Country of origin	Depth of sample (cm)	pH 1:1 in H_2O	Soil solution concentration (mM)	AEC	ECEC
					$(cmol_c\,kg^{-1})$	
Oxisol	Kenya	45–68	4.9	1.2‡	0.29	4.30
Ultisol	Thailand	80	5.7	1.3	0.01	0.13
Alfisol	Brazil	Subsoil	5.4	0.4	0.67	1.44
Oxisol	Malaysia	35–50	5.1	1.1	1.09	1.17
Inceptisol	Colombia	40–60	5.5	0.7	1.74	0.89
Ultisol	Cameroon	40–60	5.9	1.3	0.10	4.08
Ultisol	Ivory Coast	Subsoil	4.8	0.8	0.06	1.93
Ultisol	Nigeria	47–70	4.5	0.7	0.06	2.63
Ultisol	Bolivia	0–5	5.6	25.7	0	3.48
Oxisol	Malaysia	0–2	4.3	2.0	0.31	1.39

† Soil Taxonomy orders (Soil Survey Staff, 1975). A simple introduction is in Brady (1990).
‡ These values are the ionic strengths of the soil solutions and can be considered to be the concentrations of ammonium chloride which cause the soils to develop the same charge as the soil solutions.
Results obtained by M. Wong from a project funded by the Overseas Development Administration of the British Government.

phosphate are specifically adsorbed. These differences are discussed in Section 7.5. The normal use of the term 'exchange capacity' is for the charge which holds the freely exchanging ions.

There has been a development of interest in methods for determining AEC over the last 30 years, reflecting the rapid development of agriculture in the humid tropics. There is no standard method, but that given in Section 7.6 is convenient and simple in operation. It relies on the anion exchange process in which chloride is displaced by nitrate. Both AEC and CEC of Group iii soils vary with pH and solution concentration and so the method is similar to the standard method for measuring CEC in acid soils (Section 7.3).

A simple method for identifying positively charged soils

The difference in soil pH measured in water and in potassium or calcium chloride solution (Section 8.1) indicates charge characteristics. For soils with a net positive charge, pH is lower measured in water; for a net negative charge, pH is lower in the chloride solution. The soils in Table 7.4 have a net negative charge except in the lowest layer of soil B where the two pH values are equal indicating that positive and negative charge are approximately equal, and the lowest layer of soil C which has a net positive charge.

The exchange capacities of variable charge soils

Both AEC and ECEC values are small in many soils from the humid tropics (Table 7.3). The small ECEC

values reflect the absence or small amounts of 2:1 clay minerals with the charge being primarily on humus and kaolinite. Table 7.4 compares the CEC values of a Group i/ii and a Group iii soil. The ECEC of soil B is closely related to its organic C content, and its large clay content carries little charge.

There are serious difficulties in managing many Group iii soils:

- The minerals present in these soils hold phosphate very strongly, particularly at the low prevailing pH values. This leads to serious phosphate deficiencies in crops. On the other hand the ability of the AEC to hold sulphate and nitrate and reduce their leaching losses is a valuable soil property.
- The low ECEC values result in a small capacity to hold nutrient cations (K^+, Ca^{2+}, Mg^{2+}) which along with leaching caused by high rainfall may lead to deficiencies. The very low charge often present in the deeper horizons means that cations once leached from the topsoil will then move almost as freely as nitrate moves in temperate regions. Organic matter has a particularly important role in holding cations and the rapid decrease in organic matter following cultivation reduces still further their ability to hold cations (Section 13.5).
- Liming materials are often not locally available and lime is therefore expensive. Its effects are rapidly lost by leaching.
- These soils may also be easily eroded by rainstorms. Thus careful management is needed when they are cultivated (Section 13.5).

Table 7.4 Cation exchange properties of two contrasting soils

Depth (cm)	Clay (%)	Loss on ignition (%)	pH H$_2$O	1 M KCl	Exchangeable bases† (cmol$_c$ kg^{-1})	ECEC
Soil A, Group i/ii						
0–8	21	10.9	5.5	5.0	18.9	28
8–30	28	4.9	6.1	5.4	16.6	20
30–66	21	5.2	6.0	5.4	16.8	22
66–89	19	2.9	6.0	5.4	10.6	13
89–107	18	2.7	6.3	5.8	9.3	12
Soil B, Group iii		Organic C (%)				
0–17	72	2.6	5.1	4.3	0.31	0.94
17–33	79	2.1	5.2	4.5	0.23	0.43
33–51	82	1.7	5.3	4.7	0.18	0.27
51–106	86	0.7	5.2	5.2	0.08	0.10
Soil C. Group iii						
0–20	65	2.5	5.3	4.7	2.6	2.8
100–200	62	0.7	5.9	6.0	0.5	0.5

† Exchangeable bases = Ca^{2+} + Mg^{2+} + K$^+$ + Na$^+$.

Soil A. Non-calcareous groundwater gley under grass developed in alluvium over Eocene clay, Reading, UK. Data from Jarvis (1968). The clay-sized fraction is predominantly smectite. Mineralogically the soil is in Group i, but the organic matter in the surface layer gives it Group ii characteristics also.

Soil B. Red–yellow latosol under forest formed on Tertiary clay sediment over a Pre-Cambrian quartzite rock, Cerrado region, Brazil. Data from E. de Sa Mendonca. The clay-sized fraction is predominantly kaolinite and gibbsite.

Soil C. An oxisol from São Paulo, Brazil. The soil contained 19, 37 and 9% Fe$_2$O$_3$, gibbsite and kaolinite respectively in the 0–20 cm layer and 18, 33 and 11% in the 100–200 cm layer (van Raij and Peech, 1972).

FURTHER STUDIES

Ideas for projects are given in Section 7.7 and calculations in Section 7.8.

Section 7.1 Negatively charged sites on the surface of humus

Because humus is composed of very large and complex molecules (molar masses between 20 000 and 100 000 g mol^{-1}), an understanding of its properties has been obtained by both fractionation and identification of characteristic groups in the unaltered molecule (Stevenson, 1982). The latter approach has shown that carboxylic–COOH, phenolic–OH and alcoholic–OH groups are the source of negatively charged sites on humus. Only carboxylic groups are significantly charged at pH values below 7.

The dissociation of carboxylic groups

Carboxylic groups dissociate (ionize) in solution, as seen in the behaviour of ethanoic (acetic) acid.

$$CH_3COOH \rightleftharpoons CH_3COO^- + H^+ \qquad [7.1]$$
ethanoic acid   ethanoate

The extent of this dissociation into negatively charged ethanoate and hydrogen ions is a characteristic of the carboxyl group and depends on what it is attached to. For example, in methanoic acid, HCOOH, the carboxylic group has a greater tendency to dissociate than in ethanoic acid. In humus the carboxylic groups also vary in this way. The extent of dissociation also depends on pH because the presence of the hydrogen ion in solution tends to drive the reaction to the left (Note 1). Thus in soils, acid conditions favour the undissociated group and result in a small number of negatively charged sites.

Dissociation ceases when an equilibrium has developed between the undissociated acid, and the products of dissociation. For Eq. 7.1, the *dissociation constant* tells us the extent of dissociation at equilibrium. It is

$$K_a = [CH_3COO^-][H^+]/[CH_3COOH]$$

where [ ] indicates concentrations (strictly activities, Note 2) in $mol\,l^{-1}$. The values of K_a for methanoic acid and ethanoic acid are 1.6×10^{-4} and 1.7×10^{-5} respectively. They are often expressed as pK_a values where $pK_a = -\log K_a$, and in the two cases above are 3.8 and 4.8.

These numbers give a simple way of thinking about the extent of dissociation, because we can write (Note 3)

$$\log K_a = \log\{[CH_3COO^-]/[CH_3COOH]\} + \log[H]^+$$

or

$$pK_a = -\log\{[CH_3COO^-]/[CH_3COOH]\} + pH \quad [7.2]$$

When $pH = pK_a$, $-\log\{[CH_3COO^-]/[CH_3COOH]\} = 0$, $[CH_3COO^-]/[CH_3COOH] = 1$ and so $[CH_3COO^-] = [CH_3COOH]$. Thus if we know the pK_a value we also have the pH value at which the concentration of undissociated groups is equal to the concentration of charged groups, or put another way, we know the pH at which half of the total carboxylic groups are dissociated. Half of the methanoic acid molecules are therefore dissociated at pH 3.8, and for ethanoic acid this occurs at pH 4.8. The greater tendency of methanoic acid to dissociate is shown by the fact that it reaches this 'half way stage' against a larger concentration of hydrogen ions.

Carboxylic groups on humus

These can be treated in a similar way except that we are not dealing with molecules in solution

$$humus-COOH = humus-COO^- + H^+$$
$$K_a = [humus-COO^-][H^+]/[humus-COOH]$$

The unit for the concentration of the groups is now chosen for convenience as $cmol\,kg^{-1}$ humus, but because we are dealing with a ratio of dissociated to undissociated groups the unit is not important. The dissociation constant K_a is again an indication of the 'half way stage'. The carboxylic groups on humus have pK_a values ranging from about 3 to 6 reflecting differences in their local chemical environment.

It is not possible to give exact charge values for humus because of its variable characteristics. However, charge has been measured for purified fractions of humus and computer models have been developed to predict their likely charge in soils (Tipping and Hurley, 1988). The predominant sources of charge are the *humic* and *fulvic acid* fractions. These can be extracted from humus by NaOH solution leaving behind the insoluble

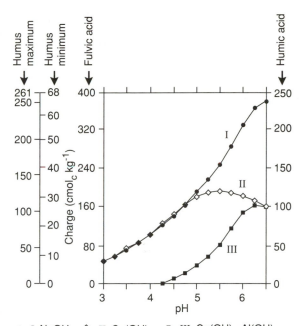

• I NaOH   ◇ II Ca(OH)₂   ■ III Ca(OH)₂-Al(OH)₃(s)

Figure 7.2 The charge free to hold exchangeable cations on purified fractions of humus, and the predicted charge on humus in soils. The humic acid data were obtained by E. Tipping based on the model for humic acid in Tipping and Hurley (1988). It has been assumed that fulvic acid behaves similarly but with increased charge. The text explains how the charge values for humus were calculated. Line I indicates the total charge developed when pH is raised with NaOH. Line II indicates the charge holding exchangeable Ca^{2+} when pH is raised with $Ca(OH)_2$. Line III indicates the charge holding exchangeable Ca^{2+} plus Al^{3+} when pH is raised by $Ca(OH)_2$ in the presence of gibbsite. In all cases the soil solution concentration is between 2.5 and 3.5 mM.

and relatively inert *humin*. Acidification of the alkaline solution precipitates the humic acid and leaves the fulvic acid in solution. Fig. 7.2 shows that the charge on humic acid increases from about 30 to 230 $cmol_c\,kg^{-1}$ when pH is raised from 3 to 6.5 with NaOH (line I). In this relatively simple system Na^+ is held exchangeably by electrostatic attraction to the charged sites:

$$humus-COOH + NaOH \rightleftharpoons humus-COO^-\,Na^+ + H_2O$$

The charge on fulvic acid probably rises from about 50 to 400 $cmol_c\,kg^{-1}$ for the same conditions. In different soils, these fractions together range from about 25 to 75 per cent (typically about 50%) of the mass of humus, and fulvic acid ranges from about 15 to 70 per cent (typically about 25%) of the mass of fulvic plus humic acid. Based on these amounts, the estimated range of charge of humus is given in Fig. 7.2. The predicted charge at pH 6.5 is between 60 and 250 $cmol_c\,kg^{-1}$ humus in agreement with measurements (Stevenson, 1982). At higher pH values, dissociation of phenolic and alcoholic groups causes further development of charge to a maximum of about 400 $cmol_c\,kg^{-1}$ humus although larger values have been reported.

The effects of bound cations

Line I in Fig. 7.2 indicates the total number of charged sites which can be developed, with exchangeable Na^+ satisfying all the sites. However, carboxylic groups react with some cations, binding them firmly by electrostatic forces and complexation reactions (Section 15.3, Note 4). The cations are no longer exchangeable.

$$humus\!\!\begin{array}{c}\diagup COOH \\ \diagdown COOH\end{array} + Ca(OH)_2 \rightleftharpoons$$

$$humus\!\!\begin{array}{c}\diagup COO^- \\ \diagdown COO^-\end{array} + Ca^{2+} + 2H_2O$$

This type of linkage is known as chelation. It is favoured by high pH, and line II shows that above pH 5 there is significant binding of Ca^{2+}, resulting in a reduction in the amount of charge which remains free to hold exchangeable cations.

Acid soils normally contain Al^{3+} released from minerals. This cation is strongly chelated to carboxylic groups, reducing the free charge still further. The effect can be modelled by assuming that the soil solution is saturated with respect to gibbsite, $Al(OH)_3$ and thus maintains the concentration of Al^{3+} (Section 8.3). Line III is therefore more typical of soil conditions and shows that the carboxylic groups are fully satisfied by bound aluminium below pH 4.25. As $Ca(OH)_2$ increases the pH, the Al^{3+} concentration in soil solution is reduced and free charge develops until at pH 6.5 Al^{3+} is having little effect because of the very low solubility of gibbsite. Bound Ca^{2+} is still present and the predicted free charge on humus is now between 25 and 100 $cmol_c\,kg^{-1}$ (typically 50) at pH 6.5. This is an estimate of the charge which is effective in terms of measurements of CEC. However, the distinction between bound and exchangeable Ca^{2+} depends on the method of extraction. The data here relate to typical conditions in soil solutions and are an indication of the data obtained by standard extractions with 1 M ammonium acetate or potassium chloride.

Chelation has other important effects. The micronutrients copper, zinc, manganese and iron are strongly held by humus in calcareous soils leading to deficiencies in crops, and potentially toxic metals are bound to the organic matter in sewage sludge, reducing their availability to plants.

The effects of solution concentration

The charge on humus increases as the concentration of the solution in contact with the surface increases. The reason for this is not so easily understood as the effect of pH, but the mechanism can be viewed as a cation exchange reaction where chemically bound H^+ exchanges for electrostatically bound Na^+:

$$humus\!-\!COOH + Na^+ = humus\!-\!COO^-\,Na^+ + H^+$$

The presence of Na^+ in solution drives the reaction to the right (Note 1). In a similar way H^+ in solution (low pH) drives the reaction to the left. Soil solution concentrations are often between 1 and 3 mM and so the effect of concentration needs to be considered in the measurement of CEC when 1 M solutions are used (Sections 7.3 and 8.2) although Fig. 8.4 suggests that this may not be a problem. The data in Fig. 7.2 relate to 3 mM solutions.

The association of humus with clay minerals and sesquioxides

The charge on clay minerals (Table 2.3), on humus (Fig. 7.2) and on sesquioxides (Fig. 7.7) are all likely to be modified by their intimate asssociation in the colloidal fraction in soils. Particularly where a charged carboxyl group is linked via a cation to charge on a sesquioxide surface, the cation is unlikely to be exchanged. There is little information on the extent of these effects.

Note 1. Chemical equilibrium For any system at equilibrium there is a relationship between the concentrations (strictly activities) of the reactants and products. For example, if a reaction is represented by the equation

$$mA + nB \rightleftharpoons pC + qD$$

then

$$[C]^p[D]^q/\{[A]^m[B]^n\} = K_c$$

where K_c is the equilibrium constant. The addition of reactants to the system (an increase in [A] or [B]) results in the reaction moving to the right (more products are formed) until the concentrations again satisfy K_c. Similarly an increase in the concentration of the products causes the reaction to move to the left.

Note 2. Activities and concentrations Chemical equilibria depend on the activities of the reactants and products. Activity is a measure of the concentration of 'free' molecules or ions in solution. Molecules (uncharged) in solution are all 'free' and so activity = concentration. As the concentration of ions (charged) increases, their 'freedom' decreases because of the proximity of other ions, and the activity becomes less than the concentration. For soil solutions which are normally dilute, the differences are small enough to be neglected for many purposes. In saline soils the differences are significant.

Note 3. Handling logarithmic equations Any equation can be changed into logarithmic form provided basic rules are followed. For example $a = b$ can be changed by taking the log of both sides of the equation to give $\log a = \log b$. Further examples are

$$a = b + c \qquad \log a = \log(b+c)$$
$$a = bc \qquad \log a = \log(bc)$$
$$\qquad = \log b + \log c$$
$$a = bc/d \qquad \log a = \log(bc/d)$$
$$\qquad = \log b + \log c - \log d$$
$$a = b^2 \qquad \log a = \log(b^2)$$
$$\qquad = 2\log b$$
$$a = b^{-2} = 1/b^2 \qquad \log a = \log b^{-2}$$
$$\qquad = -2\log b$$

To convert these to p values as in the case of K_a being converted into pK_a $(= -\log K_a)$ or $[H^+]$ into pH $(= -\log[H])$, then care must be taken to keep the signs correct. For example

$$a = bc/d \qquad \log a = \log b + \log c - \log d$$
$$\qquad -\log a = -\log b - \log c + \log d$$
$$\qquad pa = pb + pc - pd$$
$$a = b^{-2} \qquad \log a = -2\log b$$
$$\qquad -\log a = 2\log b$$
$$\qquad pa = -2pb$$

Section 7.2 The measurement of exchangeable cations and cation exchange capacity in neutral soils

The method (Fig. 7.3) involves leaching soil with 1 M ammonium ethanoate (acetate) solution at pH 7 in which the displaced exchangeable cations can be measured. The exchange capacity is filled with ammonium ions and ammonium ethanoate solution remains in the wet soil. This solution is displaced by ethanol. Acidified potassium chloride then displaces exchangeable ammonium which is measured to give the CEC. The results are expressed as $cmol_c\,kg^{-1}$.

Choice of displacing solution

The standard method was developed in temperate regions. Ammonium ethanoate is a buffered solution (Section 8.4) and raises the pH of the soil close to 7. It was considered useful to measure all soils at this pH partly because soils in these regions are limed to about pH 7 for agricultural use, and partly because the pH-dependent charge on humus is brought to a standard condition. For neutral soils the method measures the soil properties as they are in the sample but for acid soils two changes occur:

1. The exchangeable K^+, Ca^{2+} and Mg^{2+} ions present are extracted, but exchangeable H^+ and Al^{3+} ions are not. The buffered solution neutralizes H^+ and precipitates Al^{3+} as $Al(OH)_3$.
2. The CEC of the soil is increased by the rise in pH.

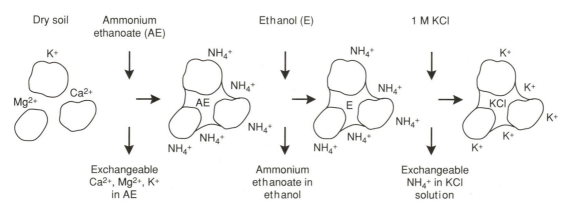

Figure 7.3 The determination of exchangeable cations and cation exchange capacity.

This may only be a small change in mineral soils with low organic matter content (Group i) but is large in organic and variable charge soils (Groups ii and iii). Modified methods for these soils are given in Sections 7.3 and 7.6.

Note that soils which contain soluble salts will release these into ammonium ethanoate. Thus calcareous soils give incorrect high values for exchangeable Ca^{2+}, and saline soils give very high values for Na^+, Ca^{2+} and Mg^{2+} (Section 14.2).

Ethanol is used to displace the excess ammonium ethanoate solution. The two solutions are miscible but exchangeable ammonium does not come off the exchange sites: there are no cations in ethanol to exchange with the NH_4^+.

The potassium chloride is acidified so that displaced NH_4^+ is not lost as ammonia.

Units of charge

Each negative charge on clay or humus is equal to that on one electron and is balanced by a positive charge on an exchangeable cation. The CEC of a soil may be in the region of 10^{23} charges kg^{-1} but a more convenient way of expressing charge is through the amount of exchangeable cations which the soil can hold. Thus for a soil having the above number of charges all satisfied by potassium there would be 10^{23} K^+ ions kg^{-1} soil. The mass of this number of ions can be calculated from the molar mass of potassium which is $39.1\ g\ mol^{-1}$. The mol is 6.02×10^{23} atoms or ions, and so there would be

$$10^{23}\ K^+ \text{ ions } kg^{-1}/6.02 \times 10^{23}\ K^+ \text{ ions } mol^{-1} = 0.166\ mol\ K^+\ kg^{-1}$$

This can also be expressed as

$$0.166\ mol\ kg^{-1} \times 39.1\ g\ mol^{-1} = 6.5\ g\ K^+\ kg^{-1}$$
(Table 7.5)

The same calculation for the soil with all the charge satisfied by Na^+ would also give $0.166\ mol\ Na^+\ kg^{-1}$ which would have a mass of 3.8 g. Thus, to express the capacity in a way which is not dependent on the ion present requires the use of $mol\ kg^{-1}$. This, however, is only satisfactory for monovalent ions, as can be seen

using the same calculation for a soil satisfied by Ca^{2+}. The 10^{23} negative charges will still be balanced by 10^{23} positive charges on Ca^{2+} ions, but only 0.5×10^{23} ions will be required due to the divalent charge. The CEC would be only $0.083\ mol\ Ca^{2+}\ kg^{-1}$. Similarly the value would be $0.055\ mol\ Al^{3+}\ kg^{-1}$.

Thus, in order to have a single value for CEC regardless of the exchangeable ion present, CEC needs to be related to the charge on the ions rather than the mass or the number of mol. This is done by using as the unit the *mole of charge* (abbreviated mol_c), which is 6.02×10^{23} charges. Thus for monovalent ions, $1\ mol = 1\ mol_c$, for divalent ions $1\ mol = 2\ mol_c$ and for trivalent ions $1\ mol = 3\ mol_c$. For Ca^{2+} where the CEC was $0.083\ mol$ $Ca^{2+}\ kg^{-1}$ we now have $0.083 \times 2 = 0.166\ mol_c\ kg^{-1}$, and for Al^{3+} we have $0.055 \times 3 = 0.166\ mol_c\ kg^{-1}$.

Not only does the CEC become a single value using this unit, it also allows cation exchange to be expressed in a satisfactory way. Using the above data, $0.166\ mol$ K^+ would exchange for $0.083\ mol\ Ca^{2+}$ if complete displacement occurred: expressed according to charge, $0.166\ mol_c\ K^+$ exchanges for $0.166\ mol_c\ Ca^{2+}$.

One final adjustment is made: because CEC values are normally between 0.02 and $0.6\ mol_c\ kg^{-1}$ we use centimole charge per kilogram ($cmol_c\ kg^{-1}$) giving values between 2 and 60. This also brings the values into line with the traditional units of charge which were milliequivalents per $100\ g$ (meq per $100\ g$). It is convenient that the numbers are the same for both units.

Preparation of extracts

Reagents

Ammonium ethanoate solution, 1 M. Dilute approximately $230\ ml$ of glacial ethanoic (acetic) acid to $1\ l$. Dilute approximately $220\ ml$ of ammonia solution (approx. 35% *m/m* NH_3) to $1\ l$. Mix the solutions together and adjust to pH 7.0 with ethanoic acid or ammonia. Dilute to $4\ l$.
Ethanol, 95 per cent.
Potassium chloride solution. Dissolve approximately $100\ g$ KCl in water and make up to $1\ l$. Add $2.5\ ml\ 1\ M$ HCl.

Extraction

Weigh $5\ g$ (±0.01) of $<2\ mm$ air-dry soil into a $100\ ml$ beaker. Add $20\ ml$ of ammonium ethanoate solution, stir and allow to stand overnight. Transfer the suspension to a filter funnel fitted with a Whatman No. 44 filter paper and standing in a $250\ ml$ volumetric flask. Thoroughly wash the beaker with ammonium ethanoate solution into the funnel. It is convenient to have a wash bottle containing ammonium ethanoate for this purpose. Leach the soil with successive $25\ ml$ volumes of ammonium ethanoate, allowing the funnel to drain

Table 7.5 Ways of expressing the charge of a soil

Number of negative charges kg^{-1}	Number of exchangeable cations kg^{-1}	Exchangeable cations		
		$(g\ kg^{-1})$	$(mol\ kg^{-1})$	$(mol_c\ kg^{-1})$
10^{23}	$1 \times 10^{23}\ K^+$	6.5	0.166	0.166
10^{23}	$0.5 \times 10^{23}\ Ca^{2+}$	3.3	0.083	0.166
10^{23}	$0.33 \times 10^{23}\ Al^{3+}$	1.5	0.055	0.166

between each addition. Continue until nearly 250 ml of filtrate has been collected. Make up to the mark with ammonium ethanoate. This extract is used for determination of potassium, calcium and magnesium.

Place the funnel in a rack over a 250 ml beaker. Wash the interior of the funnel, the soil and the paper with five 25 ml volumes of ethanol allowing the funnel to drain between washings. Discard the washings. It is essential that all the ammonium ethanoate solution is removed: the use of a wash bottle containing ethanol is advised.

Place the funnel in a 100 ml volumetric flask. Leach with successive 25 ml volumes of KCl solution, allowing the funnel to drain between each addition. Continue until nearly 100 ml has been collected and make up to the mark. This solution is used to determine the CEC.

Determination of exchangeable potassium

This is determined in the ammonium ethanoate leachate by flame analysis using a flame photometer. Gravimetric methods are the only alternative, but with less convenience and accuracy (Piper, 1947).

Reagents

Standard potassium solutions, $0-10 \mu g K^+ ml^{-1}$. These are prepared from a volumetric standard containing $1 mg K^+ ml^{-1}$ which can be purchased ready prepared or made as follows: dry potassium nitrate, KNO_3, at $105\,°C$ for 1 h and cool in a desiccator. Dissolve 1.293 g in water in a 100 ml beaker and add 1 ml of hydrochloric acid (approx. 36% *m/m* HCl) as a preservative if the solution is to be stored for more than a few days. Transfer with washings to a 500 ml volumetric flask and make up to the mark. Pipette 10 ml of this solution into a 100 ml flask and make up to the mark with ammonium ethanoate solution. This contains $100 \mu g K^+ ml^{-1}$. Pipette 0, 2, 4, 6, 8 and 10 ml of this solution into 100 ml flasks and make up to the mark with ammonium ethanoate solution. These contain 0, 2, 4, 6, 8 and $10 \mu g K^+ ml^{-1}$.

The flame photometer

The components of the photometer are shown in Fig. 7.4. Solution containing K^+ is sprayed into a gas–air mixture. In the flame the potassium ions are first converted into atoms. These gain energy from the flame and electrons are temporarily moved out into a higher energy orbit around the nucleus, a state known as 'excited' (Fig. 7.5). On falling back to their original orbit the stored energy is released (emitted) as light with a wavelength which is characteristic of the energy change. In simple terms, the colour of the light varies for different elements, a property which is used in

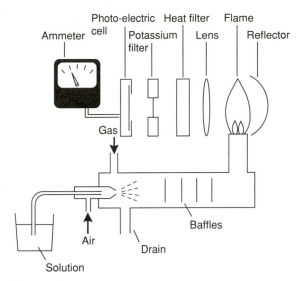

Figure 7.4 The components of a flame photometer.

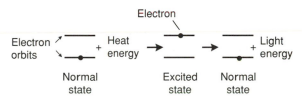

Figure 7.5 The emission of light in a flame photometer.

qualitative flame tests (Nuffield Advanced Chemistry I, Topic 4).

The amount of light emitted depends on flame conditions and on the rate at which K^+ enters the flame. Thus for a standard setting of the photometer and uniform flow rates, the emitted light is proportional to the concentration of K^+ in the solution. This light is selected from that emitted by other elements by a coloured glass filter so that it can be measured using a photoelectric cell and milliammeter. Only certain elements emit sufficient light to be measured in this way (K^+, Ca^{2+}, Na^+, Li^+) but for these the method is sensitive and simple in operation.

Measurement

Following the manufacturer's instructions and with a K^+ filter in place set the milliammeter to read zero on the zero K^+ standard, and to full scale deflection with $10 \mu g K^+ ml^{-1}$. Read all the standards. Spray into the flame the ammonium ethanoate extract (Note 1). Check the stability of the instrument by spraying the zero and maximum standard again, and remeasure if necessary.

Plot the calibration graph (Fig. 7.6). It is a straight line for small concentrations but may show curvature at

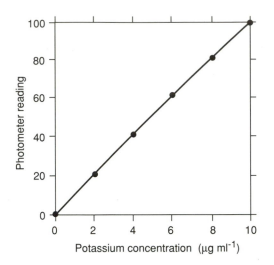

Figure 7.6 A calibration curve for the measurement of potassium by flame photometry.

larger concentrations depending on the characteristics of the photoelectric cell. Read from the curve the concentration of K^+ in the extract.

Calculation

Method summary

air-dry soil ($? \, cmol_c \, K^+ \, kg^{-1}$)
$$\downarrow$$
$$5 \, g \rightarrow 250 \, ml$$
$$\downarrow$$
$y \, (\mu g \, ml^{-1})$ measured by flame photometry

Example The concentration of the extract $y = 4.6 \, \mu g \, ml^{-1}$, and the K^+ present in 250 ml of extract is

$$4.6 \, \mu g \, ml^{-1} \times 250 \, ml = 1150 \, \mu g \text{ or } 1.150 \, mg$$

This came from 5 g of soil. Therefore from 1 kg of soil, $1.150 \times 1000/5 = 230 \, mg \, K^+$ would be extracted. The molar mass of K^+ is $39.1 \, g \, mol^{-1}$ or $39.1 \, mg \, mmol^{-1}$, and so the exchangeable K^+ is

$$230 \, mg \, kg^{-1}/39.1 \, mg \, mmol^{-1} = 5.88 \, mmol \, kg^{-1}$$

The exchangeable K^+ is therefore $0.59 \, cmol \, kg^{-1}$ (10 mmol is 1 cmol). Because K^+ is univalent, $mol_c = mol$ and the final value is expressed as $0.59 \, cmol_c \, kg^{-1}$ air-dry soil.

   This calculation simplifies to

$$cmol_c \, kg^{-1} \text{ air-dry soil} = 0.1283y \times 5/\text{soil mass (g)}.$$

If the air-dry soil contains 5 g H_2O per 100 g oven-dry soil (Section 3.3), then 1 kg air-dry soil contains 952.4 g

oven-dry soil. The exchangeable K^+ is

$$0.59 \times 1000/952.4 = 0.62 \, cmol_c \, kg^{-1} \text{ oven-dry soil}$$

Note 1 The standards and the extracts are normally placed in small beakers to allow them to be sprayed into the photometer. The volume placed in the beaker does not have any bearing on subsequent calculations. The instrument is being set with readings depending on *rate of flow* into the flame. Provided the rate of flow is the same for both standards and extracts, the concentration in the extract is measured in the same units as the standards regardless of the amount of the solution used.

DETERMINATION OF EXCHANGEABLE CALCIUM

The standard method requires an atomic absorption spectrophotometer: technical assistance should be available, and so details are not given here. It is a flame method but relies on absorption of light by calcium in a flame rather than emission. A very wide range of cations can be measured sensitively by this method. The concentration of Ca^{2+} in ammonium ethanoate extracts is between 10 and 250 $\mu g \, ml^{-1}$ and dilution is necessary to reduce the concentration to the range of measurement by atomic absorption ($0–5 \, \mu g \, ml^{-1}$). Prepare standard calcium solutions for calibration containing $0–5 \, \mu g \, Ca^{2+} \, ml^{-1}$. Prepare a releasing agent by dissolving 2.68 g of lanthanum chloride, $LaCl_3.7H_2O$, in water and diluting to 100 ml. To 20 ml of each standard and extract add 1 ml of lanthanum chloride solution and mix before measurement.

   Two other methods can be used: flame photometry and direct titration.

1. Calcium by flame photometry

The technique is less sensitive than for potassium and subject to interference but is adequate for many purposes.

Reagents

Standard calcium solutions, 0–250 $\mu g \, Ca^{2+} \, ml^{-1}$. These are prepared from a volumetric standard containing 1 mg $Ca^{2+} \, ml^{-1}$ available commercially or prepared as follows: dry anhydrous calcium nitrate $Ca(NO_3)_2$ at 105 °C for 1 h and cool in a desiccator. Dissolve 2.05 g in ammonium ethanoate solution in a 100 ml beaker and add 1 ml of HCl (approx. 36% *m/m* HCl) as a preservative. Transfer with washings to a 500 ml volumetric flask and make up to the mark with ammonium ethanoate. Pipette 0, 5, 10, 15, 20 and 25 ml into 100 ml volumetric flasks and make up to the mark with ammonium ethanoate. These contain 0, 50, 100, 150, 200 and 250 $\mu g \, Ca^{2+} \, ml^{-1}$.

Measurement

Calibrate the flame photometer using the calcium standards and a calcium filter and measure the concentration of Ca^{2+} in the soil extract.

Calculation

Follow that described for potassium, but note that the molar mass of Ca^{2+} is $40.1 \, g \, mol^{-1}$, and that 1 cmol Ca^{2+} contains 2 $cmol_c$ Ca^{2+}.

Therefore multiply cmol Ca^{2+} kg^{-1} by 2 to obtain $cmol_c$ Ca^{2+} kg^{-1}.

The calculation simplifies to

$$cmol_c \, kg^{-1} \text{ air-dry soil} = 0.249y \times 5/\text{soil mass (g)}.$$

where y is the concentration of Ca^{2+} in $\mu g \, ml^{-1}$ in the soil extract.

2. Calcium plus magnesium by titration

The method involves chelation of the cations with ethylenediaminetetra-acetic acid (abbreviated EDTA), and is the same as that used for water hardness determinations. The usual procedure is to determine Ca^{2+} and Mg^{2+} together using solochrome black indicator.

The principle

In alkaline solution solochrome black forms a red complex with Ca^{2+} and Mg^{2+}. EDTA has a stronger complexing power for these ions and so takes them from the solochrome black. Removal of all the ions in this way converts the indicator to its non-complexed form which is blue.

Reagents

Ethylenediaminetetra-acetic acid, disodium salt, 0.005 M. The molar mass is $372.24 \, g \, mol^{-1}$. Dry the salt at 105 °C for 1 h and cool in a desiccator. Weigh 1.86 g, dissolve in water in a 250 ml beaker, transfer with washings to a 1 l volumetric flask and make up to the mark.
Buffer solution. Dissolve 17.5 g of ammonium chloride, NH_4Cl, in water in a 250 ml beaker. Transfer with washings into a 250 ml volumetric flask, add 143 ml of ammonia solution (approx. 35% NH_3 *m/m*) and make up to the mark. This solution should be handled in a fume cupboard.
Solochrome black indicator. Dissolve 0.25 g of solochrome black in 190 ml of triethanolamine and 63 ml of ethanol.
Calcium solution, 250 μg Ca^{2+} ml⁻¹. Follow the procedure in the previous section, but make up the solution in water.

Method

The EDTA solution does not need to be standardized but the calcium solution should be used to familiarize yourself with the end point.

Pipette 25 ml into a 250 ml conical flask, *add 2 ml of buffer solution and a few drops of solochrome black. Titrate with 0.005 M EDTA until the colour changes from purple-red to a pale, slightly greenish blue. The end point is when the last trace of pink disappears from the blue colour. The titration should be 15.6 ml.

Titration of the soil extract Pipette 25 ml of the extract into a conical flask and continue from the asterisk above. Acid soils may require only 1–2 ml of EDTA, whereas neutral or alkaline clay soils may require up to 80 ml. Because of this wide range, titrate quickly to obtain an approximate end point, and repeat (using a smaller volume of extract if convenient) to give an accurate end point. The colour change is less clear for soil extracts than for clean solutions: an initial quick titration allows the change to be seen more easily. Add more indicator if required.

Calculation

Method summary

air-dry soil (? $cmol_c$ Ca^{2+} + Mg^{2+} kg^{-1})

$\downarrow$

$5 \, g \rightarrow 250 \, ml$

$\downarrow$

25 ml titrated against y ml of 0.005 M EDTA

The titration reaction is

$$Ca^{2+}(Mg^{2+}) + EDTA^{2-} \, Na_2^+ \rightarrow$$
$$EDTA^{2-} \, Ca^{2+}(Mg^{2+}) + 2Na^+$$

Therefore 1 mol of Ca^{2+} + Mg^{2+} reacts with 1 mol of EDTA Na_2.

Example The volume of EDTA used (y) is 12.5 ml, and so the number of mol of EDTA used in the titration is

$$0.005 \, mol \, l^{-1} \times 12.5/1000 \, l = 6.25 \times 10^{-5} \, mol$$

The amount of Ca^{2+} plus Mg^{2+} is therefore also $6.25 \times 10^{-5} \, mol$. This is in 25 ml of solution, and so in 250 ml there is $6.25 \times 10^{-4} \, mol$. This comes from 5 g of soil, and the amount that would be leached from 1 kg is

$$6.25 \times 10^{-4} \times 1000/5 = 0.125 \, mol \, kg^{-1}, \text{ or}$$
$$12.5 \, cmol \, kg^{-1}$$

For divalent Ca^{2+} and Mg^{2+} this is 25.0 $cmol_c$ kg^{-1} air-dry soil.

This calculation simplifies to

$$cmol_c \, kg^{-1} \text{ air-dry soil} = 10y \times 5/\text{soil mass (g)}.$$

Express the result in terms of oven-dry soil. (Example: 5 g H_2O per 100 g oven-dry soil, 26.2 cmol$_c$ kg^{-1} oven-dry soil.)

3. Calcium by titration

If required, Ca^{2+} can be determined alone by a similar titration procedure using murexide indicator. Calcium but not magnesium complexes with this indicator, and the colour change occurs when EDTA has removed the Ca^{2+} from the murexide.

Reagent

Murexide indicator. Shake 0.5 g of murexide in 100 ml of water for 30 min. Filter through a Whatman No. 1 paper. Prepare fresh for use.

Method and calculation

Follow the method given for Ca^{2+} + Mg^{2+} but use murexide instead of solochrome black.

Determination of exchangeable magnesium

The standard method is by atomic absorption. The concentration of Mg^{2+} in ammonium ethanoate extracts is normally between 2 and 50 μg ml^{-1} (between 0.5 and 12 cmol$_c$ kg^{-1}), and dilution is necessary to reduce the concentration to the normal range of measurement (0–1 μg ml^{-1}). Follow the method for calcium using lanthanum chloride as a releasing agent.

The titration methods described for Ca^{2+} and Mg^{2+} and for Ca^{2+} alone allow exchangeable Mg^{2+} to be determined by difference but with poor precision.

Determination of cation exchange capacity

Ammonium in the KCl extract is determined by steam distillation and titration following the method for organic N in Section 3.6.

Method

Pipette 25 ml of the KCl extract (Note 2) into the distillation unit. Distil and titrate against 0.01 M HCl.

Calculation

Method summary

air-dry soil (? cmol$_c$ NH$_4^+$ kg^{-1})
↓
  5g → 250ml
    ↓
      25 ml distilled and titrated against *y* ml of
                                    0.01 M HCl

The titration reaction is

$$NH_4OH + HCl = NH_4Cl + H_2O$$

In this reaction 1 mol HCl reacts with 1 mol NH_4OH.

Example The titration volume (y) = 15.50 ml, and so the amount of HCl used is

$$0.01\ mol\ l^{-1} \times 15.50/1000\ l = 1.55 \times 10^{-4}\ mol$$

Therefore the amount of NH_4^+ is also 1.55×10^{-4} mol. This must have been present in 25 ml of extract and so in 250 ml of extract there was 1.55×10^{-3} mol. This came from 5 g of soil. Therefore from 1 kg of soil the amount of NH_4^+ leached would have been

$$1.55 \times 10^{-3} \times 1000/5 = 0.310\ mol\ kg^{-1}$$

which is 31.0 cmol kg^{-1} or 31.0 cmol$_c$ kg^{-1} air-dry soil (univalent NH_4^+).

This calculation simplifies to

$$cmol_c\ kg^{-1}\ air\text{-}dry\ soil = 2y \times 5/soil\ mass\ (g).$$

Express the result in terms of oven-dry soil (Example: 6 g H_2O per 100 g oven-dry soil, 32.9 cmol$_c$ kg^{-1} oven-dry soil.)

Note 2 The volume pipetted into the distillation unit may have to be adjusted for soils with large or small CEC values. The calculation above shows that using 25 ml, the titration is numerically half the CEC value. The range of CEC is normally between 2 and 60 cmol$_c$ kg^{-1}. Thus for small values increase the volume to 50 ml, and for large values reduce the volume to 10 ml. Adjust the calculation.

Section 7.3 The measurement of exchangeable cations and effective cation exchange capacity in acidic soils

Acidic soils contain exchangeable hydrogen and aluminium. These are displaced along with other cations using an unbuffered salt solution (1 M KCl) which passes through the soil without changing the soil pH appreciably. H^+ and Al^{3+} are determined by titration and Ca^{2+} and Mg^{2+} by the methods of Section 7.2. K^+ present in relatively small amounts is not measured. The CEC is determined from the sum of exchangeable cations displaced. It is less than the CEC at pH 7 (which could be determined using the ammonium ethanoate method) because of the variable charge on humus and sesquioxides. For this reason it is termed the effective cation exchange capacity (ECEC). It is the exchange capacity which is effective at the natural pH of the soil.

If the ECEC is required, but the amounts of individual exchangeable cations are not of interest, then the method in Section 7.6 can be used.

Method

Reagents and equipment

Potassium chloride, 1 M. Dissolve 74.6 g of KCl in 1 l of water.

Phenolphthalein indicator. Dissolve 0.1 g phenolphthalein in 100 ml of ethanol (95% *m/m*).

Sodium fluoride solution. Dissolve 40 g of NaF in water and make up to 1 l.

Hydrochloric acid, 0.01 M.

Sodium hydroxide, 0.01 M.

pH meter, an alternative to the indicator.

Preparation of the extracts

Follow the method of Section 7.2 modified as follows. Place 10 g of soil and 30 ml of 1 M KCl in a beaker overnight. Leach with successive 10 ml volumes of KCl into a 100 ml volumetric flask. Make up to the mark. Ethanol and acidified KCl leachings are not required.

Exchangeable calcium and magnesium

Follow the methods of Section 7.2. The amounts are small in acid soils, but the ratio of soil to solution has been increased. The calculation must be adjusted to take account of the soil mass and solution volume used.

Exchangeable hydrogen and aluminium

Pipette 50 ml of the soil extract into a 250 ml conical flask. Add five drops of indicator. Titrate to the appearance of a pink colour or to pH 6.8 using 0.01 M NaOH. This measures the exchangeable $H^+ + Al^{3+}$ (Note 1).

Either: Bring the solution back to colourless by adding one drop of 0.01 M HCl. Add 10 ml of NaF solution. Titrate with 0.01 M HCl until the pink colour disappears (Note 1). This measures the exchangeable Al^{3+}.

Or: Omit the drop of HCl, add the NaF and titrate back to pH 6.8.

Principles and calculations

Exchangeable $H^+ + Al^{3+}$ = exchangeable acidity

These titrate against NaOH as follows (Note 2)

$$H^+ + NaOH = H_2O + Na^+$$
$$Al^{3+}_{(aq)} + 3NaOH_{(aq)} = Al(OH)_{3(s)} + 3Na^+_{(aq)}$$

In the first equation 1 mol of NaOH reacts with 1 mol of

H^+. One mol of NaOH therefore reacts with 1 mol_c of H^+. In the second equation 3 mol of NaOH reacts with 1 mol of Al^{3+} which contains 3 mol_c of Al^{3+}: thus 1 mol of NaOH reacts with 1 mol_c of Al^{3+}. The two reactions occur together so 1 mol of NaOH reacts with 1 mol_c of $H^+ + Al^{3+}$.

Method summary

air-dry soil ($? \, cmol_c \, H^+ + Al^{3+} \, kg^{-1}$)

$\downarrow$

$\quad 10\,g \rightarrow 100\,ml$

$\qquad \downarrow$

$\qquad$ 50 ml titrated against y ml of 0.01 M NaOH

Example The titration volume $y = 10.5$ ml and so the amount of NaOH used is

$$0.01 \, mol \, l^{-1} \times 10.5/1000 \, l = 1.05 \times 10^{-4} \, mol$$

This reacts with $1.05 \times 10^{-4} \, mol_c \, H^+ + Al^{3+}$ which was present in 50 ml of extract. Therefore in 100 ml of extract there was $2.10 \times 10^{-4} \, mol_c \, H^+ + Al^{3+}$. This came from 10 g of soil and so the amount that would be leached from 1 kg is

$$2.10 \times 10^{-4} \times 1000/10 = 2.1 \times 10^{-2} \, mol_c \, H^+ + Al^{3+}$$

The exchangeable acidity = $2.1 \, cmol_c \, H^+ + Al^{3+} \, kg^{-1}$ air-dry soil.

This calculation simplifies to

$$cmol_c \, kg^{-1} \text{ air-dry soil} = 0.2y \times 10/\text{soil mass (g)}.$$

Express the result in terms of oven-dry soil. (Example: 7 g H_2O per 100 g oven-dry soil, 2.2 $cmol_c \, H^+ + Al^{3+} \, kg^{-1}$ oven-dry soil.)

Exchangeable Al^{3+}

On completion of the first titration the addition of NaF causes the following reaction:

$$Al(OH)_{3(s)} + 3NaF_{(aq)} = 3NaOH_{(aq)} + AlF_{3(s)}$$

The released NaOH is then titrated against HCl (Note 2).

$$NaOH + HCl = NaCl + H_2O$$

Three mol of HCl reacts with the NaOH released by 1 mol of $Al(OH)_3$. This contains the aluminium which was originally exchangeable. Thus 1 mol of HCl reacts with the NaOH released from 1 $mol_c \, Al^{3+}$. The method summary is the same as that for $H^+ + Al^{3+}$ except that the final titration is z ml of 0.01 M HCl.

Example The titration volume $z = 5.3$ ml and so the number of mol of HCl used is

$$0.01 \, mol \, l^{-1} \times 5.3/1000 \, l = 5.3 \times 10^{-5}$$

which is also the number of mol of NaOH in 50 ml.

Therefore in 100 ml there were 1.06×10^{-4} mol NaOH which was released from 1.06×10^{-4} mol$_c$ Al^{3+}. The exchangeable Al^{3+} is

$$1.06 \times 10^{-2} \text{ mol}_c \text{ kg}^{-1} = 1.1 \text{ cmol}_c \text{ kg}^{-1} \text{ air-dry soil}$$

This calculation simplifies to

$$\text{cmol}_c \text{ kg}^{-1} \text{ air-dry soil} = 0.2z \times 10/\text{soil mass (g)}.$$

Express the result in terms of oven-dry soil.

Exchangeable H$^+$

This is determined by difference (exchangeable H$^+$ + Al^{3+} minus exchangeable Al^{3+}).

Note 1 Using the indicator the end points are clear but not stable. This reflects slow changes in the form of the aluminium hydroxide. Titrate until the colour is stable for 15 s. The uncertainty of the end point may lead to an overestimate of the amount of exchangeable Al^{3+} by about 0.2 cmol$_c$ kg^{-1} especially if the titration is done very slowly. The precision of the end points is much improved by using a pH meter.

Titration is the only method available for exchangeable acidity but aluminium can be determined colorimetrically or by atomic absorption. Between 0 and 5 cmol$_c$ Al^{3+} kg^{-1} is the normal range in soils giving 0–45 μg Al^{3+} ml^{-1} in the KCl extract. The calibration range of the atomic absorption spectrophotometer is 0–50 μg ml^{-1} using a nitrous oxide–acetylene flame. A releasing agent is not required.

Note 2 The aluminium displaced by KCl is predominantly Al^{3+} but other aluminium species, e.g. AlOH^{2+}, will be present (Section 8.3). In the titration

$$\text{AlOH}^{2+} + 2\text{NaOH} = \text{Al(OH)}_{3(s)} + 2\text{Na}^+$$

1 mol$_c$ AlOH^{2+} reacts with 1 mol$_c$ NaOH. Thus the calculated amount of exchangeable H$^+$ + Al^{3+} includes other aluminium species, but is correctly expressed in units of charge. After adding NaF, all the aluminium is present as Al(OH)$_3$: the second titration determines aluminium in units of charge but can also be used to calculate the mass of aluminium if required.

Effective cation exchange capacity

This is equal to the sum of exchangeable Ca^{2+} + Mg^{2+} + H$^+$ + Al^{3+} (cmol$_c$ kg^{-1}). There are no problems due to soluble salts in acidic soils because the salts have been leached away in the field. The error due to omission of exchangeable K$^+$ is small (normally less than 0.2 cmol$_c$ kg^{-1}). If the amount of exchangeable K$^+$ is required for other purposes (e.g. as an indicator of availability to crops), it can be determined by the methods of Sections 7.2 or 9.1.

Section 7.4 The measurement of the composition of soil solutions

Solution can be extracted from the pores of a moist soil by two methods: (a) centrifugal force will spin the solution out of the soil, and (b) the soil can be mixed with a liquid which has a high density. Again using a centrifuge the solution floats on top of the liquid provided the two liquids are immiscible (Kinniburgh and Miles, 1983). However, moist soils contain between 10 and 30 per cent water, and the extraction methods only remove about 30 per cent of this. Therefore from a 100 g soil sample between 3 and 10 ml of solution might be obtained.

Many of the standard analytical techniques described in this book are not adequate for these small volumes. The advent of the flame technique known as inductively coupled plasma analysis (abbreviated ICP) has made possible the analysis of many elements simultaneously with small concentrations in small volumes of liquid. Thus our knowledge of soil solutions is expected to grow rapidly in the coming years.

It is possible, however, to use the standard techniques described here if we accept the small changes in soil solution composition that occur when extra water is added to soil. Thus a suspension of soil can be prepared using 100 g of soil and 100 ml of water. This can be filtered to yield about 30 ml of solution (less from heavy-textured soils). For those ions which are well buffered by the soil (H$^+$, K$^+$, Ca^{2+}, Mg^{2+}, H$_2$PO$_4^-$, SO$_4^{2-}$) there is a small dilution by the added water. For Cl$^-$ and NO$_3^-$ in all but the variable charge soils, dilution is in proportion to the amount of water added and the expected concentration in soil solution can be calculated.

Section 14.2 gives details of the preparation of a saturation extract which is the standard method for determining the solution composition in saline soils and can be used for other soils. The ratio of soil to water is about 100 g:50 ml, so causing less dilution of the soil solution. Using either a suspension extract or a saturation extract allows comparisons to be made between soils even though absolute values are not obtained.

The soil solution concentrations (centrifuge extraction) given in Table 7.6 are the median values for a range of soils sampled several times through the year in Oxfordshire. All the soils had pH values near to pH 7 (median = 7.7).

Measurements can be made on solutions extracted from soils as follows:

- *pH* as in Section 8.1. This gives the H$^+$ and OH$^-$ concentration.

Table 7.6 The soil solution composition of soils from Oxfordshire (median values from 24 sites).

Ion	Concentration	
	(mol l^{-1})	(µg ml^{-1})
Na$^+$	4.7×10^{-4}	11
K$^+$	3.9×10^{-4}	15
Mg^{2+}	1.4×10^{-4}	3
Ca^{2+}	2.1×10^{-3}	84
NO$_3^-$	8.6×10^{-4}	12(N)
Cl$^-$	1.6×10^{-3}	57
SO$_4^{2-}$	3.3×10^{-4}	11 (S)
H$_2$PO$_4^-$	6.4×10^{-5}	2 (P)

pH 7.7, range 5.7–8.5.
From Campbell *et al.* (1989).

- *Na$^+$* by the methods of Section 14.1.
- *K$^+$* by the methods of Section 7.2, but using K$^+$ standards made up in water.
- *Ca^{2+}* and *Mg^{2+}* by the methods of Section 7.2.
- *NO$_3^-$* by the methods of Section 11.1. Nitrate in soil solution is often extracted using KCl solution and expressed as mg nitrate−N kg^{-1} soil.
- *Cl$^-$* by the method of Section 7.6.
- *SO$_4^{2-}$* by the method of Section 10.5.
- *H$_2$PO$_4^-$* by the method of Section 10.1.
- *HCO$_3^-$ and CO$_3^{2-}$* by the method of Section 14.1.
- *Al^{3+}* can be measured in acid soil solutions using atomic absorption methods (Section 7.3) or predicted from the pH value, but only in an approximate manner (Section 8.3).

The total concentration of ions in soil solution can be estimated by measuring its electrical conductivity (Section 14.1).

Solution extracted from saline soils may need diluting before measurement (Tables 7.1 and 14.2).

Section 7.5 Surface charge and the binding of anions to sesquioxide and clay surfaces

The development of charge on sesquioxides follows similar principles to that for humus (Section 7.1). As shown in Section 2.2, hydroxyl groups in the surface of iron and aluminium compounds dissociate as pH rises releasing hydrogen ions and becoming negatively charged. A dissociation constant, K_a, is used to describe the extent of dissociation. Similarly at low pH values the groups associate with hydrogen ions to become positively charged and an *association constant, K_b,* is used. For humus, pK_a indicates the pH value at which half of the reactive groups are dissociated. For

sesquioxides, however, it is more useful to indicate the behaviour of the hydroxyl groups through the *point of zero net charge* (abbreviated PZNC). This is the pH value at which the surface carries equal numbers of positive and negative charges, and so the *net charge* (= positive plus negative charge) is zero.

Fig. 7.7 shows the measured charge of pure gibbsite

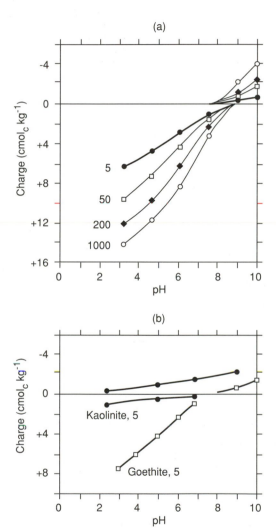

Figure 7.7 The charge on (a) gibbsite, and (b) goethite and kaolinite. The materials used were pure synthetic gibbsite and goethite, and Georgia kaolinite. Data from A. Mashali, Reading. The charge on gibbsite was measured in LiCl solutions: the concentrations in mmol l^{-1} are given against the lines. The charge on goethite and kaolinite was measured in 5 mM NaCl. The larger charge values which have been recorded for other kaolinites are probably due to impurities of 2:1 clays (negative charge) and sesquioxides (positive charge). See also Table 2.3.

and goethite. The two materials have similar amounts of charge and PZNC is about 8.

The effect of soil solution concentration

The data show that at a given pH, charge increases as concentration increases. The theoretical relationship is known: for measurements with monovalent ions, charge is expected to be proportional to the square root of concentration although this is not always found in practice. The effect on negative charge can be explained in a similar manner to that given for humus (Section 7.1), with chemically bound H^+ exchanging for electrostatically bound Na^+.

sesquioxide$-OH + Na^+ \rightleftharpoons$
$$\text{sesquioxide}-O^- \; Na^+ + H^+$$

However, the effect on a positively charged surface cannot be explained by ion exchange. An increased concentration of Cl^- or other anions apparently encourages the sorption of H^+ onto $-OH$ groups, and Cl^- becomes an exchangeable anion.

sesquioxide$-OH + Cl^- + H^+ \rightleftharpoons$
$$\text{sesquioxide}-OH_2^+ \; Cl^-$$

In this sorption reaction both Cl^- and H^+ (low pH) move the reaction to the right (Section 7.1, Note 1). An alternative way of viewing these reactions is that the development of both negative and positive sites is encouraged by increased 'availability' of counter-ions (Na^+ and Cl^- in this case).

The implications are important. The amount of pH-dependent charge in soils fluctuates as soil solution concentration varies during leaching or drying, although buffering of the solution may minimize the changes. When charge is measured, however (Section 7.6), it is essential not only that the pH of the solution in contact with the soil is similar to that in the moist soil but also that the concentration of the solution matches that in the field. For example, if a 1 M solution is used to determine AEC (as for the CEC method) charge is seriously overestimated because soils in the humid tropics have soil solutions which are around 1 mM (Table 7.3). Thus in Fig. 7.7 the 5 mM solution represents soil conditions reasonably well.

The adsorption of anions

Chloride and nitrate are attracted to positively charged surfaces by electrostatic forces. They exchange freely. Other anions in soil solution (phosphate, sulphate, silicate, bicarbonate) form a chemical bond in a similar way to the binding of polyvalent cations to humus. Fig. 7.8 gives a simplified view of the reactions for phosphate. Phosphate is bound either to a single site or very

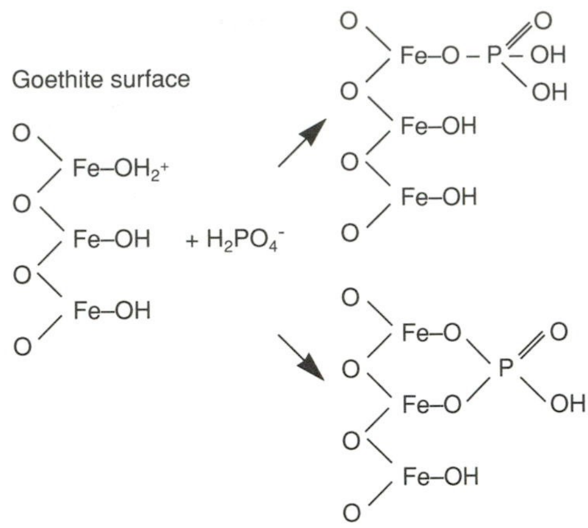

Figure 7.8 The reaction of phosphate with a goethite surface.

strongly to two adjacent sites (Section 15.3, Note 4). It adds its negative charge to the surface although in a more complex manner than shown and the surface becomes less positive (or more negative). Phosphate is very firmly bound at low pH, contributing to low availability to crops (Ch. 10). Its attraction to the surface is so strong that it becomes bound even to negatively charged surfaces in neutral and alkaline conditions although less strongly than in acid soils. Much of the fertilizer-phosphate applied to soils reacts in this way.

Sulphate is attracted to the surface less strongly than phosphate, only becoming bound at pH values below the PZNC. Thus in neutral soils sulphate is free to leach along with nitrate and chloride. Organic anions such as citrate, oxalate and ethanoate are also likely to be bound.

The chemical binding of anions means that sesquioxides in soils are more negative at a given pH than pure materials and as a result the PZNC occurs at lower pH values. In addition, clay minerals and humus contribute negative charge, so that the PZNC of the whole soil is even lower. Examples are shown in Fig. 7.9 for an oxisol from Brazil. The PZNC of the 100–200 cm layer is 6.5, reflecting the characteristics of the sesquioxides, the presence of bound anions and the kaolinite and small amount of organic matter in the soil. The 0–20 cm layer has similar clay content and mineralogy and so the increased amount of organic matter is probably the major factor increasing negative charge and reducing the PZNC to 3.8. The number of sites holding bound anions may also be increased.

Displacement of anions

Although chloride and nitrate exchange freely they will not replace chemically bound anions. However, those

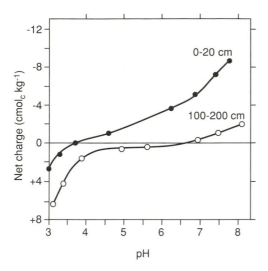

Figure 7.9 The effect of pH on the charge of a Brazilian oxisol. Data from van Raij and Peech (1972). The properties of this soil are given in Table 7.4, soil C. Measurements were made in 1 mM NaCl.

which bind into the surface will exchange with each other depending on their affinities for the surface. Thus a bicarbonate solution is used to displace phosphate in the routine method for determining phosphate availability in soils (Section 10.3), and sulphate can be displaced with a phosphate solution (Section 10.5).

Implications for the determination of AEC

The aim is to measure for the soil in its field condition the 'free' positive charge which holds chloride and nitrate. The value will depend on the characteristics of the sesquioxides including the extent to which anions are bound into the surface. To satisfy this aim, the method should not displace bound anions, otherwise the positive charge will increase (Section 7.6).

Because AEC is affected by soil pH and by bound anions, its value is dependent on soil management, particularly liming and the application of phosphate and sulphate as fertilizers.

Allophane and the 1:1 lattice clays

Allophane is an aluminium silicate which carries a variable charge and so its properties place it with the sesquioxides (Section 2.2 and Table 2.3). Allophane-rich soils, although not so widespread as those containing appreciable amounts of sesquioxides, often have appreciable AEC values.

Kaolinite clay also carries variable charge on the edges of its crystals. Fig. 7.7 shows that the amount of charge is, however, smaller than on sesquioxides and

the PZNC is lower (between 4 and 5). The weathering conditions which favour the formation of sesquioxides also favour kaolinite formation and the charge carried by soils of the humid tropics often depends on the properties of these components. Their intimate association in the clay-sized fraction affects the properties of both surfaces.

Section 7.6 The measurement of anion and cation exchange capacity in variable charge soils

Measurement of anion exchange capacity

Anion exchange capacity varies with pH and the concentration of the solution in the soil. Thus its determination has to be carried out with the extracting solution matched to these conditions. Soil pH should be measured following the methods of Section 8.1. Soil solution concentrations are normally between 0.5 and 2 mM in leached soils of the humid tropics (Table 7.3). A 2 mM solution is taken as a standard condition. If required it can be determined by the methods of Section 7.4.

Reagents and equipment

Ammonium chloride, 2 mM. Dissolve 0.107 g of NH_4Cl in water and make up to 1 l. Adjust to the soil pH using 0.05 M HCl.
Ammonium chloride, 50 mM. Dissolve 2.675 g of NH_4Cl in water and make up to 1 l. Adjust to the soil pH using 0.05 M HCl.
Potassium nitrate, 20 mM. Dissolve 2.022 g of KNO_3 in water and make up to 1 l.
Silver nitrate, 1 mM. Dissolve 0.170 g of $AgNO_3$ in water and make up to 1 l.
Standard potassium chloride, 5 mM. Dissolve 0.3727 g of KCl (dried for 1 h at 105 °C and cooled in a desiccator) in 20 mM KNO_3 and make up to 1 l with this solution.
Nitric acid, 70 per cent HNO_3 m/m. (CARE. Corrosive, handle in a fume cupboard using an Eppendorf pipette, Note 1)
Small leaching tubes. Hypodermic syringes (10 ml) are suitable (Fig. 7.10). Cut discs of nylon cloth (about 60 μm hole size) to the internal diameter of the syringe. Glue (around the edge only) a disc to the base of the tube to retain the soil. Attach about 40 cm of capillary tube (internal diameter 0.5 mm) on the syringe outlet. Place a clip on the tube.
Silver–silver chloride electrode.
Reference electrode (mercury–mercurous sulphate). A calomel electrode cannot be used because it may release chloride.

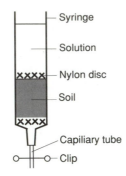

Syringe

Solution

Nylon disc

Soil

Capiliary tube

Clip

Figure 7.10 A syringe leaching tube.

Millivoltmeter. Most pH meters can be set to read millivolts.
Eppendorf pipette or similar to dispense 2 ml.

The leaching method

Weigh a dry syringe with its capillary tube and an extra nylon disc. Weigh 4.0 g ($\pm$0.01) of air-dry <2 mm soil into the syringe. Compress the soil with a suitable rubber bung attached to a rod. Place the extra nylon disc over the soil. Pour 5 ml of 2 mM NH$_4$Cl into the tube. When solution begins to run from the capillary tube, apply the clip and allow to stand overnight (cover the syringe) for thorough wetting of the soil. Leach with successive 5 ml volumes of 2 mM NH$_4$Cl over a period of 3 h using about 50 ml in total (Note 2). This solution runs to waste. Allow drainage to cease.

Weigh the syringe and wet soil to determine the volume of solution held in the soil (assume mass = volume).

Leach the soil with successive 5 ml volumes of 20 mM KNO$_3$ over a period of about 3 h, collecting the leachate in a weighed tube. Reweigh to determine the volume of leachate collected. The 40 cm length of capillary tube should give the correct flow rate. Adjust the length if necessary.

The measurement of chloride by potentiometric titration

Connect the silver–silver chloride, and the reference electrodes to the millivoltmeter following the manufacturer's instructions.

Titration A Pipette 5 ml of 5 mM KCl into a 100 ml beaker on a magnetic stirrer. Add 20 ml of water and 2 ml of nitric acid (70%) using an Eppendorf pipette. Position the electrodes in the beaker and add the stirring rod. Titrate with 1 mM AgNO$_3$ until the potential

is the same as that given by 25 ml of 20 mM KNO$_3$ with 2 ml of HNO$_3$. This standardizes the AgNO$_3$.

Titrations B and C Follow the same procedure using B, 5 ml of 2 mM NH$_4$Cl, and C, 25 ml of the 20 mM KNO$_3$ extract from the syringe, omitting the 20 ml of water.

Calculation

The soil is equilibrated with 2 mM NH$_4$Cl. In this condition the AEC is filled with Cl$^-$ and free NH$_4$Cl solution is present in the syringe. The weighings give the mass of solution held in the soil. All of the chloride is displaced with KNO$_3$ and measured by titration.

Example

	(ml)
Volume of solution held in the soil	2.37
Volume of leachate	49.38
Titrations: volume of 1 mM AgNO$_3$ used for	
A: 5 ml of 5 mM KCl	24.30
B: 5 ml of 2 mM NH$_4$Cl	10.20
C: 25 ml of KNO$_3$ extract	5.65

Steps

1. The concentration of the AgNO$_3$ solution, titration A.

$$KCl_{(aq)} + AgNO_{3(aq)} \rightarrow AgCl_{(s)} + KNO_{3(aq)}$$

Equal numbers of mol of AgNO$_3$ and KCl react together. Therefore

$$5 \times 10^{-3}\,mol\,l^{-1} \times 5/1000\,l = y \times 24.30/1000\,l$$

where y = the molarity of the AgNO$_3$.
Thus $y = 1.029 \times 10^{-3}$ M.

2. The concentration of the NH$_4$Cl solution, titration B.

$$AgNO_{3(aq)} + NH_4Cl_{(aq)} \rightarrow AgCl_{(s)} + NH_4NO_{3(aq)}$$

Following the same principles,

$$1.029 \times 10^{-3}\,mol\,l^{-1} \times 10.20/1000\,l = z \times 5/1000$$

where z is the molarity of the NH$_4$Cl. Thus $z = 2.099 \times 10^{-3}$ M.

3. The amount of Cl$^-$ held in the solution in the soil. The amount of Cl$^-$ in 2.37 ml of this solution is

$$2.099 \times 10^{-3}\,mol\,l^{-1} \times 2.37/1000\,l = 4.975 \times 10^{-6}\,mol$$

4. The amount of Cl$^-$ in the soil extract, titration C. The number of mol of Cl$^-$ in 25 ml of soil extract is equal to the number of mol of AgNO$_3$ used in the titration. This is

$1.029 \times 10^{-3}\,\mathrm{mol\,l^{-1}} \times 5.65/1000\,l =$
$$5.814 \times 10^{-6}\,\mathrm{mol\ Cl^-}$$

In the whole extract there is $5.814 \times 10^{-6} \times 49.38/25 = 1.148 \times 10^{-5}\,\mathrm{mol}$.

5. The amount of exchangeable Cl^-. This is determined as the difference between the Cl^- in the extract and that held in the solution in the soil, which is

$$1.148 \times 10^{-5} - 4.975 \times 10^{-6} = 6.505 \times 10^{-6}\,\mathrm{mol}$$
$$(\text{Note 3})$$

This was in 4 g of soil (use your measured mass of soil) and so the AEC is

$$6.505 \times 10^{-6} \times 1000/4 = 1.626 \times 10^{-3}\,\mathrm{mol\,kg^{-1}}\ \text{or}$$
$$0.16\,\mathrm{cmol_c\,kg^{-1}}\ \text{air-dry soil.}$$

Express the result in terms of oven-dry soil. (Example: 7 g H_2O per 100 g oven-dry soil, $0.17\,\mathrm{cmol_c\,kg^{-1}}$ oven-dry soil.)

Note 1 An Eppendorf pipette allows the safe handling of a standard volume of concentrated HNO_3. If this pipette is not available the concentration of acid can be reduced for safer handling: use 2 ml of 1 M HNO_3, prepared by adding 63 ml of HNO_3 (70% *m/m*) to water and making up to 1 l.

Note 2 The amount of Cl^- in the NH_4Cl leaching solution must be adequate to fill the AEC with Cl^-. A larger volume could be used but sulphate is displaced and the measured AEC increases. The amount of Cl^- in 50 ml of 2 mM NH_4Cl is 10^{-4} mol. In 4 g of soil with an AEC of $1\,\mathrm{cmol_c\,kg^{-1}}$ the required Cl^- is 4×10^{-5} mol. Thus 50 ml is about the minimum possible for this purpose.

Note 3 In this soil the amount of Cl^- in the solution in the soil is about half of that appearing in the extract. As the AEC becomes smaller this proportion increases, and errors increase because of the small difference between the two amounts.

The measurement of the effective cation exchange capacity

Section 7.3 describes the determination of ECEC by the 'sum of exchangeable cations' method. If the amounts of each of the exchangeable cations are not required an adapted AEC method can be used. During the NH_4Cl leaching the negative charge is satisfied by NH_4^+ at a solution concentration and pH which match those of the soil in the field and is measured after displacement.

Method

Follow the procedure for AEC but include a prelimi-nary leaching of the soil with 25 ml of 50 mM NH_4Cl at the pH of the soil. This ensures that the ECEC (which may be much larger than the AEC) is filled with NH_4^+. Follow this with the 2 mM NH_4Cl at the soil pH and extract with KNO_3.

Determine the concentration of NH_4^+ in 25 ml of both the 2 mM NH_4Cl and the KNO_3 extract by steam distillation and titration against 0.01 M HCl (Section 3.6). The ammonium chloride will require a titration of about 20 ml, and the extract will require about 5 ml for each $1\,\mathrm{cmol_c\,kg^{-1}}$ of exchange capacity.

Calculation steps

1. Calculate the concentration of NH_4^+ in the NH_4Cl solution.
2. Calculate the amount of NH_4^+ in the solution held in the soil.
3. Calculate the total amount of NH_4^+ extracted from the soil in the KNO_3 solution.
4. Calculate the ECEC by difference as $\mathrm{cmol_c\,kg^{-1}}$ oven-dry soil.

Section 7.7 Projects

1. From standard textbooks collect data on CEC, clay content and organic matter content (or loss on ignition) of soils. Plot the relationships between CEC and clay, CEC and organic matter, and clay and organic matter. Fit the best straight line through the points by eye or using linear regression analysis (Section 9.6). Think about the reasons for the scatter of points around the lines.
2. Obtain soils from local areas where you know there are differences in parent material. As far as possible use sites under similar vegetation, and sample both topsoil and subsoil. Measure CEC, exchangeable $Ca^{2+} + Mg^{2+}$, C percentage (or loss on ignition), pH (Section 8.1) and clay percentage (or texture by the finger method which gives a possible range of clay percentage through Fig. 2.6). Attempt to explain the differences between the samples. Use Section 8.2 when considering effects of acidity.
3. On field or garden plots (Section 11.6, Project 4 and Section 13.1) set up an experiment to compare the effects of manures and fertilizers on the Cl^- concentration in soil solution. Suggested application rates are as follows: the Rothamsted Broadbalk Experiment used 35 t of cattle manure ha^{-1} which is $3.5\,\mathrm{kg\,m^{-2}}$, and KCl inputs of 37 kg K^+ ha^{-1} which is $3.7\,\mathrm{g\ K^+\,m^{-2}}$ or $7\,\mathrm{g\ KCl\,m^{-2}}$. Spread the manure evenly over the plot and mix into the top few inches. Dissolve the KCl in water (2 l) and apply evenly over the area with a watering can. Apply an equal volume

of water to the manure plots. After a few days sample the topsoil (0–20 cm), extract the soil solution and measure Cl^- concentration using the methods of Sections 7.4 and 7.6. Think about the sampling procedure and required replication (Sections 1.1 and 1.3). Measure the changes resulting from leaching with rain or more applied water.

4. Soils with positive charge are not easily obtainable in temperate regions. They can, however, be simulated using a mixture of sandy soil and anion exchange resin. A suitable resin is Dowex 1–XB (available from BDH Ltd). It has an AEC of about 350 cmol$_c$ kg^{-1} of moist resin and contains about 43 per cent water. Prepare a mixture of resin and soil such that the AEC of the mixture is about 2 cmol$_c$ kg^{-1}. The mixture should be made from air-dry <2 mm soil and air-dry resin. Thorough mixing of the dry materials is required (shake together in a polythene bag). As supplied the resin has Cl^- on its exchange sites.

Leach a variable charge soil or the soil-resin mixture with solutions containing various ratios of Cl^- and NO_3^- following the method of Section 7.6 and then measure the exchangeable Cl^-. The leaching solutions should be mixtures of NH_4Cl and NH_4NO_3 such that $[Cl^-] + [NO_3^-] = 2$ mM. Include 2 mM Cl^- + 0 mM NO_3^-.

Plot the ratio of exchangeable Cl^-/exchangeable NO_3^- against the solution concentration ratio, $[Cl^-]/[NO_3^-]$. The slope of the line is the exchange coefficient, G, for these two ions, and the relationship is described by the equation

$$\text{exchangeable } Cl^-/\text{exchangeable } NO_3^- = G[Cl^-]/[NO_3^-]$$

If $G = 1$ then the sites have an equal affinity for the ions. This project is extended in Section 11.6, Project 5.

Section 7.8 Calculations

1. A soil contains 15 per cent smectite and 3 per cent organic matter. Assuming that the surfaces act independently, calculate possible values for the CEC of the soil at pH 6.5 using the data in Table 2.3 and Fig. 7.2. (*Ans.* 15.8–18.1 cmol$_c$ kg^{-1})

2. A soil from the humid tropics has 5 per cent goethite, 20 per cent kaolinite, 2 per cent organic matter and a pH of 4. Assuming that the surfaces act independently, calculate possible values for AEC and CEC using the data in Figs. 7.2 (humus) and 7.7 (goethite and kaolinite). Compare your values to those in Table 7.3. (*Ans.* AEC 0.5, CEC 0.1 cmol$_c$ kg^{-1})

3. To the Malaysian topsoil in Table 7.3 imagine that sulphate is added as a fertilizer at a rate of 60 kg S ha^{-1} (2000 t soil ha^{-1}) and is adsorbed by the sesquioxides such that 1 mol$_c$ of adsorbed SO_4^{2-} reduces the positive charge by 1 mol$_c$. Calculate the new AEC. The molar mass of S is 32.1 g mol^{-1} (*Ans.* 0.29 cmol$_c$ kg^{-1}) It is more likely that charge is only reduced by about 0.5 mol$_c$ for each mol$_c$ of SO_4^{2-} adsorbed.

4. Table 7.6 gives the average composition of soil solutions from neutral soils in Oxfordshire. Calculate the total anion and cation charge. (*Ans.* 3.2 and 5.3 mmol$_c$ l^{-1}). Which important ions have been omitted?

5. Lime is applied to acid soils in amounts up to 10 t $CaCO_3$ ha^{-1}, and potassium fertilizer in amounts up to 60 kg K$^+$ ha^{-1}. Convert these amounts into cmol$_c$ kg^{-1} assuming that there are 2000 t soil ha^{-1}. If a soil contains 10 cmol$_c$ Ca^{2+} kg^{-1} and 0.5 cmol$_c$ K$^+$ kg^{-1}, what is the percentage increase in these amounts if all the added Ca^{2+} and K$^+$ becomes exchangeable? (*Ans.* 100 and 15)

If the K$^+$ is added as KCl, and all the Cl$^-$ remains in soil solution, what is the increase in Cl$^-$ concentration in the soil solution if the moist soil contains 30 per cent water m/m? (*Ans.* 2.6 mM) Compare your result to that in Section 9.4. Molar masses are in Appendix 2.

6. A soil has an AEC of 1.0 cmol$_c$ kg^{-1}, half of which is occupied by NO_3^-. Calculate the amount of NO_3^-–N present ha^{-1}$_{(2000 t)}$. The molar mass of N = 14 g mol^{-1}. (*Ans.* 140 kg)

7. Imagine that pure humus has 65 cmol carboxylic groups kg^{-1}, all having p$K_a = 4.0$. Calculate the fraction of the groups which are dissociated at pH 3, 4, 5 and 6 using Eq. 7.2. From the fractions calculate the charge associated with the humus at each pH. (*Ans.* 5.9, 32.5, 59.1, 64.4 cmol$_c$ kg^{-1}) Plot the data. Compare your graph to line I in Fig. 7.2. Now assume that one quarter of the groups have a p$K_a = 3.0$ and three quarters have a p$K_a = 5.5$. Recalculate and plot charge against pH. This is one of the procedures adopted in modelling charge–pH relationships. Adjust the pK_a values, the number of groups in each fraction and the number of fractions to give the best fit to the data.

8. The upper limit recommended for copper additions to soils from sewage wastes in Britain is 135 mg kg^{-1} of soil. If a soil contained 3 per cent humus with a negative charge of 60 cmol$_c$ kg^{-1} humus, and all the copper became strongly bound to the humus by chelation, what percentage of the humus charge would be neutralized? The molar mass of copper = 63.5 g mol^{-1}, and 1 mol Cu^{2+} carries 2 mol charge. (*Ans.* 24).

CHAPTER 8

Soil Acidity and Alkalinity

Climate has a dominating influence on soil properties. Differences between soil profile characteristics in the various major climatic regions were the initial stimulus to the study of pedology and to the development of systems of soil classification. Rainfall and temperature control the intensity of leaching and the weathering of soil minerals (Section 2.1), thus having a major influence on the chemical properties of soils particularly, acidity, alkalinity and salinity.

Acidity is associated with leached soils, whereas alkalinity occurs predominantly in drier regions. Within a given region the extent to which soils become acidic depends on the inputs of acidity from vegetation, the microbial biomass and the atmosphere and the ability of the primary minerals to resist the acidifying effects of leaching. The development of alkalinity also depends on local parent materials, vegetation and hydrology.

Our management of soil alters the balance of these factors. Thus liming reverses the natural trend towards soil acidity whereas conifer plantations may increase the rate of acidification. Irrigation of soils in arid regions may if carefully managed reduce the tendency to alkalinity, but history shows that this has often not been the case. On a broader scale, atmospheric pollution increases inputs of acidity and the possibility of change in climate due to the greenhouse effect also has implications in terms of acidification.

obtained is primarily related to the solution pH. However, hydrogen ions are also present on cation exchange sites and have an effect on the measurement. Also as soils become more acidic (pH $7 \rightarrow 3$), there are associated changes in the following properties:

- The amounts of exchangeable Ca^{2+} and Mg^{2+} decrease. These together with exchangeable K^+, Na^{2+} and NH_4^+ are known as the *basic* cations: their total amount is often expressed as a percentage of the CEC which is termed the *percentage base saturation* (Section 8.2).
- The amount of exchangeable Al^{3+} increases and is often expressed as the *percentage aluminium saturation* of the ECEC (Section 8.2).
- The negative charge on humus decreases and the positive charge on sesquioxides increases (Sections 7.1 and 7.5).
- The availability of plant nutrients is changed. For example, phosphate solubility is reduced (Ch. 10).
- The availability of toxic elements is changed. For example, aluminium and manganese become more soluble in acid soils (Section 8.3).
- The activity of many soil organisms is reduced resulting in an accumulation of organic matter, reduced mineralization and a lower availability of N, P and S.

THE NATURE OF SOIL ACIDITY

Acidity strictly defined depends on the concentration of hydrogen ions in a solution. It is measured as a pH value which is $-\log[H^+]$ where $[H^+]$ is the hydrogen ion concentration (strictly its activity, Section 7.1, Note 2) in $mol\,l^{-1}$. In acid solutions pH is less than 7, neutral solutions have a pH of 7 and in alkaline solutions pH is greater than 7. Section 8.1 explains the terms involved. It should be noted that when applied to soils, 'neutral' is given a slightly different meaning, being a range from about pH 6.5 to 7.

Soil acidity involves more than just the pH of the soil solution. This is still the main principle and the measurement of soil pH (Section 8.1) is normally made in a suspension of soil in water such that the value

THE DEVELOPMENT OF SOIL ACIDITY

In pure water the concentration of H^+ ions is $10^{-7}\,mol\,l^{-1}$ and the pH is 7. When in contact with the atmospheric concentration of CO_2 a dilute carbonic acid solution is formed with a pH of 5.6. Distilled or deionized water in the laboratory therefore has a pH of about 5.6. For the pH to differ from this value some other acid or base must be present. Thus 'acid rain' contains nitric and sulphuric acid dissolved from the atmosphere (or ammonia and oxides of N and S which can form these acids). Its pH is below 5.6; the average pH of rain over eastern Britain is about 4.4 (DOE, 1990). Even in unpolluted air rain picks up small amounts of naturally occurring acid and has a pH of about 5. Ammonia and oxides of N and S are also

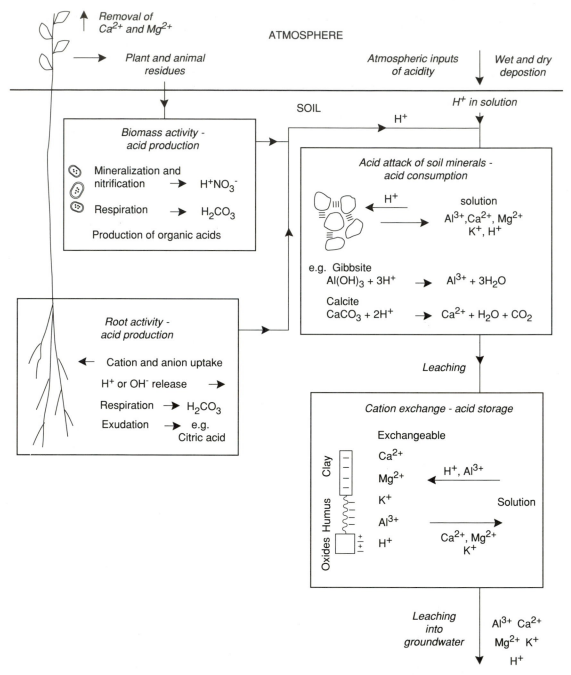

Figure 8.1 The process of soil acidification. Notes: (1) Acidity production by nitrification is increased by inputs of ammonium fertilizers; (2) respiration is only an acidifying process above about pH 5. Below this value the dissociation of H_2CO_3 to release H^+ is small; (3) calcite can be a naturally occurring soil mineral or is added as lime.

deposited dry on vegetation and soil and are washed into the soil by rain where they produce acidity. Thus the atmosphere is an external source of acidity (Fig. 8.1).

Soils also have internal sources of acidity:

- Root and microbial respiration produce CO_2 which dissolves to produce carbonic acid in soil solution (Section 8.1). This weak acid is significantly dissociated only above pH 5, and in neutral and alkaline soils becomes a major source of H^+.

- H^+ is released during the decomposition of soil organic matter as the result of mineralization, nitrification and leaching.
- Organic acids are released from vegetation, soil organic matter and plant roots.
- Roots release H^+ or OH^- to maintain electrical neutrality at their surfaces during the uptake of nutrient ions. Thus they can be a source of either acid or base.
- Pedogenic minerals are normally acidic, releasing H^+ during their weathering by soil water.

Not all the acidity entering or produced in soil remains in soil solution because of reactions which can be considered as acidity 'sinks':

- Many inherited minerals are basic, reacting with water to give a pH near to or above 7.
- Ion exchange reactions on clays, sesquioxides and organic matter remove H^+ from solution in exchange for basic cations.
- Leaching removes H^+ from the soil.

Management of soils alters the balance between sources and sinks, the principal effects being as follows:

- The clearing of natural vegetation causes accelerated organic matter decomposition, nitrate leaching and acidity production.
- The nitrification of ammonium fertilizers produces H^+.
- Lime is a basic mineral which reacts with acidity.

Because of the presence of acidity sinks in soils, the decrease in pH caused by a given input of H^+ is less than it would be if the same amount of H^+ was added to water. This is known as buffering (Section 8.4), a term already used in Ch. 7 for the control of ion concentrations in solution by ion exchange. Soils are buffered to different extents depending on their composition. For example, a calcareous soil holds the pH at about 8 as long as calcium carbonate ($CaCO_3$) is present to be dissolved (Section 8.3) and clay and organic matter buffer the solution through H^+ exchange and adsorption reactions. Thus soils vary in their sensitivity to inputs of acid or base with sandy soils being much more prone to the development of acidity than heavy-textured soils. Consideration of the damaging effects of acid rain takes this into account and has led to the concept of a *critical load* of acidity which is the quantity of acidity which can fall on a given area per year without producing adverse effects: the critical load is based primarily on buffer capacity (DOE, 1991). Buffering is also important in agriculture, with sandy soils requiring less lime to raise pH than heavy-textured soils (Section 8.5).

The buffering of soil acidity is measured as the *buffer capacity* which is the amount of acid (or base) required to lower (or raise) the pH by one unit kg^{-1} of soil (Section 8.4). The buffering of nutrient ions in soil solution is measured as the *buffer power* (Section 9.4).

One further aspect of the soil system has to be considered in relation to the development of acidity. Any input of H^+ causes reactions to occur which release products. Further reaction is restricted unless the products are removed. This is a basic chemical principle known as Le Châtelier's principle (Nuffield Advanced Chemistry II, Topic 12). For example, $CaCO_3$ is attacked by acid as follows:

$$CaCO_3 + 2H^+ \rightleftharpoons Ca^{2+} + H_2O + CO_2$$

Unless the Ca^{2+} is removed or the CO_2 can escape, the reaction is restricted. In soil, leaching removes the products of reaction; the effects are seen in Fig. 8.1. Of particular importance is the leaching away of exchangeable Ca^{2+} and Mg^{2+} ions when H^+ and dissolved Al^{3+} take their place. If $Ca^{2+} + Mg^{2+}$ were not removed further inputs of H^+ would have restricted effects. Thus the development of soil acidity requires both acidic inputs and leaching.

Summary

Soils become acidic as a result of atmospheric inputs of H^+ or compounds which produce H^+ together with internal production of H^+. On reaction with soil minerals, the amounts of exchangeable H^+ and Al^{3+} ions are increased, and soluble products are lost by leaching. The pH change depends on the magnitude of the external and internal inputs and the extent to which H^+ reacts with the soil and products are leached. The resistance of soil to change of pH is known as the buffer capacity.

ACIDIFICATION UNDER NATURAL VEGETATION

The buffer capacities of soils between pH 7 and 4 vary depending on texture and organic matter content, with a range from about 10 to 100 mmol $H^+ kg^{-1} pH^{-1}$ (= mg $H^+ kg^{-1} pH^{-1}$). Inputs of acidity are normally given in relation to an area of land. For this purpose, the buffer capacity is between about 25 and 250 kg $H^+ ha^{-1} pH^{-1}$ (1 ha to 20 cm, 2500 t soil).

When soil pH reaches about 4 it tends to stabilize despite continuing inputs of acidity. Minerals dissolve to neutralize some of the acidity; the remainder is either leached down into the parent material where it is neutralized, or flows laterally over hard rocks into streams as acidic water. Thus the above values of buffer capacity do not apply to very acidic soils.

Inputs to soils under natural vegetation have recently been the subject of detailed study because of concern regarding the effects of acid rain on forests, lakes, rivers and soils. Inputs from the atmosphere in Britain have been reviewed by DOE (1990) and are shown in Colour Plates 13(a) and (b).

Internal production is more difficult to measure. Table 8.1 gives examples. In agricultural soils it is large compared to atmospheric inputs but in acidic soils under natural vegetation the latter become more im-

Table 8.1 Examples of acidity inputs and production in soils

	(kg H$^+$ ha^{-1} a^{-1})
(a) Average Danish agricultural soils (Petersen, 1986)	
Atmospheric inputs	1.2
Internal production by:	
respiration	4.5
nitrification	>2.1
nutrient uptake	0.8

Soil pH in H$_2$O varies from an average of 5.75 in the western part of Denmark to 7.15 in the east. Rainfall varies from about 800 to 500 mm a^{-1}, evapotranspiration is 350–450 mm a^{-1} and leaching >400–<100 mm a^{-1}. Average use of fertilizer-N is about 140 kg ha^{-1}, mostly as NH$_4$NO$_3$. Acidity produced by nitrification is calculated from the average leaching loss of NO$_3^-$ (30 kg N ha^{-1}): for NH$_4$NO$_3$, 1 mol of H$^+$ remains in the soil per mol of NO$_3^-$ leached (Reuss and Johnson, 1986).

(b) A soil under Sitka Spruce
Mature forest catchment:

	(kg H$^+$ ha^{-1} a^{-1})
H$^+$ in the rain, pH 4.4, 2717 mm a^{-1}	0.92
Acidity due to atmospheric N inputs and N transformations in the soil†	1.07
Total acidity inputs	1.99
Output of H$^+$ in the streams	0.89
Output of NO$_3^-$ in the streams	11–16 kg N

An adjacent felled area, second year after felling:

H$^+$ in the rain	0.86
Acidity due to atmospheric N inputs and N transformations in the soil	4.90
Total acidity inputs	5.76
Output of H$^+$ from the C horizon	0.82
Output of NO$_3^-$ from the C horizon	72 kg N

† Calculated from the differences in amounts of NH$_4^+$ and NO$_3^-$ entering and leaving the mature forest catchment (Stevens *et al.*, 1989), or entering and leaving the profile in the felled area (Stevens, unpublished).
The catchments are at Beddgelert in N. Wales. The soils are stagnopodzols with an average pH of 3.1–4.1 through the profile. Acidity production due to S inputs and S transformations is small and has been discounted.

portant. Internal production can be changed significantly by changes in land management. In particular, large flushes of acidity are released as a result of decomposition of organic matter and leaching of nitrate when an area is deforested (Table 8.1(b) and Section 13.5) or when grassland is ploughed (Fig. 11.7). The Hubbard Brook Experimental Forest in the USA has given valuable data on changes after deforestation (Likens *et al.*, 1970) and Table 8.1(b) is from a British study described by Stevens *et al.* (1989).

To assess the environmental impact of atmospheric inputs or of changes in land management, *acidity budgets* of the type shown above are needed comprising the atmospheric input, internal production, internal consumption and drainage loss. Drainage loss is normally a small component of the budget, increasing as through-put of water increases. Because of the complexity of the real system, quantitative chemical models are used to predict long-term effects of changes in atmospheric inputs (Tipping and Hurley, 1988).

THE EFFECTS OF ACIDIFICATION

Although the effects are often measured as a change of soil pH, the implications are primarily in relation to an increase of solution- and exchangeable aluminium (Sections 8.2 and 8.3) leading to toxicities in sensitive plants, and a decrease in solution- and exchangeable calcium and magnesium leading to deficiencies, with the ratio of Al^{3+}/Ca^{2+} + Mg^{2+} often controlling the effects (Colour Plate 12). Normally aluminium toxicity is the main factor, with direct effects on plant metabolism, including an interference with the transfer of ions and water through root cell membranes. Roots become shortened and thickened, further affecting the plant's ability to take up water and nutrients, particularly phosphate (Colour Plate 15). Acidity is a primary yield-limiting factor in the humid tropics. It is also a primary factor governing the natural distribution of plants because of their varying sensitivity (Nuffield Advanced Biology II, Ch. 26.5).

The implications of soil acidity in relation to tree decline in forests are still unclear, although increased uptake of aluminium and decreased uptake of calcium and magnesium may be contributing factors (Roberts *et al.*, 1989). The composition of drainage waters leaving the soil influences streams and lakes (Howells and Dalziel, 1992). Again increased concentrations of Al^{3+} ions and increased Al^{3+}/Ca^{2+} + Mg^{2+} ratios seem to be the main problem in relation to fish survival. The aluminium eventually reaches drinking water supplies with possible implications in relation to Alzheimer's disease.

ACIDIFICATION IN AGRICULTURAL SYSTEMS

Agriculture introduces into acidity budgets the effects of fertilizers and livestock. Probably of most significance is the use of N fertilizers, particularly those containing ammonium ions. The nitrification of NH_4^+ to NO_3^- occurs within a few weeks in agricultural soils:

$$NH_4NO_3 + 2O_2 \rightarrow 2NO_3^- + 2H^+ + H_2O.$$

In this reaction 1 kg N (as NH_4^+) releases 0.14 kg H^+. The increased uptake of NO_3^- by the crop causes a release of OH^- to maintain electrical neutrality at the root surface, depending however on the extent to which cation uptake is also increased.

Field measurements in England and Wales between 1965 and 1971 showed that the net acidity produced by the various forms of N fertilizers in use was between one-half and two-thirds of the theoretical value, about $0.08 \, kg \, H^+ \, kg^{-1}$ of NH_4^+-N applied (Gasser, 1973). Because uptake of NO_3^- counteracts acidity production, NO_3^- lost by leaching rather than NH_4^+ applied primarily determines acidification. A simple model shows that for NH_4NO_3 and urea, $0.07 \, kg \, H^+$ is produced $kg^{-1} \, NO_3^-$-N leached and for $(NH_4)_2SO_4$ the same value applies plus $0.06 \, kg \, H^+ \, kg^{-1} \, SO_4^{2-}$-S leached (Reuss and Johnson, 1986). Similarly for KCl, $0.03 \, kg \, H^+$ is produced $kg^{-1} \, Cl^-$ leached if all the K^+ is taken up. Taking the average values for winter wheat in England and Wales (Table 11.1(b)), 61 kg NO_3^--N ha^{-1} is leached with a net production of 4.3 kg $H^+ \, ha^{-1}$, which is about twice the value given for Danish soils in Table 8.1(a) and is similar to the inputs from all other sources. The situation is not likely to be improved by using organic farming systems unless winter leaching losses can be controlled (Section 13.3) because the same amount of acidity remains per kg NO_3^--N leached for both NH_4NO_3-N and mineralized N (Reuss and Johnson, 1986).

If a light-textured soil has a buffer capacity of $100 \, kg \, H^+ \, ha^{-1} \, pH^{-1}$ (Table 8.5) and a total input of $9 \, kg \, H^+ \, ha^{-1} \, a^{-1}$ (Table 8.1(a)) the fall in pH is given by $9/100 = 0.09$ pH units a^{-1}. In heavy-textured soil pH falls more slowly. Examples are given in Fig. 8.2 for soils at Rothamsted and Woburn. The addition of 5 t $CaCO_3 \, ha^{-1}$ increased pH by about 1.1 units, indicating that despite textural differences both soils apparently had buffer capacities of about $4.5 \, t \, CaCO_3 \, ha^{-1} \, pH^{-1}$ ($90 \, kg \, H^+ \, ha^{-1} \, pH^{-1}$). At Rothamsted the lime was applied to the soil surface whereas at Woburn cultivation followed application. The apparent buffer capacity of the Rothamsted soil is lower than expected (Table 8.5). Subsequent acidification occurred at between 0.05 and 0.1 pH units a^{-1}. The lower rate at Rothamsted may reflect a smaller input of acidity,

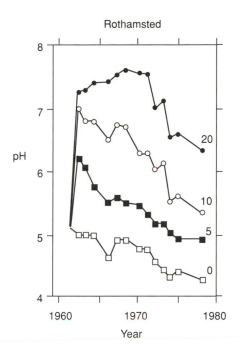

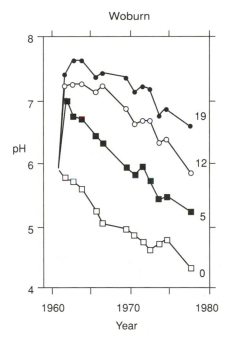

Figure 8.2 The effects of lime and subsequent acidification on the pH of soils. Rothamsted silty clay loam under permanent grass. Woburn sandy loam under an arable rotation. The numbers against the lines are lime applications (t CaCO$_3$ ha^{-1}) made in 1962. Soil samples were taken from 0 to 23 cm. From Goulding *et al.* (1989).

smaller leaching losses due to a larger water-holding capacity or a larger actual buffer capacity.

LIMING PRACTICE

Lime is usually applied as $CaCO_3$. Normally the aim is to raise pH to a target value, but increasingly in the humid tropics the aim is to reduce the percentage aluminium saturation to a satisfactory value (Sanchez, 1976). The target pH value is normally between 6.5 and 7.0 for arable crops in temperate regions but lower in the humid tropics. It is related to the sensitivity of the crop to acidity (Table 8.2), the organic matter content of the soil, the cost of liming and the rate at which soils become acidic. Thus in the humid tropics, the local crops and varieties are more tolerant to acidity, acidification is rapid because of excessive leaching and lime costs are high; the target pH is between 4.5 and 5.5. For grass crops in the hill lands of the west and north of Britain a pH of 5.5–6.0 is the target (tolerant grass species, high rates of acidification) whereas in the east of England, sugar beet and barley are sensitive to acidity, rates of acidification are slow and a target pH of 7 may be economically worth while. This is well above the sensitivity values given in Table 8.2, but reflects the need for (a) the pH of all parts of a spatially variable field to be raised above the sensitivity value and (b) a lime treatment to last for several years during which time pH will be falling.

Table 8.2 (a) The sensitivity of crops grown in temperate regions to acidity expressed as the soil pH below which crop growth is adversely affected (MAFF, 1981)

Crop	pH	Crop	pH
Lucerne (alfalfa)	6.2	Wheat	5.5
Beans	6.0	Oats	5.3
Barley	5.9	Potato	4.9
Sugar beet	5.9	Rye	4.9
Peas	5.9	Wild white clover	4.7
Red clover	5.9	Fescues	4.7
Maize	5.5	Ryegrass	4.7

(b) The tolerance of tropical crops to acidity expressed as a range of aluminium saturation percentages above which growth may be adversely affected (Caudle, 1991)

Crop	Tolerance	Crop	Tolerance
Maize	0–40	Cassava	70–100
Soybean	0–70	Cowpea	40–100
Sorghum	0–70	Bean	0–70
Millet	40–100	Mungbean	0–40
Groundnut	40–70	Wheat	0–70
Rice	40–100	Cotton	0–40

Experience and data from field experiments are needed for the development of a satisfactory system for determining liming requirements. Note that the higher the soil pH, the greater the production of acidity due to the increased dissociation of carbonic acid (Table 8.1) and the fall in pH will be more rapid. It is therefore poor management practice to apply more than the recommended amount of lime.

Lime requirements are normally determined by measuring the buffer capacity of the soil expressed as the amount of $CaCO_3$ required to raise the pH of 1 kg of soil (or 1 ha to a given depth) by 1 pH unit. Buffer capacity multiplied by the required change in pH gives the lime requirement. The measurement of buffer capacity is described in Section 8.4 and routine methods for determining lime requirements are in Section 8.5.

The removal of the direct effect of aluminium on plant metabolism is probably the main factor causing crops to respond to lime, but root growth is improved which allows the crop to take up larger amounts of nutrients and water and the rate of mineralization may be increased. This is an example of interactions between various growth limiting factors (Ch. 13).

ALKALINE SOILS

For a soil to have a pH above 7, it must either be calcareous (contain calcite, $CaCO_3$), dolomitic (contain dolomite, $CaCO_3.MgCO_3$) or sodic (contain Na_2CO_3). Section 8.3 explains the chemistry of calcareous soils. They are very strongly buffered by calcite. A soil containing 5 per cent $CaCO_3$ will neutralize $1 \, g \, H^+ \, kg^{-1}$ ($2500 \, kg \, H^+ \, ha^{-1}$ for the 0–20 cm layer) in dissolving the calcite. Only then will pH fall below 7 and other buffering mechanisms become dominant.

The effects of alkalinity

As soils become more alkaline (pH 7–8.5), the concentration of OH^- increases with associated increases in bicarbonate which becomes the dominant anion (Section 8.3). Carbonate also increases but its concentration is much lower than bicarbonate. There are no significant changes in the exchangeable cations present because the soil is already base saturated at pH 7.

It is not known whether bicarbonate has any direct effect on plant growth. However alkaline conditions cause a range of plant disorders and have an important influence on the natural distribution of plant species. The most striking symptom is chlorosis of leaves of sensitive plants resulting from their inability to utilize iron and manganese (Colour Plates 14(a) and (b)). Yields of tree and bush fruit crops can be reduced by a combination of low solubility of minerals containing iron and manganese, strong binding of the metals on humus, and an interference of bicarbonate with iron and manganese metabolism. Deficiencies of copper and zinc may also occur for similar reasons.

Phosphate deficiencies are common on calcareous soils but can be remedied relatively easily by fertilizer use. Low solubility of phosphate minerals, and uptake reduced by bicarbonate are likely reasons. Plant tolerance varies greatly: wheat, for example, grows normally in fertilized calcareous soils in which tomato growth is seriously inhibited.

Soils containing very large $CaCO_3$ contents, e.g. shallow soils overlying chalk, have a poor ability to hold potassium due to leaching losses (Section 9.5). Similarly, their small content of organic matter reduces the amount of available mineralized N (Section 11.4).

The management of alkaline soils

Where alkalinity is due to the presence of calcite or dolomite, no attempts are made to neutralize these except on a localized scale for horticulture because of the large amounts of acid that would be required. Choice of tolerant crops or varieties together with suitable fertilization and if necessary foliar sprays of micronutrients are the primary management techniques. The management of alkalinity in sodic soils is discussed in Ch. 14.

FURTHER STUDIES

Ideas for projects are given in Section 8.6 and calculations in Section 8.7.

Section 8.1 The meaning of pH, and its measurement

The pH of a solution is defined as $-\log (H^+)$ where (H^+) is the activity of hydrogen ions in the solution. In dilute solutions the activity is approximately equal to the concentration $[H^+]$ in $mol\,l^{-1}$ (Section 7.1, Note 2). Thus for all except saline soil solutions we can write

$$pH \simeq -\log [H^+]$$

Water dissociates to release hydrogen and hydroxyl ions

$$H_2O \rightleftharpoons H^+ + OH^-$$

The dissociation constant K_w is $(H^+)(OH^-)$ and has a value of $10^{-14} (mol\,l^{-1})^2$ at 25 °C. In pure water $(H^+) = (OH^-) = 10^{-7}\,mol\,l^{-1}$ and the pH is 7 which is neutral. An input of acid (as HNO_3 for example) raises the H^+ concentration, causes some association of H^+ and OH^- to form water and leaves the OH^- concentration $<10^{-7}$, $[H^+]>10^{-7}$ and pH < 7. An input of base (as NaOH for example) causes an association of H^+ and OH^- and leaves the concentration of $OH^- > 10^{-7}$, $[H^+] < 10^{-7}$ and pH > 7. The range of pH in soil solutions is from about 3 to 10 $(10^{-3}$–$10^{-10}\,mol\,H^+\,l^{-1})$.

Note that an acid is a substance that is a source of H^+ in solution, whereas a base is a substance that can combine with H^+. Thus HNO_3 is an acid releasing H^+ in solution, and lowering the pH, whereas NaOH is a base, the OH^- combining with H^+ to form water and so raising the pH.

Water will absorb CO_2 from the atmosphere which reacts to form carbonic acid, H_2CO_3. This is a weak acid (Note 1) and dissociates to release H^+, HCO_3^- and CO_3^{2-}:

$$
\begin{aligned}
CO_2 + H_2O &\rightleftharpoons H_2CO_3 \\
H_2CO_3 &\rightleftharpoons HCO_3^- + H^+ \\
HCO_3^- &\rightleftharpoons CO_3^{2-} + H^+
\end{aligned}
$$

Pure water in equilibrium with CO_2 at the atmospheric concentration (0.03% *v/v*) has a pH of 5.6. A soil solution has a pH value lower or higher than 5.6 if the soil is acting as either an acid or a base.

Note 1 A weak acid is only partially dissociated to release H^+. A strong acid is almost completely dissociated. The terms do not apply to the concentration of acid in solution but to the extent to which the acid molecules dissociate (Nuffield Advanced Chemistry II, Topic 12.4). For example, nitric acid is a strong acid and is almost completely dissociated into H^+ and NO_3^-, whereas carbonic acid tends to remain as undissociated H_2CO_3. Thus for the same concentration of acid, nitric acid gives a higher concentration of H^+ and a lower pH than carbonic acid.

The measurement of pH in solution

pH is measured using a glass electrode (Fig. 8.3), a reference electrode (Fig. 6.14) and a pH meter which is a millivoltmeter. When there is a difference of pH between the solution inside and outside the bulb of the glass electrode, a potential difference develops across the glass membrane. This can only be measured relative to another potential and a calomel reference electrode is normally used. The potential of the calomel electrode does not depend on pH. The potential difference between the electrodes is directly proportional to pH, and so the meter can be calibrated to give a pH reading. The two electrodes are often manufactured as a single combination electrode (Fig. 8.3).

Equipment and reagents

The pH meter is calibrated using two buffer solutions normally with pH values of 4 and 7 or of 7 and 9, depending on the values to be measured. These solutions are well buffered, i.e. they have pH values which are stable even if traces of acid or base contaminate them (Nuffield Advanced Chemistry II, Topic 12.5 and Section 8.4). Buffer solutions are prepared from buffer

STANDARD GLASS
ELECTRODE

COMBINATION pH
ELECTRODE

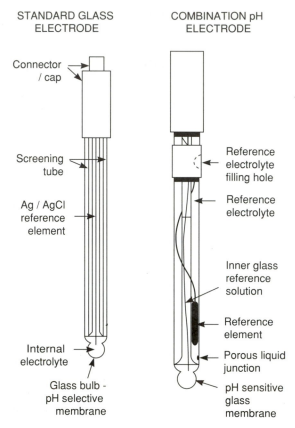

Connector
/ cap

Screening
tube

Ag / AgCl
reference
element

Reference
electrolyte
filling hole

Reference
electrolyte

Inner glass
reference
solution

Reference
element

Internal
electrolyte

Porous liquid
junction

Glass bulb -
pH selective
membrane

pH sensitive
glass
membrane

Figure 8.3 A glass electrode and a combination pH electrode.

tablets dissolved in distilled water following supplier's instructions, or as follows.

Buffer solution, pH 4.0. Dry potassium hydrogen phthalate, $COOHC_6H_4COOK$, at 105 °C for 1 h and cool in a desiccator. Dissolve 10.21 g in water and dilute to 1 l.

Buffer solution, pH 6.9. Dry potassium dihydrogen orthophosphate, KH_2PO_4, and disodium hydrogen orthophosphate, Na_2HPO_4, at 105 °C for 1 h and cool in a desiccator. Dissolve 3.39 g of KH_2PO_4 in water, add 3.53 g of Na_2HPO_4, dissolve and make up to 1 l.

Calibration of the pH meter

The pH values of buffer solutions are temperature dependent and are supplied with the buffer tablets. For routine purposes, the above solutions can be used and variations with temperature neglected. Set the temperature control of the pH meter to the ambient temperature.

Wash the electrodes with water from a wash bottle. Gently dry the electrode with a tissue. Place the electrodes in the pH 4 buffer solution and gently swirl the solution. This reduces the time required for the reading to stabilize. Adjust the meter to pH 4.0 using the buffer control. Remove the electrodes, rinse again in water, dry and insert into the pH 6.9 buffer. When the reading is stable adjust the slope control until the reading is pH 6.9. Repeat these readings until they are correctly set for both buffer solutions.

The pH meter is now calibrated, and can be used to measure the pH in any other solution following the above procedure. If a series of measurements is being made, periodically check that the correct pH reading is obtained with one of the buffer solutions. When not in use keep the bulb of the glass electrode and the lower part of the calomel electrode in water.

The measurement of soil pH

This is normally made using a suspension of soil. Weigh 10 ± 0.1 g air-dry <2 mm soil (Note 2) into a bottle with a screw cap. Add 25 ml of water from a measuring cylinder. Shake for 15 min on a shaking machine, or occasionally by hand over a 15 min period.

Stir the suspension, insert the electrodes, swirl the suspensions over the electrodes, and record the pH after 30 s (Note 3).

Note 2 For routine purposes a 10 ml scoop (Figure 9.5) can be used to dispense the soil. It will contain approximately 10 g air-dry <2 mm soil. Small variations in mass between samples do not alter the pH reading.

Note 3 In neutral and alkaline soil suspensions, the reading is slow to stabilize. For routine measurements, the value after 30 s is accepted. For precise chemical work, a longer period is needed .

The implications of adding water to the soil

Soil solutions are well buffered. When distilled water which has a pH of about 5.6 is added to a soil having a different pH, the pH of the water changes to be close to that of the solution in the moist soil. This is brought about in acid soils by Ca^{2+} and Al^{3+} in soil solution moving on to the exchange sites to displace H^+ in response to the dilution of the soil solution. In calcareous soils pH is controlled primarily by the dissolution of $CaCO_3$ so again the solution is buffered. By using a standard procedure for the measurement of pH, comparisons between soils can be made with confidence even though absolute values are difficult to interpret. Differences in pH caused by adding different volumes of water to soil are known as *suspension effects*.

Measurements can be made in the surface layer of a

moist soil provided sufficient water is present to make liquid contact between the electrodes (Section 6.7).

The implications of measuring soil pH in a salt solution

In an attempt to standardize conditions further, measurements can be made in $0.01\,M$ $CaCl_2$ (or $0.1\,M$ KCl) solution. For acidic soils with a negative charge (Group i and ii soils, Ch. 7) this causes cation exchange with H^+ being displaced into solution. The measured pH is about 0.5 units lower than that of soil in water.

For acidic soils with a net positive charge (variable charge, Group iii soils) the increase in salt concentration causes H^+ to be adsorbed on to reactive sites (Section 7.5). The solution pH therefore increases by up to 0.5 units. These differences in pH are known as *salt effects* and are measured as

$$\Delta pH = \text{soil pH in salt solution} - \text{soil pH in water}$$

where Δ means 'change in'. Δ pH values are positive in soils with a net positive charge and negative in soils with a net negative charge, and its magnitude increases as net charge increases.

For calcareous soils, Ca^{2+} is a *common ion* in the dissolution of $CaCO_3$ (Section 8.3) and reduces the amount of carbonate which dissolves, again depressing pH by about 0.5 units.

Section 8.2 The measurement of percentage base saturation and percentage aluminium saturation

Acidic and basic cations

In neutral and alkaline soils the CEC is satisfied by Ca^{2+} Mg^{2+}, K^+, Na^+ and NH_4^+. As acidity develops, the amounts of these ions decrease, and H^+ and Al^{3+} occupy an increasing proportion of the exchange sites. The need to distinguish between the two groups of ions led to the use of the terms *basic* and *acidic* cations. The latter is a correct use of the term because H^+ is the acidic ion, and Al^{3+} is an acid because it can react with water to produce H^+ and hydroxy aluminium ions, $AlOH^{2+}$ being an example. However, the term basic cations is not strictly correct because they do not accept H^+ or release OH^- on reaction with water. They are almost neutral except for NH_4^+ which is slightly acidic. All that can be said in defence of the term is that basic cations are less acidic than H^+ and Al^{3+}. Historically the term basic cations was used because they appeared to counteract the acidic cations. However, although the term is questionable from a chemical viewpoint it conveniently distinguishes the two sets of cations.

Percentage base saturation

Definition

Percentage base saturation

$$= 100 \times \text{exchangeable } (Ca^{2+} + Mg^{2+} + K^+ + NH_4^+ + Na^+)/CEC$$

Methods

Exchangeable cations and CEC are measured as in Section 7.2. Exchangeable Na^+ and NH_4^+ are normally not measured because of their small concentrations in acidic soils where base saturation is of interest.

Interpretation of results

The above definition of percentage base saturation is superficially unambiguous. However, the following points need to be considered:

1. The property was first measured for soils of temperate regions dominated by permanent charge (Group i soils). The CEC was measured using $1\,M$ ammonium ethanoate buffered at pH 7, abbreviated here CEC_7. When limed to pH 7 these soils were found to be 100 per cent base saturated. With decreasing pH, exchangeable H^+ and Al^{3+} increasingly occupied exchange sites reducing the saturation.

2. For Group i soils, CEC can be considered to be almost independent of pH and concentration of solution provided that organic matter and sesquioxide contents are small (Sections 7.1 and 7.5). Group ii soils, however, carry pH-dependent charge on humus. This causes two problems:
   (a) The ECEC decreases as pH falls. Although the basic cations may be saturating the exchange sites at, for example pH 6, the saturation calculated relative to CEC_7 will be less than 100 per cent. This can be termed a *pH–charge effect*;
   (b) The charge on humus will increase as the ionic concentration in solution increases. The measurement of CEC_7 involves washing the soil with a $1\,M$ solution of ammonium ethanoate, whereas the soil solution is likely to be about $3\,mM$ giving an expected increase in charge. The effects of the subsequent ethanol wash are not well understood. However, the fact that, when expressed relative to CEC_7, soils at pH 7 are found to be 100 per cent base saturated shows that the method also measures (perhaps fortuitously) the ECEC at pH 7.

3. The charge on the minerals in Group iii soils is

dependent both on pH and solution concentration which, together with the charge on humus, leads to the problems outlined above becoming more pronounced. In this case CEC_7 is found to be greater than ECEC at pH 7 which can be termed a *salt–charge effect*. It may be due to the effect of the increased ionic concentration of the ammonium ethanoate or to the adsorption of ethanoate on to the sesquioxide surfaces. Thus even though exchangeable bases saturate the charge at pH 7, the base saturation expressed relative to CEC_7 will be less than 100 per cent.

These points lead to an alternative approach in which the saturation of exchange sites at any pH value can be determined by expressing the amount of basic cations as a percentage of the ECEC. A comparison of these values with those obtained using CEC_7 will indicate the extent of the pH– and salt–charge effects. Examples are given in Table 8.3 and Fig. 8.4. The acidic soils were incubated with varying amounts of $Ca(OH)_2$ to produce the range of pH values and the methods of Sections 8.1, 7.2 and 7.3 were used. The following conclusions can be drawn:

- The gley shows a pH–charge effect only. The lack of a salt-charge effect shows that $CEC_7 = ECEC$.
- The ultisol shows both a pH– and a salt–charge effect.
- Below pH 5.5 the saturation of exchange sites (values relative to ECEC) differs greatly for the two soils. The same effect is seen for percentage aluminium saturations in Fig. 8.5 and is discussed below.

Note that because percentage base saturation can be calculated using both CEC_7 and ECEC values care is needed when using published data to establish which measurement of charge has been used.

The use of percentage base saturation values

Because calcium, magnesium and potassium are nutrients, and aluminium is a toxic element, percentage base saturation has been used as a general indicator of soil fertility. In addition, because soils dominated by variable charge minerals have lower saturations at any given pH than soils dominated by permanent charge minerals, the property is used in the classification of soils in the American system of Soil Taxonomy (Soil Survey Staff, 1975).

The measurement and use of percentage aluminium saturation

Exchangeable aluminium and ECEC are measured by the methods in Section 7.3.

Table 8.3 Percentage base saturations of soils containing variable and permanent charge

pH	Exchangeable bases	Sum of cations = ECEC	Base saturation relative to	
			CEC_7	ECEC
	(cmol$_c$ kg^{-1})		(%)	
(a) Ultisol. Onne, Nigeria				
4.24	0.35	2.39	5.5	14.6
4.87	1.69	2.41	26.5	70.1
5.52	2.89	2.96	45.0	97.6
6.70	3.52	3.53	55.0	99.7
7.55	4.59	4.59	71.9	100.0
		$CEC_7 = 6.38$		
(b) Gley. Reading, England				
3.93	9.74	14.55	39.7	66.9
4.31	12.61	13.25	59.0	95.2
4.76	14.86	15.01	72.4	99.0
5.49	16.80	16.90	81.2	99.5
6.36	18.60	18.70	89.5	99.5
7.03	20.75	20.88	109.4	99.4
		$CEC_7 = 19.52$		

The ultisol (0–20 cm) contained 2.5% organic matter and 15% clay, predominantly kaolinite with a little goethite, and is a Group iii soil. The gley (15–30 cm) contained 6.2% organic matter and 13% clay, predominantly smectite with mica and kaolinite, and is a Group i/ii soil.
Data by A. Bantirgu, A. Dudley, and C. Guest, Reading.

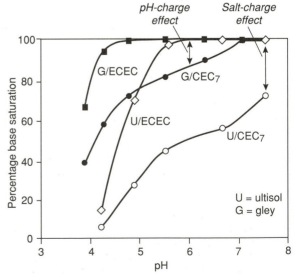

Figure 8.4 The relationship between percentage base saturation and pH. Table 8.3 gives the soil properties.

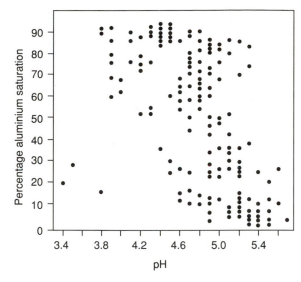

Figure 8.5 The relationship between pH and percentage aluminium saturation for soils from South America, mostly Colombia. Exchangeable aluminium is expressed as a percentage of the ECEC. Data from Dr S. McKean, measured at Centro Internacional de Agricultura Tropical, Colombia. Samples from 0 to 20 cm depth.

Definition

Percentage aluminium saturation =

$$100 \times \text{exchangeable Al}^{3+}/\text{ECEC}$$

In contrast to percentage base saturation, its meaning is clear: it is a measure of Al^{3+} as a percentage of the total cations that are removed by leaching with 1 M KCl solution. However some of the older literature expresses it relative to CEC at pH 7.

The use of data

The relationship between percentage aluminium saturation and pH is shown in Fig. 8.5. The general trend is the reverse of that for base saturation, being low above pH 5.5 because of the effect of pH on the solubility of gibbsite and other minerals (Section 8.3). Humus binds aluminium strongly and reduces the percentage aluminium saturation; this and differences in mineralogy are probably the main reasons for the variability of the data at any given pH.

Aluminium is the predominant toxic element in acidic soils. Its concentration in solution governs its effect on plants, but is difficult to measure. Solution aluminium is buffered by exchangeable aluminium, and percentage aluminium saturation is probably the best index of aluminium toxicity that can be routinely measured. Figs. 8.5 and 8.7 show that pH is unlikely to be as good an index.

Data on the tolerance of tropical crops in terms of aluminium saturation are only slowly becoming available. The range of values for each crop in Table 8.2(b) reflects uncertainties due to soil variability and differences between crop varieties. Acidity management is now being considered in terms of liming to reduce the percentage aluminium saturation to satisfactory levels (Section 8.5), and plant breeding for these regions includes the need to produce varieties tolerant to high aluminium saturations.

Most of the crops grown in temperate regions are sensitive to small amounts of soluble aluminium. Percentage aluminium saturation is no longer a useful index and a pH of between 6 and 7 is the normal target (Section 8.5).

Section 8.3 The chemistry of acidity and alkalinity in soils

SOIL ACIDITY

The properties of aluminium are central to an understanding of soil acidity. Its chemistry is, however, complex and there are difficulties both in understanding and in measuring its properties in soils (Lindsay, 1979).

Aluminium in soil solutions

The predominant minerals which appear to control the solubility of aluminium in mineral soils are gibbsite and kaolinite. In organic soils its ability to complex with reactive sites on the surface of humus is also important.

Gibbsite dissolves to release aluminium and hydroxyl ions into solution:

$$Al(OH)_{3(s)} \rightleftharpoons Al^{3+}_{(aq)} + 3OH^-_{(aq)} \qquad [8.1]$$

The *solubility product* $K_{sp} = (Al^{3+})(OH^-)^3 = 10^{-34}$ (Note 1). Using this equation, the relationship between activity of Al^{3+} and pH can be found, and is shown in Fig. 8.6.

Al^{3+} reacts with water to produce the hydroxy−Al species $AlOH^{2+}$ and $Al(OH)_2^+$ in solution by the following reactions (Note 2):

$$Al^{3+} + H_2O \rightleftharpoons AlOH^{2+} + H^+ \qquad [8.2]$$
$$AlOH^{2+} + H_2O \rightleftharpoons Al(OH)_2^+ + H^+ \qquad [8.3]$$

Their activities are shown in Fig. 8.6. Also present in solution are polymers (hydroxy aluminium species linked into larger units) and aluminium bound to organic molecules. Below pH 4.5, Al^{3+} is the dominant species in mineral soils.

Fig. 8.7 shows measured concentrations of aluminium in acidic soil solutions. The calculated

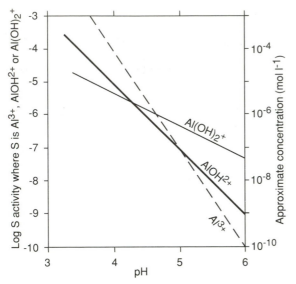

Figure 8.6 The relationship between activity of aluminium species and pH in solution in equilibrium with gibbsite.

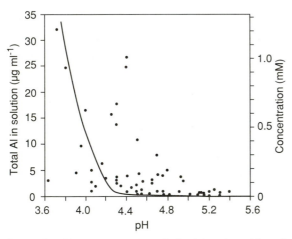

Figure 8.7 The concentration of aluminium in acidic soil solutions. The full line is the theoretical total aluminium concentration in equilibrium with gibbsite. The concentrations (right hand axis) have been calculated assuming that only Al^{3+} was present in solution. From Wild (1988).

concentration of $Al^{3+} + AlOH^{2+} + Al(OH)_2^+$ in equilibrium with gibbsite is also shown. Most of the data lie above the line primarily because polymers and organically complexed aluminium are measured in solution but excluded from the calculations. Little aluminium is present in solution above pH 5.5, and at any given pH a wide range of aluminium concentrations can be found, reflecting the range of exchangeable aluminium already seen in Fig. 8.5. The significance of soluble aluminium in relation to plant growth has been discussed in Section 8.2.

Aluminium on particle surfaces

Exchangeable aluminium is present as Al^{3+}, $AlOH^{2+}$ and $Al(OH_2)^+$. When displaced using 1 M KCl, their determination by titration is related to the charge on the ions (Section 7.3). Atomic absorption or colorimetric methods, however, give a mass of aluminium.

Increase in pH favours the formation in solution and on particle surfaces of hydroxy aluminium polymers which can be considered as intermediates in the formation of gibbsite. Aluminium bound to humus (Section 7.1) is not exchangeable with 1 M KCl. A range of extracting solutions can be used to fractionate soil aluminium (Jarvis, 1986).

Hydrogen on particle surfaces

Exchangeable H^+ is held on permanent charge sites of clay minerals. This together with exchangeable Al^{3+} is known as exchangeable acidity (Section 7.3). Hydrogen present in hydroxyl groups on sesquioxides and kaolinite and in carboxyl groups on humus can be released as pH rises and is neutralized by the added base. Similarly, some aluminium bound to humus is also released and neutralized. Together these are known as *non-exchangeable acidity*. The resulting negatively charged sites hold exchangeable cations and the CEC is increased (Fig. 8.12). When lime is the added base, Ca^{2+} becomes bound on to some of these sites (Section 7.1), so giving a smaller increase in CEC than when NaOH is added (Fig. 7.2). An example of these changes is shown in Table 8.4. An addition of 6.20 cmol$_c$ kg^{-1} of Ca^{2+} increased the exchangeable Ca^{2+} by 3.21 cmol$_c$ kg^{-1}, the remaining 2.99 cmol$_c$ kg^{-1} becoming non-exchangeable. The decrease in exchangeable Mg^{2+} indicates that 0.05 cmol$_c$ kg^{-1} became bound as charge increased. The addition of 6.2 cmol$_c$ kg^{-1} of

Table 8.4 The properties of a Nigerian ultisol treated with lime

Added Ca(OH)$_2$ (cmol$_c$ kg^{-1})	pH	Exchangeable cations (cmol$_c$ kg^{-1})					
		Ca^{2+}	Mg^{2+}	K^+	Al^{3+}	H^+	Sum of cations = ECEC
0	4.24	0.19	0.07	0.09	1.49	0.55	2.39
1.55	4.87	1.52	0.07	0.10	0.43	0.29	2.41
3.10	5.52	2.74	0.06	0.09	0	0.07	2.96
6.20	6.70	3.40	0.02	0.10	0	0.01	3.53

This is the same soil as in Table 8.3(a), where its properties are described.
Data by A. Bantirgu, Reading.

$Ca(OH)_2$ neutralized $2.03 \, cmol_c \, kg^{-1}$ of exchangeable acidity, the remaining $4.17 \, cmol_c \, kg^{-1}$ neutralizing non-exchangeable acidity. The increase in charge was $1.14 \, cmol_c \, kg^{-1}$, showing that $3.03 \, cmol_c \, kg^{-1}$ of the non-exchangeable acidity which was neutralized did not remain as charged sites, this being approximately equal to the $2.99 \, cmol_c \, kg^{-1}$ of bound Ca^{2+} + $0.05 \, cmol_c \, kg^{-1}$ of bound Mg^{2+}.

Note 1 Equation 8.1 is an example of an equilibrium reaction (Section 7.1, Note 1). The gibbsite dissolves to produce a saturated solution. Because there is excess gibbsite present, its amount has no effect on the composition of the solution, and the equilibrium state can be expressed simply through the activities of the products of the reaction. The solubility product is a constant for any salt or mineral. Because soil solutions are normally dilute, activities and concentrations are similar and Fig. 8.6 indicates the approximate concentrations of the aluminium species. The difference between activity, a, and concentration, c, depends on the total ionic concentration of the solution and the valency of the ion. For example, in a typical (non-saline) soil solution, $c \simeq 1.1a$ for monovalent ions, $c \simeq 1.4a$ for divalent ions and $c \simeq 2a$ for trivalent ions. Although at first sight these differences appear to be large, particularly for Al^{3+}, the range of concentrations being considered (a factor of 10^6 in Fig. 8.6) is very large in comparison.

Note 2 It is sometimes said that Al^{3+} is a source of acidity in soils. Equations 8.2 and 8.3 illustrate this, showing that Al^{3+} is an acid. However, it is not satisfactory to isolate these reactions from overall changes in the soil system. The source of the Al^{3+} also has to be considered. Equation 8.1 shows one of its sources and clay mineral dissolution is another. In these reactions H^+ is used to dissolve the minerals. Equation 8.1 can be rewritten as

$$Al(OH)_3 + 3H^+ = Al^{3+} + 3H_2O \qquad [8.4]$$

When Eqs 8.4, 8.2 and 8.3 are coupled together there is a net loss of H^+ from solution; acid is being used to dissolve the Al^{3+} and it is therefore not correct *in this context* to consider Al^{3+} as a source of acidity. However, when pH is raised by liming Al^{3+} does behave as an acid.

SOIL ALKALINITY

Calcareous soils

Calcite has a low solubility, but its reactions with water have a profound effect on soil properties through the control it imposes on soil pH. The chemical reactions are given below:

$$CaCO_{3(s)} \rightleftharpoons Ca^{2+}_{(aq)} + CO_3^{2-}_{(aq)} \qquad [8.5]$$
$$CO_3^{2-}_{(aq)} + H_2O \rightleftharpoons HCO_3^-_{(aq)} + OH^-_{(aq)} \qquad [8.6]$$
$$HCO_3^-_{(aq)} + H_2O \rightleftharpoons H_2CO_{3(aq)} + OH^-_{(aq)} \qquad [8.7]$$
$$H_2CO_{3(aq)} \rightleftharpoons CO_{2(aq)} + H_2O \qquad [8.8]$$
$$CO_{2(aq)} \rightleftharpoons CO_{2(g)} \qquad [8.9]$$

Equation 8.5 describes the initial dissolution of calcite. Equations 8.6 and 8.7 show that carbonate reacts with water (known as hydrolysis) to form bicarbonate and carbonic acid, raising the pH. Thus carbonate and bicarbonate are acting as bases. Equation 8.8 shows that carbonic acid and dissolved CO_2 come to equilibrium in water, and Eq. 8.9 describes the equilibrium between CO_2 in the soil atmosphere and that dissolved in the water.

When calcite, water and air are in contact all these reactions tend to come to equilibrium. Using these equations and their equilibrium constants the following equation can be obtained:

$$2pH = 9.6 - \log(Ca^{2+}) - \log P_{CO_2} \qquad [8.10]$$

where P_{CO_2} is the partial pressure of CO_2 in the air and (Ca^{2+}) is the activity of Ca^{2+} in solution. The atmosphere contains 0.03 per cent CO_2 *v/v* and so $P_{CO_2} = 0.0003$. In soil air P_{CO_2} may be up to 100 times this value. This equation is plotted in Fig. 8.8 to show the range of pH values likely to be present in calcareous soils.

The following conclusions can be drawn:

- Water in equilibrium with calcite and the atmosphere has a pH of 8.4.
- An increase in the concentration of CO_2 causes a decrease in pH.
- An increase in the concentration of Ca^{2+} (from some source other than calcite and therefore a common ion) causes pH to decrease.
- The solubility of soil calcium carbonates differs from that of calcite due to impurities and coatings on the surface interfering with dissolution. A lower solubility gives a lower equilibrium pH value.
- Bicarbonate is the dominating anion in solution in the range of pH which is of interest in calcareous soils (7–8.5).

The pH of a calcareous soil is therefore variable over time depending on the concentration of CO_3^{2-} and Ca^{2+}. Measurement of pH in 10 mM $CaCl_2$ solution tends to standardize these conditions. The measured pH then reflects differences in the solubility and to some extent the amount of $CaCO_3$ present because when small amounts occur in soils or where the particles have a low surface area (a few large particles) the rate of dissolution of $CaCO_3$ into the solution is slow compared to the standard 15 min shaking time (Section 8.1).

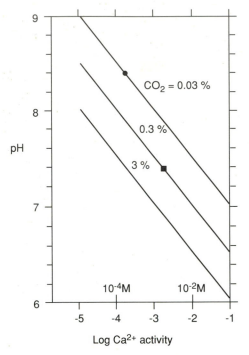

- The solution properties when calcite and water equilibrate in the atmosphere: (Ca^{2+}) = 0.2 mM, CO_2 = 0.03 %. pH = 8.4

■ An example of the solution properties in a calcareous soil: (Ca^{2+}) = 2 mM, CO_2 = 0.3 %. pH = 7.4

Figure 8.8 The relationship between pH and activity of Ca^{2+} in solutions in equilibrium with calcite.

Sodic soils

In arid regions Na_2CO_3 tends to accumulate in soils. It is a soluble salt. Dissolved CO_3^{2-} reacts with water and CO_2 as shown in Eqs 8.6–8.9 and pH rises to give values up to 10.5 (Ch. 14).

Section 8.4 The buffer capacity of soils and its measurement

A buffer solution is one which resists pH change when acid or base is added. The chemical mechanism is explained in Section 8.5, but the principles can be seen by comparing the effects of adding acid to water and a soil suspension as shown in Fig. 8.9.

If to 25 ml of water 1 ml of 10^{-1} M HCl is added, the amount of H^+ present is

$$10^{-1} \, mol \, l^{-1} \times 1/1000 \, l = 10^{-4} \, mol \, H^+$$

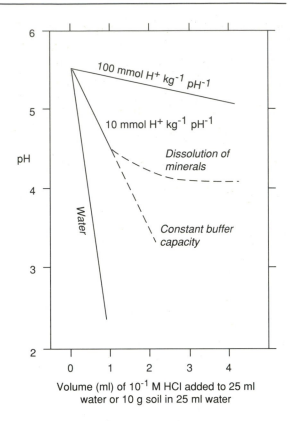

Figure 8.9 The buffering of pH by water and soil.

The contribution of H^+ initially present in the water is insignificant. The H^+ ions are in a volume of $26/1000$ l. The concentration of H^+ in $mol \, l^{-1}$ is therefore 3.8×10^{-3} and

$$pH = -\log (H^+) \simeq -\log (3.8 \times 10^{-3}) = 2.4 \quad \text{(Section 8.3, Note 1)}$$

Because of the reactions which occur between added H^+ and soil particles (Fig. 8.1) the decrease in pH of a soil treated in the same way is much less than the water, with soil buffer capacities of between about 10 and 100 mmol $H^+ \, kg^{-1} pH^{-1}$ (Table 8.5). Again the pH change can be calculated: an input of 10^{-4} mol H^+ to 10 g of soil is 10 mmol $H^+ \, kg^{-1}$. Thus for the soil with a buffer capacity of 10 mmol $H^+ \, kg^{-1} pH^{-1}$ the change of pH is 1 unit, and for a buffer capacity of 100 mmol $kg^{-1} pH^{-1}$ it is 0.1 unit. Further additions of acid cause pH to fall linearly if the buffer capacity is constant. In practice, however, it increases at about pH 4 due to mineral dissolution, so that provided time is allowed for the soil particles to react with the added acid, the pH is eventually held at about this value, i.e. the buffer capacity becomes very large.

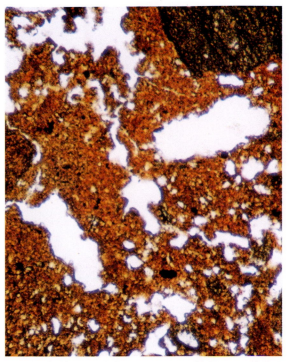

Plate I Micrographs of thin sections of rocks and soils (field of view = 3 mm). (a) an Ah horizon of a brown earth (Denbigh Series, North Wales) with an open spongy fabric produced primarily by faunal activity;

(b) a dense glacial till with rock fragments in a clay matrix and a few pores;

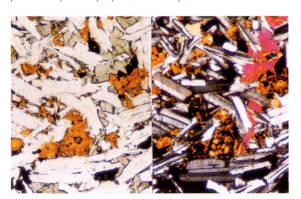

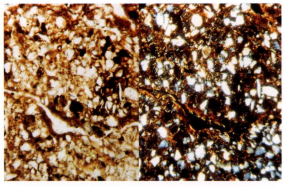

(c) A solid crystalline igneous rock (olivine dolerite) seen in plane polarized light (left) and between crossed polarizers (right);

(d) a Bt horizon of an argillic brown earth (Burseldon Series, Southern England) showing concentrations of clay (argillans) lining channels and filling pores: in plane polarized light (left) and between crossed polarizers (right). Photographs by D. Jenkinson, University College of North Wales, and K. Gutteridge.

Plate 2 Lichen growing on the surface of rocks above Val d'Isère, France, probably *Xanthoria elegans*, a species widespread in extreme environments in arctic and alpine regions. Photograph by D. Rowell.

Plate 3 Equipment with large, low pressure 'Terra' tyres being used to prepare a seedbed at the Scottish Centre of Agricultural Engineering. Photograph by D. Campbell.

(a)

(b)

Plate 4 A vertisol at Qedma, Israel, showing structural features resulting from shrinkage and swelling. (a) The profile down to 2 m; (b) a closer view of the pressure faces (slickensides) formed when adjacent aggregates are forced together during swelling. Field of view = 10 cm. Photographs by D. Rowell.

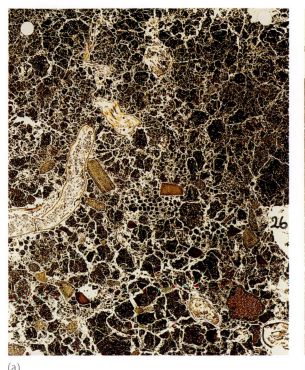
(a)

(b)

Plate 5 Large thin sections of the A₁ horizon of a Linhope Series silty clay loam from the Southern Uplands of Scotland. The vertical length of the sections is 10 cm. (a) A soil under grazed unimproved grassland, cleared from forest at least 200 years ago. The structure is fine crumb becoming blocky with worm faecal pellets, a prominent bracken root and a porosity of about 0.6 cm³ cm⁻³; (b) the same after tillage, pasture re-establishment and compaction by roller, sheep and cattle. The structure has been completely degraded and the porosity is much reduced. The darker colour on the left is due to the uneven thickness of the slide. Photographs by R. MacEwan.

Plate 6 Winter wheat being sprayed with pesticide using a 'tramline' system of management. After drilling, sprays and fertilizers are spread with the tractor wheels running only along the tramlines where no crop grows, thus minimizing crop damage and restricting the area of soil subject to compaction. Photograph by Holt Studios.

Plate 7 Compaction by tractor wheels in a cultivated sandy loam soil at the Scottish Centre of Agricultural Engineering. Photograph by D. Campbell.

(a)

(b)

Plate 8 The vegetation at ICRISAT, near Niamey, Niger, photographed during (a) the wet season and (b) the dry season. The foreground is dominated by *Guiera senegalensis* bushes and shallow rooted annual grasses which establish when a cultivated area (in the distance) reverts to fallow. Photograph by S. Allen, Institute of Hydrology.

Plate 9 Rice production near Rangkasbitung in Western Java, Indonesia. The buffalo is pulling a puddling plough prior to planting. Young buffaloes follow their mother. Adjacent fields have been recently planted. Banana trees, as in the foreground, are commonly grown on the banks around the paddy fields. Photograph by G. Warren.

Plate 10 A soil in the Gower Peninsular, South Wales showing the effects of almost permanently waterlogged conditions. Peat has accumulated over a strongly reduced mineral horizon (grey) showing patches of more oxidized soil (brown) probably associated with old root channels. Depth to water = 65 cm. Photograph by D. Rowell.

Plate 11 Concretions of gypsum formed in an aridisol in the Arava in the south of Israel. Gypsum is dissolved, moved and deposited by the occasional rainfall. The soil surface is covered by a layer of stones known as a desert pavement. The rod is 60 cm long. Photograph by Pinhas Fine, The Volcani Center.

Plate 12 The effects of calcium deficiency and aluminium toxicity on the growth of sorghum. The plants were grown in soils containing varying amounts of exchangeable Ca^{2+} and Al^{3+}, expressed as meq per 100 g soil (= $cmol_c$ kg^{-1}). Photograph by K. D. Ritchey, EMBRAPA.

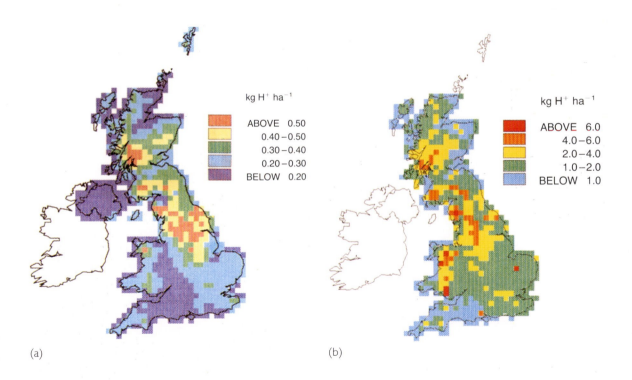

(a)

(b)

Plate 13 Annual deposition of acidity from the atmosphere onto soils in Britain (1986–1988 mean). (a) The amount of H^+ deposited in the rain and (b) the potential acidity calculated from the total N and S entering the soil by both wet and dry deposition minus the $Ca^{2+} + Mg^{2+}$ input. By DOE, Warren Spring Laboratory.

(a)   (b)

Plate 14 Tomato leaves showing inter-veinal chlorosis induced by high pH. The plants were grown in sand culture supplied with a complete nutrient solution with pH adjusted to (a) 5.6 and (b) 9.3 with Na_2CO_3. The uptake and utilization of iron and manganese is inhibited by the large bicarbonate concentration, interfering with the production of chlorophyll. Photographs by A. Masshady and D. Rowell.

(a)   (b)

Plate 15 The effect of phosphate, applied before sowing, on root growth of a grass, *Andropogon guyanus* in an acidic oxisol near Planaltina, Brazil. The scale is given by the book which is 20 cm high. (a) Without P and (b) with P. Photographs by P. Le Mare.

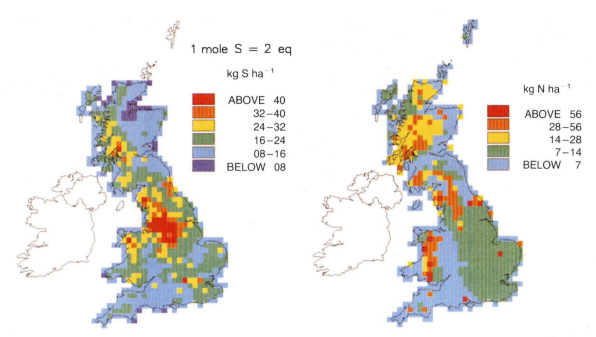

1 mole S = 2 eq

kg S ha^{-1}

ABOVE	40
	32−40
	24−32
	16−24
	08−16
BELOW	08

kg N ha^{-1}

ABOVE	56
	28−56
	14−28
	7−14
BELOW	7

Plate 16 Total annual deposition of sulphur from the atmosphere onto soils in Britain (wet deposited sulphate + dry deposited SO_2 + sulphate deposited in cloud water, 1986−1988 mean). By DOE, Warren Spring Laboratory.

Plate 17 Total annual deposition of nitrogen from the atmosphere onto soils in Britain (wet NO_3^- + wet NH_4^+ + dry HNO_3 + dry NO_2 + dry NH_3, 1986−1988 mean). By DOE, Warren Spring Laboratory.

(a)   (b)

Plate 18 The growth of winter wheat in June at the MAFF Bridgets Experimental Husbandry Farm, near Winchester on a soil which had previously grown arable crops and had a nitrogen Index 0. (a) Without added fertilizer−N and (b) with 200 kg fertilizer−N ha^{-1} applied in the spring. Photographs by D. Rowell.

Plate 19 An aerial view of the Broadbalk (centre) and Hoosfield (foreground) classical experiments at Rothamsted Experimental Station, Harpenden. A modern multifactorial experiment is in Pastures Field (upper right) and Rothamsted Manor can just be seen in the trees (upper centre). Reproduced by permission of Rothamsted Experimental Station.

Plate 20 A field trial to determine the biological dose response curve of wheat for a herbicide. There were five rates of application (including the control) and four replicates of each treatment. The maximum treatment was equivalent to nearly one thousand times the normal field application. Photograph courtesy of M. Lane, Zeneca Agrochemicals.

(a)

(b)

Plate 21 Soil salinity. (a) A dried estuarine mud at Yarmouth, Isle of Wight, showing the extensive shrinkage of clays dominated by exchangeable sodium from sea water, and the accumulation of salts on the soil surface; (b) the upper 25 cm of soil adjacent to the Dead Sea, West Bank, Israel, showing salt crystals throughout the soil and on the surface. The accumulation of salt forms and fills a honeycomb of pores within the soil. Photographs by D. Rowell.

Table 8.5 Buffer capacities of soils expressed in various units

Soil type	A	B	C	D	E
Sand	16	0.8	0.6	2	40
Light	48	2.4	1.8	6	120
Medium	56	2.8	2.1	7	140
Heavy	64	3.2	2.4	8	160
Clay	80	4.0	3.0	10	200
Organic	80	4.0	3.0	10	200
Peat	128	6.4	4.7	16	320

A = $mmol\ H^+\ kg^{-1}\ pH^{-1}$;
B = $g\ CaCO_3\ kg^{-1}\ pH^{-1}$;
C = $g\ Ca(OH)_2\ kg^{-1}\ pH^{-1}$;
D = $t\ CaCO_3\ ha^{-1}\ pH^{-1}$ (2600 t soil ha^{-1} to 20 cm);
E = $kg\ H^+\ ha^{-1}\ pH^{-1}$ (2600 t ha^{-1}).
Except for sand and clay, the values are based on those used by ADAS in Britain given in Table 8.6 where the soil types are explained. Because of the range of texture within each type the more extreme values for sand and clay have been added. These are based on Helyar *et al.* (1990) who used the following equation to obtain buffer capacity values for Australian soils from their clay and organic matter (OM) contents;
buffer capacity $(kg\ H^+\ ha^{-1}_{2500t}\ pH^{-1} = (7.5 \times OM\%)\ +$
$(3.6 \times clay\ \%)$.
The equation only gives approximate values primarily because the mineralogy of the clay-sized fraction is not taken into account. A similar relationship has been obtained by Aitken *et al.* (1990) for variable charge soils.

The measurement of soil buffer capacity: addition of acid

Equipment and reagents

Hydrochloric acid, 0.1 M.
Sodium hydroxide, 0.1 M.
Potassium chloride solution. 0.1 M. Dissolve 7.5 g of KCl in water and make up to 1 l.
pH meter and electrodes.
Magnetic stirrer.

Method

Place 10 g of air-dry <2 mm soil in a 100 ml beaker. Add 25 ml of 0.1 M KCl solution. Place on a magnetic stirrer and insert the stirring bar. Calibrate the pH meter (Section 8.1) and measure the pH. Add a volume of acid from a burette (see below), stir for 15 min and measure pH. Repeat the addition and measurement until the required pH range has been covered. Plot the buffer curve (pH against added acid) and calculate the buffer capacity (see below).

Volumes of acid Sandy soils have small buffer capaci-

ties. Use 0.5 ml additions of acid. For the larger buffer capacities of heavy textured soils use 2.5 ml additions.

Time of equilibration On adding acid the pH will fall sharply and then rise slowly as reactions proceed. It may take many hours to approach an equilibrium value. The time allowed will depend on the purpose of the experiment. Where, for example, a 24 h equilibration time is required, carry out a preliminary experiment as described above using 15 min equilibration times. On the basis of the results, set up a series of beakers containing soil and solution, and add volumes of acid to cover the range required. Stir occasionally by hand over 24 h and then measure pH. Alternatively place the soil suspension and acid in a capped bottle and shake overnight on a mechanical shaker before measuring pH. (Caution. Do not use this procedure for a calcareous soil; CO_2 will be produced and may break the bottle.)

The use of 0.1 M KCl solution This experiment could be carried out in water rather than KCl solution, but the addition of HCl would progressively increase the chloride (salt) concentration in solution (Fig. 8.10) introducing errors due to the salt effect (Section 8.1). By placing the soil in 0.1 M KCl solution, the salt concentration remains constant during the experiment provided that 0.1 M HCl is used. Fig. 8.11 gives examples of the curves obtained.

Note that there is a change in the ratio of soil to solution during the experiment so small errors due to the suspension effect may be involved (Section 8.1). Additions of 1 M HCl using an Eppendorf pipette would reduce the suspension effect but introduce a salt effect. In the field anions are leached and the soil solution concentration does not change significantly. Thus there are difficulties in choosing the best method.

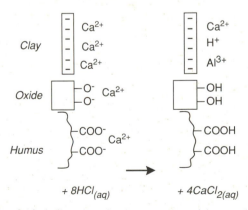

Figure 8.10 A diagrammatic representation of the reaction of acid with a neutral soil. Gibbsite and other minerals release Al^{3+} as pH decreases. Other ions will also be present in solution and on the exchange sites.

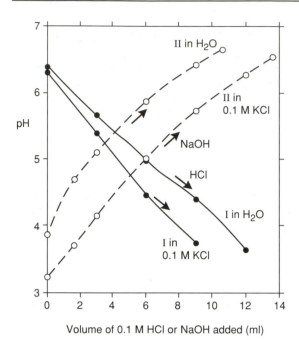

Figure 8.11 Buffer curves obtained by adding NaOH or HCl to soil suspensions. Ten g of soil was placed in 25 ml of H₂O or 0.1 M KCl, NaOH or HCl was added and pH measured after overnight equilibration. Soil I: A brown earth, 3.0 per cent organic matter, 18 per cent clay, predominantly smectite with illite and kaolinite, CEC = 13.2 cmol_c kg⁻¹. Soil II: The gley described in Table 8.3.

The measurement of soil buffer capacity: addition of base

Addition in soil suspensions

The above method can be used, but with additions of 0.1 M NaOH. Exchangeable H^+ and Al^{3+} are neutralized by OH^- to form water and precipitate $Al(OH)_3$, and the added Na^+ takes their places as the exchangeable cation (Fig. 8.12). If variable charge is present, the rise in pH causes sites to become negatively charged as H^+ is released, and again Na^+ takes its place on these new sites. Thus there is no increase in salt concentration in solution and so the experiment can be carried out in either water or 0.1 M KCl. The pH values will be lower in KCl solution if the soil is negatively charged (Section 8.1), as shown in Fig. 8.11 .

Incubation of soil with CaCO₃

The main reason for measuring the buffer capacity when pH is raised is in relation to liming. It is therefore

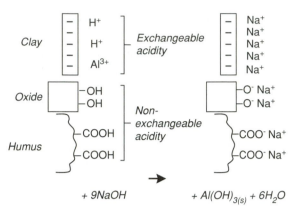

Figure 8.12 A diagrammatic representation of the reaction of a base with an acidic soil. Other ions will also be present in solution and on the exchange sites.

logical to mix soil with $CaCO_3$ and measure the pH after a suitable period of incubation.

The amounts of lime required Buffer capacities of between 10 and 100 mmol H^+ kg⁻¹ pH⁻¹ can be converted into lime treatments as follows. Assuming reversibility of reactions a depression of 1 pH unit for the above range of H^+ inputs can be reversed by adding 5–50 mmol $CaCO_3$ kg⁻¹ pH⁻¹, since

$$2H^+ + CaCO_3 = Ca^{2+} + H_2O + CO_2$$

The molar mass of $CaCO_3$ is 100 g mol⁻¹ and so between 0.5 and 5 g $CaCO_3$ kg⁻¹ is required to raise the pH by 1 unit (Table 8.5). For laboratory incubation studies $Ca(OH)_2$ reacts more rapidly than $CaCO_3$, its molar mass is 74 g mol⁻¹ and so between 0.4 and 4 g $Ca(OH)_2$ kg⁻¹ pH⁻¹ is required.

Method

Measure the pH of the soil (Section 8.1). Decide on the pH range you wish to cover in the experiment. Table 8.5 gives expected buffer capacities in relation to texture and organic matter content. Using these values decide on a range of treatments in terms of g $Ca(OH)_2$ per 100 g soil. For example, a light-textured soil at pH 4 would theoretically have its pH raised to 7 by adding 3×1.8 g $Ca(OH)_2$ kg⁻¹ or 0.54 g per 100 g soil, and so treatments of between 0 and 0.6 g per 100 g soil in 0.1 g increments might be suitable.

Weigh into polythene bags 100 g samples of <2 mm air-dry soil. Add $Ca(OH)_2$ as required. Shake to mix thoroughly. Add water to bring the soil to 40 per cent of its water holding capacity (Table 5.2 and Section 3.2,

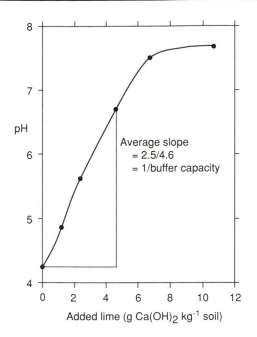

Figure 8.13 The buffer curve of a Nigerian ultisol obtained by incubating moist soil with $Ca(OH)_2$. Data from A. Bantirgu for the soil described in Tables 8.3 and 8.4.

Note 2). Store for 1 week with the bag loosely folded, opening the bag and shaking occasionally to ensure good aeration.

At the end of the incubation period measure the pH of each soil sample (Section 8.1, 10 g soil: 25 ml H_2O).

The buffer curve Plot your results as in Fig. 8.13. The slope of the line $= pH/g\,Ca(OH)_2\,kg^{-1}$ is the reciprocal of the buffer capacity. The results are unlikely to give a straight line so an average value for a given range can be recorded. In this case the buffer capacity between pH 4.2 and 6.7 is

$$1/(2.5/4.6) = 1.8\,Ca(OH)_2\,kg^{-1}\,pH^{-1}$$

Alternatively the value at a given pH value can be found by drawing a tangent to the line. The above buffer capacity is that given for light textured soils in Table 8.5. The limiting pH of about 7.7 indicates that $Ca(OH)_2$ has been converted to $CaCO_3$ which remains in the soil.

The reversibility of pH changes

Buffer capacities measured by addition of acid are approximately equal to those measured by addition of base over the pH range 4.5–6.5. Mineral dissolution at low pH values and the presence of $CaCO_3$ at high pH values limit this range.

Section 8.5 The determination of lime requirements

The lime requirement of a soil is the amount of lime (normally $CaCO_3$) required to raise the pH of the soil in the field to a given target value.

The texture–pH method

A combination of a knowledge of crop sensitivity (Table 8.2) and buffer capacity based on extensive laboratory measurements and field experience has led to simple guidelines being developed by ADAS in Britain. Table 8.6 gives the details for soils to be used for arable cropping.

Table 8.6 Lime requirements of soils used for arable crops

Soil type	Recommended pH	Target pH	Buffer capacity† ($t\,CaCO_3\,ha^{-1}\,pH^{-1}$)
Light	6.5	6.7	6
Medium	6.5	6.7	7
Heavy	6.5	6.7	8
Organic	6.2	6.4	10
Peat	5.8	6.0	16

Soil type	Texture (UK system) and organic matter
Light	Sands and loamy sands
Medium	Loams except clay loam
Heavy	Clay loams and clays
Organic	10–25% organic matter
Peat	Greater than 25% organic matter

† The buffer capacity is for a 20 cm depth of soil (2600 t ha^{-1}) and is called a *lime factor* by ADAS.
Recommended and target pH. Table 8.2 shows that at a pH of 6.0 some crops begin to be affected. To allow for the decrease in pH over a few years, 6.5 is the recommended pH for mineral soils, and to allow for variations in pH over the field and uneven application the target pH is 0.2 above this.
Organic soils and peat. Aluminium is strongly bound to organic matter, and so lower pH values are tolerated than in mineral soils. Large amounts of lime are needed to change the pH of these soils: overliming leads to (a) the development of micronutrient deficiencies (Section 7.1) and (b) high ratios of Ca^{2+}/K^+ and Ca^{2+}/Mg^{2+} which may cause K^+ and Mg^{2+} deficiencies.
From MAFF (1986b).

Method

Soil is sampled from the field following the methods given in Section 1.3. It is important that this sample represents the whole field to a depth of 20 cm. If there is any reason for thinking that pH might vary systematically over the field (change of soil type or management) then a sample should be obtained from each area and measured separately. Prepare a <2 mm air-dry sample. Measure soil pH (1:2.5 in H_2O, Section 8.1). Determine the soil texture (Section 1.2) and its buffer capacity from Table 8.6. Calculate the lime requirement as follows:

$$\text{lime requirement} = (\text{target pH} - \text{soil pH}) \times \text{buffer capacity}$$

For grass crops, the target values are 0.5 pH units lower than for arable crops and recommendations are made for a reduced depth of soil (15 cm) because the land will not be ploughed after application and large applications would take time to be washed into the soil. Buffer capacities expressed as $t\,CaCO_3\,ha^{-1}\,pH^{-1}$ are therefore $0.75 \times$ those for arable sites.

A buffer solution method

The lime requirement of a soil is measured most directly by field trials. These are expensive and time consuming. In the laboratory, incubation studies (Section 8.4) allow soil and lime to react together without the complications of soil variability and uneven mixing which occur in the field. Incubation experiments are therefore a direct way of measuring lime requirements but are also time consuming for routine purposes. For this reason buffer solution methods have been developed which are economical and simple to use (Woodruff, 1948; Shoemaker et al., 1961).

The principle

A buffer solution is one in which pH change is resisted (buffered) when acid or base is added. Normally it consists of a weak acid and its salt (Nuffield Advanced Chemistry II, Topic 12.5). An example is ethanoic acid and ammonium ethanoate. Its use in determining CEC at pH 7 has already been discussed (Section 7.2). Ethanoic acid is a weak acid: it is only partially dissociated in water:

$$CH_3COOH \rightleftharpoons CH_3COO^- + H^+$$

Hydrochloric acid added to this solution is a strong acid which is almost completely dissociated:

$$HCl \rightleftharpoons H^+ + Cl^-$$

Some of the added H^+ associates with CH_3COO^- to increase the concentration of CH_3COOH. The ammonium ethanoate is a source of CH_3COO^- so that more H^+ can be removed in this way. As a result the decrease in pH is less than it would be in water.

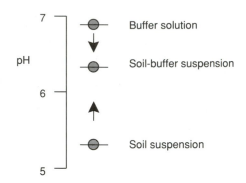

Figure 8.14 The pH values measured in the buffer solution method for determining lime requirement.

The method involves mixing a buffer solution at pH 7 with an acid soil. The two buffered systems react together, the pH of the buffer solution decreasing and the pH of the soil increasing until both are the same (Fig. 8.14). The decrease in the solution pH will depend on the amount of soil acidity reacting with the buffer as soil pH increases to the equilibrium value. The buffer capacity of the buffer solution and its decrease in pH can therefore be used to measure the buffer capacity of the soil. In effect the buffer solution is being used as a liming material to raise the pH of the soil.

The method described below is based on that by Woodruff (1948) and MAFF (1986a) modified to obtain a buffer capacity for the soil.

Reagents and equipment

The buffer solution. At 105 °C, dry calcium ethanoate, $(CH_3COO)_2Ca$, for 1 h and cool in a desiccator. Weigh 40.0 g and add to about 900 ml of water, together with 8.0 g of 4-nitrophenol, $NO_2C_6H_4OH$, (CAUTION, poison) and 0.6 g of magnesium oxide (light), MgO. Warm to dissolve the solids. Cool and dilute to 1 l. Filter if the solution is not clear. The pH should be between 6.9 and 7.1. If necessary adjust to this range by adding concentrated HCl or MgO.
pH meter, electrodes and buffers.
Bottles and caps.

The method

The soil should be sampled as for the texture–pH method above.

Calibrate the pH meter. Prepare a soil suspension using 10 g (or 10 ml) of soil and 25 ml of water and measure its pH (Section 8.1). If the pH is above the recommended pH in Table 8.6 it does not need lime. In the example below, soil pH = 5.25.

For acid soils use the above suspension for the measurement of lime requirement. Pipette 20 ml of the buffer solution into the suspension. Cap the bottle and shake for 5 min (Note 1). Mix also 25 ml of water and 20 ml of the buffer solution in a beaker.

Measure the pH of the buffer–soil suspension (for example 6.34) and the buffer–water mixture (for example 6.92). If the pH of the buffer–soil suspension is below 6.0 repeat the experiment using 5 g (or 5 ml) of soil.

Calculation

The reaction can be summarized as follows:

buffer–OH^- + soil–H^+ = soil–buffer suspension + H_2O

The amount of OH^- reacting is given by the buffer capacity of the buffer solution multiplied by the change in pH of the buffer solution called Δ buffer pH (Δ meaning 'change in'). The amount of H^+ reacting is given by the soil buffer capacity × Δ soil pH. When expressed in mol, the two amounts must be equal.

The buffer capacity of 20 ml of buffer solution with 25 ml of water is shown in graphical form in Fig. 8.15 (Note 2): 0.86 mmol H^+ per 45 ml solution is required to depress the pH by 1 unit. If the solution initially has a pH of 7, the full line applies: the dotted line shows the

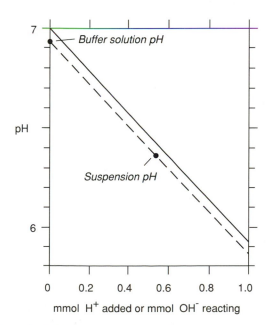

Figure 8.15 The changes in pH produced by adding acid to the buffer solution (20 ml buffer solution + 25 ml water).

behaviour of the buffer solution initially at pH 6.92 as in the example above.

Subtract the soil–buffer suspension pH from the buffer solution pH to give Δ buffer pH. Its value is 6.92 − 6.34 = 0.58. The amount of buffer–OH^- which has reacted can be determined from the graph, or calculated as

$$0.86 \, \text{mmol} \, OH^- \, \text{pH}^{-1} \times 0.58 \, \text{pH} = 0.50 \, \text{mmol} \, OH^-$$

Thus the amount of soil–H^+ which has reacted is 0.50 mmol H^+ per 10 g soil or 50 mmol kg^{-1}. The value of Δ soil pH is 6.34 − 5.25 = 1.09. The soil's buffer capacity is therefore

$$50 \, \text{mmol} \, \text{kg}^{-1}/1.09 \, \text{pH} = 45.8 \, \text{mmol} \, \text{kg}^{-1} \, \text{pH}^{-1}$$

The use of buffer capacity to determine a field recommendation

In the field acidity is normally neutralized by $CaCO_3$. Since

$$2H^+ + CaCO_3 = Ca^{2+} + CO_2 + H_2O$$

only 0.5 mol of $CaCO_3$ is required to neutralize 1 mol of H^+ in the soil. Therefore the buffer capacity can be expressed as

$$45.8/2 = 22.9 \, \text{mmol} \, CaCO_3 \, \text{kg}^{-1} \, \text{pH}^{-1}$$

The molar mass of $CaCO_3$ is 100 mg mmol^{-1}. Therefore the buffer capacity is

$$22.9 \times 100 \, \text{mg} \, \text{kg}^{-1} \, \text{pH}^{-1}$$

or

$$2.29 \, \text{g} \, CaCO_3 \, \text{kg}^{-1} \, \text{pH}^{-1} \, (= \text{kg} \, CaCO_3 \, \text{t}^{-1} \, \text{pH}^{-1})$$

Assuming 2600 t soil ha^{-1} to 20 cm depth, the buffer capacity is

$$2.29 \, \text{kg} \, \text{t}^{-1} \, \text{pH}^{-1} \times 2600 \, \text{t} \, \text{ha}^{-1} = 5954 \, \text{kg} \, \text{ha}^{-1} \, \text{pH}^{-1} = 5.95 \, \text{t} \, CaCO_3 \, \text{ha}^{-1} \, \text{pH}^{-1}$$

These calculations simplify to

buffer capacity (t $CaCO_3$ ha^{-1} pH^{-1}) = 11.18 × Δ buffer pH/Δ soil pH

If 5 g soil is used the factor is 22.36 instead of 11.18. Adjustments have to be made to the calculation if a different depth of soil is considered (15 cm, 1950 t) or if the soil has a bulk density very different from 1.3 g cm^{-3} (Section 4.2).

As shown on previous page, the recommended application of lime is

(target pH − soil pH) × buffer capacity

For the example above (assuming arable cropping and a mineral soil, Table 8.6) the recommendation is

$$(6.7 - 5.25) \times 5.95 = 8.6 \, \text{t} \, CaCO_3 \, \text{ha}^{-1}$$

A simplified method For routine advisory purposes a field recommendation (2600 t soil ha^{-1}) can be calculated as

$$t\,CaCO_3\,ha^{-1} = 11.2 \times \Delta\,buffer\,pH$$

This is based on the original method of Woodruff (1948) and makes the assumption that the soil–buffer suspension pH is approximately equal to the target pH, hence the need for the former to be above pH 6.

Note 1 This shaking time is very short compared to the time of reaction of lime in the field. Particularly for organic soils it may underestimate the buffer capacity.

Note 2 The graph is a straight line between pH 7 and 6 but becomes curved below pH 6. The validity of this graph can be checked by measuring the pH values of the following solutions: to 20 ml of buffer solution add 5, 10, 15 and 20 ml of 0.05 M HCl with 20, 15, 10 and 5 ml of water respectively. These solutions contain 0.25, 0.5, 0.75 and 1 mmol HCl.

The liming effect of hard water

Hardwater contains calcium and magnesium carbonate and bicarbonate, normally dissolved from limestone. The concentration (expressed as $Ca(HCO_3)_2$) is about 3 mM and can be measured using the methods of Section 14.1.

Example

Water containing 3 mM $Ca(HCO_3)_2$
($= 6$ mM HCO_3^-)
Soil buffer capacity
50 mmol OH^- (or HCO_3^-) kg^{-1} pH^{-1}

One litre of water contains 6 mmol of HCO_3^-. When added to 1 kg of soil, the pH change is 6 mmol/50 mmol pH^{-1} = 0.12 pH units.

A single irrigation event in the field or in a pot designed to bring soil to field capacity may add 200 ml kg^{-1} giving a pH change of 0.024. These effects are small compared to liming of acid soils for arable crops, but for horticulture hard water should not be used for lime-sensitive plants such as azaleas. For this purpose, water is acidified to pH 5.8–6.0 to remove most of the bicarbonate.

A method for determining the lime requirement of soils from the humid tropics

In the humid tropics, it is not necessary to lime to target pH values between 6 and 7: the crops are less sensitive to acidity, yields can be depressed and rates of lime loss

are high. Methods have been developed with lower target pH values (Shoemaker *et al.*, 1961) or which aim to neutralize part or all of the exchangeable aluminium, so reducing the amount of soluble aluminium below its toxic concentration (Section 8.2 and Table 8.2(b)).

Kamprath (1970) has shown that in order to reduce the amount of exchangeable aluminium the following empirical relationship can be used:

OH^- required (cmol$_c$ kg^{-1}) = 1.5 × the amount of exchangeable Al^{3+} to be neutralized (cmol$_c$ kg^{-1})

The 1.5 factor takes account of other forms of acidity in the soil apart from exchangeable Al^{3+} which have to be neutralized (Fig. 8.12) and applies only to mineral soils similar to those used by Kamprath (oxisols and ultisols). Organic soils have very much higher factors. This method is still being developed (Caudle, 1991; Cochrane *et al.*, 1980).

Section 8.6 Projects

1. Obtain an acidic soil (woodland soils are normally acidic) and measure or predict its buffer capacity (Section 8.4). Calculate suitable $Ca(OH)_2$ treatments to raise its pH to a range of values up to 6.5. Apply $Ca(OH)_2$ to the soil (100 g samples), and incubate moist in polythene bags. After 1 week measure pH and exchangeable Ca^{2+}. Calculate the increases in the amount of exchangeable Ca^{2+} and compare to the amounts added. The difference is predominantly due to Ca^{2+} which has become bound on to humus in the soil (Table 8.4). Compare your measured pH values to those predicted using the chosen buffer capacity. Calculate the buffer capacity from the incubation experiment and compare to the predicted value.

2. Obtain a neutral soil (limed agricultural soils are normally between pH 6 and 7). Incubate 100 g samples moist with added ammonium sulphate reagent for several weeks (Section 11.3). Measure pH to determine the acidifying effect of the nitrification of ammonium fertilizers.

   Suggested treatments. A wheat crop may receive 150 kg N ha^{-1}. This would be 0.28 g $(NH_4)_2SO_4$ kg^{-1} assuming 2500 t ha^{-1}. Thus applications from 0.28 to 2.8 g kg^{-1} soil would give between 1 and 10 times the normal field rates. Apply the reagent to dry soil, shake well and moisten to 40 per cent of its water holding capacity (Table 5.2 and Section 3.2, Note 2).

   Each mol of NH_4^+ theoretically releases 2 mol H^+. The treatment with 0.28 g kg^{-1} is 4.2 mmol NH_4^+ kg^{-1} and releases 8.4 mmol H^+ kg^{-1}. If this occurs in a sandy soil with a buffer capacity of

$10 \, mmol \, H^+ \, kg^{-1} \, pH^{-1}$, the expected decrease in pH is 0.84 units. For a heavy-textured soil with a buffer capacity about 10 times this value, the pH change is about 0.1.

Extension of the project. Measure the buffer capacity of the soil (Section 8.4). From the pH changes measured in the incubation experiment calculate the amounts of acid produced by the fertilizer. Repeat the experiment with NH_4NO_3 and urea.

3. Set up a pot experiment to determine the effects of acidity on plants with different sensitivities to acidity, or to examine competition between these plants.

Obtain an acidic (woodland) soil with a pH of between 4 and 5. Air dry and pass through a garden sieve. From Table 8.5 determine the lime requirement to raise its pH to 6.5. Obtain barley and rye seeds (or other suitable species) from your local agricultural supplier. Decide on lime treatments up to the full requirement, using perhaps four different treatments including zero lime. Set up three pots per crop per treatment including adequate N, P, and K^+ additions. Section 9.2 gives the methods. Germinate the seeds in a seed compost, and transplant three per pot when they are about 3 cm high. Harvest the plants after a few weeks growth, and determine dry matter yield. Measure soil pH. Plot mean yield per pot against pH. Is there any indication that the critical pH values for field crops in Table 8.2 apply to pot-grown plants?

A mixture of red clover and ryegrass can be used instead of a single species. Weigh 0.25 g of ryegrass seed and sprinkle uniformly into each pot. Place 10 clover seeds equally spaced on the soil in each pot. Cover with a 5 mm depth of soil. At harvest cut at soil level, separate the grass and clover, dry and weigh. Plot dry matter yield of each species against soil pH (see also Section 13.4).

Use a simple statistical analysis (Section 3.8) to help you decide whether mean values for treatments differ from one another.

4. *Data analysis project.* The relationship between buffer capacity, organic matter and clay contents of soils is given by Helyar *et al.* (1990):

   (a) Check that the equation given by Helyar (p. 526 in his paper) has been correctly converted to give the equation under Table 8.5.
   (b) Rewrite the equation to give buffer capacities expressed as $mmol \, H^+ \, kg^{-1} \, pH^{-1}$.
   (c) What is the buffer capacity of the organic matter according to this equation? Compare your value to Fig. 7.2. Does the organic matter in these Australian soils have similar properties to the humus used to obtain the figure?

   (d) Using the data in Helyar's Table 7, Site 6, calculate the acidity input associated with NH_4NO_3. What might be the reason for the negative input of acidity (a development of alkalinity) on some plots?

5. Obtain soil samples from various sites in chalk and limestone areas. Measure their $CaCO_3$ contents (Section 2.4) and their pH values. Plot your results to show that there is little relationship between these two properties. On a soil over chalk, sample down through the profile to include the underlying chalk. In this case, the pH values should increase with the increase in $CaCO_3$ content (Schinas and Rowell, 1977).

6. Obtain samples of liming materials from an agricultural merchant or garden centre. Measure their neutralizing ability (Section 2.4), expressing your results as an equivalent amount of pure $CaCO_3$ per kilogram of material. For example, hydrated (slaked lime) is mostly $Ca(OH)_2$. The molar masses are 100 and $74 \, g \, mol^{-1}$ respectively and so 1 kg $Ca(OH)_2$ has the neutralizing ability of $100/74 = 1.35$ kg $CaCO_3$. Note that in the UK, *neutralizing value* has a different meaning, i.e. the mass of CaO (molar mass $56 \, g \, mol^{-1}$) which has the same neutralizing ability as 100 kg of the material.

Section 8.7 Calculations

1. For a soil with a buffer capacity of $20 \, mmol \, H^+ \, kg^{-1} \, pH^{-1}$:

   (a) calculate the buffer capacity as $kg \, H^+ \, ha^{-1} \, pH^{-1}$ assuming 2500 t soil ha^{-1}; (*Ans.* 50)
   (b) plot a graph showing the relationship between pH and added H^+ for the pH range 3 to 7 with pH on the ordinate and added H^+ in $kg \, ha^{-1}$ on the abscissa, and
   (c) label the abscissa also in terms of $t \, CaCO_3 \, ha^{-1}$ assuming that the soil reactions are reversible (the molar mass of $CaCO_3$ is $100 \, g \, mol^{-1}$ and 1 mol $CaCO_3$ reacts with 2 mol H^+).

      If the soil initially has a pH of 5.0 determine from the graph the pH of the soil after

   (d) H^+ inputs from the atmosphere and produced within the soil amounting to $4 \, kg \, H^+ \, ha^{-1} \, a^{-1}$ for 10 years (*Ans.* 4.2), followed by
   (e) an application of $3 \, t \, ha^{-1}$ of powdered limestone having 75 per cent of the neutralizing ability of $CaCO_3$. (*Ans.* 5.1)

2. A soil has a CEC of $20 \, cmol_c \, kg^{-1}$ of which 60 per cent is satisfied by exchangeable $H^+ + Al^+$. Calculate the amount of lime (g $CaCO_3 \, kg^{-1}$), required to neutralize this exchangeable acidity. (*Ans.* 6).

3. A mineral soil at pH 5.2 had a lime requirement of

$9\,t\,CaCO_3\,ha^{-1}$ for the production of arable crops (Table 8.6). Using the data in Table 8.1(a) calculate how long it would be after the application of this amount of lime before a further lime application would be needed assuming that the farmer allowed the pH to fall to 6.0. (*Ans.* 10 years)

Suggestion: Calculate the buffer capacity of the soil as $t\,CaCO_3\,ha^{-1}\,pH^{-1}$ knowing that pH is raised to 6.7. Convert to $kg\,H^+\,ha^{-1}\,pH^{-1}$. Calculate the input of H^+ $(kg\,ha^{-1})$, needed to reduce the pH to 6.0 and divide by the annual input $(kg\,H^+\,ha^-\,a^{-1})$, to give time in years.

4. Use the data in Colour Plates 13(a) and (b) and the buffer capacities given in Table 8.5 to calculate expected depressions in pH over a 10-year period resulting from low and high atmospheric inputs of acidity (1 and $6\,kg\,H^+\,ha^{-1}\,a^{-1}$) for a sand. Assume that all the acidity reacts in the top 20 cm of soil $(2600\,t\,ha^{-1})$. (*Ans.* 0.25 and 1.5) If the area of high input receives 2000 mm rain a^{-1} with a H^+ concentration of $20\,\mu g\,H^+\,l^{-1}$ (pH 4.7), calculate the acidity input from this source in $kg\,H^+\,ha^{-1}\,a^{-1}$ $(1\,m^3$ rain $= 1000\,l)$. (*Ans.* 0.4) Calculate the

expected 10-year pH depression from this source. (*Ans.* 0.1)

5. Use the data in Fig. 8.2 for the following calculations:

(a) Calculate the apparent buffer capacity of each soil from the change in pH produced by the $5\,t\,ha^{-1}$ lime application to confirm the value given on page 157. (*Ans.* Rothamsted 4.35, Woburn $4.55\,t\,CaCO_3\,ha^{-1}\,pH^{-1}$)

(b) Calculate the inputs of acidity $(kg\,H^+\,ha^{-1}\,a^{-1})$ required to cause the observed decreases in pH; use the data for Rothamsted + 5 t lime ha^{-1} and Woburn + 0t lime ha^{-1} so that changes are compared starting at about pH 6 on both sites. Compare your answers to Table 8.1 and Colour Plates 13(a) and (b).

Suggestion. Calculate the average change in pH a^{-1} between 1962 and 1978. (*Ans.* 0.08 and 0.11) Using the buffer capacity values calculate the annual loss of $CaCO_3$. (*Ans.* 0.36 and 0.48 $t\,ha^{-1}$) Convert your answers to $kg\,H^+\,ha^{-1}\,a^{-1}$. (*Ans.* 7 and 10).

CHAPTER 9

The Availability of Plant Nutrients – Potassium, Calcium and Magnesium

The growth of plants depends on inputs from both the soil and the atmosphere. The elements in plant material are obtained in the following ways:

- Carbon and oxygen are obtained from CO_2 in the atmosphere, being assimilated by photosynthesis in the leaves.
- Hydrogen and oxygen are taken up as water through the roots.
- Potassium, calcium and magnesium are taken up by roots from the soil solution as the cations K^+, Ca^{2+} and Mg^{2+}.
- Nitrogen, phosphorus and sulphur are taken up from soil solution as the anions, NO_3^-, $H_2PO_4^-$ and SO_4^{2-}. Nitrogen is also taken up as NH_4^+, and is obtained indirectly from the soil atmosphere through N fixation by bacteria.
- Other elements are acquired in small amounts (micronutrients) from soil solution.

Based on the composition of dry crop materials in Table 9.1, Table 9.2 gives typical amounts of macronutrient elements that have to be supplied to various crops. In general, N is required in the largest amounts,

Table 9.1 Typical concentrations of macronutrients in crop dry matter (g kg^{-1} = kg t^{-1})†

Crop	N	P	K$^+$	Ca^{2+}	Mg^{2+}	S
Cereals, grain	20	4	6	0.6	1.5	1.5
straw	7	0.8	8	3.5	0.9	1.1
Potato, tubers	14	1.8	22	0.9	0.9	1.4
Ryegrass	25	3	18	4	1.2	1.2
Oilseed rape, seed	36	7	10	4	2.5	10

† Nutrient concentrations in the fresh crop are smaller depending on the water content of the plant material (Table 9.2). For example, cereal grain with 85% dry matter typically contains 17 g N kg^{-1}.
A range of values applies to all the data. The N content of ryegrass is particularly variable (16–32 g kg^{-1}) depending on the stage of growth.
From Archer (1988).

followed by potassium, with the remaining nutrients all required in smaller amounts but with much variation between crops.

Table 9.2 Typical amounts of macronutrients in crops at harvest

Crop	Percentage dry matter†	N	P	K$^+$	Ca^{2+}	Mg^{2+}	S
				(kg ha^{-1})			
Cereals							
6 t grain	85	100	20	30	3	8	8
4.5 t straw‡	85	30	3	30	13	3	4
Potatoes							
50 t tubers	22	150	20	240	10	10	15
Grass							
50 t silage	20	250§	30	180	40	12	12
Oilseed rape							
3 t seed	92	100	20	30	11	7	30

Amounts removed from the soil¶							
(mg kg^{-1})							
Cereals							
6 t grain + 4.5 t straw		50	9	24	6	4	5
(cmol$_c$ kg^{-1})							
				0.06	0.03	0.04	

† See Section 9.3.
‡ The ratio of harvested straw to grain is normally between 0.6 and 0.75 for wheat and barley depending on variety, straw length and height of cutting.
§ The range is from 160 to 320 kg ha^{-1}.
¶ Calculated assuming that all the nutrients are removed from the top 20 cm of soil (2500 t ha^{-1}).

Although the nutrient elements are taken from the soil solution, little of each is present in solution at any given time. They are maintained in solution by various processes:

- Mineral weathering is the primary source of all the nutrients except for C, H, O and N.

- Nitrogen gas is fixed in the soil by bacteria and is subsequently released (mineralized) into the soil solution in the form of NH_4^+ and NO_3^-.
- Plant residues entering the soil rapidly release K^+, Ca^{2+} and Mg^{2+} but only slowly release P, S and N through mineralization, with large amounts stored in soil organic matter. Fig. 3.2 illustrates the processes.
- Clay minerals, sesquioxides and humus give the soil its cation exchange properties. K^+, Ca^{2+} and Mg^{2+} are held as exchangeable cations and buffer (maintain) their concentrations in solution.
- Phosphate is held on sesquioxide surfaces, and as precipitated phosphate minerals. Its release by desorption or dissolution buffers the solution.
- Sulphate held on sesquioxide surfaces in acidic soils buffers the solution. Most of the SO_4^{2-} is free in the solution in neutral soils (Section 7.5). In alkaline soils of arid regions SO_4^{2-} in gypsum may dissolve to maintain solution concentrations.
- Nitrogen is held temporarily as exchangeable NH_4^+. When this is nitrified, all the NO_3^- is free in soil solution except in soils with positive charge (Group iii soils, Ch. 7) which hold exchangeable NO_3^-.
- Compounds (mainly oxides) of N and S together with salts of Ca^{2+}, Mg^{2+} and K^+ enter the soil solution in rain or via dry deposition from the atmosphere.

It would be logical to begin a discussion of plant nutrients with N. It is required in the largest amounts and is the nutrient which most commonly limits growth. However, it is relatively much simpler to understand the principles governing the supply of potassium, calcium and magnesium. These principles can then be applied to the more complex systems.

POTASSIUM AVAILABILITY

The forms and average amounts of potassium in 20 British soils are shown in Fig. 9.1 (Arnold and Close, 1961). There are very large reserves of potassium in soil minerals, particularly in heavy-textured soils. The amount of exchangeable K^+ is between 10 and 100 times larger than the amount in solution which it effectively buffers. Slowly exchangeable K^+ ions are held at the edges of illite clay minerals (Fig. 9.2 and Plate 9.1) and have properties intermediate between exchangeable K^+ and that in the soil minerals. The amount which can be released to plants depends on mineralogy, the weather, the crop and its growth period.

A typical wheat crop may remove 60 kg K^+ ha^{-1} in the grain and straw (Table 9.2) and may contain up to three times this amount during growth (Fig. 11.3). To supply 200 kg K^+ ha^{-1} from the upper 20 cm of soil (2500 t ha^{-1}) requires 80 mg K^+ kg^{-1} soil in a form which is available to the roots during the growing sea-

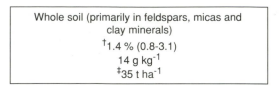

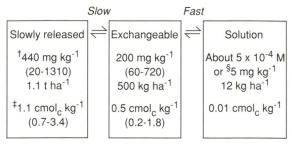

† The average value for the 20 soils studied by Arnold and Close (1961) is followed by the range in brackets.

‡ In the upper 20 cm (2500 t ha^{-1})

§ Assuming 250 g H_2O kg^{-1} soil

Figure 9.1 The amounts of potassium in 20 British soils.

son. There appears to be plenty of exchangeable K^+ in the Arnold and Close soils for the needs of a wheat crop provided the roots can take it up, although this is not true for all soils. The term *nutrient availability* is used in this context. In its broadest sense it is the ability of a soil to supply K^+ to growing plants and thus *available potassium* is the amount which can be taken up. It is dependent on (a) the amount of K^+ in the chemical fractions from which uptake can occur, i.e. solution, exchangeable and some slowly exchangeable K^+, and (b) its accessibility to roots. In its narrowest sense it is the amount of K^+ which can be extracted from the soil by a routine chemical procedure, normally the exchangeable K^+.

There are two aspects to accessibility. Firstly, there may be physical barriers to root growth such as a plough pan or waterlogged soil below a certain depth. The potassium in only part of the soil is therefore accessible. Secondly, the normal distribution of roots (Plate 3.1) means that K^+ has to move through the soil to the root surface. This requires time and so accessibility will depend on root spacing, ease of movement of K^+ through the soil, and the period of growth. Thus the accessibility of K^+ to plants is very restricted in dry soils because there is insufficient water to allow the K^+ to move easily. Also in a badly compacted soil roots may be far apart so restricting uptake. Accessibility is a more serious problem in relation to phosphate which moves very slowly in soils.

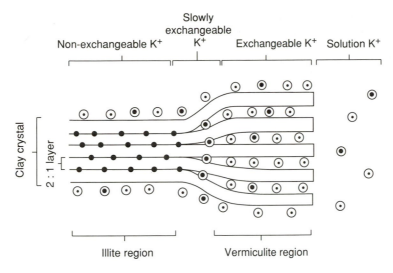

Figure 9.2 The forms of potassium in soils. Vermiculite is the weathered part of the clay crystal. Potassium that can be slowly released is in the illite adjacent to the expanded region. As exchangeable K^+ diffuses out of the vermiculite region to maintain the solution K^+ concentration other cations (Ca^{2+} and Mg^{2+}) enter the wedge-shaped positions and force more of the crystal to expand. This releases more K^+. A large concentration of fertilizer–K^+ can reverse this process, causing K^+ to become non-exchangeable.

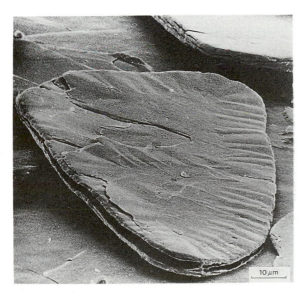

Plate 9.1 A flake of biotite mica after partial removal of interlayer potassium by artificial weathering using a barium chloride solution. The expanded edge is similar to the vermiculite region in Figure 9.2. Unweathered mica (the central region) normally holds its potassium more firmly than illite. From Smart and Tovey (1981).

The measurement of available potassium

Chemical extraction

The simplest measurement is through routine methods which extract exchangeable (plus solution) K^+ or in some cases exchangeable plus some slowly exchangeable K^+ (Section 9.1). The commonly used 30 min extraction period is short compared to the period of growth of crops and so may not extract all the K^+ which could become available during a growing season. On the other hand, unlike roots, it extracts from all the soil efficiently and uniformly and so normally overestimates the amount the crop can take up in one season. *Extractable potassium* is therefore best considered simply to rate soils in order of availability rather than to measure the amount available to a crop.

Pot experiments

Pot experiments allow the measurement of K^+ uptake by plants over time periods similar to those in the field. Thus availability can be measured in a way which allows for the slow release of K^+. However, there are unlikely to be accessibility problems provided the experiment is managed correctly because of the large number of roots constrained in a small volume of soil. For the same reason, the demand for K^+ is large and more K^+ may be slowly released than over a similar period in the field. Thus pot experiments are not a good simulation of field conditions, although they normally give a better estimate of availability than a chemical extraction method. Soil cores (15 cm diameter, 1 m deep) with undisturbed structure simulate field conditions better than pots of soil (Ogunkunle and Beckett, 1988).

The management of pot experiments is described in Section 9.2, and Section 9.3 gives methods for the analysis of the nutrient content of plants.

Pot experiments have many uses in the study of nutrient supply to plants. The results of a classical experiment published in 1961 by Arnold and Close will be used here as an example. The potassium-supplying abilities of soils were compared by growing ryegrass (*Lolium perenne*) in pots in a glasshouse. Water and all nutrients apart from potassium were supplied as required. The grass was cut at intervals and its potassium content was measured over a period of nearly 2 years. Ryegrass is a very suitable plant for pot experiments in that it survives repeated cutting and continues to grow slowly even when nutrients have been seriously depleted.

Fig. 9.3 shows the uptake from four soils, together with the amounts of exchangeable K^+ initially present. These graphs are termed *cumulative uptake curves* because they are obtained by summing the successive K^+ removals to accumulate a total amount. If the amount removed in each cut were plotted separately the graphs would fall rapidly over the first six cuts and then level off at low uptake values. The initial rapid uptake of exchangeable K^+ is followed by removal of slowly exchangeable K^+.

The amount of slowly exchangeable K^+ taken up over any time period can be found from the *potassium budget*. For a closed system where there are no other gains and losses, the sum of potassium removed in the crop and potassium remaining in the soil must be constant. Thus

plant K^+ + exchangeable K^+ + non-exchangeable K^+
$$= \text{a constant}$$

The term *non-exchangeable K^+* has now been introduced. This is all of the soil K^+ apart from exchangeable K^+ (which here includes solution K^+ because of the method used for its measurement). Over a period of growth, plant K^+ must have come from soil K^+ and so

plant K^+ = decrease in exchangeable K^+ + release of
non-exchangeable K^+

If plant uptake and exchangeable K^+ at the beginning and end of the experiment are measured, then the

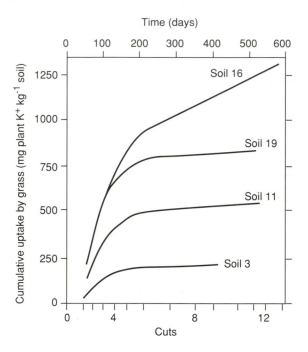

	Soil			
	3	11	19	16
		(mg kg^{-1})		
Exchangeable K^+, initial	90	140	335	720
Exchangeable K^+, final	47	69	84	110
Total uptake by the crop	208	548	818	1313
Release of non-exchangeable K^+	165	477	567	703

Figure 9.3 Cumulative uptake of potassium by ryegrass grown in pots.

amount of non-exchangeable K^+ which has been released can be calculated. Amounts are shown in Fig. 9.3. The graphs suggest that there may be three fractions of potassium available to plants: exchangeable K^+, slowly released K^+, and K^+ which weathers very slowly from soil primary minerals when the solution concentration is small giving the final straight line part of the graphs.

Correlation

The precision with which the initial exchangeable K^+ predicts availability in the Arnold and Close pot experiment is shown in Fig. 9.4. The relationship is clear but with much variability. The amount of exchangeable K^+

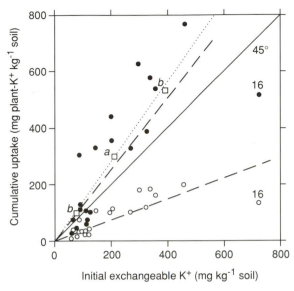

Figure 9.4 The relationship between exchangeable K^+ and uptake by ryegrass in the Arnold and Close (1961) pot experiment. The fitted lines and symbols are explained in the text and in Section 9.6.

o After 50 d
● After 105 d

therefore gives a useful indication of the amount of available K^+. The 45° line shows where cumulative uptake equals initial exchangeable K^+. After 50 days' growth about half of the exchangeable K^+ has been taken up. After 105 days some soils have released significant amounts of non-exchangeable K^+ to the grass. For this set of soils, those which are initially well supplied with exchangeable K^+ are also good releasers with one exception where a poor releaser had presumably recently been fertilized.

The movement of potassium to roots

Because only a small fraction of the soil is in direct contact with roots (Section 3.1) uptake involves movement of nutrient ions through the soil to the root surface.

Two processes are involved.

1. Potassium ions in solution are carried to the root as water is taken up to supply transpiration needs. This is called *mass flow*.
2. If insufficient K^+ arrives at the root by mass flow to satisfy plant needs, the concentration of K^+ in the soil solution adjacent to the root is reduced by plant uptake. This produces a concentration gradient in the soil, and *diffusion* of ions then occurs.

The amounts of K^+ supplied by mass flow and diffusion can be determined in pot experiments. Water use is measured (Section 9.2) and the concentration of K^+ in soil solution is found using the method in Section 7.4 or an *exchange isotherm* described in Section 9.4. The amount of K^+ supplied by mass flow is

$$\text{mg } K^+ = \text{volume of } H_2O \text{ (l)} \times \text{concentration of } K^+ \text{ (mg l}^{-1}\text{)}$$

The amount supplied by diffusive flow is simply total uptake minus the amount supplied by mass flow.

Experiments have shown that all or most of the nitrate, calcium and magnesium can often be supplied by mass flow but only about 20 per cent of the potassium and about 1 per cent of the phosphate. Thus soil structure which influences root spacing has an important effect on the availability of K^+ and P as do soil properties which influence the rate at which ions can diffuse through the soil. The most important of these are the water content (diffusion occurs almost entirely through the water) and the ability of soil to maintain the concentration of ions in soil solution by ion exchange (its buffer power, Section 9.4).

These mechanisms are now well understood (Nye and Tinker, 1977). It is possible to predict the rate of supply of nutrients to plants using mathematical methods based on the measurement of soil and root characteristics. This, together with increased knowledge of the effects of atmospheric conditions on plant growth, has significantly improved our understanding of crop production.

The availability of potassium in the field

Measurement of a *potassium availability index* is described in Section 9.1. This is a simple rating of K^+ availability (0–4 for most arable soils) based on the amount of extractable K^+. In field experiments the potassium index has been measured and responses of crops to applied K^+ have been determined. Fig. 9.5 shows a *response curve* for potatoes which have a large K^+ requirement (Table 9.2). The optimum K^+ input and therefore the optimum yield depend on the cost of the fertilizer and the value of the crop. It is less than the input required to give maximum yield, because as maximum yield is approached, the returns from a given input become smaller and eventually become less than the cost of the extra fertilizer. The typical shape of a response curve is rarely seen for potassium fertilization of cereals, because of adequate availability and a smaller demand than potatoes. For cereals, maintenance applications are normally used to balance crop removals with no direct response being expected. In contrast, N applications normally give direct responses, and typical response curves are obtained.

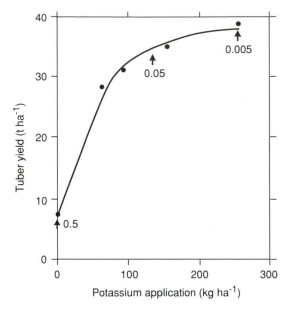

Figure 9.5 A response curve, showing the effect of potassium applications on potato yield. Response is the increase in yield resulting from an input of fertilizer and is numerically equal to the slope which is the increase in yield per kilogram of (extra) K^+ applied. Values are marked on the graph. From Henkens (1986).

Table 9.3 Recommendations of fertilizer needs: potassium for cereal crops

Extractable K^+				
(mg l^{-1} soil)†	0–60	61–120	121–240	Over 240
(kg ha^{-1})†	0–150	151–300	301–600	Over 600
Index	0‡	1	2	Over 2
	Recommended application (kg K^+ ha^{-1})§			
Straw ploughed in or burnt				
Expected yield,				
6 t grain ha^{-1}	71	30	30 M¶	Nil
10 t grain ha^{-1}	91	50	50 M	Nil
Straw removed‖				
Expected yield,				
6 t grain ha^{-1}	101	60	60 M	Nil
10 t grain ha^{-1}	141	100	100 M	Nil

† Assuming 2500 t ha^{-1}, and a soil density in the measuring cup of 1 kg l^{-1} (Section 9.1).
‡ For soils with very small availability, large inputs of K^+ are needed over a number of years to raise the index to 2, except on sandy soils where excessive leaching would result. For example to raise the K^+ from 50 mg kg^{-1} (index 0) to 200 mg kg^{-1} (index 2) in 2500 t of soil, 375 kg K^+ ha^{-1} is needed, or even more depending on how much enters the non-exchangeable fraction.
§ Application rates are given as kg K^+ to facilitate comparison to crop and soil data. Fertilizers are still marketed with K^+ content given as kg K_2O although there is no K_2O in the fertilizer, which is normally KCl. The use of K_2O relates to the old conventions for expressing results from gravimetric analyses. To obtain kg K_2O, multiply kg K^+ by 1.2.
¶ M is a maintenance application. It supplies the K^+ removed in the crop and is based on a K^+ content of 5 kg t^{-1} grain, and the amount is doubled if the straw is removed. A response to K^+ is not expected at index 2. Zero application would result in a decrease in extractable K^+ each year or would deplete reserves of non-exchangeable K^+. Most clay soils have reserves of slowly exchangeable K^+ and maintenance applications may not be needed. Measurements of extractable K^+ after several years will indicate whether availability is being maintained (Section 9.5).
‖ Estimates of the K^+ content of the straw vary. ADAS work on the basis of 5 kg K^+ in the straw for each tonne of grain produced, but the content may be larger (Table 9.2). From MAFF (1988).

Based on field experiments fertilizer recommendations are made taking account of the potassium index, the expected yield, the soil type and management. An example is shown in Table 9.3 for cereals in Britain.

In order to supply the needs of a crop the soil has to contain more extractable K^+ than is taken up. This is because not all of the extractable K^+ is accessible to roots during the growing season and during the period of rapid growth a high rate of supply is needed (Fig. 11.3). For example, a potassium index of 2 is recommended for the production of wheat. The crop may contain up to 200 kg K^+ ha^{-1} (Fig. 11.3) although less is removed from the field. Soils with index 2 have between 300 and 600 kg extractable K^+ ha^{-1} (2500 t soil ha^{-1}). Thus the topsoil alone initially needs to contain at least 1.5 times the maximum amount of K^+ subsequently taken up by the crop and at least 10 times the amount harvested in the grain, with additional large amounts of extractable K^+ in the subsoil. At this index a maintenance application is recommended. At index 1 (150–300 kg ha^{-1}) there may not be adequate K^+, and a yield depression might occur if fertilizer were not applied.

The precision with which exchangeable K^+ can predict fertilizer requirements is shown in Fig. 9.6. There is even more variability than in pot experiments (Fig. 9.4) resulting from local soil and weather conditions and soil and crop management. In these experiments many index 1 soils showed little response to inputs and there is clearly a need to distinguish more effectively the responsive sites. Similar data for P fertilization of sugar beet are given by Cooke (1982, Fig. 23).

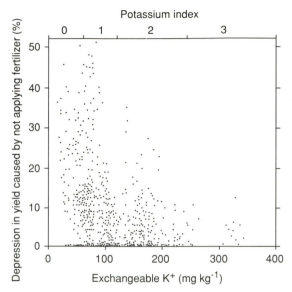

Figure 9.6 The effect of the availability of soil potassium on the response of rotation crops to potassium fertilizer in long-term experiments in northern France. Between 1962 and 1983, a rotation of arable crops was grown which varied slightly between sites. An example is wheat (9), barley (2), maize (4), sugar beet (2), lucerne (2) and peas (1): the brackets indicate the number of years each crop was grown in the 20-year period. Yield (Y_1) with 100 kg fertilizer–K^+ ha^{-1} a^{-1} was compared to Y_0 with no application. For each crop the depression in yield resulting from omission of the fertilizer, $100(Y_1 - Y_0)/Y_1$, was calculated and plotted against the amount of exchangeable K^+ present. Data from Julien (1989).

Field budgets

Potassium budgets are difficult to obtain for field conditions. However, crop removals, fertilizer inputs and changes in exchangeable K^+ can be measured (Section 9.5). These show that in many clay soils significant amounts of K^+ are released each year from non-exchangeable forms and little fertilizer is needed. For example the Boulder Clays of eastern England have supplied all the K^+ needed by crops for many years with no fertilizer inputs. In contrast, only small amounts of non-exchangeable K^+ are released in sandy and some chalky soils which, together with significant leaching losses during the winter, causes difficulties in maintaining adequate levels of available K^+. Chemical methods have been developed to estimate the contribution of non-exchangeable K^+ to crop uptake. Neither these, nor measurements of total soil potassium significantly improve estimates of availability.

CALCIUM AVAILABILITY

Table 9.2 gives the amounts of Ca^{2+} removed by crops: a wheat crop contains about 20 kg Ca^{2+} ha^{-1}. This is small compared to the amounts of exchangeable Ca^{2+} in limed soils where the CEC is mostly filled with Ca^{2+} and Mg^{2+}. A soil with 20 $cmol_c$ Ca^{2+} kg^{-1} has 10 t exchangeable Ca^{2+} ha^{-1} in the plough horizon (2500 t). Thus calcium deficiencies do not occur in limed soils.

In acidic soils of the humid tropics, exchangeable Ca^{2+} levels may be very low. Deficiencies occur together with the toxic effects of aluminium. A critical Ca^{2+} level of 0.2 $cmol_c$ kg^{-1} has been suggested. Other factors, particularly the ratio of Ca^{2+} to Al^{3+} and Ca^{2+} to Mg^{2+}, influence the development of calcium deficiencies (Colour Plate 12).

In terms of amounts of Ca^{2+} in the field 0.2 $cmol_c$ kg^{-1} is 100 kg ha^{-1}. Thus the ratio of the critical level of exchangeable Ca^{2+} to crop needs (Table 9.2) is about 5:1. Not all of the Ca^{2+} may be accessible to plant roots during the growing season, a problem made worse by the fact that root growth is directly related to Ca^{2+} supply. Thus with insufficient available Ca^{2+}, root growth is poor and less of the Ca^{2+} is accessible.

In soils of the humid tropics there is often very little exchangeable Ca^{2+} in the subsoil. This causes particular problems because unlike other nutrients, Ca^{2+} is not easily translocated down through the root system. Thus root growth is inhibited in the subsoil and liming of the topsoil does not immediately rectify the problem because of the slow downward movement of Ca^{2+}. Shallow rooting makes the crop liable to the effects of drought, because even in humid regions there may be only a few days' supply of water in the topsoil. This is an example of an interaction between Ca^{2+} supply and water availability. Interactions are discussed in Ch. 13.

In a calcium budget for limed soils, crop removals are small compared to leaching losses. The latter are closely related to acidity inputs which displace Ca^{2+} (and K^+ and Mg^+ in smaller amounts) by cation exchange. Two mol of H^+ (2 g) displace 1 mol of Ca^{2+} (40 g), and so an input of 10 kg H^+ ha^{-1} a^{-1} (Table 8.1) theoretically displaces 200 kg Ca^{2+} ha^{-1} a^{-1} which is equivalent to 500 kg $CaCO_3$. Fig. 9.7 gives examples. Acidity inputs are related to rainfall (the throughput of water) and the extent of the exchange reaction depends on both pH (the amount of exchangeable Ca^{2+}) and drainage which removes the products of the reaction. Losses are more serious in the humid tropics.

MAGNESIUM AVAILABILITY

Deficiencies of magnesium are more common in temperate climatic regions than are calcium deficiencies.

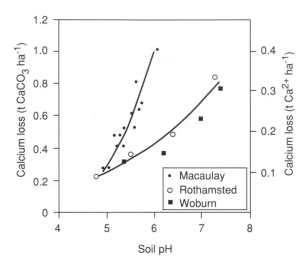

Figure 9.7 Annual calcium losses from cropped soils. Both the Rothamsted and Woburn data (Bolton, 1972) and the Macaulay (Aberdeen) data (Reith, 1962) are calculated from the measured changes in exchangeable Ca^{2+} in the top 23 cm of soil (3000 t soil ha^{-1}). Soil pH decreased during the experiments. Median values are plotted. The mean annual rainfall at Woburn, Rothamsted and Macaulay is 632, 686 and 893 mm respectively.

This reflects the small inputs of Mg^{2+} in lime unless a dolomitic limestone is used. The chemical extraction of Mg^{2+} from soils is carried out using the same method as for K^+ and an index system is also used (Section 9.1). For many crops, fertilizer applications are only likely to be needed if the extractable Mg^{2+} is less than about 15 mgl^{-1} of soil (the lower part of index 0) which is 0.12 cmol$_c$ kg^{-1} and is about half the critical level for calcium. It is also about 40 kg ha^{-1} and again taking wheat as an example (Table 9.2), between 10 and 15 kg is taken up by the crop giving a ratio of required extractable Mg^{2+} to crop Mg^{2+} of about 3:1. For potatoes, sugar beet and grass crops, magnesium applications may be beneficial on index 1 soils.

Particularly in tropical regions Mg^{2+} may be supplied by mineral weathering in larger amounts than Ca^{2+}, reflecting the composition of the parent material. A ratio of Mg^{2+}:Ca^{2+} greater than 1 may lead to calcium deficiencies even though exchangeable Ca^{2+} is greater than 0.2 cmol$_c$ kg^{-1}.

FURTHER STUDIES

Ideas for projects are in Section 9.7 and calculations in Section 9.8.

Section 9.1 The chemical extraction of available potassium, magnesium and calcium

POTASSIUM

There are a number of methods in use. All extract exchangeable and solution K^+ together with varying amounts of the slowly exchangeable fraction. The method described here is the routine procedure for British soils, and uses 1 M NH_4NO_3 as the extracting solution. Alternative extractants are 1 M ammonium ethanoate (acetate), or a mixture of HCl and ammonium fluoride known as the Bray solution (Page, 1982).

Reagents and equipment

Ammonium nitrate, 1 M. Dissolve 80 g of ammonium nitrate, NH_4NO_3, in water and make up to 1 l.
Standard potassium solutions. Dry potassium nitrate, KNO_3, at 105 °C for 1 h and cool in a desiccator. Dissolve 1.293 g in water, add 1 ml of HCl (approx. 36% *m/m* HCl) as a preservative if required and make up to 500 ml. This solution contains 1 mg K^+ ml^{-1}. Pipette 0, 1, 2, 3, 4, and 5 ml into 100 ml flasks and make up to the mark with 1 M NH_4NO_3. These contain 0, 10, 20, 30, 40, and 50 μg K^+ ml^{-1}.
Bottles with caps.

Method

Transfer 10.0 g (or 10 ml, see below) of air-dry <2 mm soil to a bottle. Add 50 ml of 1 M NH_4NO_3. Cap and shake for 30 min on a shaking machine or occasionally by hand for 30 min. Filter and retain the filtrate for measurement of potassium using a flame photometer (or atomic absorption spectrophotometer) following the methods of Section 7.2.

Calculation

Method summary

air-dry soil (? mg K^+ kg^{-1} or mg l^{-1} of soil)
↓
10 g (or 10 ml soil) → 50 ml solution
↓
y μg K^+ ml^{-1}

Example The extract concentration $y = 14.3$ μg ml^{-1} and so in 50 ml of solution there is $14.3 \times 50 = 715$ μg K^+. This came from 10 g of soil. Therefore from 1 kg of soil, $715 \times 1000/10\,\mu g$ or $71.5\,mg\ K^+$ would be extracted. This calculation simplifies to

$$mg\,K^+\,kg^{-1}\ \text{air-dry soil} = 5y$$

Express in terms of oven-dry soil if necessary.

The use of soil volume

Crops are rooting to about the same depth (using the same soil volume) in a range of soils and so volume is a more appropriate way of measuring the index of available potassium. Its use also removes the need to take account of the difference between air-dry and oven-dry soil. For routine analysis, 10 ml of soil is rapidly dispensed using a measuring cup (Fig. 9.8). Scoop the soil

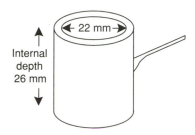

Figure 9.8 Measuring cup.

into the cup and without tapping, remove excess soil using a straight edge (a ruler). The calculation is unchanged numerically but the answer is given as

$$mg\,K^+\,l^{-1}\ \text{of soil} = 5y$$

The bulk density of the soil in the cup (often about $1\,g\,cm^{-3}$) is normally less than the bulk density in the field. However, the values vary roughly in proportion to field values.

The potassium availability index

In Britain soils are placed in classes and allocated an index according to their content of extractable K^+ (Table 9.4). For many arable crops indices 0, 1 and 2

Table 9.4 Potassium availability indices (MAFF, 1988)

Index	Extractable K^+ (mg l^{-1} of soil)
0	0–60
1	61–120
2	121–240
3	241–400
4	400–600

are deficient, low and medium respectively. Index 2 soils have adequate K^+ to supply the needs of cereal crops, but inputs of K^+ are required to maintain the availability of K^+. Index 3 soils have enough K^+ to supply needs for a few years without further inputs. The use of broad classes is an indication of the uncertainties involved when measurements of extractable K^+ are used to indicate availability in the field. Apart from the problems of accessibility, the method cannot assess the availability of slowly released K^+. Therefore even though the chemical method gives a precise value, this cannot be used as an index of availability with the same precision.

Stony soils

The index is determined on <2 mm soil. If a soil contains an appreciable quantity of stones, then extractable K^+ expressed in terms of <2 mm soil overestimates availability in the field. An estimate of stone volume is likely to be approximate (Section 1.2) particularly because stone content is often variable over a field, and so an approximate correction is adequate. If in the field a soil contains 30 per cent v/v stones, then 70 per cent of the soil volume is <2 mm soil and the extractable K^+ in the example given above is $71.5 \times 0.7 = 50.5\,mg\,l^{-1}$ of soil including stones.

MAGNESIUM

The extraction procedure is the same as for potassium. The magnesium in the extract is measured using an atomic absorption spectrophotometer with standards between 0 and $1\,\mu g\,Mg^{2+}\,ml^{-1}$. The extract contains between 0 and $50\,\mu g\,ml^{-1}$ and has to be diluted by 50 (2 ml made up to 100 ml) to bring it into the range for measurement. A releasing agent is needed to prevent interference in the measurement; dissolve 2.68 g of lanthanum chloride, $LaCl_3.6H_2O$, in water and make up to 100 ml. To 20 ml of each standard and extract add 1 ml of releasing agent and mix before measurement.

Alternatively an EDTA titration can be carried out on the undiluted extract (Section 7.2) but with much reduced precision.

Table 9.5 Magnesium availability indices (MAFF, 1988)

Index	Extractable Mg^{2+} (mg l^{-1} of soil)
0	0–25
1	26–50
2	51–100
3	101–175
4	176–250

Classes of extractable Mg^{2+} and availability indices are shown in Table 9.5.

CALCIUM

Deficiencies of calcium are only likely to occur on very acidic soils. Extractable Ca^{2+} need only be determined if the soil pH (Section 8.1) indicates that it might be deficient. It can be extracted following the method for magnesium, with standards between 0 and $5 \, \mu g$ $Ca^{2+} \, ml^{-1}$. Critical levels are discussed on p. 181. The determination of the lime requirement of a soil (Section 8.5) estimates the lime needed to raise the soil pH to a target value which will contain a very much greater amount of Ca^{2+} than that needed as a fertilizer.

Section 9.2 Pot experiments – techniques and management

There are advantages in using pot exeriments rather than field experiments for the study of nutrient availability in soil:

- They are normally easier to organize and take less time.
- Soil variability is eliminated by thoroughly mixing the soil.
- All pots are placed in an identical environment.
- Rain can be excluded, allowing control of the water supply.
- A 'closed' system can be used so that no leaching losses occur.
- The whole plant including its roots can be harvested and measured.
- Nutrient treatments can be applied to a known mass of soil.

There are, however, disadvantages:

- The roots grow in a confined volume.
- Watering has to be carefully controlled to avoid waterlogging.
- Competition between adjacent plants is dissimilar to field conditions.
- In a heated glasshouse, low light intensity in the winter may cause plants to become etiolated (unnaturally elongated).

The preparation of the soil

Soil dug from a field site should be allowed to dry in the air until it can be pressed through a garden sieve (approximately a 1 cm aperture). It can then be dried further as required. Unless there are special reasons, soil should not be passed through a 2 mm sieve. This destroys much of its structure and causes problems of aeration and waterlogging. If chemical measurements are to be made the following techniques can be adopted:

1. A subsample of the 1 cm sieved soil should be air dried, weighed and then passed through a 2 mm sieve and weighed again. From this the mass of $<2 \, mm$ soil kg^{-1} of $<1 \, cm$ soil in the pot can be obtained, and chemical determinations made on the $<2 \, mm$ soil can be related to the mass of soil in the pot.
2. The pot experiment can be carried out using $<2 \, mm$ soil but special control of water supply to the pots is needed (see below).

Size of pots and amount of soil

The amount of soil required for the range of commonly used pots depends on bulk density after filling the pot. This is often about $1 \, g \, cm^{-3}$ and is less than field bulk densities because of sieving and repacking. Approximate values are given in Table 9.6. Manufacturers normally give pot sizes as the diameter of the top of the pot in inches.

Table 9.6 The approximate amounts of soil needed for commonly used pots

Pot diameter (inches)	Dry soil mass (kg)
6	1.5
5	1.0
4.5	0.6
4.0	0.4
3.5	0.3
3.0	0.2
'Drainpipe' pots	
15 cm diam × 50 cm	8
10 cm diam × 25 cm	2

Application of lime

Table 8.5 gives buffer capacities for soils of varying texture.

Measure soil pH (Section 8.1) and determine texture by the finger method (Section 1.2). On the basis of the required pH values calculate the lime required per kilogram of soil. This can be applied as $CaCO_3$ powder, or as $Ca(OH)_2$ powder if a rapid reaction with the soil is required. Place dry soil in a large polythene bag, add the required mass of lime and mix well by shaking the bag before putting the soil in the pot. Treat each pot of soil separately because bulk treatment leads to uneven distribution between pots and less exact replication.

After remoistening the soil, allow time for the lime to react (1 or 2 weeks).

Application of nutrients

Because the masses of nutrients required per pot are small it is convenient to make up solutions and apply the necessary volume to each pot of soil. It is also quicker if many pots have to be treated. The solutions should be added to the soil in the polythene bag after the lime has been mixed. Sprinkle the solution over the soil with shaking to ensure as uniform a distribution as possible. Alternatively place the soil in a domestic food mixer, and stir slowly while adding the solution.

Nitrogen

Application rates can be based on field recommendations or on expected crop removals (Tables 9.2 and 11.5). Taking as an example $100 \, kg \, N \, ha^{-1}$, the application can be calculated on the basis of the area of the soil surface in the pot or the mass of soil:

- *Area of soil.* Assuming that a 5 inch diameter pot is used for 1 kg of soil, the area of the soil surface is $\pi(0.063 \, m)^2 = 0.012 \, m^2$. Since $1 \, ha = 10^4 \, m^2$, the application rate is 100 kg N per $10^4 \, m^2$. Therefore to obtain the same application rate per unit area $100 \times 0.012/10^4 \, kg = 0.12 \, g \, N$ is required per pot. This is normally supplied as NH_4NO_3. One mol (80 g) contains 28 g N. Therefore the required application is $0.12 \times 80/28 = 0.34 \, g \, NH_4NO_3$ per pot.
  Make up a solution containing $13.6 \, g \, NH_4NO_3 \, l^{-1}$ and apply $25 \, ml \, kg^{-1}$ of soil (per pot).
- *Mass of soil.* The standard value for the mass of soil per hectare can be used (2500 t to 20 cm depth), or a reduced value taking account of the depth of soil in the pot (about 12 cm for a 5 inch pot). An application of $100 \, kg \, N \, ha^{-1} = 100 \, kg$ per 2500 t or $0.04 \, g \, N \, kg^{-1}$ or $0.114 \, g \, NH_4NO_3 \, kg^{-1}$ (per pot). Make up a solution containing $4.6 \, g \, l^{-1}$ and apply $25 \, ml \, kg^{-1}$ of soil.
- *The difference.* Note that the first rate is about three times the second. This is because there is more soil under a given area in the field than under the same area in the pot. Normally the higher rate of application is chosen because the soil in the pot has to supply the needs of the crop whose growth rate is dependent on the input of light per unit area of soil surface. However, there may be situations where the lower rate should be applied.

The disturbance and preparation of the soil leads to more rapid mineralization of N in the pots than would occur under field conditions. Particularly in soils taken from woodland or grassland areas with large organic matter contents, there may be a good supply of N without the use of fertilizer.

Other macronutrients

Suggested treatments are given in Table 9.7 based on the area of soil ($0.012 \, m^2 \, pot^{-1}$). Alternatively, phosphorus and potassium can be added together. An application of 25 ml of a solution containing $6.60 \, g \, K_2HPO_4 \, l^{-1}$ gives the equivalent of 24 kg P and $60 \, kg \, K^+ \, ha^{-1}$. Similarly 25 ml of a solution containing $7.68 \, g \, MgSO_4.7H_2O \, l^{-1}$ will supply the equivalent of $20 \, kg \, S$ and $15 \, kg \, Mg^{2+} \, ha^{-1}$.

If a soil has only a small amount of a major nutrient more may be needed. Measurements of extractable nutrients together with a knowledge of critical levels (MAFF, 1988) allow suitable application rates to be calculated.

Note that seeds contain nutrients, and their amounts may have to be measured as an input to the pot if a nutrient budget is being determined (Section 9.3).

Table 9.7 Suggested nutrient applications for pot experiments

	Nutrient				
	N	K^+	Mg^{2+}	P	S
Field application ($kg \, ha^{-1}$)	100	60	10	20	20
Pot application, (mg element per pot)	120	72	12	24	24
Pot application, (g reagent per pot)†	0.34	0.14	0.10	0.11	0.11
Solution, 25 ml applied per pot (g reagent l^{-1})	13.6	5.6	4.0	4.4	4.4

† N as NH_4NO_3, K^+ as KCl, Mg^{2+} as $MgCl_2.6H_2O$, P as $CaHPO_4$ and S as Na_2SO_4.
Pot applications are for a 5 inch pot (about 1 kg soil) to give the same input per unit area of soil surface as in the field.

Micronutrients

Soils normally contain sufficient available micronutrients for plant needs in pot experiments. Section 13.1 gives a method for determining whether micronutrients are deficient together with suggested application rates.

Application of water

The normal technique used by a gardener would be to apply water each day (or as required) to the soil surface and allow the excess to drain away. This, however, causes leaching of nitrate and other nutrients.

A 'closed' system can be established if the pot stands in a plastic saucer. Excess water drains into the saucer but can then be taken back into the soil by capillary action as the plant transpires. However, this means that soil at the bottom of the pot is standing in water, and anaerobic conditions may develop. Provided soil structure is very good, this system is satisfactory. It should be remembered, however, that pores 1 cm above the water level need to be 3 mm in diameter to remain air filled (p. 85). This is the reason why a coarse sieve only should be used to prepare the soil. Subsoils with small organic matter contents have weak structure, which may collapse when the soil is watered. Anaerobic conditions lead to poor root development and loss of nitrate by denitrification.

Drainage can be improved by mixing coarse sand, peat, expanded vermiculite or 'Perlite' with the soil. Peat and vermiculite also change the chemical properties of the soil whereas sand and Perlite are relatively inert.

Watering can be controlled if necessary by adding daily to maintain the soil near to the field capacity. Tables 5.2 and 12.1 give values for soils of varying texture. For an experiment with standard pot size and mass of soil per pot, a simple pair of scales makes maintenance easy. Fill a bottle with dry sand until its mass is the same as the required mass of pot and moist soil and place it on one of the pans. Each day place the pots in turn on the other pan and add water to re-establish the mass. Growing plants add extra mass to the pot and adjustment of watering to allow for this may be necessary. Unless very large plants are grown, their mass is small compared to the mass of water in the pot.

Water use by the plants can be measured using this technique. Add water from a measuring cylinder and record the volume. Alternatively, the pots can be weighed on a normal balance and the loss of water recorded. Control pots without plants are required if evaporation from the soil surface needs to be subtracted from total water loss to give transpiration.

Germination

Seeds should be germinated in a tray of soil, expanded vermiculite or Perlite. Seedlings of uniform size should be selected for transplanting into the pots. As a precaution plant one or two extra seedlings in each pot and thin out to the required number when they are established. The number of plants per pot depends on pot size and the final plant size. For example, cereals are normally grown at about 300 plants m^{-2}. In a 5 inch pot with a soil surface of $0.012\,m^2$, three or four seedlings per pot gives a field density.

Grass seedlings are difficult to handle. It is convenient to sow the seeds directly into the pot and to allow many seedlings to grow so that yield depends on the size of the pot rather than the number of plants. A field seeding rate for ryegrass is about $3\,g\,m^{-2}$ or $0.04\,g$ per 5 inch pot. Five times this rate allows a full cover to be established rapidly. Germination is more uniform if the seeds are covered with a few mm of coarse sand.

Seeds of common garden plants are easily obtained from garden centres. Crop seeds can be obtained from agricultural merchants; weed seeds are also available. Addresses are given in Appendix 1.

Biological variability and replication

The statistical analysis of variability resulting from soil sampling and analysis is discussed in Sections 1.3 and 3.8. Plant experiments introduce biological variability and it is even more important to take this into account when differences in growth and nutrient uptake are being examined. Normally at least three replicates per treatment are required. Mean values can then be compared using standard methods. Statistical advice at the planning stage of a pot experiment is valuable. Differences between plants are minimized by selecting uniform seedlings or establishing many plants as in the case of grasses.

Differences in growth may occur if the environment varies over the bench in the glasshouse. Light, temperature and draughts may all be involved. For this reason randomize the pots over the bench and change their positions occasionally.

Pests and diseases

Standard sprays can be used. The most common problem in Britain is the development of fungal infections on cereals. This can be easily controlled using Benlate (ICI) or a similar compound.

Harvesting

Cut the tops of the plants at soil level, place in a paper bag and dry at 100 °C. Both fresh weight and the mass of dry matter can be determined.

Roots can be separated from soil using the methods given in Section 3.1. Fresh weight, root length and the mass of dry matter can be determined.

The nutrient content of the dry matter can be determined using the methods in Section 9.3.

Section 9.3 The analysis of plant material

The macronutrient elements are converted into their oxides and carbonates by ashing dry plant material at between 450 and 500 °C. The ash is dissolved in HCl and potassium, calcium, magnesium and phosphorus are determined by standard methods. Most of the N and much of the S are lost in the ashing process as gaseous oxides. An acid digestion (Section 11.2) is used for N, and a modified dry ashing method for S (Section 10.5). An alternative acid digestion method for potassium, calcium, magnesium, and phosphorus is given in Page (1982) and MAFF (1986a). The effects of temperature during dry ashing are discussed by Isaac and Jones (1972).

Burning vegetation in the field is similar to laboratory ashing. At the higher temperatures of the fire, however, some P is also lost. The other elements remain to dissolve in rain water and are washed into the soil. For this reason bonfire ash is a good source of nutrients other than N and S.

Determination of percentage dry matter

Plant materials when harvested contain variable water contents. Table 9.2 gives examples expressed as a percentage of dry matter in the crop. The mass of the crop can be expressed as a mass of fresh material (fresh weight), a mass of air-dry material or as dry matter after drying to constant weight at 100 °C which may take between 20 and 40 h depending on the plant material. The water content of fresh or air-dry material is normally expressed as a percentage: g H_2O per 100 g of fresh or air-dry material. The dry matter content is also expressed as a percentage: g dry matter per 100 g of fresh or air-dry material. Note that this differs from soil water contents expressed as g H_2O per 100 g oven-dry soil (Section 3.3).

Preparation of plant dry matter

Dry the plant material at 100 °C and grind to pass through a 1 mm sieve. Alternatively, fine chopping with scissors or a food chopper prepares a material from which a representative subsample can be taken.

Dry ashing and the preparation of the plant extract

Reagents and equipment

Concentrated hydrochloric acid, approximately 36 per cent m/m HCl.

Hydrochloric acid, approximately 6 M HCl. Mix equal volumes of concentrated HCl and water. (Pour acid into the water – CARE).
Evaporating basins, 20 ml capacity.
Muffle furnace.
Water bath.

Method (Also see Note 1)

Weigh 2 g (± 0.01) of plant dry matter into an evaporating basin. Heat at 450–500 °C overnight in a muffle furnace to give a grey ash. If black C remains, moisten with water, dry at 105 °C and heat again at 450–500 °C. After cooling add 10 ml of 6 M HCl and cover with a watch glass. When effervescence is complete, rinse splashed material off the watch glass into the basin, place on a boiling water bath and evaporate to dryness. Continue heating for a further hour on the bath or in an oven at 105 °C. Moisten the residue with 2 ml of concentrated HCl, cover with a watch glass and boil on the water bath for 2 min. Add 10 ml of water and boil again. Remove and rinse the watch glass into the basin. Quantitatively transfer the contents with washings into a 50 ml volumetric flask. Make up to the mark. Filter the solution through a Whatman No. 541 paper rejecting the first few ml. Retain the remainder of the plant extract for analysis of potassium (below), phosphorus (Section 10.4) and sulphur (Section 10.5). Calcium, magnesium and sodium can also be determined by flame emission or absorption methods.

At the same time carry out the above procedure omitting the plant material. This gives a blank solution which should be analysed in the same way as the plant extract.

THE DETERMINATION OF POTASSIUM IN THE PLANT EXTRACT

The concentration of potassium in the diluted plant extract is determined using a flame photometer.

Reagents

Standard potassium solutions. Dry potassium chloride, KCl, at 105 °C for 1 h and cool in a desiccator. Dissolve 0.954 g in water, add 1 ml of concentrated HCl (approx. 36% *m/m* HCl) if required as a preservative, and dilute to 500 ml in a volumetric flask. This solution contains 1 mg K^+ ml^{-1}. Pipette 0, 1, 2, 3, 4 and 5 ml into 100 ml flasks and make up to the mark to give standards containing 0, 10, 20, 30, 40, and 50 μg K^+ ml^{-1}.

Method

Plant dry matter is likely to contain between 5 and

$30 \, mg \, K^+ \, g^{-1}$ (Table 9.1). When 2 g is extracted into 50 ml, the concentration in solution is between 0.2 and $1.2 \, mg \, ml^{-1}$. This is too large for direct measurement. Pipette 2 ml of the extract into a 100 ml flask and make up to the mark with water. The concentration will be between 4 and $24 \, \mu g \, K^+ \, ml^{-1}$.

Calibrate the flame photometer following the methods in Section 7.2 and determine the concentration of K^+ in the plant extract.

Calculation

Method summary

plant dry matter ($? \, g \, K^+ \, kg^{-1}$)
↓
2 g → 50 ml plant extract
↓
2 ml → 100 ml diluted extract
↓
$y \, \mu g \, K^+ \, ml^{-1}$

Example The diluted extract has a concentration $y = 15.2 \, \mu g \, ml^{-1}$. In the 100 ml of diluted extract there is $100 \times 15.2 \, \mu g$ or $1.52 \, mg \, K^+$. This was in 2 ml of the undiluted extract. Therefore in 50 ml of undiluted extract there are $1.52 \times 50/2 = 38.0 \, mg \, K^+$. This came from 2 g of dry matter (use the actual mass here), and so in 1 kg of dry matter there is $38.0 \times 1000/2 \, mg$ or $19.0 \, g$ K^+. Expressed as a percentage in the dry matter this is 1.9 per cent.

The calculation simplifies to

$g \, K^+ \, kg^{-1}$ plant dry matter = 2.5y/mass of plant dry matter

If a blank solution was prepared, subtract its concentration from that of the diluted extract before making the calculation.

Note 1 If potassium only is to be determined the procedure can be simplified because K^+ is easily released from plant material. An example of this is the loss of K^+ from the leaves of cereals into rain water during the ripening period (Fig. 11.3).

Prepare approximately 1 M HCl by diluting 25 ml of concentrated HCl (approx. 36% *m/m* HCl) to 500 ml. Place 2 g of plant dry matter in a beaker and soak in 25 ml of 1 M HCl overnight (cover with a watch glass). Filter through a Whatman No. 541 paper into a 100 ml flask, rinsing the beaker and filter paper with 1 M HCl. Make up to the mark. Pipette 4 ml into a 100 ml flask and make up to the mark with water. This diluted extract is used for the determination of potassium.

The detail of the calculation is changed but it simplifies into the same relationship as that given for the ash extract.

Section 9.4 The measurement and use of potassium exchange isotherms

The term 'isotherm' applies to a line joining points (on a map or a graph) at *the same temperature*. In studies of the properties of surfaces it applies to the graph showing the relationship between the amount of an ion or a molecule sorbed on to the surface and the amount in the air or solution in contact with the surface. This *sorption isotherm* is determined at a given temperature and so the graph joins points *determined at the same temperature*. Depending on the process being studied, the terms adsorption-, desorption- or exchange isotherm are also used.

A potassium exchange isotherm is shown in Fig. 9.9 for a simple system. A clay is washed with a $CaCl_2$ solution until the exchange capacity is filled only with Ca^{2+}: it is said to be calcium saturated. Into a suspension of this clay in 10 mM $CaCl_2$ solution, KCl solution is added in increasing amounts. Potassium exchanges with Ca^{2+} and an equilibrium is established for each addition.

$$clay^{2-} \, Ca^{2+} + 2KCl_{(aq)} = clay^{2-} \, K_2^+ + CaCl_{2(aq)}$$

The concentration of K^+ remaining in solution at equilibrium is measured. The difference between the amount of K^+ added and the amount which remains in solution is the amount on the exchange sites. A relationship can be plotted between the amount of exchangeable K^+ and the solution concentration.

Initially K^+ ions are strongly attracted to the 'wedge'

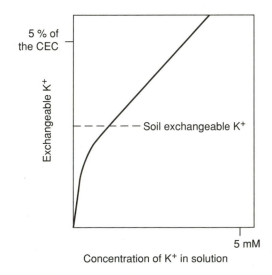

Figure 9.9 A potassium–calcium exchange isotherm.

sites shown in Fig. 9.2 and then more weakly on to sites on the exposed clay surface. Thus the graph is curvilinear, the slope decreasing as the amount of exchangeable K^+ increases.

The amounts of exchangeable K^+ in soils are normally less than 5 per cent of the CEC, and the soil solution concentrations are less than 5 mM. For practical purposes therefore only this part of the isotherm is studied.

A soil is a less simple system but the same principles apply. As sampled from the field it already contains some exchangeable and solution K^+. The dotted line in the figure shows this starting point. The graph above this line shows K^+ being adsorbed on to exchange sites, and below the line K^+ is being desorbed.

A further complication has to be considered. The extent to which K^+ is exchanged depends on the concentration of the other competing cations in the system. Limed agricultural soils in temperate regions have their exchange sites dominated by Ca^{2+}, and soil solution concentrations of about 3 mM (Table 7.6). The exchange isotherm is therefore measured using a $CaCl_2$ solution. The concentration chosen is often 10 mM which imposes a standard condition on all soils despite differences in their natural concentrations. For some purposes it is necessary to measure the Ca^+ in soil solution and use this concentration.

The isotherm is normally plotted with the units of $cmol_c\,kg^{-1}$ soil and $mmol\,l^{-1}$ solution for exchangeable and solution K^+ respectively. Alternatively $mg\,kg^{-1}$ and $mg\,l^{-1}$ can be used.

The measurement of the isotherm

Reagents and equipment

Calcium chloride, 0.4 M. Dissolve 14.7 g $CaCl_2.2H_2O$ in water and make up to 250 ml.
Potassium chloride, 0.1 M. Dry KCl at 105 °C for 1 h and cool in a desiccator. Dissolve 3.727 g in water and make up to 500 ml.
Standard potassium chloride solutions in 10 mM calcium chloride. Into six 1 l volumetric flasks pipette 25 ml of 0.4 M $CaCl_2$. Into the same flasks pipette 0, 5, 10, 20, 30 and 50 ml of 0.1 M KCl solution. Make up to the mark. These solutions contain 0, 0.5, 1, 2, 3 and 5 mM K^+ in 10 mM $CaCl_2$.

If the alternative method with K^+ in mg rather than mmol is to be used, adjust the preparation of solutions as follows. Dissolve 3.813 g of KCl in water and make up to 500 ml. This contains 4 g $K^+\,l^{-1}$. Dilute as above to give 0, 20, 40, 80, 120 and 200 mg $K^+\,l^{-1}$ in 10 mM $CaCl_2$.
Shaking bottles with caps to hold 50 ml.
Flame photometer.

Method

Weigh into six bottles 2.5 g of air-dry <2 mm soil. Pipette 25 ml of the standard KCl solutions into the bottles and place on a shaker for 30 min, or shake occasionally by hand over this period of time. Filter the suspensions through a Whatman No. 41 paper into 25 ml beakers rejecting the first few ml.

Calibrate the flame photometer using the standard KCl solutions following the methods of Section 7.2, and measure the K^+ concentration in the six filtrates.

Calculation

Example The calculation below is worked through for a solution which was initally 0.5 mM, and after shaking was 0.65 mM. Values are also given on the right-hand side for the 5 mM solution which after shaking was 4.40 mM.

(a) The concentration of K^+ before shaking was 0.50 $mmol\,l^{-1}$	5.00
(b) The amount of K^+ in 25 ml of solution before shaking was $0.50 \times 25/1000 = 0.0125$ mmol	0.125
(c) The concentration of K^+ after shaking was 0.65 $mmol\,l^{-1}$	4.40
(d) The amount of K^+ in 25 ml after shaking was $0.65 \times 25/1000 = 0.01625$ mmol	0.110
(e) The change in the amount of K^+ in 25 ml of solution during shaking was $0.01625 - 0.0125 = +0.00375$ mmol The positive sign indicates a gain, negative a loss.	−0.015
(f) The change in the amount of K^+ in 2.5 g of soil during shaking was −0.00375 mmol	+0.015
(g) The gain or loss by the soil is converted into standard units. Thus −0.00375 mmol per 2.5 g $= -0.00375 \times 1000/2.5$ $= -1.5\,mmol\,kg^{-1}$ $= -0.15\,cmol_c\,kg^{-1}$ air-dry soil	+0.60

The calculation simplifies to:

gain or loss by the soil ($cmol_c\,kg^{-1}$ air-dry soil)
    = concentration before shaking − concentration after shaking ($mmol\,l^{-1}$)

If required, express in terms of oven-dry soil. *Example:* the soil contains 5 g H_2O per 100 g oven-dry soil giving −0.16 and 0.63 $cmol_c\,kg^{-1}$ respectively.

Using the alternative method with solution concentrations in mg $K^+\,l^{-1}$, the calculation simplifies to:

gain or loss by the soil $(\text{mg kg}^{-1}$ air-dry soil)

= 10 × (concentration before shaking − concentration after shaking) (mg l^{-1}).

It can be seen that when a solution with a small concentration of K^+ is shaken with the soil, K^+ is desorbed. Adsorption occurs with an initially large concentration. The values which have been calculated are plotted in Fig. 9.10(a) as closed circles. The four open circles are values from the other solutions.

The isotherm can be replotted to show the relationship between exchangeable K^+ and solution concentration (Fig. 9.10(b)). Measure exchangeable K^+ using the standard method of Section 7.2. For this soil its value might be 0.45 $\text{cmol}_c\,\text{kg}^{-1}$. To the calculated gains and losses by the soil add 0.45 and replot the values. The dotted horizontal line marks the condition of the soil as sampled from the field. Note that the two methods are not strictly comparable since K^+ is being adsorbed and desorbed in 10 mM $CaCl_2$ but is extracted in 1 M ammonium ethanoate.

The results can, however, be usefully displayed in this way.

Errors

The points do not lie perfectly on the line. This results from analytical errors, which are particularly serious when a small difference between two measurements is the basis of the calculation. For example, if there is a 5 per cent error in the measurement of both the 0.5 and 0.65 mM solutions (0.5 ± 0.025 and 0.65 ± 0.0325) the difference may lie between 0.525 − 0.6175 = −0.0925 and 0.475 − 0.6825 = −0.2075, and the error in the calculated gain or loss by the soil would be 0.15 ± 0.0575 which is ±38 per cent. This is marked on Fig. 9.10(a).

There is a general principle here: *the error of the difference between two measured values is always greater than the error of a single measurement.* This is seen in a statistical context when the difference between two mean values is determined (Section 3.8). The errors can be reduced by measuring the standards and extracts in pairs to ensure that the calibration of the flame photometer does not change during the measurements.

The use of the isotherm

The soil solution concentration

The point where the isotherm crosses the abscissa (Fig. 9.10, point c) is the concentration in solution causing no adsorption or desorption of K^+. Provided the Ca^{2+} concentration is the same as that in the field, then this point gives the soil solution concentration of K^+ (1.1 mM in the example here).

A simple adjustment can be made if the measurements were determined at a Ca^{2+} concentration other than that in soil solution. For example, the approximate

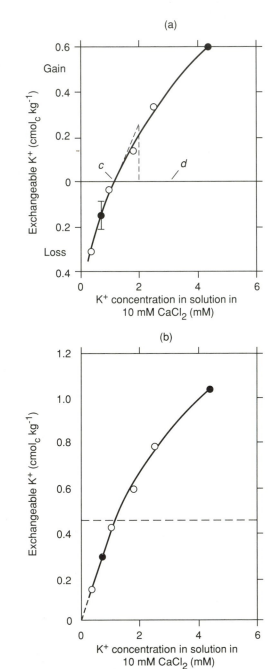

Figure 9.10 Potassium exchange isotherms (a) showing adsorption and desorption, and (b) replotted knowing the amount of exchangeable K^+ in the soil.

K^+ concentration, $[K^+]$, can be calculated at a Ca^{2+} concentration, $[Ca^{2+}]$, of 3 mM. The ratio $[K^+]/[Ca^{2+}]^{1/2}$ remains constant despite changes in $[Ca^{2+}]$ for a given ratio of exchangeable K^+/exchangeable Ca^{2+}. This is known as the *Ratio Law*. Thus if $[K^+]$ = 1.1 mM and $[Ca^{2+}]$ = 10 mM, then

$$[K^+]/[Ca^{2+}]^{1/2} = 1.1 \times 10^{-3}/(10 \times 10^{-3})^{1/2} = 0.011$$

This must equal $y/(3 \times 10^{-3})^{1/2}$ where y is the value of $[K^+]$ at $[Ca^{2+}]$ = 3 mM. Thus y = 0.6 mM. The Ratio Law is related to the Gapon exchange equation (Section 9.8, Calculation 4).

The buffering of potassium in soil solution

The isotherm (Fig. 9.10(a)) illustrates the meaning of buffering in relation to the K^+ concentration in soil solution. The moist soil may contain 300 ml of water in 1 kg of soil. The soil solution concentration is 1.1 mM, and so in 300 ml there is

$$1.1 \times 300/1000 = 0.33 \text{ mmol or } 0.033 \text{ cmol}_c \text{ } K^+$$

If the solution were not in contact with soil particles removal of this amount of K^+ from solution would reduce its concentration to zero. However, when in contact with soil, K^+ is released by exchange in response to a decrease in concentration.

Release of this amount of K^+ (0.033 cmol$_c$) from 1 kg of soil would cause only a small change in the total amount of exchangeable K^+ (0.45 cmol$_c$ kg^{-1}). Thus the solution concentration is maintained close to its initial value.

Buffering and fertilizer use

The significance of buffering is seen in relation to normal fertilization practice. Wheat may receive 60 kg fertilizer-K^+ ha^{-1} (Table 9.3) which is subsequently removed in the crop. This input can be expressed in the units used for the isotherm. Assuming that the 0–20 cm layer contains 2500 t soil ha^{-1}, 60 kg ha^{-1} = 24 mg K^+ kg^{-1} which is 0.06 cmol$_c$ kg^{-1} (molar mass = 39.1 g mol^{-1}). Thus the fertilizer application would raise the solution concentration from 1.1 to 1.3 mM, and the concentration would return to 1.1 mM during cropping. Near to a fertilizer granule the concentration will be higher, but for the whole soil layer the average change is very small. A mean value of 0.4 mM has been found for soils in Oxfordshire (Table 7.6) with a range from 0.01 to 2 mM.

These principles can be applied to one of the tenets of organic farming systems where the use of KCl as a fertilizer is not allowed partly on the grounds that the concentration of the dissolved salt, particularly the chloride, is raised to an unacceptable level leading to leaching losses. In relation to K^+ there is clearly very little change because of the buffering of the solution. Even if almost all the added K^+ remained in soil solution in a very poorly buffered sandy soil then the concentration of K^+ would only be raised by about 2 mM (300 ml solution kg^{-1} soil).

The input of Cl$^-$ would be 54 kg ha^{-1} for the above application. This is not buffered in solution and making the same assumptions the concentration would also be raised by about 2 mM. The Oxfordshire soils had a mean value of 1.6 mM and a range from 0.1 to 10 mM. Rainfall has a concentration of 0.05–0.5 mM depending on proximity to the sea. Thus even in an unbuffered system, the change in concentration is small apart from around a fertilizer granule (see also Section 13.3.) It is unlikely that there are detrimental effects on soil or crop. The Cl$^-$ is leached out of the soil each winter and so does not accumulate as in an arid environment.

Buffer power

The term 'buffer capacity' applied to soils strictly only applies to the buffering of soil pH (Section 8.4). It is sometimes used in relation to nutrients in solution although the alternative term *buffer power* is preferred. Buffer power for soil potassium can be defined as the mass of K^+ required per kilogram of soil to change the solution concentration by one unit. Based on the isotherm, the unit of concentration is mmol l^{-1} and buffer power is given by the slope expressed as cmol$_c$ kg^{-1} (mmol l^{-1})$^{-1}$.

The curvature of the isotherm indicates that buffer power varies depending on the amount of potassium present. The steeper the slope, the more strongly buffered is the solution. The buffer power of the soil in its field state is found at the point where the isotherm crosses the abscissa by drawing a tangent to the curve. Its slope here (the dashed line in Figure 9.10(a)) is 0.22/0.9 = 0.244 expressed in the above units.

Apart from the nature of the exchange sites, the main factor influencing buffer power is the CEC. This is most easily understood by reference to the isotherm again. To raise the concentration from c to d, 0.4 cmol$_c$ kg^{-1} has to be added to the soil. If the exchange capacity were doubled (a larger clay or organic matter content) 0.8 cmol$_c$ kg^{-1} would have to be added to give the same change of concentration, and the slope of the isotherm would be doubled.

Using buffer power values, changes in solution concentration can be calculated for known inputs of potassium. For approximate calculations, assume that all the added K^+ becomes exchangeable. Thus in the case of the wheat crop above an input of 0.06 cmol$_c$ kg^{-1} will give in this soil a concentration change of

$$0.06/0.244 \text{ cmol}_c \text{ kg}^{-1}/\text{cmol}_c \text{ kg}^{-1} (\text{mmol l}^{-1})^{-1}$$
$$= 0.026 \text{ mM}$$

Buffer power values can also be used to calculate the proportion of added fertilizer-K^+ which remains in solution. For this purpose it is useful to express buffer power as $mmol_c kg^{-1} (mmol l^{-1})$ which for the above example is 2.44 with the units simplifying to $l kg^{-1}$. The reciprocal of this value (0.41) is the ratio of the amount of added K^+ which remains in solution (per litre) to the amount which becomes exchangeable (per kilogram). In a moist soil with 0.3 l solution kg^{-1} soil the ratio is $(0.41 \times 0.3):1 = 0.12:1$. Thus the proportion of the added K^+ which remains in solution is $0.12/(1 + 0.12) = 0.11$ or 11 per cent. Buffer power values range from about 1 to 20 $l kg^{-1}$, showing that between 25 and 1 per cent of the added K^+ may be in solution. Leaching losses will be related to these proportions.

A similar approach is used in Section 15.3 to express the way in which pesticides distribute between particle surface and soil solution. In this case the adsorption coefficient is often expressed as $cm^3 g^{-1}$.

Buffer power and leaching losses

Soil solution is displaced by rain water, and the K^+ it contains is lost from the soil. The isotherm shows the relationship between changes (losses) of exchangeable K^+ and soil solution concentration, and can be used to predict losses if the amount of water moving through the soil is known.

Example In the Reading area the annual rainfall is about 600 mm a^{-1}. During the 6 months from October to March the excess of rainfall over evapotranspiration is about 300 mm. A soil at its wilting point at the end of September, requires rain to bring it back to field capacity, and any extra will then leach through. For a sandy loam with a wilting point of 0.05 $cm^3 H_2O\, cm^{-3}$ soil and a field capacity of 0.3 $cm^3 cm^{-3}$ (Table 4.3), the top 20 cm of soil needs 50 mm of water to bring it back to field capacity. Thus of the 300 mm, about 250 mm moves through this surface layer during the winter. This volume of water ha^{-1} is

$$0.25\, m \times 10^4\, m^2 = 2.5 \times 10^3\, m^3 \text{ or } 2.5 \times 10^6\, l.$$

If the K^+ concentration is maintained at 1.1 mM, the amount of K^+ leached to below 20 cm is

$$1.1 \times 10^{-3} \times 2.5 \times 10^6 = 2750\, mol \text{ or } 108\, kg\, K^+\, ha^{-1}$$

(molar mass = 39.1 $g mol^{-1}$). A more typical K^+ concentration in soil solution (Table 7.6) is 0.4 mM, giving a predicted loss of 39 kg ha^{-1}. Measured losses into drains (>60 cm depth) are about 20 kg ha^{-1} from sandy soils (e.g. Woburn 14 kg ha^{-1}, Table 10.3) and about 5 kg ha^{-1} from clay soils compared to fertilizer inputs and crop removals of between 30 and 60 kg $K^+ ha^{-1}$.

Other factors influencing losses are as follows:

- The solution concentration will decrease as K^+ is removed from exchange sites. The isotherm shows the relationship. A loss of 108 kg $K^+ ha^{-1}$ is 43 mg kg^{-1} (2500 t ha^{-1}) or 0.11 $cmol_c kg^{-1}$. The isotherm shows that the concentration will decrease from 1.1 to about 0.8 mM, a change which will reduce the calculated loss by between 10 and 20 kg.

- In order for the K^+ to be exchanged to maintain the solution concentration, Ca^{2+} or some other cation has to take its place. In neutral and acid soils it is likely that the release of K^+ will be restricted by the small amounts of other cations in soil solution. It is therefore not possible at present to predict these losses. However, in calcareous soils, $CaCO_3$ will dissolve to maintain the supply of Ca^{2+} and HCO_3^-, and large K^+ losses are predicted. Field measurements are presented in Section 9.5 which support these predictions.

- Even though K^+ may be leached out of the surface soil it may be trapped in the subsoil. Particularly in heavy textured subsoils the large exchange capacity and ability to hold water will restrict losses. Roots of cereals penetrate deeply (Fig. 3.1) and K^+ will be taken up from these deeper layers. A sandy, stony or chalky subsoil will have a poor ability to retain K^+.

- Sandy soils have small buffer powers. Thus for a given amount of exchangeable K^+, the solution concentration will be larger than in heavy textured soils. This together with a smaller capacity to hold water, will lead to greater leaching losses.

- Large losses occur in stony soils for two reasons. Firstly, the water content at field capacity is reduced by the volume of stones, and more water will pass through the soil during the winter. Secondly, the effective buffer power of the whole soil is reduced. For example, on the isotherm the ordinate has the units $cmol_c kg^{-1}$ of <2 mm soil. The presence of say 50 per cent of the soil mass as stones means that the values on the ordinate have to be halved if the buffer power is expressed relative to the whole soil. Thus for a given amount of exchangeable K^+ in the whole soil, the solution concentration is larger in a stony soil.

- In regions of the world receiving large amounts of rainfall, there may be serious losses of K^+ and other cations especially if the soils have small exchange capacities. This leads to difficulties in the management of fertilizers and lime.

Section 9.5 Potassium budgets in the field

The components of a field budget are shown in Fig. 9.11. If a budget is to be drawn up for a crop of wheat, the exchangeable K^+ can be determined by extraction

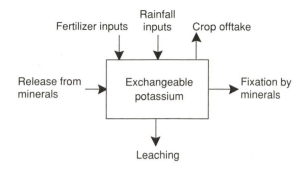

Figure 9.11 Inputs and losses in a potassium budget.

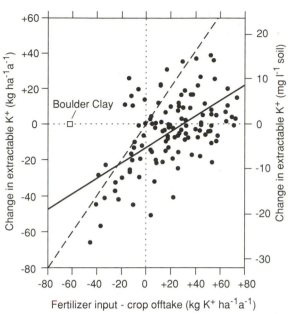

Figure 9.12 The potassium budget for cereal crops on chalky soils. Exchangeable K^+ in the plough horizon was measured as $mg\,l^{-1}$ by NH_4NO_3 extraction (Section 9.1): values on the ordinate are also given as $kg\,ha^{-1}$, calculated assuming $2500\,t\,ha^{-1}$ in the plough horizon, and that the density of dry $<2\,mm$ soil in the measuring cup is $1\,kg\,l^{-1}$. The regression equation plotted as the full line is $y = -13 + 0.42x$, $r = 0.62$.
Data from Soil Services Ltd, Swindon.

with NH_4NO_3 solution in September just before the crop is sown, and again a year later. The change in exchangeable K^+ is therefore measured and the budget can be written:

change in exchangeable K^+
= (atmospheric input + fertilizer input + release)
  − (crop offtake + fixation + leaching)

Release is the movement of K^+ from non-exchangeable to exchangeable forms, and fixation is the reverse process (Fig. 9.2). Atmospheric inputs are small (Table 10.3). Fertilizer inputs and crop offtakes can be determined. Rearranging the budget equation gives:

change in exchangeable K^+
= fertilizer input − crop offtake
  + (release − fixation + atmospheric input − leaching)

The terms in the brackets are not easily measured. A useful way of displaying this equation is shown in Fig. 9.12. Field measurements were made on chalky soils in southern England for the seasons 1985–1990. Input minus offtake is plotted on the abscissa. The change in the amount of exchangeable K^+ is shown on the ordinate. A dotted line has been drawn showing what would happen if input minus crop offtake = change in exchangeable K^+ in the plough horizon. Data points on this line show that the net effect of fixation, release, atmospheric inputs and leaching is zero. However the majority of the data lie below this line and the fitted regression line (Section 9.6) indicates that on average fertilizer input must exceed crop removal by about $31\,kg\,ha^{-1}$ to prevent a decrease in extractable K^+. There is unlikely to be net fixation or release since these soils have been fertilized and cropped for many years. Atmospheric inputs are about $4\,kg\,ha^{-1}a^{-1}$ and so leaching losses appear to be about $35\,kg\,ha^{-1}a^{-1}$ in soils where extractable K^+ is being maintained. The loss

predicted on the opposite page is $39\,kg\,ha^{-1}$ from the top $20\,cm$ of soil based on winter leaching and a soil solution concentration of $0.4\,mM$. Not surprisingly, losses tend to increase as input minus offtake increases.

The Boulder Clays of East Anglia often receive no K^+ fertilizer, crop removals are about $60\,kg\,ha^{-1}a^{-1}$ and yet there is no change in exchangeable K^+ from year to year. A point has been plotted for these soils. In this case, fertilizer input minus crop offtake = $-60\,kg\,ha^{-1}$. If loss by leaching on these heavy-textured soils is about equal to atmospheric inputs, then there must be a net release of about $60\,kg\,ha^{-1}a^{-1}$.

It is of interest that the Boulder Clays maintain a potassium index of 2–3 without fertilizer applications, whereas it is sometimes difficult to maintain the index of chalky soils above 1 even with large inputs. On soils which are easily leached, spring rather than autumn applications would reduce the losses. If autumn applications are necessary, top dressing would be more beneficial than cultivating the fertilizer into the soil.

Section 9.6 Regression and correlation

The measurement of nutrient availability in soils and its relationship to yield or nutrient uptake by a crop leads to the questions 'What is the mathematical relationship between availability and yield (or uptake)?' and 'How good is this relationship?' An example is in Fig. 9.4 where the K^+ uptake by ryegrass is clearly related to exchangeable K^+ but the relationship is not perfect. Statistical analysis helps to answer these questions. Straight line relationships only will be dealt with here.

Straight line graphs

The equation of a straight line graph is

$$y = A + Bx$$

where y and x are the two variables, A is the intercept on the y axis and B is the slope (Fig. 9.13). If the equation is known, the line can be plotted by substituting values of x and y. Only two points are needed to position the line. If the graph is available, the equation can be obtained from the intercept and the slope. The slope can be found by drawing a triangle as shown and dividing the number of y units by the number of x units $= (y_1 - y_2)/(x_1 - x_2)$. Note that these are not the lengths of the lines, but the values in the units given on the axes. Graphs (a) and (b) in Fig. 9.13 have positive gradients and (c) has a negative gradient.

Plotting data

When data are obtained from an experiment, possible relationships between variables can be examined by plotting them on a scattergraph (Fig. 9.14). It is normal to plot the independent variable (the one you control) on the x-axis and the dependent variable (the one that is affected by altering x) on the y-axis. If dependency is not known, the variables can be plotted on either axis.

Fitting lines to data

Where there appears to be a relationship then a line can be drawn to display this relationship. This is the line of best fit. Four simple methods can be used to obtain this line:

1. Place a transparent ruler on the scattergraph and move it until the edge passes through the points leaving roughly the same number of points on each side of the edge.
2. Draw two parallel lines to enclose most or all of the points. 'Stray' points can be excluded. Draw a straight line equidistant from the two parallel lines.

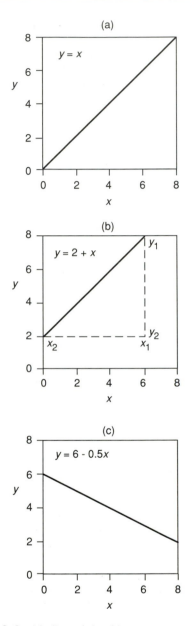

Figure 9.13 Straight line relationships.

3. Calculate the mean value of the x values ($\bar{x}$) and the mean of the y values ($\bar{y}$) (Section 3.8). Plot the point $\bar{x}, \bar{y}$. If you know that the line passes through the origin, join $\bar{x}, \bar{y}$ to the origin. The data in Fig. 9.4 and Table 9.8 for the relationship between exchangeable K^+ and uptake after 105 d give $\bar{x} = 210$ and $\bar{y} = 296$. The dotted line on Fig. 9.4 passes through this point (marked a) and the origin. The slope of the line is $296/210 = 1.41$, the intercept $= 0$

Table 9.8 Uptake of potassium by ryegrass in the pot experiment of Arnold and Close (1961)

| Soil number | Exchangeable K^+ | | Cumulative uptake after days: | | | Release of non-exchangeable K^+ after 600 d |
| | Initial | After 600 d | 50† | 105 | 600 | |
	($mg\ kg^{-1}$)			($mg\ plant\ K^+\ kg^{-1}\ soil$)		($mg\ kg^{-1}$)
	120	50	43	107	202	132‡
2	110	40	42	110	179	109
3	90	47	35	110	208	165
4	115	37	28	71	129	51
5	60	31	13	30	57	28
5A	75	42	17	43	62	29
5B	110	49	30	62	92	31
6	65	21	36	78	108	64
6A	90	33	61	130	182	125
8	200	107	101	360	685	592
10	265	126	103	323	637	498
11	140	69	107	332	548	477
12	85	39	75	303	525	479
13	325	135	116	388	696	506
14	455	125	198	773	1643	1313
15	295	100	180	626	1179	984
16	720	110	137	518	1313	703
17	355	220	163	541	1158	1023
18	195	115	104	440	1075	995
19	335	84	181	571	818	567

† The values given are for the K^+ in the tops at 50 and 105 d and in the tops + roots at 600 d.
‡ Release of non−exchangeable K^+ = uptake after 600 d minus decrease in exchangeable K^+.

and so the equation of the line is $y = 1.41x$. Similarly for uptake after 50 d the equation is $y = 0.42x$.

4. If you do not know that the line of best fit passes through the origin, the Bartlett method can be used. Arrange the data (105 d) in ascending order with respect to one of the variables; for this purpose exchangeable K^+ is chosen because this is the independent variable (Table 9.8). Divide the data into three equal groups, or if this is not possible Groups 1 and 3 should be equal as in this case. Calculate $\bar{x}$ (210) and $\bar{y}$ (296) for all the data (point *a* on Fig. 9.4). Calculate $\bar{x}$ and $\bar{y}$ separately for Group 1 and Group 3. Plot these points (marked *b* on Fig. 9.4) and draw a straight line through them and point *a*. The points may not lie exactly on a straight line. This line (not plotted) is moved slightly downwards. Its slope is unchanged, and the intercept is −5, giving $y = -5 + 1.41x$.

Regression

The use of statistical methods to determine the line of best fit through a set of data is known as *regression*, the line is then called a *regression line* and the equation of the line is called a *regression equation*. A scientific calcu-

Table 9.9 The Bartlett method for determining a line of best fit

| Group 1 | | Group 2 | | Group 3 | |
x	y	x	y	x	y
60	30	110	110	265	323
65	78	115	71	295	626
75	43	120	107	325	388
85	303	140	332	335	571
90	110	195	440	355	541
90	130	200	360	455	773
110	62			720	518
$\bar{x} = 82$	$\bar{y} = 108$			$\bar{x} = 393$	$\bar{y} = 534$

lator with a linear regression mode can be used to obtain the required parameters. Always plot a scattergraph and fit a line by eye before calculating a regression equation. This serves as a rough check on your calculations.

If the line passes through the origin, the slope is given by $\sum xy / \sum x^2$ and calculation using the 50 d and 105 d data gives $y = 0.37x$ and $y = 1.27x$ as the regression equations respectively. These are plotted (dashed) in Fig. 9.4.

If the line is not known to pass through the origin the slope B is

$$\sum(x - \bar{x})(y - \bar{y})/\sum(x - \bar{x})^2$$

and the intercept A is $\bar{y} - B\bar{x}$. Calculation now gives $y = 31.1 + 0.273x$ and $y = 72.3 + 1.06x$ respectively. Statistics does not help us to decide whether the line should pass through the origin. In this experiment if an imaginary soil had no exchangeable potassium, then measured uptake would be very small due only to atmospheric inputs. The potassium in the seed has already been subtracted from the measured K^+ in the plants to give the uptake value, and so at most the intercept should be only a few mg: the lines passing through the origin are therefore probably the best lines.

It is tempting to reject soil No. 16 from the regression on the grounds that it is a stray point. The fit of the lines would be improved and the slopes increased slightly. However, there is no basis for rejecting this point other than its position on the graph, and so it has been included.

Correlation

Fig. 9.14 shows data that fit to regression lines with various degrees of fit. *Correlation* is the statistical term for the degree of fit. Where the slope is positive, then there is a positive correlation. Correlation is quantified through the *correlation coefficient* (r). The diagrams show examples ranging from a perfect positive correlation where $r = +1$, through data where there is no relationship (r is almost zero) to a perfect negative correlation where $r = -1$. These diagrams can be used to give an approximate (but useful) estimate of the correlation coefficient. For example, the 105 d data in Fig. 9.4 appear to have a correlation coefficient of between 0.5 and 0.8.

The correlation coefficient can be calculated using the following formula:

$$r = \sum[(x - \bar{x})(y - \bar{y})]/[\sum(x - \bar{x})^2 \sum(y - \bar{y})^2]^{1/2}$$

Using a calculator, $r = 0.78$ for the 105 d data and 0.77 for the 50 d data. It is always worthwhile to use a scattergraph to estimate r (Fig. 9.14) before calculating its value.

Is the correlation significant?

In Section 3.8 the significance of a mean value was determined by calculating 95 per cent confidence limits. In a similar way the significance of correlation can be determined. As the number of data points

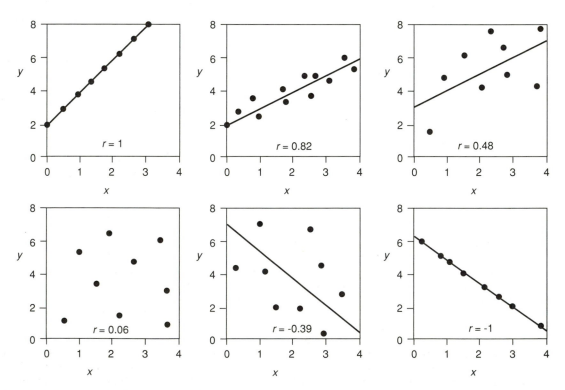

Figure 9.14 Scattergraphs. From Garvin (1986).

Table 9.10 A table of correlation coefficients

df	r	df	r	df	r
1	0.997	11	0.55	25	0.38
2	0.95	12	0.53	30	0.35
3	0.88	13	0.51	35	0.33
4	0.81	14	0.50	40	0.30
5	0.76	15	0.48	45	0.29
6	0.71	16	0.47	50	0.27
7	0.67	17	0.46	60	0.25
8	0.63	18	0.44	70	0.23
9	0.60	19	0.43	80	0.22
10	0.58	20	0.42	90	0.21
				100	0.20

increases a significant correlation is attained at lower values of r. Table 9.10 gives the r values above which we can be 95 per cent certain that the correlation is significant. The number of degrees of freedom (df) is 2 less than the number of data points. For the potassium results (105 d), 20 data points give 18 degrees of freedom and r must be >0.44 to be significant. There is thus a statistically significant correlation between exchangeable K^+ and K^+ uptake.

The significance of correlation can also be considered through the value of r^2, known as the *coefficient of determination*, which is a measure of the proportion of the variability in y accounted for by variations in x. In the above case (105 d data), $r^2 = 0.61$, indicating that 61 per cent of the variability in K^+ uptake was accounted for by the differences in exchangeable K^+.

Section 9.7 Projects

1. To a soil sample add KCl at rates up to 600 kg K^+ ha^{-1} including a zero addition (up to 10 times the normal field application, Table 9.3). Incubate moist for 1 or 2 weeks and measure the K^+ extractable in 1 M NH_4NO_3 following the methods of Section 9.1. Calculate the amount of the added K^+ which is not extractable and has therefore been fixed for each treatment.

2. Treat a sandy soil and a clay soil with KCl equivalent to 60 kg K^+ ha^{-1} (Table 9.7). Carry out a pot experiment (Section 9.2) on each using ryegrass, following the methods used by Arnold and Close (p. 178) to measure the cumulative K^+ uptake. At the end of the experiment measure the K^+ extractable from the soil in 1 M NH_4NO_3 following the methods of Section 9.1. Also measure the extractable K^+ in a freshly treated soil and in a soil sample treated at the beginning of the pot experiment but not cropped. By calculating K^+ budgets, determine whether K^+

moves into or out of the non-exchangeable fraction during the experiment. This project requires at least several months.

3. Determine the K^+ exchange isotherms of soils with a range of textures following the methods of Section 9.4. Measure exchangeable K^+ and CEC (Section 7.2). Buffer power increases with CEC, but because of the curvature of the isotherm, buffer power decreases with increased K^+ saturation of the CEC. Plot buffer power aginst CEC, and against exchangeable K^+/CEC. Determine the regression equations and the correlation coefficients (Section 9.6). Imagine that to each soil was added 60 mg K^+ kg^{-1}. Determine from the graphs or calculate from the buffer powers the expected increase in soil solution concentration.

   Assume that this application rate was used in the field (2500 t soil ha^{-1}, 0–20 cm layer). If 300 mm of drainage water passed through each of these treated soils, and the soil solution concentration did not change, calculate the leaching loss as kg ha^{-1} from each soil. In reality the solution concentration decreases during leaching: devise a way of calculating the losses to include this effect.

4. Leaching losses for the soils in Project 3 can be measured using methods given in Section 11.5 for nitrate leaching. The columns should be leached with 10 mM $CaCl_2$ if comparison is to be made to losses predicted from the isotherms. Leaching with water will remove much less potassium. In field soils the calcium concentration is maintained at about 3 mM, although it may not be in the more rapid leaching of laboratory experiments. An isotherm determined in 3 mM $CaCl_2$, and leaching also with this solution is probably the best way of simulating field conditions.

5. Prepare nine pots each containing 1 kg of soil uniformly treated with lime and nutrients (Section 9.2). Germinate leek plants in compost and transplant three seedlings into each pot. Using another three seedlings, determine root length (Section 3.1), oven dry the tops and roots and retain as a combined sample for analysis. After 2 weeks, harvest three pots treating the three plants from each pot together as one sample. Separate roots from soil, washing with water. Measure root length, oven dry the tops and roots and retain for analysis. Harvest three pots 2 weeks later and the final three pots after another 2 weeks, repeating the measurements. You may need to increase these time periods if conditions for growth are poor.

   Combine the tops and roots from each pot and determine the mass of dry matter and the K^+ content in each sample (Section 9.3). Include the seedling sample.

Plot yield (mg dry matter per pot) and K^+ uptake ($\mu g\,K^+$ per pot) against time taking the transplanting date as the zero time.

A useful measure of uptake rate of K^+ is known as inflow, the amount of K^+ entering unit length of root in unit time. From the graph determine the uptake rate per day from the slope of the graph at 2, 4 and 6 weeks after planting. Calculate inflow at these times. Values between 50 and $500\,\mu g\,K^+\,m^{-1}\,d^{-1}$ have been found in pot experiments.

From Fig. 11.3 calculate the maximum uptake rate (during May) of K^+ by a wheat crop in $kg\,K^+\,ha^{-1}\,d^{-1}$ and convert into $g\,K^+\,m^{-2}\,d^{-1}$. Fig. 3.1 gives a root length of 180 cm under $1\,cm^2$ of soil surface for this crop in June; from this calculate the root length under $1\,m^2$ of soil, and using the above uptake rate calculate the inflow as $\mu g\,m^{-1}\,d^{-1}$. There are approximations in this calculation: it only gives an order of magnitude. Compare to your value for leeks: it has generally been found that young plants take up nutrients at about 10 times the rate of older plants. Leeks are useful experimental plants because they have a simple root system which allows easy root length measurements.

6. Extend Project 5 by measuring the water use by the leeks (Section 9.2) and the concentration of K^+ in soil solution (Section 9.4 or 7.4). Calculate the contributions of mass flow and diffusion to K^+ uptake.

Section 9.8 Calculations

1. Using the average values in Fig. 9.1 calculate the percentage of the total soil K^+ held in the various fractions. (*Ans.* Slowly released 3.1, exchangeable 1.4, solution 0.03) What percentage of the total offtake of a 6 t wheat crop including straw (Table 9.2) is held in solution? (*Ans.* 20)

2. Ten ml of soil was shaken with 50 ml of $1\,M$ NH_4NO_3 solution for 30 min. The suspension was filtered and the solution was found to contain $20\,\mu g\,K^+\,ml^{-1}$. What potassium index would be ascribed to this soil (Section 9.1)? (*Ans.* $100\,mg\,l^{-1}$ soil, index 1) How much fertilizer-K^+ would be required to raise the extractable K^+ of the 0–20 cm layer of this soil to $180\,mg\,l^{-1}$ soil (index 2)? Assume that all the added K^+ remains in an extractable form, that the dry soil in the measuring cup has a density of $1\,g\,cm^{-3}$ and that there are 2500 t soil ha^{-1}. (*Ans.* $200\,kg\,ha^{-1}$)

3. Calculate the amount of KCl required to supply $50\,kg\,K^+\,ha^{-1}$ (Table 9.3). (*Ans.* $95\,kg\,ha^{-1}$) What would be the appplication in $kg\,K_2O\,ha^{-1}$? (*Ans.* 60)

4. Six samples of 5 g of soil were weighed into bottles

Table 9.11 Experimental data for Calculation 4

Bottle	Concentration of K^+ (mM)	
	Before shaking	After shaking
1	0	0.30
2	0.5	0.65
3	1.0	1.05
4	2.0	1.90
5	3.0	2.80
6	4.0	3.65

and each was shaken for 10 min with 50 ml of solution containing KCl in 10 mM $CaCl_2$. The suspensions were then filtered and the concentrations of K^+ determined in the filtrates. The data are given in Table 9.11. Plot the exchange isotherm (Graph 1) as gain or loss of K^+ by the soil ($cmol_c\,kg^{-1}$) against solution concentration (mM). Draw a smooth curve through the points. From the graph determine the equilibrium concentration of K^+ in the soil solution, and the soil's buffer power. (*Ans.* 1.3 mM, $0.2\,cmol_c\,kg^{-1}/mmol\,l^{-1}$)

The exchangeable K^+ extracted in $1\,M$ NH_4NO_3 is $0.40\,cmol_c\,kg^{-1}$. Replot the isotherm as exchangeable K^+ against solution concentration (Graph 2).

The CEC of the soil is $15\,cmol_c\,kg^{-1}$. Assuming that Ca^{2+} is the only other exchangeable cation present and that the Ca^{2+} concentration is solution remains at 10 mM, plot the exchange isotherm as exchangeable K^+/exchangeable Ca^{2+} against $[K^+]/[Ca^{2+}]^{1/2}$ where [ ] indicates concentration in $mol\,l^{-1}$. This gives the data in a form which allows the *Gapon exchange coefficient* to be found. The Gapon exchange equation is

$$\text{exchangeable } K^+/\text{exchangeable } Ca^{2+} = G[K^+]/[Ca^{2+}]^{1/2}$$

where G is the exchange coefficient in $(mol\,l^{-1})^{-1/2}$. Determine the value of G for the soil as sampled from the field. (*Ans.* 1.4)

In a similar experiment, bottle 6 was taken before filtration. It was centrifuged, and the solution poured off and the concentration measured. The remaining soil contained 1.12 g of solution. To this was added 50 ml of 10 mM $CaCl_2$ containing no potassium, the bottle was shaken and centrifuged and the solution was found to contain 0.55 mM K^+. Determine the amount of K^+ desorbed from the soil during this shaking. (*Note:* calculate the amount of K^+ in the 1.12 g (= ml) of solution. Its concentration is 3.65 mM. Then calculate the amount of K^+ in the 50 + 1.12 ml of solution after shaking. Its concentration is 0.55 mM. The difference can be used to calculate the amount of K^+ desorbed per kilogram of soil). Plot this desorption point on Graph 2. (*Ans.* Coordinates 0.55, 0.27) Is the exchange process reversible?

5. Using the initial form of the exchange isotherm in Calculation 4 (Graph 1), determine the fertilizer addition (kg K^+ ha^{-1}) that would be required to raise the soil solution concentration in the 0–20 cm layer to 3.5 mM (2500 t ha^{-1}). (*Ans.* 293) How much exchangeable K^+ does this soil hold before fertilization expressed as kg K^+ ha^{-1}? (*Ans.* 391)

6. In the experiments by Arnold and Close (Table 9.8), the yield for soils 5A, 14 and 19 were 26.3, 114.3 and 62.3 g dry matter kg^{-1} soil respectively after 600 d. Calculate the K^+ concentration in the plant dry matter. (*Ans.* 2.4, 14.3 and 13.1 mg g^{-1}) Compare your values to Table 9.1. Was the ryegrass well supplied with potassium?

7. Six t of lime ha^{-1} was applied to a field in the form of dolomitic limestone. It contained 95 per cent CaMg(CO$_3$)$_2$. It was cultivated into the top 20 cm of soil (2500 t ha^{-1}). Calculate the input as (a) t Ca^{2+} ha^{-1} and t Mg^{2+} ha^{-1} and (b) cmol$_c$ Ca^{2+} kg^{-1} and cmol$_c$ Mg^{2+} kg^{-1} (molar masses are in Appendix 2). (*Ans.* (a) 1.8 and 1.1, (b) 3.7 of both). Table 9.2 gives the calcium and magnesium requirements of various crops. Calculate the percentage of the Ca^{2+} and Mg^{2+} in the 6 t of lime which would be taken up each year by a 6 t ha^{-1} barley crop. (*Ans.* 0.9 and 1.0)

8. Use the data in Table 9.8 to determine the regression line and correlation coefficient for the relationship between exchangeable K^+ and uptake after 600 d. The regression analyses for the 50 and 105 d data are in Section 9.6. Which set of data gives the best fit?

Consider the hypothesis that those soils which have large amounts of exchangeable K^+ are also able to release appreciable amounts of non-exchangeable K^+. Use the data in Table 9.8 to examine the statistical validity of this hypothesis: plot initial exchangeable K^+ (x axis) against the release of non-exchangeable K^+ after 600 d, and carry out a regression analysis. (*Ans.* $y = 97.9 + 1.64x$, $r = 0.69$)

CHAPTER 10

Phosphorus and Sulphur

PHOSPHORUS

Phosphorus occurs both as inorganic phosphate and in the organic matter in soils. The dominant characteristics of soil phosphate are very low solubility of phosphate minerals and its strong binding on to particle surfaces giving small soil solution concentrations. As a result phosphate deficiencies in crops are common. A large research effort has been devoted to this problem. From an environmental viewpoint, however, the small solubility of phosphate means that drainage waters have very small concentrations, and pollution of water supplies is rarely associated with phosphate from soils.

Amounts of phosphorus in soils

The distribution of P between various fractions is shown in Fig. 10.1.

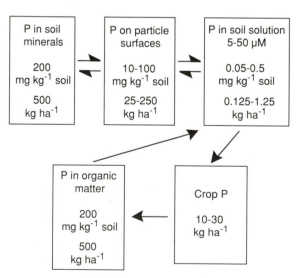

Figure 10.1 Typical values for the distribution of phosphorus in a reasonably fertile soil. Values are given for the 0–20 cm layer, 2500 t soil ha⁻¹. Organic P values are for a soil with 1.7 per cent organic matter and 12 g P kg⁻¹ organic matter. Soil-solution P is calculated assuming that 1 kg of soil contains 300 ml of solution. Arrows indicate movement between fractions. Based on Wild (1988).

- *Phosphate minerals*. These are primarily calcium, iron and aluminium phosphates.
- *Particle surfaces*. Phosphate is bound on to sesquioxides by chelation (Section 7.5). Clay minerals hold phosphate but the mechanisms are not well understood. They include binding with −OH groups at the edges of kaolinite particles and the formation on clay surfaces of very small particles of mineral phosphates. Phosphate seems not to be held on 2:1 clay minerals unless the surface is contaminated with sesquioxides. In calcareous soils phosphate is adsorbed on the surface of calcite forming calcium phosphate. Phosphate also appears to bind to aluminium on humus surfaces.
- *Soil solution*. Phosphate is present predominantly as $H_2PO_4^-$ and HPO_4^{2-} over the usual range of soil pH values.
- *Plant phosphorus*. This is mostly in the form of organic esters in which P is bonded to C via O.
- *Organic matter*. The form of organic P reflects that in plants and micro-organisms, being present primarily as esters.

Inorganic phosphorus

Solution phosphate

The strong adsorption of phosphate on to particle surfaces and the small solubility of phosphate minerals leads to concentrations in soil solutions between 1 and 50 μM. These are about one-tenth of those for potassium and one-hundredth of those for nitrate. Solution phosphate is normally measured using a spectrophotometer or colorimeter after the development of a blue-coloured phosphomolybdate complex (Section 10.1).

Reactions between particle surfaces and solution phosphate

Adsorption isotherms have been extensively used to study these reactions. The principles are the same as those described for potassium (Section 9.4). The most important differences are the stronger adsorption and smaller solution concentrations for phosphate. The reaction experiments are normally carried out over 24 h in a 10 mM $CaCl_2$ solution and are described in Section 10.2. In contrast to potassium experiments where K^+ and Ca^{2+} compete for the same exchange surfaces, Cl^- does not compete with $H_2PO_4^-$ for adsorption sites. The $CaCl_2$ solution simply prevents the clay from dispersing and maintains a solution concentration slightly above that commonly found in neutral soils (Table 7.6).

One of the results of the strong binding of phosphate, known as *fixation*, is that its subsequent removal from particle surfaces is difficult. However, fixation is not a satisfactory term, firstly because it is used in a very different way in relation to soil-N, and secondly, because the binding process is reversible with slow release providing phosphate for crop use over many years. The more general term *sorption* is preferred.

Although chloride does not displace adsorbed phosphate, bicarbonate competes for the same sites (Section 7.5). For this reason 0.5 M sodium bicarbonate solution is often used to measure the amount of extractable P as a way of indicating its availability (Section 10.3). However, even this solution extracts only part of the phosphate that may be available.

The role of phosphate minerals

Surface reactions include the formation of phosphate minerals with small solubility: their rate of formation is normally slow compared to adsorption on to sesquioxides. Thus an adsorption isotherm determined from a 24 h equilibration of soil and solution may not be influenced much by the precipitation of these minerals. Under field conditions, however, their formation is more important, and P can be slowly released again by dissolution, when solution concentrations are reduced by plant uptake. Adsorption–desorption and precipitation–dissolution reactions are illustrated below. Although surface-P and mineral-P are considered separately surface phosphate forms a continuum from that held on sesquioxides through to discrete mineral particles.

$$\begin{array}{ccc} \text{strong adsorption} & & \text{slow precipitation} \\ \text{surface-P} \leftrightharpoons & \text{solution-P} & \rightleftharpoons \text{mineral-P} \\ \text{restricted desorption} & & \text{slow dissolution} \end{array}$$

Organic phosphorus

Phosphorus is added to soil in plant residues and is released from soil organic matter by mineralization. For a soil in which the amount of organic matter is not changing, e.g. a continuous arable site, the turnover is between 4 and 8 kg P ha^{-1} a^{-1}. Supply from this source is greater under conditions where large amounts of organic matter are decomposing. For example, after

clearing a secondary forest in Nigeria, the organic P in the top 10 cm of soil decreased from 194 to 147 mg kg^{-1} during 22 months (Nye and Greenland, 1960). Assuming 1250 t of soil ha^{-1} in this layer, 59 kg P ha^{-1} was released. In a situation where organic matter levels are increasing, e.g. forest regrowth or grass after arable, there is a net accumulation of organic P, but each year there is a small turnover also. These principles will be developed further for N in Chapter 11. They are the basis of organic farming and shifting cultivation systems (Ch. 13).

The availability of phosphorus

The principles involved in measuring nutrient availability have been developed for potassium in Ch. 9, and are applied here to phosphorus.

Extractable phosphate

Various extraction solutions are used including 0.5 M sodium bicarbonate at pH 8.5 known as Olsen's solution which extracts phosphate which is exchangeable with bicarbonate and some readily soluble calcium phosphate (Section 10.3).

Exchangeable phosphate

Phosphate labelled with the radioactive isotope ^{32}P has been used to measure the amount of *isotopically exchangeable phosphate* in a soil. Labelled phosphate exchanges with soil phosphate so avoiding the need to use another anion such as bicarbonate in the measurement. The method is beyond the scope of this book, but results show that in many soils the amounts of bicarbonate-extractable phosphate and isotopically exchangeable phosphate are similar.

The low mobility of phosphate and the role of mycorrhizae

Because of strong adsorption and small solution concentrations, the contribution of mass flow to uptake of phosphate by roots is very small and diffusion rates are slow. Thus, roots take up phosphate only from their immediate vicinity. Using isotopic labelling techniques (Plate 10.1), the extent of the depleted region (normally called the *depletion zone* has been measured (Fig. 10.2). As an example, most of the P needs of a cereal crop are supplied during the rapid growth period (April, May and June in Britain, Fig. 11.3) and it is likely that depletion zones of only 2–3 mm radius around the roots are involved. An extensive root system is therefore important for adequate uptake.

The significance of the low mobility of phosphate was

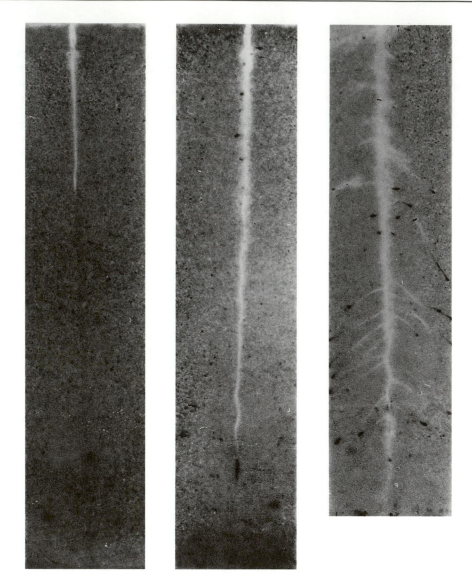

Plate 10.1 Autoradiographs of phosphate depletion regions around roots of oilseed rape (*Brassica napis*) grown in soil blocks labelled with ^{33}P. The blocks were 9 cm long × 2 cm wide × 0.5 cm thick and a <0.15 mm fraction of a sand-textured soil was used. The dark background is caused by radiation from the isotope and the white areas show soil depleted by uptake of phosphate. The darkened tips of the primary root and some laterals indicate accumulation of P in the roots. (a) After 2 days' growth, (b) after 4 days and (c) after 7 days. Photograph by E. Owusu-Bennoah, Reading.

demonstrated in a pot experiment carried out at Roth-amsted in 1980. Ten soils with low phosphate status were each treated with five rates of P fertilizer. Bicarbonate-extractable P was measured. Each soil sample was divided in two. Half the soil was sterilized by gamma irradiation and half left in its natural state. The radiation killed all micro-organisms in the soil. Leeks were then grown with an adequate supply of all nutri-

ents except P and their yield is shown in Fig. 10.3 leading to the following conclusions:

- In the sterilized soils (Fig. 10.3(a)) bicarbonate-extractable P is a good indicator of yield despite the low mobility of phosphate. However, in the unsteri-lized soils (Fig. 10.3(b)) yield is less clearly related to extractable P. The reason for this difference lies in

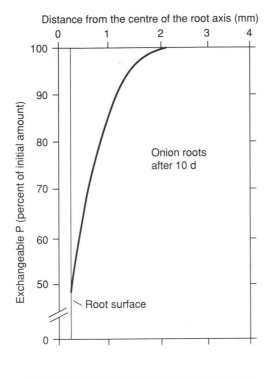

Figure 10.2 Depletion zones around roots. Onions were grown in a sandy loam and the depletion measured using autoradiography after a 10 d uptake period. The depletion zone is normally considered to be the region with less than 90 per cent of the initial amount of nutrient. From Bhat and Nye (1973, 1974).

the ability of roots to form a symbiotic relationship with mycorrhizal fungi. *Symbiosis* is the close association of two organisms, often to their mutual benefit. The mycorrhizae grow out from the roots into the soil, take up P (and other nutrients) and transport it to the root in return for a supply of carbohydrate from the plant. In some of the soils mycorrhizal infection was well developed and P supply was improved. In others the roots had to rely on diffusion alone.

- In soils with large concentrations of extractable P, mycorrhizae do not increase uptake. The root demand is easily satisfied by diffusion.

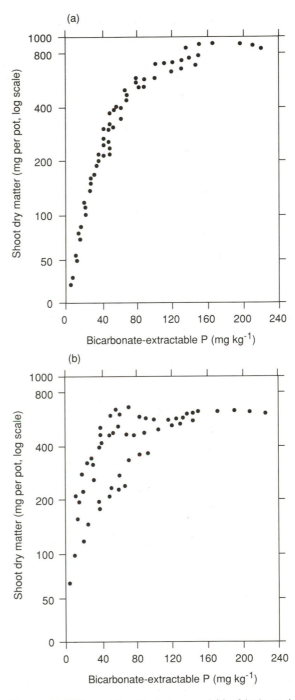

Figure 10.3 The relationship between yield of leeks and bicarbonate-extractable phosphorus. (a) Soils sterilized by gamma irradiation to kill the mycorrhizae; (b) plants naturally infected with mycorrhizae in unsterilized soils. From Stribley *et al.* (1980).

• Maximum yield is greater in the sterilized soils possibly due to removal of pathogens or to mineralization of N and P following sterilization (Section 3.2). In this context, mycorrhizae may be acting as parasites, taking carbohydrates from the host plant and reducing yield.

As a result of these and similar experiments there has been much interest in the use of mycorrhizae to improve the utilization of P by crops. However, it has proved difficult to inoculate seed to ensure that an effective symbiosis develops. A more effective way of overcoming the low mobility of P is to place fertilizer granules near to the seed rather than mixing the fertilizer into the soil during cultivation. The roots then make good use of the small volume of enriched soil. However in intensive agriculture, the P status of the whole soil is raised so that enough P is available despite its low mobility.

Pot experiments

The above discussion shows the value of pot experiments in studying the availability of P. However, as for potassium, many of the variables which influence the utilization of P in the field are excluded. The determination of the P content of plant dry matter is described in Section 10.4.

Availability of phosphorus in the field

An availability index

The extraction of phosphate by bicarbonate is used to allocate a P index to a soil (Section 10.3). Table 10.1 gives the recommended fertilizer applications for cereal crops in Britain. Visual symptoms of deficiency occur in extreme cases only.

Although the principles involved in making P recommendations are similar to those for potassium, two important differences should be noted:

1. Clay soils often have large reserves of non-exchangeable K^+ in minerals which can supply some or all of the crop needs from year to year. However, there are very few soils which have large reserves of P in soil minerals.

   The limitations imposed on the availability of nutrients resulting from low mobility were discussed in Chapter 8 for potassium. Phosphate is much less mobile, and it would be expected that only a small fraction of the extractable P in soils would be accessible. A wheat crop yielding 6 t of grain takes up about 30 kg P ha^{-1} compared to up to 200 kg K^+ ha^{-1} (Fig. 11.3). To avoid yield constraints both P and K^+ need to be maintained at index 2 (39–63 kg P ha^{-1} and 301–600 kg K^+ ha^{-1}

Table 10.1 Fertilizer recommendations: phosphorus for cereal crops

Extracted P					
(mg l^{-1} soil)	0–9	10–15	16–25	26–45	Over 45
(kg ha^{-1})	0–23	24–38	39–63	64–113	Over 113
Index	0	1	2	3	Over 3

Recommended applications (kg P ha^{-1})					
Expected yield,					
6 t grain ha^{-1}	47	21	21M	21M	Nil
10 t grain ha^{-1}	57	35	35M	35M	Nil

1. The extracted P has been converted into an amount ha^{-1} assuming 2500 t soil ha^{-1} and a soil density in the measuring cup of 1 kg l^{-1} (Sections 9.1 and 10.3).
2. Application rates are given as kg P ha^{-1} to facilitate comparison to crop and soil data. To convert to the units used by the fertilizer manufacturers, kg P × 2.29 = kg P_2O_5.
3. Only small amounts of P are removed in straw (Table 9.2) and so there is no difference in the recommendation according to whether straw remains in the field or is removed, in contrast to potassium (Table 9.3).
4. M indicates a maintenance application. No response is expected, but crop removals are balanced by applications based on a P content in the grain of 3.5 kg P t^{-1} which is slightly smaller than that in Table 9.1. Recommendations for other yields can be calculated on this basis.
5. For soils with index 0, large applications of P are needed over a number of years to raise the index to 2. Much of the added P is held strongly and is not extractable in the bicarbonate solution.

From MAFF (1988)

respectively). Based on these values, the topsoil alone needs to contain at least 1.3 times the amount of P the crop needs, and 1.5 times as much K^+: on this evidence there is no indication that accessibility of P is a particular problem. However, the fact that adequate amounts of P can be taken up from a small fraction of the soil in the depletion zones suggests that much more P is available in the soil than is extractable by bicarbonate; this can be confirmed by a second extraction of the soil sample with bicarbonate which removes about half as much P as the first extraction, whereas a second extraction with ammonium nitrate removes very little K^+. Thus the bicarbonate method does not measure the total amount that could be extracted by roots, but simply rates soils according to their ability to supply P; hence the use of the term 'index of availability'.

The residual value of phosphorus fertilizers

The rapid adsorption of P on to particle surfaces is followed by a slower conversion into less available

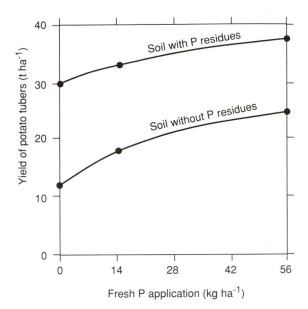

root growth and function (Colour Plate 15(a) and (b)) and the combination of poor root growth and low mobility of P may lead to deficiencies even if extractable P is present in what would otherwise be adequate amounts. In addition, poor root growth particularly in acidic subsoils may lead to a lack of available water adding a further constraint to growth.

These effects are seen in Table 10.2 which shows the response of maize to both lime and P applications in an acidic soil in Ghana. Without added P, yield was raised from 1 to $3.1\,t\,ha^{-1}$ by lime. Without added lime, yield was raised to $2.8\,t\,ha^{-1}$ by the fertilizer. When applied together a maximum yield of $3.6\,t\,ha^{-1}$ was recorded. Thus the soil had almost enough P for the crop provided that constraints due to acidity were removed.

Table 10.2 The response of maize to lime and phosphate applications on an acidic soil in Ghana

Lime application ($t\,CaCO_3\,ha^{-1}$)	Soil pH	P application ($kg\,ha^{-1}$)				
		0	11	22	33	44
		(t grain ha^{-1})				
0	4.5	1.05	2.63	2.92	2.90	2.83
1	5.0	1.38	2.79	3.10	3.13	3.44
2	5.7	2.23	2.70	2.99	3.20	3.61
4	6.0	3.08	3.10	3.42	3.06	3.33

From Lathwell (1979).

Figure 10.4 The residual value of phosphate fertilizers at Agdell Field, Rothamsted Experimental Station, 1960. From Johnston *et al.* (1970).

forms including mineral phosphates. Thus fertilizer-P is most available in the first season after application, but some remains available over long periods of time. Responses after the first season are termed *residual effects*.

The importance of residual effects is shown in Fig. 10.4. Two plots in Agdell field at Rothamsted have received no P since 1848, and two have received $37\,kg\,P\,ha^{-1}$ once every 4 years. In 1960 the plots were split into subplots which were given varying fertilizer treatments. With no fresh P, the residual P more than doubled the yield of potato tubers, and the large additions of fresh P did not match the combined effects of residual and fresh P. The special value of residual P is probably related to its uniform distribution through the soil: the whole root system is in contact with soil which is well supplied with P. Freshly applied fertilizer stays in a small volume of soil around each granule and even large concentrations in these volumes do not supply adequately the crop's needs.

Constraints due to soil acidity

The production of arable crops in temperate climatic regions normally involves liming as a first step in soil management. However in the tropics the expense of lime, and in the wetter areas the rapid reacidification of limed soils, results in crops being grown in soils with low pH values. These often have very small amounts of extractable P, and applications of fertilizer-P are needed. However, acidity has deleterious effects on

Leaching of phosphate

Soil solution concentrations of between 0.2 and $2\,mg\,l^{-1}$ (5 and $50\,\mu M$, Fig. 10.1) imply that very little P is lost from soils by leaching. The concentration entering drains and watercourses may be even smaller because of P adsorption as soil water drains through the subsoil. Table 10.3 gives data collected at Woburn

Table 10.3 Concentration of phosphorus, potassium and sulphur in water collected at Woburn, 1970–74

	Mean concentration ($mg\,l^{-1}$)			*Mean annual content* ($kg\,ha^{-1}$)		
	K^+	P	S	K^+	P	S
Rainfall, $548\,mm\,a^{-1}$	0.39	0.03	2.59	2.1	0.16	14.2
Drainage,† $250\,mm\,a^{-1}$	5.0	0.13	55.0	12.5	0.3	138
Stream	4.3	0.15	35.5			
Lake	8.8	0.03	55.0			

† The amount of drainage is an estimate. Annual lysimeter drainage between 1988 and 1990 was 235 mm (K. Goulding) and undisturbed soils may give slightly more.
From Williams (1976).

Experimental Station. The soil is a sandy loam over Lower Greensand parent material and so leaches relatively easily. The concentrations of P in rain-water and drainage water are very small, being 0.03 and 0.13 mg l^{-1} respectively. On heavier soils at Saxmundham Experimental Station, the concentration in drainage water from some plots is smaller than that in the rain indicating that the soil has adsorbed P from the rainwater (Williams, 1976). The above concentrations are for soluble P: suspended mineral particles containing P may greatly increase the total P burden of the water.

The concentrations of soluble P in local streams and lakes at Woburn are similar to those of the drainage water. However, the major rivers in Britain have concentrations up to 1 mg l^{-1}. The source of this extra P is probably sewage, slurry effluent and domestic washing machines. Raised concentrations of P may stimulate the growth of algae in rivers and lakes (Birch and Moss, 1990). There is much uncertainty regarding the concentrations at which algal blooms may form. If N is not limiting algal growth, then between 0.01 and 0.1 mg P l^{-1} may be needed; if P is not limiting growth then between 0.1 and 1 mg N l^{-1} may be needed. In most natural waters P rather than N is the limiting nutrient (Ch. 11).

SULPHUR

The total S content of soils varies widely (20 μg to 50 g kg^{-1}) depending on the amount in the parent material and that held in organic matter (about 5 g S kg^{-1} organic matter). Soils with large contents of S are found in arid regions where gypsum accumulates and in coastal marshes where pyrite, FeS_2, is often present.

Sulphur is supplied to plants as sulphate released by mineralization of organic sulphur (between 2 and 4 kg S ha^{-1} a^{-1}) or entering the soil as dry-deposited SO_2 and wet-deposited SO_4^{2-} from the atmosphere. In acidic soils sulphate is bound to sesquioxides and kaolinite in a similar manner to phosphate (Section 7.5). However, in neutral soils it behaves like nitrate: it is not bound and is easily leached. Its concentration in soil solution is similar to that of K^+ (about 0.3 mM in southern England, Table 7.6), but with wide variability depending on atmospheric inputs.

Sulphur budgets in agricultural soils

Crop requirements are given in Table 9.2. Between 10 and 15 kg ha^{-1} is removed in most crops although brassicas require larger amounts. In Britain these requirements have been met from two sources:

1. *Atmospheric inputs.* These are shown in Colour Plate 16 and have been measured as part of the current interest in acid rain. Over most of the country the

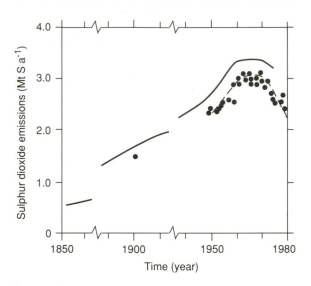

Figure 10.5 Changes in sulphur dioxide emissions in the UK from 1850. The data are from Warren Spring Laboratory together with historical trends from other sources (full line). From DOE (1990).

annual input is adequate for crop needs. However, reductions in S emissions since the mid-1960s have led to a steady decline in inputs to soil (Fig. 10.5) and the west and north of the country are not now being supplied with what is needed by crops. Excess sulphate at the end of a growing season is mostly leached away during the winter as is the case for nitrate.

2. *Fertilizers.* Superphosphate which contains about 12 per cent S has been the main source of fertilizer-P since its development in 1840 and has thus been a major source of S also. In recent years however, ammonium phosphate has been used increasingly in compound fertilizers, so that S inputs have decreased to low levels.

The S budget points to the need for S fertilizers for some crops in the north and west of Britain, and field experiments are beginning to show responses. The area of the country requiring fertilizer inputs will increase as S emissions continue to decrease. Sulphur deficiencies have been recorded in many parts of the world where atmospheric inputs are low.

The data in Table 10.3 for Woburn show that much more S was lost in drainage water than entered in the rain, the excess coming from dry deposition and fertilizers.

Sulphur budgets in non-agricultural soils

In Britain, the main areas of non-agricultural soils are

in the hill lands of the north and west. These acidic soils bind sulphate and so there has been an increase in soil sulphate over the last 150 years. In contrast inputs of nitrate are lost by leaching each year.

Acidification resulting from atmospheric inputs has been discussed in Ch. 8. Soluble aluminium is a product of acidification and is leached from soils together with nitrate and sulphate. Its loss into drainage water can only occur if anions are present, and so acidification of streams and lakes is closely related to atmospheric inputs of N and S. The accumulation of sulphate in soils means that there is a large potential for aluminium to be leached in the future despite the lower inputs resulting from reduced emissions of S.

Measurements of sulphur in soils and plants

The methods used for the determination of total S and sulphate-S in water, soils and plants are not entirely satisfactory, and further development of simple and reliable techniques is required. Standard methods are given in Section 10.5.

FURTHER STUDIES

Ideas for projects are given in Section 10.6 and calculations are in Section 10.7.

Section 10.1 The measurement of phosphate in solution

Soil solutions and drainage waters normally have phosphate concentrations between about 1 and 50 μM. (0.03–2 mg P l^{-1}). Measurement therefore requires a technique which is sensitive for small concentrations. A spectrophotometric method is used, in which phosphate and ammonium molybdate form a complex which is reduced with ascorbic acid to produce a blue colour in solution. Red light passing through the solution is absorbed, the amount depending on the concentration of phosphate present. A spectrophotometer or colorimeter is used to measure light absorption. Because of the small concentrations involved, especially clean analytical procedures are required.

The use of a spectrophotometer

The components of the spectrophotometer are shown in Fig. 10.6. Light passes through a diffraction grating. A given wavelength band is selected and passes through a shutter which allows the intensity to be adjusted. For P a wavelength of 880 nm is required. Light passing through the blue solution in an optical cell falls on to a photoelectric cell which converts light energy into elec-

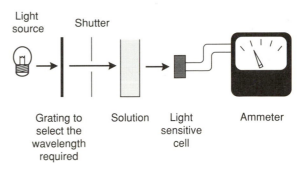

Figure 10.6 The components of a spectrophotometer.

trical energy which is measured with an ammeter. The current is directly proportional to the light intensity provided the intensity is low.

The relationship between absorbed light and concentration of the complex in solution (Fig. 10.7) is given by the Beer–Lambert Law:

$$\log(L_0/L_t) = kcx$$

where L_0 is the incident light intensity, L_t is the transmitted light intensity, c is the concentration, x is the optical path length, often 10 mm, and k is a constant. If the current, I, is directly proportional to L_t, then the relationship beween I and c is shown in Fig. 10.8. This does not give a calibration curve which is easy to use. However, $\log(L_0/L_t)$ is linearly related to c and is known as the *absorbance* of the solution. It has a value of zero when $c = 0$ and $L_t = L_0$ because $\log(L_t/L_0) = \log 1 = 0$. This graph is ideal as a calibration curve. The spectrophotometer scale is calibrated in absorbance units so that the reading is linearly related to concentration. The sensitive range for measurement is between 0 and 1 absorbance units or between 0 and 90 per cent of the light absorbed.

Note the meaning of the words used here:

- *absorb* or *absorption* = taken *into* a solution or a solid.
- *adsorb* or *adsorption* = taken *on to* a surface.
- *absorbance* is the logarithmic function in the Beer–Lambert equation.

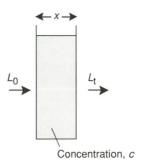

Figure 10.7 Light absorption in an optical cell.

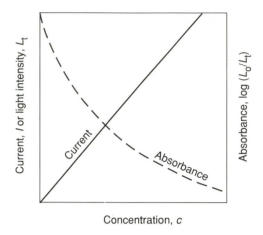

Figure 10.8 A graphical illustration of the Beer–Lambert Law.

A colorimeter is a simple spectrophotometer. Light is selected by a filter (red for the P determination) and the optical cell is a glass tube.

The calibration of the spectrophotometer

Reagents

Ammonium molybdate solution. Dissolve 12 g of powdered ammonium molybdate, $(NH_4)_6Mo_7O_{24}.4H_2O$, and 0.3 g of antimony potassium tartrate, $KSbO-C_4H_4O_6$, in 600 ml of water. Slowly add 148 ml of concentrated sulphuric acid (approx. 98 per cent *m/m* H_2SO_4) (CARE), and dilute to 1 l. Mix 125 ml of this solution with 875 ml of water. Store in a cool place.

Ascorbic acid solution. Dissolve 1.5 g of ascorbic acid, $C_6H_8O_6$, in water and make up to 100 ml. Prepare a fresh solution each day.

Standard phosphate solution. Dry potassium dihydrogen orthophosphate, KH_2PO_4, at 105 °C for 1 h and cool in a desiccator. Dissolve 1.099 g in water, add 1 ml of concentrated hydrochloric acid, (approx. 36% *m/m* HCl) and make up to 250 ml. This contains $1 \, mg \, P \, ml^{-1}$. Add one drop of toluene if the solution is to be stored for more than a few days.

Clean glassware. Pipettes and volumetric flasks must be carefully washed. Soak either in concentrated H_2SO_4 overnight or in 'Decon' or an equivalent cleaning agent. Thoroughly rinse with tap-water and demineralized water.

Method

Pipette 10 ml of the $1 \, mg \, P \, ml^{-1}$ solution into a 100 ml flask and make up to the mark to give a solution containing $100 \, \mu g \, P \, ml^{-1}$. Pipette 10 ml of this into a 1 l flask and make up to the mark to give a solution containing $1 \, \mu g \, P \, ml^{-1}$.

Into six 100 ml volumetric flasks pipette 0, 5, 10, 15, 20 and 30 ml of the $1 \, \mu g \, P \, ml^{-1}$ solution. Dilute to about 80 ml with water and mix. Into each pipette 8 ml of ammonium molybdate solution and 8 ml of ascorbic acid solution. Make up to the mark, mix and allow to stand for 20 min for full colour development. These solutions contain 0, 5, 10, 15, 20 and 30 μg P in the 100 ml flasks. Measure the absorbance of each solution using a 10 mm optical cell after setting to zero absorbance with water (Note 1) in a matched 10 mm cell. Check the zero (and if necessary readjust) before each measurement.

Plot the calibration curve (Fig. 10.9). It is common for the first set of measurements to be erratic. If points are seriously off the line repeat the experiment until analytical problems have been overcome.

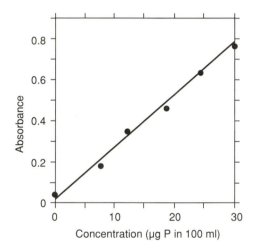

Figure 10.9 A spectrophotometer calibration graph for phosphate.

Measurement of an unknown solution sample

Soil solution (Section 7.4), or other water samples can be measured against this calibration.

Pipette 5 ml (or any volume up to 80 ml) of the sample into a 100 ml volumetric flask at the same time as the standards are prepared. Dilute to 80 ml with water and develop the colour as for the standards. Read from the calibration curve the concentration of P in solution (μg in 100 ml).

Calculation

Method summary

water sample ($?\,\mu g\,P\,ml^{-1}$)

$\downarrow$

$5\,ml \rightarrow 100\,ml : \mu g\,P$ in $100\,ml$

Example The measured concentration from the calibration curve is $2.5\,\mu g\,P$ in $100\,ml$. This came from a $5\,ml$ water sample. Therefore this sample has $2.5/5 = 0.5\,\mu g\,P\,ml^{-1}$. This can be expressed as a molarity; it is

$0.5 \times 10^{-3}\,g\,l^{-1}/31\,g\,mol^{-1} =$
$$1.6 \times 10^{-5}\,mol\,P\,l^{-1} \text{ or } 16\,\mu M$$

Phosphate in soil solution

If the above sample was a soil water extract prepared by the methods of Section 7.4 ($100\,g$ of air-dry soil + $100\,ml$ of water), the concentration of this extract would be $16\,\mu M$ P present as phosphate. Phosphate is well buffered in soil solutions and it can be assumed that the concentration in the extract is similar to that in the solution in the moist soil. If the air-dry soil contained $5\,g\,H_2O$ per $100\,g$ oven-dry soil, then the suspension contained $95.24\,g$ of oven-dry soil and $104.76\,ml$ of water, compared to about $30\,ml$ of water in a moist soil. Thus the soil solution would have been diluted by a factor of about 3, and the P concentration by between 1 and 3 depending on the extent of the buffering.

Expressing the concentration If a field moist soil contains $30\,ml$ of soil solution per $100\,g$ of oven-dry soil and the solution has a concentration of $0.5\,\mu g\,P\,ml^{-1}$, the amount of P in the soil solution is $0.5 \times 30 = 15\,\mu g$. This is often expressed relative to the mass of soil. There are $15\,\mu g$ P in solution in $100\,g$ of oven-dry soil, or

$$15 \times 1000/100 = 150\,\mu g \text{ solution-P }kg^{-1}\text{ soil}$$

Note 1 A choice can be made regarding the solution in the matched cell. If water, which is optically completely stable, is used a small reading will normally be obtained with the zero standard, and the calibration graph will not pass through the origin. Alternatively the zero standard can be used. The matched cell is needed because the zero setting has to be checked between each measurement to allow for slow changes in the zero reading because of changing lamp intensity or electronic drift.

Section 10.2 Adsorption and desorption isotherms for phosphate

The principles of isotherm measurement and use are developed in Section 9.4 to give a potassium exchange isotherm. Phosphate is adsorbed or desorbed on sesquioxide and clay surfaces, and in this case the isotherm is often called a *sorption isotherm*.

Reagents and equipment

Calcium chloride, 0.4 M. Dissolve $14.7\,g\,CaCl_2.2H_2O$ in water and make up to $250\,ml$.

Standard phosphate solution. Dry potassium dihydrogen orthophosphate, KH_2PO_4, at $105\,^\circ C$ for $1\,h$ and cool in a desiccator. Dissolve $2.197\,g$ in water and make up to $250\,ml$. Pipette $25\,ml$ into a $500\,ml$ flask and make up to the mark. This contains $100\,\mu g\,P\,ml^{-1}$.

Standard phosphate solutions in 10 mM calcium chloride. Into six $1\,l$ volumetric flasks pipette 0, 5, 10, 20, 30 and $50\,ml$ of the $100\,\mu g\,P\,ml^{-1}$ solution. Add to each $25\,ml$ of $0.4\,M\,CaCl_2$ solution and make up to the mark. These contain 0, 0.5, 1, 2, 3 and $5\,\mu g\,P\,ml^{-1}$ in $10\,mM\,CaCl_2$.

Shaking bottles with caps.

Spectrophotometer with matched 10 mm optical cells (or colorimeter).

Method

Into six bottles weigh $2.5\,g$ (± 0.01) of $<2\,mm$ air-dry soil. Pipette $25\,ml$ of the standard phosphate solutions into the bottles. Cap and place on a shaker for $24\,h$ or shake occasionally by hand over this period. Filter the suspensions through Whatman No. 41 papers, rejecting the first few ml of solution.

Following the methods of Section 10.1 prepare a calibration curve for the phosphomolybdate method. Use the above standard phosphate solutions ($0–5\,\mu g\,P\,ml^{-1}$) and take $5\,ml$ samples for colour development. The $100\,ml$ flasks of blue solution will contain 0, 2.5, 5, 10, 15 and $25\,\mu g\,P$.

Measure the phosphate concentrations of the six filtered solutions again using $5\,ml$ samples for colour development.

Calculation

For each bottle, calculate the final concentration in solution ($\mu g\,P\,ml^{-1}$) and the amount of P adsorbed or desorbed during shaking ($\mu g\,P\,g^{-1}$ soil).

Example A solution initially contained $5\,\mu g\,P\,ml^{-1}$. After shaking, the blue complex was developed and $100\,ml$ of coloured solution contained $5.2\,\mu g$.

1. The final concentration: $5.2 \,\mu g\,P$ was in 5 ml of filtered solution. Therefore its concentration was $5.2/5 = 1.04 \,\mu g\,P\,ml^{-1}$.
2. The amount of P initially in 25 ml of solution was

$$5 \,\mu g\,ml^{-1} \times 25\,ml = 125 \,\mu g\,P$$

The amount finally present was $1.04 \times 25 = 26 \,\mu g\,P$.
3. The amount of P removed from solution was $125 - 26 = 99 \,\mu g\,P$. This had been adsorbed on to 2.5 g of soil. Therefore

$$\text{adsorbed P} = 99/2.5 = 39.6 \,\mu g\,P\,g^{-1} \quad \text{(Note 1)}$$

The calculation simplifies to

adsorbed $P\,(\mu g\,g^{-1}) =$
$10 \times$ (initial concentration $-$ final concentration)
$$(\mu g\,P\,ml^{-1})$$

and a negative result indicates desorption.

Note 1 These calculations relate to adsorption from 25 ml of added solution on to 2.5 g air-dry soil. If results are required on an oven-dry basis, then determine the water content of the soil (e.g. 5 g H_2O per 100 g oven-dry soil). The mass of oven-dry soil used is 2.38 g and the calculation can be adjusted at (3). The amount of water in the air-dry soil (0.12 ml per 2.5 g) is small compared to the 25 ml added, and is not normally considered in the calculation.

Isotherms

Fig. 10.10 shows results obtained for three soils.

Oxisol, Brazil This uncultivated soil has its clay-sized fraction dominated by iron oxides and kaolinite. It has a very small soil solution concentration $(0.001 \,\mu g\,P\,ml^{-1})$, a very large phosphate buffer power and a very small amount of bicarbonate-extractable phosphate $(0.3 \,mg\,P\,kg^{-1}$, index 0, Section 10.3). There are serious problems in raising the P status of soils of this type. The amount of available P is limited and there is the added problem of the very small solution concentration which limits uptake rates. Even to raise the concentration to $0.1 \,\mu g\,ml^{-1}$ in this soil (which is often considered the lower limit for crop production) requires an input of $210 \,mg\,kg^{-1}$ or about $500 \,kg\,P\,ha^{-1}$ and more may be needed in the field because of continued slow adsorption.

Upper Greensand soil, Oxfordshire This has a small buffer power and has been well fertilized, which has raised its solution concentration and given it an assumed P index of 2 $(20 \,\mu g\,g^{-1})$.

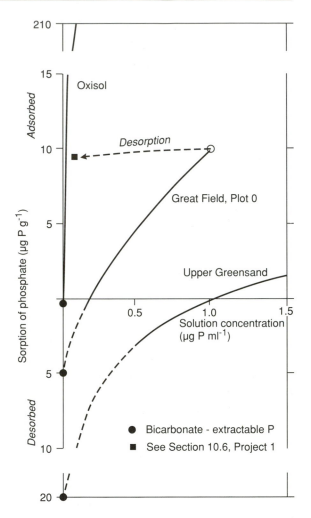

Figure 10.10 Adsorption–desorption isotherms for soil phosphate. Oxisol: G. Warren, Reading; Great Field: P. Le Mare and Reading students; Upper Greensand: White and Beckett (1964). For Upper Greensand the bicarbonate-extracable P is an assumed value for an index 2 soil.

Great Field soil, Rothamsted This soil was taken from plot 0 of the Great Field Experiment. Its buffer power reflects its silty clay loam texture, and its low solution concentration and P index 0 reflect its management. It received no fertilizer or manure between 1959 and 1985 when it was sampled. During that time it had been cropped with swedes, potatoes, barley and a grass-clover ley. From other plots in the same experiment which received phosphate (Table 10.4), crop uptake of P was estimated to be about 20 per cent of the inputs. Thus in plot 4, about 341 of the $426 \,mg\,kg^{-1}$ remain in the soil. Only 37 mg was extracted by the sodium bicarbonate solution (Section 10.3), but much of the

Table 10.4 The Great Field Experiment, Rothamsted

Plot	Total P inputs (1959–1985)		Bicarbonate−extractable P	Soil−solution concentration†
	(kg ha^{-1})	(mg kg^{-1})‡	(mg kg^{-1})	(μg P ml^{-1})
0	0	0	5	0.15
1	278	89	8	0.20
2	552	177	9	0.25
3	942	301	19	0.60
4	1332	426	37	1.65

† Soil solution concentrations were determined from the intercepts of sorption isotherms with the abscissa.
‡ Soil samples were taken down to 25 cm (3125 t ha^{-1}).
Data from E.A. Johnston and Reading students

of the 341 mg kg^{-1} will be available because this plot is now an index 3 soil. The soil solution concentration has been raised to a normal level for agricultural soils of temperate regions (Table 7.6).

Section 10.3 The extraction of soil phosphate with sodium bicarbonate solution (Olsen's method) and the determination of an availability index

Phosphate extracted from soil by sodium bicarbonate solution is measured using the phosphomolybdate method described in Section 10.1. The amount of extractable P is used to allocate to the soil a P index in a similar way to that described for potassium in Section 9.1. The Bray solution (30 mM NH_4F + 25 mM HCl) is also commonly used for extraction, and isotopic dilution techniques are used to measure exchangeable phosphate (Page, 1982).

Reagents and equipment

Polyacrylamide solution (Note 1). Dissolve 0.5 g of polyacrylamide in about 600 ml of water by stirring for several hours. Make up to 1 l.
Sodium bicarbonate solution, 0.5 M. Dissolve 84 g of sodium bicarbonate, $NaHCO_3$, in water, add 10 ml of polyacrylamide solution and dilute to 2 l. Adjust the pH to 8.50 (Section 8.1) by adding drops of NaOH solution (50 g NaOH dissolved in 100 ml of water).
Sulphuric acid, approximately 1.5 M. Add 80 ml of concentrated H_2SO_4 (approx. 98% *m/m* H_2SO_4) to water (CAUTION) and dilute to 1 l.
Shaking bottles with caps.

Spectrophotometer with matched 10 mm optical cells (or colorimeter).

Extraction

Weigh 5 g (±0.05) of <2 mm air-dry soil into a bottle. Add 100 ml of sodium bicarbonate solution. Cap and shake on a shaking machine for 30 min at 20 °C (Note 2). Filter through a Whatman No. 125 paper rejecting the first few ml of filtrate. Retain the remainder for the determination of phosphate (Note 3).

Measurement

Prepare a calibration curve following the methods in Section 10.1. An adjustment to the method is required: after pipetting 0–30 ml of the 1 μg P ml^{-1} standard solution into the 100 ml flasks, add to each 5 ml of the sodium bicarbonate solution. Then add to each flask 1 ml of approximately 1.5 M H_2SO_4 and swirl the flasks to release the CO_2. Do not put stoppers in the flasks until the gas has escaped. Develop the blue colour and measure absorbance as before.

Measure the P concentration in the extract as follows. *Pipette 5 ml (Note 3) into a 100 ml flask at the same time as the standards are prepared. Add 1 ml of 1.5 M H_2SO_4 and then proceed as above.

The adjustment to the method neutralizes the bicarbonate in the extract so that the conditions for colour development are similar to those in the water extracts of Section 10.1. The standards have to be matched with the extracts. Hence these also contain sodium bicarbonate and acid.

Calculation

Method summary

air-dry soil (? mg P kg^{-1})
$\downarrow$
5 g $\rightarrow$ 100 ml soil extract
$\downarrow$
5 ml $\rightarrow$ 100 ml blue extract:
μg P in 100 ml

Example The measured amount of P in 100 ml of blue solution is 10.5 μg. This came from 5 ml of soil extract. In 100 ml of extract there is 10.5 × 100/5 = 210 μg P. This came from 5 g of soil and so there is

210 × 1000/5 = 42 000 μg or 42 mg P kg^{-1} air-dry soil

The calculation simplifies to

mg P kg^{-1} air-dry soil =
4 × (μg P in 100 ml of blue solution)

If the air-dry soil contains 5 g H_2O per 100 g oven-dry soil then 4.76 g of oven-dry soil was used. The extractable P is therefore

$$210 \times 1000/4.76 = 44\,118\,\mu g \text{ or } 44\,\text{mg P kg}^{-1} \text{ oven-dry soil}$$

The use of soil volume

For routine analytical purposes a 5 ml measuring cup filled with soil is used instead of 5 g. The cup dimensions are 18 mm internal diameter by 19 mm deep (Fig. 9.8). The procedure is otherwise unaltered. The calculation simplifies to

$$\text{mg P l}^{-1} \text{ soil} = 4 \times (\mu g \text{ P in 100 ml of blue solution})$$

Section 9.1 explains the use of soil volume.

Note 1 The polyacrylamide is used to prevent clay dispersing when the soil is shaken with sodium bicarbonate solution. It is a large molecular weight organic polymer which binds to clay particles.

Note 2 The amount of phosphate extracted is temperature dependent. If a constant temperature room is not available, work in an area as near to 20 °C as possible.

Note 3 If the extract is highly coloured (brown) due to dissolved organic matter, proceed with the calibration as described. Prepare two extract samples at the stage marked with an asterisk. Treat one as described, but in the second do not add the ascorbic acid: simply make up to the mark with water after adding the ammonium molybdate. Thus the blue colour does not develop but an absorbance reading will be given by the brown solution. The difference in the concentration readings for the two solutions gives the amount of P present.

The phosphorus availability index

An index system is shown in Table 10.5. For cereal

Table 10.5 The phosphorus availability indices used in Britain by ADAS

Index	Extractable phosphate (mg P l^{-1} of soil)†
0	0–9
1	10–15
2	16–25
3	26–45
4	46–70
5	71–100

† 1 litre of <2 mm air−dry soil.
From MAFF (1988).

crops fertilizer applications are recommended up to index 3 (Table 10.1). Some horticultural crops receive fertilizer applications up to index 5.

Section 10.4 The measurement of phosphorus in plant material

PLANT PHOSPHORUS

Plant dry matter is ashed to convert organic P into phosphate which is then dissolved in HCl and measured using the phosphomolybdate method. The alternative phospho-vanado-molybdate method can be used (Page, 1982; MAFF, 1986a).

Method

Following the methods of Section 9.3, 2 g of plant dry matter is ashed at 450–500 °C, and the phosphate dissolved in 50 ml of solution to give a plant extract. The P content of the dry matter is about 3 g P kg^{-1} (Table 9.1) and so there is about 6 mg of P in the 50 ml of extract.

The concentration of P is measured using the phosphomolybdate method (Section 10.1). The range of measurement for the method is between 0 and 30 μg P in a 5 ml sample and so the above extract (600 μg P per 5 ml) has to be diluted to bring it into the range for measurement.

Pipette 5 ml of the extract into a 250 ml volumetric flask. Make up to the mark with water. Develop the blue complex using 5 ml of this diluted plant solution, and measure its absorbance.

Calculation

Method summary

plant dry matter (? g P kg^{-1})
$\downarrow$
2 g $\rightarrow$ 50 ml plant extract
$\downarrow$
5 ml $\rightarrow$ 250 ml diluted solution
$\downarrow$
5 ml $\rightarrow$ 100 ml : μg P in 100 ml

Example There is 12.5 μg P in 100 ml of blue solution. This came from 5 ml of diluted solution and so in 250 ml of diluted solution there is $12.5 \times 250/5 = 625$ μg P. This came from 5 ml of the plant extract, and so in 50 ml of this there is 6250 μg P or 6.25 mg. This came from 2 g of plant dry matter, and its content is

$$6.25 \times 1000/2 = 3125 \text{ mg or 3.1 g P kg}^{-1}$$

The calculation simplifies to

$g P kg^{-1}$ dry matter =
$$(\mu g \text{ P in } 100 \text{ ml of blue solution})/4$$

Section 10.5 The measurement of sulphur in soils and plants

Sulphate in soil solution can be extracted in water or in 10 mM $CaCl_2$ solution. Chloride may displace a small amount of adsorbed sulphate. These solutions can be used to measure available S in neutral soils which contain little adsorbed sulphate. Solution plus adsorbed sulphate can be extracted with calcium phosphate solution, and in acidic soils this is a measure of availability. Plant S is determined by ashing to convert organic S to sulphate.

Sulphate in these extracts is determined turbidimetrically, a technique that is simple but may be unreliable when first used. Other methods are available but are more difficult for general use.

The turbidimetric determination of sulphate

The addition of barium chloride to a solution containing sulphate causes barium sulphate to be precipitated. The absorbance of light in the turbid solution is measured under standard conditions with a spectrophotometer. The use of a spectrophotometer is described in Section 10.1.

Reagents and equipment

Barium chloride. Sieve the salt, retaining the 425–600 μm crystal size (Note 1).
Hydrochloric acid, approximately 1 M. Dilute 25 ml of concentrated HCl (approx. 36% *m/m*) to 250 ml with water.
Standard sulphate solutions, 0–50 μg S ml^{-1}. Dry potassium sulphate, K_2SO_4, at 105 °C for 1 h and cool in a desiccator. Dissolve 0.544 g in water and dilute to 1 l. This contains 100 μg S ml^{-1}. Into 100 ml flasks pipette 0, 5, 10, 20, 30, 40 and 50 ml of this solution. Into each flask pipette 10 ml of concentrated HCl, (approx. 36% *m/m*) and make up to the mark. These contain 0 to 50 μg S ml^{-1} in approximately 1 M HCl.
Spectrophotometer with 40 mm optical cells.
Measuring cylinders, 50 ml with stoppers, or 250 ml conical flasks with bungs.

Method

Pipette 5 ml of each standard into a 50 ml measuring cylinder (or conical flask) and dilute to 20 ml with water. Add 0.25 g of barium chloride, stopper the cylinder and shake for 30 s. Stand for 30 min.

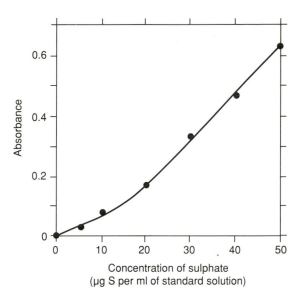

Figure 10.11 A calibration curve for sulphate using the turbidimetric method.

*Invert one cylinder several times and without delay measure the absorbance of the suspension in a 40 mm optical cell at 480 nm. Repeat for each cylnder in turn (Note 1). Plot a calibration curve relating absorbance to concentration in the standard solutions (Fig. 10.11).

Note that the calibration curve relates absorbance to the concentration of S (μg ml^{-1}) in the 5 ml of standard solution and not to the final concentration in the 20 ml of turbid solution. The final concentration could be used, but for both soil and plant extracts, 5 ml is placed in the measuring cylinder and so it is more convenient to relate the calibration to this volume.

Note 1 The reliability of the method depends on carefully standardizing the conditions. Absorbance depends on a uniform and representative distribution of barium sulphate particles in the optical cell. It may be necessary to perform the experiment several times so that a routine develops in handling the solutions. This routine must then be used for the extracts. The reason for sieving the barium chloride is to ensure that it dissolves in a standardized way in each sample. The calibration graph is not linear below 20 μg ml^{-1}.

Soil solution sulphate

Soluble sulphate is extracted in 10 mM $CaCl_2$ solution which prevents clay dispersion but does not displace adsorbed sulphate. Activated charcoal is also present to decolorize the extract and to prevent soluble organic matter from interfering in the subsequent measurement. The standard sulphate solutions used in the turbidimetric analysis are modified to match the extracting solution.

Reagents and equipment

Activated charcoal. This has to be purified to remove sulphate. To 100 g of charcoal in a 1 l beaker add 150 ml of water and 250 ml of concentrated HCl (approx. 36% *m/m*). Boil for 15 min, stir well and allow to settle for 30 min. Pour off the liquid, add 500 ml of water, mix well and allow to stand for 30 min. Repeat this water wash twice more. Add 500 ml of water and boil for 15 min. Wash with water three more times. Dry the charcoal at 105 °C overnight. Store in a tightly capped jar.
Calcium chloride, 10 mM. Dissolve 1.47 g $CaCl_2.2H_2O$ in water and make up to 1 l.
Calcium chloride, 100 mM. Dissolve 1.47 g $CaCl_2.2H_2O$ in water and make up to 100 ml.
Standard sulphate solutions. Prepare as described under 'Turbidimetric determination of sulphate', but before making up to 100 ml, pipette into each flask 10 ml of 100 mM $CaCl_2$ solution. The standard solutions then contain 10 mM $CaCl_2$.
Bottles with caps, 125 ml, or 250 ml conical flasks with bungs.

Extraction and sulphate measurement

To 25 g ($\pm$0.1) of <2 mm air-dry soil in a bottle or conical flask add approximately 1 g of charcoal and 50 ml of 10 mM $CaCl_2$ solution. Cap and shake for 30 min on a shaking machine (or occasionally by hand over this period). Filter through a Whatman No. 40 paper and retain the filtrate.

Carry out a blank determination omitting the soil. This determines the absorbance resulting from any remaining sulphate extracted from the charcoal.

Pipette 5 ml of each filtrate into a 50 ml measuring cylinder. Add 5 ml of 1 M HCl, 10 ml of water and 0.25 g of barium chloride. Stopper the cylinder, shake for 30 s and stand for 30 min. Measure the absorbance following exactly (from the asterisk) the procedure used for calibration above.

Measure the background absorbance using the soil extract and acid, omitting the barium chloride and making up to 20 ml with water. This absorbance is the result of fine particles of clay passing through the filter paper.

Calculation

From the calibration curve above determine the concentration of sulphate equivalent to the absorbance in the soil extract, the blank and the background. Sum the blank and background concentrations.

Example

Extract	4.5 µg S ml^{-1}
Blank + background	0.3 µg S ml^{-1}
Extracted sulphate	4.2 µg S ml^{-1}

Method summary

air-dry soil (? mg S kg^{-1})
$$\downarrow$$
$$25\,g \rightarrow 50\,ml$$
$$\downarrow$$
$$\mu g\,S\,ml^{-1}$$

In the above example, 50 ml of extract contains 4.2 × 50 = 210 µg S. This came from 25 g of soil. Therefore from 1 kg there would be

$$210 \times 1000/25 = 8400\ \mu g\ \text{or}\ 8.4\ mg\,S\,kg^{-1}\ \text{air-dry soil}$$

The calculation simplifies to

mg S kg^{-1} air-dry soil =
$$2 \times (\text{extract concentration}, \mu g\,S\,ml^{-1})$$

If the air-dry soil contains 5 g H_2O per 100 g oven-dry soil (952.4 g oven-dry soil kg^{-1} air-dry soil), the result is

$$8.4 \times 1000/952.4 = 8.8\ mg\,S\,kg^{-1}\ \text{oven-dry soil}$$

This can be converted into a molarity in soil solution as follows. Assuming that the water-holding capacity of the soil is 30 per cent *m/m*, 1 kg of oven-dry soil would have, when moist, 300 ml of soil solution. In this there would be

$$8.8\ mg\,S\ \text{or}\ 8.8 \times 1000/300 = 29\ mg\,l^{-1}$$

The molar mass of S is 32.1 g mol^{-1} and so the molarity is

$$29 \times 10^{-3}\,g\,l^{-1}/32.1\,g\,mol^{-1} =$$
$$9.0 \times 10^{-4}\,mol\,l^{-1}\ \text{or}\ 0.9\ mM$$

(Table 7.6 gives 0.3 mM for Oxfordshire soils). Note that the calculation assumes that all the extracted sulphate has come from the 300 ml of soil solution. This is probably the case in a neutral soil. If adsorbed sulphate is present in an acidic soil some desorption may occur during extraction.

Extractable sulphate in soils

Both sodium bicarbonate and calcium tetrahydrogen phosphate solutions are commonly used (Page, 1982). The latter gives clear extracts with most soils and is the method described here. Phosphate competes strongly with sulphate for adsorption sites, displacing it into solution. The standard sulphate solutions used in the turbidimetric analysis are modified to match the extracting solution.

Reagents

Extracting solution. Dissolve 2.52 g $CaH_4(PO_4)_2.H_2O$ in water and make up to 1 l.

Standard sulphate solutions. Prepare as described under 'Turbidimetric determination of sulphate' but before making up to 100 ml, pipette into each flask 10 ml of a solution containing 2.52 g $CaH_4(PO_4)_2.H_2O$ dissolved in water and made up to 100 ml. The standard solutions then contain the same calcium phosphate concentration as the extracting solution.

Extraction and measurement

Follow the method under 'Soil solution sulphate' modified as follows. Extract 10 g of soil with 1 g of charcoal and 100 ml of extracting solution for 30 min. Measure the S concentration in 5 ml of extract.

Calculation

Adjust the calculation to take account of the altered soil to solution ratio. The calculation simplifies to

mg S kg^{-1} air-dry soil =
$$10 \times (\text{extract concentration, } \mu g\, S\, ml^{-1})$$

and the result should be expressed in terms of oven-dry soil.

Critical levels of extractable sulphate

Between 10 and 13 mg of extractable sulphate-S kg^{-1} (25–33 kg ha^{-1}, 2500 t ha^{-1}) is sometimes considered to be the critical soil concentration below which S availability may restrict crop growth. This concentration is hardly measurable using the above method (1–1.3 $\mu g\, ml^{-1}$ in the extract). Crop requirements (Table 9.2) and atmospheric inputs (Colour Plate 16) vary greatly and an index-recommendation system has not yet been developed in Britain.

Sulphur in plant material

Plant S is released as sulphate by ashing. The procedure of Section 9.3 is modified to ensure that organic S is oxidized to SO_4^{2-} which is then determined turbidimetrically.

Reagents and equipment

Magnesium nitrate solution. Dissolve 71.3 g $Mg(NO_3)_2.6H_2O$ in water and make up to 100 ml.
Electrical hot plate.

Extraction

To 1 g of plant dry matter (Section 9.3) in an evaporating basin add 10 ml of magnesium nitrate solution over the whole surface of the sample. Heat to 180 °C on a hot plate, and when dry raise to 280 °C. Organic S and sulphide have now been oxidized to sulphate. When the colour of the residue changes from brown to yellow, heat at 450 °C in a muffle furnace overnight until a white ash remains. Cool and cover with a watch glass. Moisten with water, add 10 ml of concentrated HCl (approx. 36% *m/m*) and gently boil for 2 min. Add 10 ml of water, remove and rinse the watch glass into the basin. Quantitatively transfer the contents of the basin with washings into a 100 ml volumetric flask and make up to the mark. Filter the solution through a Whatman No. 541 paper, rejecting the first few ml. Retain the remainder for analysis.

Measurement of sulphate

The method under 'Turbidimetric determination of sulphate' is used. The plant extract has an acid concentration equal to that in the standards. Pipette 5 ml of extract into a 50 ml measuring cylinder, make up to 20 ml with water and add 0.25 g of barium chloride. Then proceed as before.

Calculation

Method summary

plant dry matter $(?\, g\, S\, kg^{-1})$
$$\downarrow$$
$$1\, g \rightarrow 100\, ml$$
$$\downarrow$$
$$\mu g\, S\, ml^{-1}$$

Example The measured concentration is 20.5 $\mu g\, S\, ml^{-1}$. In 100 ml of extract there is $20.5 \times 100 = 2050\, \mu g$ or 2.05 mg S. This came from 1 g of plant dry matter. The S content is 2.05 g S kg^{-1} dry matter.

Section 10.6 Projects

1. Determine the extent to which adsorbed phosphate is not desorbed by a $CaCl_2$ solution. Follow the methods of Section 10.2 and determine the sorption isotherms for a soil or for soils with differing textures. The only modification to the standard method is that the soils and solutions should be shaken in centrifuge tubes.

Weigh the centrifuge tube before adding the soil and solution. After shaking, centrifuge and pour off the liquid which is retained for the measurement of phosphate. Weigh the tube containing the wet soil to give the mass of solution left in the soil (known as the entrained solution). Add 25 ml of 10 mM $CaCl_2$ solution, shake for 24 h, centrifuge and measure the concentration of P in the solution.

An example follows based on the Great Field soil isotherm in Fig. 10.10. The maximum point on the isotherm $(10 \, \mu g \, g^{-1}, 1 \, \mu g \, ml^{-1})$ is taken through the above procedure. The entrained solution was 0.6 g, and the final concentration $0.1 \, \mu g \, ml^{-1}$. The amount of P in the entrained solution is $1 \times 0.6 = 0.6 \, \mu g$. After the final shaking the amount of P in 25 + 0.6 ml of solution is $0.1 \times 25.6 = 2.56 \, \mu g$. Therefore the P desorbed into the 10 mM $CaCl_2$ solution is $2.56 - 0.6 = 1.96 \, \mu g$ P. This came from 2.5 g of soil and so $1.96/2.5 = 0.78 \, \mu g \, P \, g^{-1}$ has been desorbed. This point has been plotted on the isotherm, and shows the extent of irreversibility of the reaction. The desorbed P is 8 per cent of that which was initially adsorbed.

For soils of differing texture is there any relationship between texture and percentage desorption? Compare your result to similar experiments with K^+ (Section 9.8, Calculation 4).

2. Carry out a pot experiment to determine the response of ryegrass to applications of lime and phosphate in an acidic soil. Follow the methods given in Section 9.2. Obtain an acidic soil (from a woodland area) with a pH of about 4. Calculate suitable lime treatments to raise the pH to 6. Apply N and K^+ to all pots. Phosphate should be applied to half the pots with three replicates per treatment. The plan is therefore

L$_0$ (pH 4)   P$_0$ (no P)
L$_1$ (pH 6)   P$_1$ (with P)

Treatments: 3 pots each for L_0P_0, L_0P_1, L_1P_0 and L_1P_1 making 12 pots in total.

The main aim of the experiment is to determine whether growth in the acidic soil is primarily constrained by acidity or by a shortage of P.

The experiment could be extended by determining the bicarbonate-extractable P in your soils (Section 10.3). Does liming increase the amount of extractable P? What fraction of the added fertilizer-P is extractable, and does liming influence this?

3. Treat samples of a soil with various amounts of phosphate following the methods given in Section 9.2. Incubate them moist for 1 month. Measure their sorption isotherms, including the untreated soil.

Plot the equilibrium concentration of P in soil

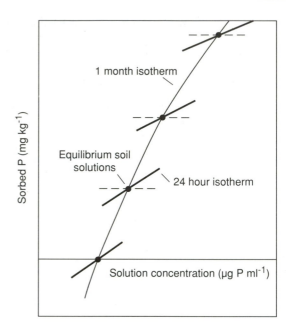

Figure 10.12 Isotherms obtained in Project 3.

solution against sorbed P. At each point plot the sorption isotherm (Fig. 10.12).

What causes the difference in slope between the 24 h isotherms and the 1 month isotherm?

4. *Data analysis project.* Using the data in Fig. 3.1, calculate the percentage of the soil volume in the 0–15 cm layer which would be within the P depletion zone around wheat roots in March and in June. Assume that the root radius is 0.15 mm and that the extent of the depletion zone is 2 mm (Fig. 10.2). (*Ans.* 19 and 37) Repeat the calculation for the 75–100 cm layer. (*Ans.* 0 and 1.4)

Extend your calculations to take account of the different uptake times associated with different parts of the root system. You will have to make a few assumptions, i.e. that root length doubles in each 6 week period from 31 December to 30 June, that there is no depletion zone at the end of December, and that the zone extends such that the radius is proportional to the square root of time (radius in mm = $0.32 \, t^{1/2}$ where t is the uptake time in days). Compare your depletion volume to the total soil volume to 1 m depth. A computer program could be written to make these calculations for you. It would involve a stepwise calculation, with new roots growing each week and then developing a depletion zone during the remainder of the growth period.

5. The literature suggests that in neutral soils nearly all the sulphate is free in soil solution. In acidic soils, particularly those containing appreciable sesquiox-

ide and kaolinite contents, large amounts of adsorbed sulphate may be present (Section 7.5). Measure solution and adsorbed sulphate in a range of soils to check these statements. Collect soils with a range of pH values and composition. Red soils often have large sesquioxide contents. Prepare <2 mm air-dry samples. Following the methods of Section 10.5 extract sulphate with calcium chloride and calcium phosphate solutions. The concentrations of S may be low in leached soils. If so, add sulphate to each soil and remeasure.

Suggested additions. Add 50 mg S kg^{-1} soil, equivalent to 125 kg ha^{-1} (2500 t soil ha^{-1}). To do this dissolve 1.357 g K_2SO_4 in water and make up to 1 l. The application of 20 ml to 100 g soil adds 27.14 mg K_2SO_4 or 5 mg S. Mix the soil and solution thoroughly and incubate moist for a few days before drying and extracting. If all this S is extracted in $CaCl_2$ solution, the extract concentration increases by 25 μg ml^{-1}. If all is extracted in calcium phosphate solution the concentration increases by 5 μg ml^{-1}.

Section 10.7 Calculations

1. Using the data in Fig. 10.1 calculate the approximate range of values for total soil P. (*Ans.* 400–500 mg P kg^{-1}) Using the larger of these two values express as a percentage of the total P in the soil the following: solution-P, P on particle surfaces, P in soil minerals, organic P. (*Ans.* 0.1, 20, 40, 40)

   To supply crop needs, P moves from soil solution into roots. Using the data for the wheat crop in Table 9.2, express the solution-P as a percentage of the total uptake. (*Ans.* 0.5–5)

   Using Fig. 11.3 calculate the approximate maximum rate of uptake of P per day by the crop. (*Ans.* 0.9 kg ha^{-1}) Is there enough P in soil solution to supply one day's needs?

2. Fertilizer was applied to a field plot at the rate of 50 kg P ha^{-1} and was cultivated into the top 20 cm. Assuming a bulk density of 1.3 t m^{-3}, calculate the equivalent application of P in a laboratory experiment in mg P kg^{-1}. (*Ans.* 19) What mass of KH_2PO_4 would be needed to give this application of P? Appendix 2 gives molar masses. (*Ans.* 84 mg kg^{-1})

What would be the field application expressed as kg P_2O_5 ha^{-1}? (*Ans.* 115)

3. The radius of the depletion zone of phosphate around a root increases approximately in proportion to the square root of time. Using the values in Fig. 10.2, write an equation relating radius (r, mm) to time (t, d). (*Ans.* $r = 0.32 t^{1/2}$) From this calculate the time required to develop a 2 mm depletion zone. (*Ans.* 39 d) If there is 4 cm of root cm^{-3} of soil, calculate the percentage of the soil volume which is within a 2 mm depletion zone neglecting the root volume. (*Ans.* 50) Fig. 3.1 gives root lengths for various depths in the soil. Repeat the calculation for the 75–100 cm layer in June assuming a 3 mm zone. (*Ans.* 14)

4. Of a phosphate solution containing 0.5 g P l^{-1}, 3 ml was thoroughly mixed with 40 g of soil and incubated moist for 1 week. Of this soil, 5 g was shaken with 50 ml of 10 mM $CaCl_2$ solution. The suspension was filtered and the P concentration in the solution was found to be 1 μg P ml^{-1}. How much of the added P had been adsorbed on to particle surfaces? (*Ans.* 1.1 mg P)

5. Using Table 10.2 calculate for each plot the difference between the amount of P added and removed. Assume that the maize from all the plots contained 4 kg P t^{-1}. Compile a new table showing these values as an indication of the amounts of residual P which might give responses in a subsequent crop. (*Example:* yield = 3.33 t ha^{-1}, P in the grain = 3.33 × 4 = 13.3 kg, added P = 44 kg, residual P = 30.7 kg)

6. Calculate the loss of P, S and N per hectare in 300 mm of drainage water if their concentrations were 0.01, 3 and 10 mg l^{-1} respectively (Table 10.3 and Fig. 11.7). (*Ans.* 0.03, 9 and 30 kg)

7. The annual rainfall at a site in northern England is 1200 mm, and its total S concentration is 2 mg l^{-1}. Calculate the input as kg S ha^{-1} a^{-1}. (*Ans.* 24) If the input of sulphur dioxide deposited dry and in the rain is 10 kg S ha^{-1} a^{-1}, and its oxidation in the soil is as follows,

$$2SO_2 + O_2 + 2H_2O = 2H_2SO_4,$$

calculate the input of acidity associated with SO_2 as kg H$^+$ ha^{-1} a^{-1}. Molar masses are in Appendix 2. (*Ans.* 0.63)

CHAPTER 11

Nitrogen in Soils

Soil nitrogen is derived primarily from atmospheric nitrogen gas, N_2. Soil micro-organisms, both free living and symbiotically associated with plants, fix N_2 to produce *organic nitrogen* in the form of amino groups, $-NH_2$, in proteins. This becomes part of the soil organic matter. The decomposition of organic matter converts some organic N into *mineral-nitrogen*, a term applied to ammonium, NH_4^+, nitrite, NO_2^- and nitrate, NO_3^-. Note the use of the word 'mineral' here for ions containing N (and S and P) in contrast to its use for soil minerals (feldspars, clays etc. Ch. 2). Nitrogen is also present in the soil atmosphere, but this is not considered to be soil-N because the gases are free to move out of the soil. Mineral-N is taken up by plants and micro-organisms and converted into organic N. Plant-N may be a large part of the total in the soil-plant system, particularly in forests.

A DYNAMIC SYSTEM

Soil-N is continuously moving from one form to another as a result of the activities of plants and micro-organisms. The processes are shown in Fig. 11.1.

- *Mineralization* is the microbial conversion of organic N to mineral-N:

$$\text{organic-}NH_2 \rightarrow NH_4^+$$

- *Nitrification* is the oxidation of ammonium-N to nitrite and nitrate by specific micro-organisms:

$$NH_4^+ \rightarrow NO_2^- \rightarrow NO_3^-$$

- *Immobilization* is the conversion of mineral-N to organic N. It occurs when micro-organisms cannot satisfy their N needs from the organic materials on which they are feeding. As a result they take up mineral-N:

$$NH_4^+ \text{ and } NO_3^- \rightarrow \text{organic-}NH_2$$

- *Volatilization* is the loss of ammonia gas from the soil. Under alkaline conditions ammonium ions are converted to ammonia molecules in solution which can then be released into the soil atmosphere:

$$NH_4^+ + OH^- \rightleftharpoons NH_3 + H_2O$$

- *Denitrification* is the loss of nitrogen and nitrous oxide gas from the soil under anaerobic conditions. Nitrate and nitrite are reduced to these gases by micro-organisms (Section 6.6):

$$NO_3^- \text{ and } NO_2^- \rightarrow N_2O \rightarrow N_2$$

- *Nitrogen fixation* is the conversion of N_2 in the soil atmosphere into NH_4^+ by specialized groups of micro-organisms. The NH_4^+ is then assimilated as organic N

$$N_2 \rightarrow NH_4^+ \rightarrow \text{organic-}NH_2$$

- *Net mineralization.* Because mineralization and immobilization occur at the same time it is difficult to separate them. Normally the change in the amount of mineral-N is measured over a given period of time, losses by leaching, denitrification and volatilization are taken into account and a net effect calculated. There may be a gain or loss of mineral-N, the latter being a *net immobilization*.

- *Nitrate leaching* is the process whereby nitrate is lost from the soil in drainage waters. Nitrate is not adsorbed on to soil particle surfaces unless they carry positive charge (Section 7.5). Thus nitrate is freely leached except in acidic soils of the humid tropics.

The N cycle is completed by plant uptake from the soil, by additions directly from the atmosphere (as nitrate, ammonia and gaseous oxides of N which are converted into nitrate in the soil) and addition of fertilizers, manures and animal excreta.

What can be measured?

Soil organic N is measured using the methods in Section 3.6. The analysis determines organic N plus ammonium-N but because mineral-N is very small, the results are referred to as organic N or total soil-N.

Mineral-N is determined by extracting soil with potassium chloride or sulphate solution. Ammonium and nitrate can be measured separately or together (Section 11.1). The same methods are used to measure mineral-N in drainage waters.

Plant-N is determined by decomposing plant ma-

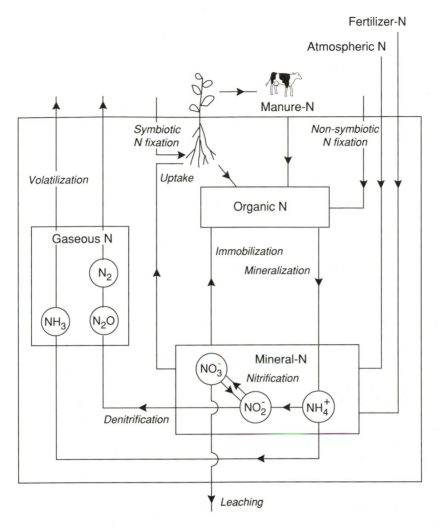

Figure 11.1 The nitrogen cycle.

terial in a concentrated acid to release ammonium-N (Section 11.2). The method is similar to that used for soil-N.

Inputs of N from the atmosphere and gaseous losses require techniques of measurement which are beyond the scope of this book. They are, however, important in determining the soil-N budget. Current estimates of atmospheric inputs in the UK are shown in Colour Plate 17.

The N cycle has been redrawn in Fig. 11.2 in terms of those fractions which can be measured by the above methods and shows the amounts of N in the mineral and organic fractions. The amount of organic N is simply related to the amount of organic matter (Ch. 3): for each 1 per cent of organic matter in the soil there is about 0.06 per cent N if the C:N ratio is 10:1. Organic

matter contents are low in soils of arid regions and so their organic N content is also small. If no organic N is mineralized, the amount of mineral-N will reflect the atmospheric inputs. In these regions the rain may contain $0.1\,mg\,N\,l^{-1}$ which in a soil containing 30 per cent m/m of this water is equivalent to $0.03\,mg\,N\,kg^{-1}$ or $<0.1\,kg\,ha^{-1}$ in the topsoil. The largest amounts of mineral-N occur where rapid mineralization has taken place or where fertilizer or manure has been applied. Inputs and losses of N are considered below in the section on N budgets.

The supply of nitrogen to plants

Plants take up mineral-N. Thus under natural vegetation, mineralized N with small amounts of mineral-N

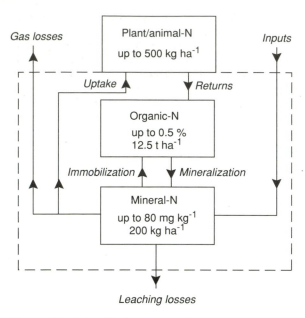

Figure 11.2 A simplified nitrogen cycle.

from the atmosphere supply plant needs and thus have a controlling effect on the productivity of natural systems. In agricultural systems crops also rely on these sources together with inputs from fertilizers and manures (Colour Plates 18(a) and (b)). The accurate assessment of fertilizer-N requirements depends on a knowledge of mineralization rates and crop needs.

The factors influencing the rate of mineralization are as follows:

- *The amount of soil organic matter and its N content.* This is the substrate (food) for microbial activity. In general, the larger the amount of organic matter, the greater will be the activity with much of the mineralized N coming from the decomposition of fresh crop residues. However, for there to be a net mineralization, the C:N ratio of the decomposing organic matter must be less than about 30:1 (more than 1.8% N): straw residues result in net immobilization, at least initially, whereas legume residues give net mineralization.
- *Water content.* The soil must be moist for microorganisms to be active. After remoistening a dry soil, there is a flush of microbial activity.
- *Temperature.* The rate of metabolism of micro-organisms increases by a factor of about 3 for each 10 °C rise in temperature up to an optimum (Section 6.1).
- *pH.* Acidic conditions reduce the rate of decomposition of organic matter and the release of mineral-N resulting in the accumulation of partially decomposed plant residues on the surface of acidic soils.

Liming increases the rate of mineralization and improves the supply of mineral-N to plants.

- *Aeration.* Anaerobic conditions also decrease microbial activity causing the accumulation of peat in flooded areas and a reduced supply of mineral-N.
- *Cultivation.* A combination of factors is involved. Plants are killed and incorporated into soil along with crop residues, and aeration may be improved. Normally the net mineralization rate is increased.

Maximum mineralization rates therefore occur in warm, moist, organic soils. The following situations are important:

- the rewetting of warm soils in the autumn, particularly if cultivated;
- the warming of moist soils in the spring; and
- the input of large amounts of fresh plant material.

The ploughing of well-fertilized or legume-rich grassland in temperate regions causes particularly high mineralization rates and, along with manures, has been the traditional basis for crop production in a mixed farming system. Clearing forest has a similar effect and is the basis for shifting cultivation systems of agriculture (Section 13.5). Organic farming systems rely to a large extent on mineralized N for crop production (Section 13.3).

Predicting mineralization rates in the field

Laboratory studies to examine the factors controlling mineralization rates have led to proposals for routine methods which measure the amount of mineral-N released during the incubation of soil under standard conditions (Section 11.3). However, because weather and soil conditions cannot be predicted, it has proved impossible to predict the amounts of mineral-N that will become available during a growing season based simply on a laboratory measurement. Current research makes use of computer models to predict the day-to-day changes in soil- and crop-N, building on the principles obtained from laboratory studies. There is also interest in the use of soil biomass as an indicator of the turnover of soil-N (Section 3.2). However, current advisory methods to assess N availability and fertilizer needs do not use a laboratory measured index of availability, in contrast to the methods for K^+ and P.

SUPPLY AND DEMAND: NITROGEN BUDGETS

A knowledge of N budgets is the basis for predicting the fertilizer needs of crops. The type of crop and its variety, soil conditions and climate control potential yield and therefore the potential N requirement.

Crop demand

Table 9.2 shows that crops in Britain remove between 100 and 250 kg N ha^{-1} a^{-1}. This must be supplied from various sources. However, because mineralization occurs throughout the growing season it is important to consider the distribution of demand by crops for N over the whole period of growth. In contrast, potassium and phosphorus supply were considered in Chs. 9 and 10

primarily in terms of the total amount available at the beginning of the growing season.

Fig. 11.3 shows the uptake of nutrients by winter wheat between sowing in October and harvest in August at the Nottingham University School of Agriculture. Only small amounts of N are needed in the autumn, and in the winter the crop is almost dormant. Uptake slowly increases in the spring and then in May and June the rapid growth of the crop demands on average about 1.6 kg N ha^{-1} d^{-1}, although on warm sunny days giving rapid growth, rates up to 6 kg N ha^{-1} d^{-1} may be taken up. This crop, including the roots, finally contained about 120 kg N ha^{-1}. Ideally the supply of N should match this distribution of demand.

There is a similar high demand for the other nutrients in May and June which is met primarily from the available nutrients in the soil at the beginning of the season except for S which is supplied from the atmosphere throughout the year.

Fig. 11.4 shows the changes in dry matter in the various part of the crop. By harvest the yield was 6.45 t ha^{-1} of grain, 6 t of straw and 1 t of roots, values which are close to the average for the UK.

The root distribution of the wheat crop has been shown in Fig. 3.1. Although the demand for nutrients is small in the autumn, the uptake rate per unit length of root known as *inflow* is high during this period, and is again high in May and June. To supply crop needs in the autumn the nutrients need to be in the topsoil where the roots are growing. Nitrogen is mineralized in

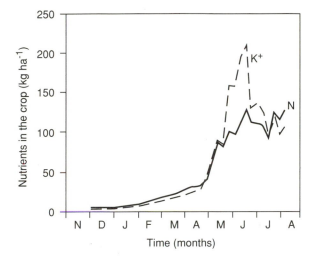

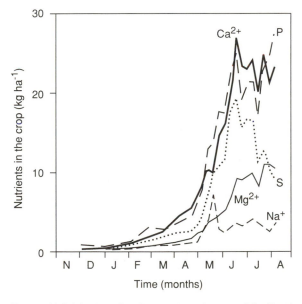

Figure 11.3 Uptake of major nutrients by a well-fertilized winter wheat crop grown on a sandy loam soil in Leicestershire, UK. The data are for the whole crop including roots and the grain yield was 6.45 t ha^{-1} at 15 per cent moisture content. From Gregory *et al.* (1979).

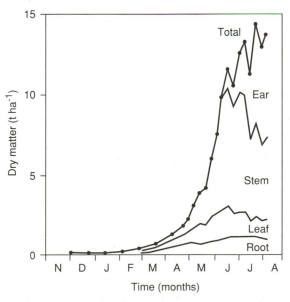

Figure 11.4 Dry matter production by a winter wheat crop: details as in Figure 11.3. From Gregory *et al.* (1979).

this soil layer which also contains the largest amounts of available phosphorus and potassium. As a result there are unlikely to be shortages of nutrients during the autumn.

The dry matter harvested in an irrigated ryegrass crop (the offtake) growing in southern England and well supplied with nutrients is shown in Fig. 11.5. The crop was cut monthly. As for wheat there is a period of rapid growth in May but in contrast growth of this perennial crop continues through the summer and autumn. The N offtake has been estimated from the yield data.

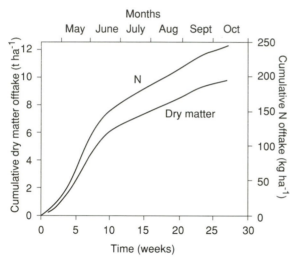

Figure 11.5 The dry matter and N content of the harvested part of a ryegrass crop subjected to repeated cuttings. Based on Robson *et al.* (1989), calculated assuming that dry matter = 1.11 × crop organic matter and N offtake is 25 kg N t^{-1} dry matter offtake.

More normal conditions for the growth and utilization of this crop are as follows:

- There are usually water deficits in the summer giving slower growth rates, but followed by a second period of rapid growth when autumn rain and mineralized N become available.
- Ryegrass is normally grazed. Under these conditions the gross uptake of N by the crop is between 600 and 700 kg ha^{-1} a^{-1}, of which up to 300 kg is returned to the soil via dung and urine, and another 300 kg returns via the senescence of leaves which are not grazed. The net offtake by the animals is 30–40 kg N ha^{-1} a^{-1} in meat, milk and wool (Parsons *et al.*, 1990). The average daily uptake of N by the crop (200 d growing season) is about 3 kg ha^{-1}.

Nitrogen supply

Measurements of the amounts of N mineralized throughout the growing season are needed to improve assessments of fertilizer needs. Measurements of changes in mineral-N give the net effect of all the components of the N budget. Thus over a given time period

Δ mineral-N =
(net mineralized N + fertilizer inputs + atmospheric inputs) minus (crop uptake + leaching loss + gaseous loss)

where Δ means 'change in the amount of'. At any given time the mineral-N is the available N which may be subsequently taken up or lost from the soil.

Fig. 11.6 shows the amounts of mineral-N present in a soil growing winter wheat at ICI Jealott's Hill Research Station. The measurements were made using the methods given in Section 11.1. It is very time consuming to obtain data of this sort. Reliable sets of data are only slowly becoming available for a range of crops, soils and climatic conditions.

The field had previously grown ryegrass and had an ADAS N index 2 (see below). It was ploughed in August and the crop was sown at the beginning of October 1982. In January 1983 there was about 10 kg mineral-N ha^{-1} in the top 30 cm of soil with presumably more in the subsoil. There was not much leaching in January and February. In late February 33 kg of

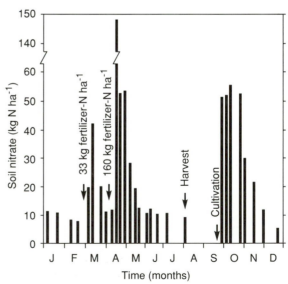

Figure 11.6 Nitrate-nitrogen in the 0–30 cm layer of soil under winter wheat: clay loam over London Clay, Berkshire, UK. The times at which fertilizer was applied are marked on the figure. ICI data, by D. Barraclough.

fertilizer-N (NH_4NO_3) was applied. Some of this may have been leached from the topsoil by early March.

Mineralization followed as soil temperatures increased. Crop uptake, leaching and denitrification account for the decreases in late March and early April. The main fertilizer application on 12 April was not immediately washed into the soil by rain, but was recovered in the mid-April sampling. Crop uptake then reduced mineral-N to about $10 \, kg \, ha^{-1}$ by harvest. The yield was $10 \, t$ grain ha^{-1} with a large N concentration in the grain resulting from the large amount of available N. An index 2 soil would not normally be given so much fertilizer-N unless wheat of bread-making quality was being grown.

Following cultivations in mid-September there was a flush of mineralization (about $50 \, kg \, ha^{-1}$ in the topsoil), probably followed by further mineralization through the autumn. Gaseous loss and leaching were the main causes of the decrease through to the end of 1983, the next crop only taking up about $10 \, kg \, N \, ha^{-1}$ over this period.

The budget

The matching of demand and supply for the winter wheat crop is good between January and August, but the N mineralized in the autumn is of little use to the crop. Much is lost by leaching and denitrification. The budget for the Jealott's Hill crop is summarized in Table 11.1(a).

For fields which have grown continuous arable crops for many years it has been estimated that there is no net mineralization, leaching losses are between 15 and $65 \, kg \, ha^{-1} a^{-1}$, and gaseous loss is between 5 and $30 \, kg \, ha^{-1} a^{-1}$. The average values for the winter wheat crop in the UK are given in Table 11.1(b). Further examples of N budgets for wheat are given in Section 11.4 and in Table 13.4.

Grass crops are more efficient in their use of available N, primarily because of their ability to take up the N mineralized in the autumn, although if grazed or fed to animals considerable losses may occur.

A nitrogen availability index

Because laboratory measurements of mineralization do not satisfactorily predict the availability of N to a crop, predictions are based on an understanding of the way in which the management of the soil in previous years influences the amount of soil-derived mineral-N which is likely to become available during the growing season. In the UK this has led to a *nitrogen availability index* (Section 11.4). Those aspects of management which are important include the ploughing of leys (grass and legume crops) and previous applications of manure and

Table 11.1 The nitrogen budget for the winter wheat crop in the UK, ($kg \, N \, ha^{-1}$)

(a) A crop which yielded $10 \, t$ grain ha^{-1} grown at ICI Jealott's Hill Research Station

Inputs		Removals	
Fertilizer	193	Crop	335
Atmospheric inputs	20	Leaching	64
Seed	5	Denitrification	12
Net mineralized	171		
Biological N fixation	4		
Total	393	Total	411

(b) The average winter wheat crop in England and Wales (1985–88) assuming the soil is under steady state conditions (no net mineralization). The average yield is $6.75 \, t$ grain ha^{-1} at 14% moisture content.

Inputs		Removals	
Fertilizer	181	Grain	123
Farmyard manure	17	Straw	23
Atmospheric inputs and seed	20	Denitrification	15
Biological N fixation	4		
Total	222		161
		Leaching by difference	61

From House of Lords (1989), Appendix 2

fertilizer. Essentially the index is a rating of the amount of mineral-N present as a residue from the previous crop plus the amount of organic N in soil organic matter, in crop residues and in manures which will mineralize in the coming season (Colour Plate 18). Any practice which increases the availability of soil-derived mineral-N can be balanced by a decreased input of fertilizer-N.

Although the index system described here is that used in Britain, the principles apply under any conditions. Crop needs are met by soil-derived mineral-N and by fertilizer-N; efficient management of fertilizer-N requires that crop needs are known and that soil-derived N can be predicted.

The leaching of nitrate

Efficient use of soil-N by crops requires minimal losses by leaching (DOE, 1986; House of Lords, 1989; NRA, 1992). All soils lose a small amount of nitrate into groundwaters: between 1 and $5 \, mg$ nitrate-$N \, l^{-1}$ is the normal concentration range for water in British aquifers compared to rain-water with up to about $1.5 \, mg \, N \, l^{-1}$. Water moves through soils during the winter and even under natural vegetation it will contain nitrate.

Whenever the concentration of nitrate increases as a result of mineralization or fertilizer application there is the potential for increased loss by leaching. Because leaching occurs during the winter in Britain, the nitrate which is at risk is that left in the soil at the end of a cropping season together with that which is mineralized during the autumn and winter (Addiscott *et al.*, 1991).

Nitrate moves freely in solution in soils of temperate regions: it is not adsorbed on to particle surfaces. In light-textured soil, solution is displaced by incoming rain, the rain mixing with the soil solution causing dilution. Section 11.5 describes laboratory experiments to demonstrate these principles. Nitrate leaching in heavy-textured soils is more complex, because water moves through large cracks and channels leaving nitrate behind in the soil aggregates where little water movement occurs. The study of these complex systems has been aided by computer models.

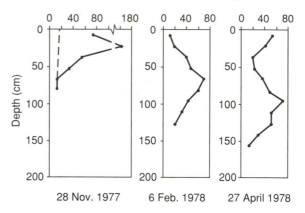

Nitrate concentration in soil solution (mg N l^{-1})

28 Nov. 1977   6 Feb. 1978   27 April 1978

Figure 11.7 The mineralization and leaching of nitrate following the ploughing of grassland. The dotted line indicates the probable concentration of NO_3^--N before ploughing in September. From Cameron and Wild (1984).

Examples of nitrate leaching in the field

Aquifers beneath the chalk hills of southern and eastern England are important sources of water. As part of a programme to study the effects of agriculture on groundwater quality, a field at the University of Reading Churn Farm in Berkshire was ploughed in September 1977. Previously it had grown ryegrass for 3 years, had received fertilizer and been grazed by sheep. The field was sown with winter wheat. Mineral-N was measured in the following November, February and April.

Fig. 11.7 shows the release of mineral-N during the autumn, and its movement by leaching. About 100 kg nitrate-N ha^{-1} was produced by the flush of mineralization in the ploughed layer, and this pulse of nitrate was moved about 1 m down through the soil during the winter. In the spring a further release occurred in the ploughed layer which is seen in the figure combined with fertilizer-N applied in March. Nitrate leached below about 1 m during the winter is not accessible to the wheat root system (Fig. 3.1): 70 kg ha^{-1} of autumn-mineralized N was below 90 cm when sampled in April.

On a similar site at Bridgets Experimental Husbandry Farm in Hampshire, deep core samples were taken in 1975 by the Water Research Centre. The results are shown in Fig. 11.8. Two effects can be seen:

1. A series of peaks are associated with the pulses of mineralization which occurred each time a grass crop was ploughed for arable crops. For example, in 1948 the field was ploughed after being in grass for as long as records existed. Mineralization released a very large pulse of nitrate; over 27 years (1948–1975) this had moved to about 30 m depth in the

chalk, approximately 1 m per year. The most recent ploughing of grass was in 1972, and by 1975 the pulse of nitrate had moved to about 3 m depth.

2. There is a steady decrease in concentration from about 20 mg l^{-1} in the top 10 m to about 3 mg l^{-1} in the solution in the aquifer. The ploughing of grassland is clearly responsible for part of this trend but increased inputs of fertilizer-N over this period are also likely to be involved (Fig. 11.9). Other factors also need to be considered: a pulse of mineralized N becomes diluted as it mixes with water around it as shown in Fig. 11.7, and there may be denitrification in the chalk. Thus it is not certain whether water with 20 mg nitrate-N l^{-1} will eventually (around the year 2025) reach the aquifer. Assuming that it will, this water will be mixed with water from surrounding areas of non-agricultural land, and the changes in the aquifer concentration will reflect land use over the whole area.

Data on the leaching of nitrate into field drains, ditches and streams in Britain can be found in MAFF (1976). Areas with field drains are normally heavy-textured land from which less nitrate is leached than from light-textured soils.

ENVIRONMENTAL IMPLICATIONS

Water quality

Large nitrate concentrations in drinking water have been blamed for the disease called methaemoglobinaemia in babies. Cases of the disease are rare and are

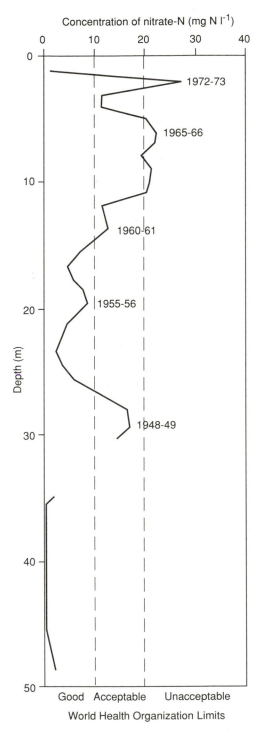

Figure 11.8 The nitrate distribution in the Hampshire Chalk under agricultural land: concentrations in water extracted from the chalk. Dates when grass was ploughed are marked against the peaks. From Young *et al.* (1976). The water table is at 56 m.

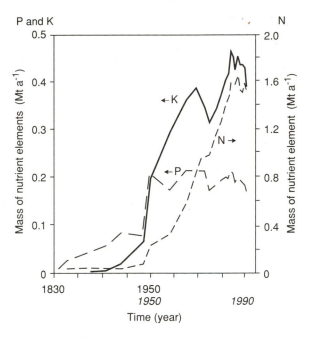

Figure 11.9 Changes in fertilizer use in Britain between 1837 and 1991. Note the different scale for nitrogen. From FMA (1981) and direct communication.

normally associated with water contaminated with both nitrate and sewage bacteria. There is also concern that the incidence of stomach cancer may be related to nitrate in drinking water. The World Health Organization has set a limit of 11.3 mg nitrate-Nl^{-1} for good quality water, and lower limits are proposed by the European Economic Community. It is clearly important to understand the effect of agricultural practices on nitrate concentrations in groundwater and rivers. The delay in nitrate entering an aquifer (Fig. 11.8) does not occur in drainage to rivers. The nitrate contents of six British rivers are shown in Fig. 11.10 and increases since 1950 are now being reversed in many areas. The possible link between nitrate and the formation of algal blooms on rivers and lakes has been discussed on p. 206.

Gaseous losses

Denitrification causes a loss of soil-N in the form of N_2 and nitrous oxide gases. Both are economically undesirable, although the former is of no concern environmentally. Denitrification occurs under warm wet conditions when all or part of the soil is anaerobic. Reduction of losses requires the maintenance of good soil structure.

Volatilization of ammonia is favoured by high pH and warm conditions. Of particular significance are losses from the excreta of grazing cattle. Recent

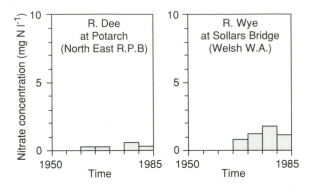

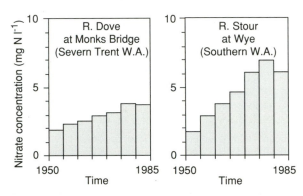

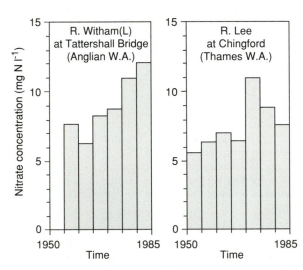

Figure 11.10 Changes in nitrate concentrations in river waters between 1950 and 1985. Five-year mean values. From DOE (1989).

estimates for grazed grassland show that an amount of N equivalent to between 9 and 15 per cent of the fertilizer applied is removed in useful outputs (milk, meat). The implication is that up to 90 per cent of the N is either stored in the organic matter or is lost. Losses from grazed grassland, like those from cut grass, depend on inputs from fertilizer and recycled excreta (Jarvis, 1992; Barraclough *et al.*, 1992; Cuttle *et al.*, 1992).

Losses of ammonia also occur from slurry applications, and from urea applied as fertilizer to calcareous soils. Apart from the financial loss, the smell can be unpleasant, and increased soil acidification results from the raised inputs of N to surrounding areas. This is a particular problem in the Netherlands with its high density of dairy cattle.

FURTHER STUDIES

Ideas for projects are in Section 11.6 and calculations are in Section 11.7.

Section 11.1 The measurement of mineral-nitrogen in soils

Soils sampled moist from the field are extracted immediately or stored at low temperature without drying or freezing. The standard extracting solution is potassium chloride which will remove soluble and exchangeable nitrate and ammonium. The method below uses 2 M KCl but a 1 M solution is almost as efficient and is cheaper. Nitrate is reduced to ammonium before being steam distilled as ammonia and titrated against standard acid.

Field sampling

The sampling method will depend on the purposes of the measurements:

- *Using a Jarrett auger*. Samples are taken from known depths following the methods of Section 1.3. The amount of mineral-N is determined per unit mass of soil. In sandy soils, particularly when dry, soil falls into the hole and it is difficult to obtain a 'clean' sample from a given depth.
- *Using a sampling tube*. A metal sampling tube typically between 4.5 and 15 cm diameter is driven into the soil and removed. The depth of the hole is measured. This is used to correct the length of the soil core for compression. The core is pushed out of the tube and sections of known length are cut and placed in polythene bags. From the length of the section and the diameter of the tube the volume of soil is known. The mass of soil in the section is also determined. The amount of mineral-N per unit mass of soil can be related to soil volume, and using the compression calculated from the hole depth the total mineral-N to a given depth in the field can be found.

A simple metal tube is adequate for sampling to 30 cm, but special sampling equipment and expertise may be needed for deeper cores (Eijkelkamp Agrisearch Equipment).

- *The number of samples.* The distribution of mineral-N can be very variable in the field, particularly in well-structured soils. To obtain a measurement which represents a given area replication is essential. For routine crop measurements guidelines are as follows: take 13–15 cores per plot with samples bulked if necessary to give 4 or 5 replicates for analysis.

Sample storage and preparation

Extraction should be carried out on the moist soil immediately after sampling (within a few hours). In emergencies samples can be stored by chilling to $+2\,^{\circ}$C to restrict mineralization. Once in KCl solution, extracts can be stored for up to about 2 months in a refrigerator at $2\,^{\circ}$C before analysis.

Gently crumble the soil and, leaving out the stones, take a representative 40 g sample for analysis. Determine the water content of a 10 g sample following the methods of Section 3.3 and express as g H_2O per 100 g oven-dry soil.

Extraction

Reagent

Potassium chloride, 2 M. Dissolve 149 g of KCl in water and make up to 1 l.

Method

Place the 40 g sample of moist soil in a 250 ml conical flask for extraction. The mass of dry soil in this sample can be calculated from its water content. Add 200 ml of 2 M KCl and shake for 1 h. Filter the suspension. The mineral-N content of this extract is determined by steam distillation.

Steam distillation

Reagents

Magnesium oxide, ignited. Heat MgO in a muffle furnace at $800\,^{\circ}$C overnight. This ensures that any $MgCO_3$ present is converted to MgO. Cool and store in a tightly stoppered bottle in a desiccator. After a few days ignite again to ensure that fresh MgO is used.
Devarda's alloy. Grind finely in a pestle and mortar.
Octan-2-ol. This is used to prevent excessive frothing during distillation.

Other reagents and the equipment are described in Section 3.6.

Method (Note 1)

Ammonium-N. Pipette 50 ml of extract into the distillation flask. Place 10 ml of fresh boric acid solution in the receiving flask and insert it under the condenser. Add 1 drop of octan-2-ol and 0.5 g of MgO to the extract. Pass steam and collect about 50 ml of distillate. Remove the receiving flask and retain for titration. Disconnect the steam supply. Place another receiving flask under the condenser for the analysis of nitrate-N.
Nitrate-N. Add 0.5 g of Devarda's alloy to the extract in the distillation flask. Immediately reconnect the steam line and distil a further 50 ml of distillate.
Mineral-N. If NH_4^+-N and NO_3^--N are not required separately, mineral-N can be determined using only one distillation. To the extract in the distillation flask add 1 drop of octan-2-ol, 0.5 g of Devarda's alloy followed by 0.5 g of MgO and distil.

Titrate each distillate against 0.01 M HCl using methyl red–bromocresol green indicator (Section 3.6). Carry out a blank determination using 50 ml of 2 M KCl solution. The titration obtained should be subtracted from the sample titration to give a corrected titration volume.

Calculations

Ammonium is distilled at the high pH produced by adding MgO to the extract. Nitrate is reduced to ammonium by the nascent hydrogen produced by the reaction of the alloy with the alkaline solution. The calculations below apply to NH_4^+-N, NO_3^--N and mineral-N.

Method summary

Moist soil ($?\,mg\,N\,kg^{-1}$ oven-dry soil)
$$\downarrow$$
$\quad$ 40 g + 200 ml 2 M KCl
$$\downarrow$$
$\quad\quad$ 50 ml extract distilled and titrated against 0.01 M HCl

Example The water content of the moist soil was 23.2 g H_2O per 100 g oven-dry soil. In 40 g moist soil there was $40 \times 100/123.2 = 32.47$ g of oven-dry soil and 7.53 g (= ml) of water. The total liquid volume during shaking was 207.53 ml. From this, after filtration, 50 ml was taken for distillation giving a titration volume (corrected for the blank titration) of 1.25 ml.

The titration reaction is

$$NH_4OH + HCl = NH_4Cl + H_2O$$

The amount of HCl used in the titration is

$$0.01\,mol\,l^{-1} \times 1.25/1000\,l = 1.25 \times 10^{-5}\,mol$$

The same number of mol of NH_4OH must have reacted in the titration. Each mol of NH_4OH contains 14 g N, and so the mass of N is

$$1.25 \times 10^{-5}\,mol \times 14\,g\,mol^{-1} = 1.75 \times 10^{-4}\,g\ or\ 0.175\,mg$$

This amount of N must have been in 50 ml of extract, and so the total extracted from the soil is $0.175 \times 207.53/50 = 0.726$ mg. This came from 32.47 g of oven-dry soil, giving

$$0.726 \times 1000/32.47 = 22.4\,mg\,N\,kg^{-1}\ oven\text{-}dry\ soil$$

The calculation simplifies to

$$mg\,N\,kg^{-1} = 2.8 \times corrected\ titration\ volume\ (ml) \times total\ liquid\ volume/oven\text{-}dry\ soil\ mass$$

As an approximate guide, each ml of titration volume is equivalent to 18 mg N kg^{-1}.

Conversion to field data

If the above example relates to a sample taken from the 0–20 cm soil layer in the field, and the bulk density is 1.35 g oven-dry soil cm^{-3}, calculate as follows.

The bulk density can also be expressed as 1.35 t m^{-3}. In 1 ha to 20 cm depth there is $0.2 \times 10^4\,m^3$ of soil, and so the mass of soil ha^{-1} is

$$1.35 \times 0.2 \times 10^4 = 2700\,t$$

The N content is 22.4 mg kg^{-1} ($=$ g t^{-1}). Therefore there is

$$22.4 \times 2700 = 6.05 \times 10^4\,g\,ha^{-1}\ or\ 61\,kg\,N\,ha^{-1}$$

Stony soils These are difficult to sample. If a stony sample is used, taking the 40 g sample for analysis rejects the stones. The stone content should be measured (Section 1.2) and expressed as (for example) 210 g of stones kg^{-1} of whole soil, or 21 per cent *m/m*. Thus in 1 kg of whole soil there is 790 g of soil excluding stones. The N content of the whole soil is therefore

$$22.4 \times 790/1000 = 17.7\,mg\,N\,kg^{-1}\ whole\ dry\ soil$$

This value together with the bulk density will give the N content as kg ha^{-1}. The bulk density in this case is the mass of whole soil (including stones) per unit volume.

Conversion to a soil solution concentration

Assuming that all the nitrate-N is in soil solution, the concentration is simply related to the amount of nitrate extracted and the soil water content. If in the above example (22.4 mg N kg^{-1}) the water content at field capacity is 30 g H_2O per 100 g oven-dry soil, then in 1 kg of soil there is 300 ml of solution. Thus the N concentration is 22.4 mg per 300 ml or 75 mg N l^{-1}. As an approximate guide, the soil solution concentration is three times the extracted N expressed in the above units.

Note that field capacity is used as a basis for expressing soil solution concentration; the concentration will be greater in a drier soil, but interest centres on the leaching of solution into drainage water. Solution normally moves through soil at a water content slightly above field capacity (Section 11.5), and so the concentration of the moving solution is slightly smaller than that calculated.

Determination of mineral-nitrogen in drainage waters

The above methods are used. A 50 ml sample of water is pipetted into the distillation unit (Note 1).

Method summary

drainage water ($?$ mg N l^{-1})
$$\downarrow$$
50 ml distilled and titrated against 0.01 M HCl

Example The titration volume was 1.55 ml. The amount of HCl used in the titration was $0.01 \times 1.55/1000 = 1.55 \times 10^{-5}$ mol. This reacts with

$$1.55 \times 10^{-5} \times 14 = 2.17 \times 10^{-4}\,g\ or\ 0.217\,mg\,N$$

This was in 50 ml and so the concentration of N in the water is

$$0.217 \times 1000/50 = 4.3\,mg\,l^{-1}$$

The calculation simplifies to

$$mg\,N\,l^{-1} = 2.8 \times the\ titration\ volume$$

Note 1 The use of a standard N solution is advised as a means of checking the analytical procedure. Section 3.6, Note 4 gives details of a standard containing 0.28 mg NH_4^+-N ml^{-1} which can be used to check the ammonium determination. A 5 ml sample when distilled should titrate against 10 ml of 0.01 M HCl.

A suitable nitrate sample containing 0.28 mg NO_3^- N ml^{-1} is made by dissolving 2.022 g of potassium nitrate, KNO_3, in water and diluting to 1 l. A 5 ml sample when distilled should titrate against 10 ml of 0.01 M HCl.

Section 11.2 The measurement of nitrogen in plant material

The method follows closely that used for organic N in soils (Section 3.6). Plant dry matter is decomposed in concentrated sulphuric acid with a catalyst. Plant-N is converted to ammonium-N which is then distilled and titrated.

Reagents and equipment

These are given in Section 3.6.

Preparation of the plant dry matter

Fresh plant material should be dried at 105 °C overnight, and ground to pass a 1 mm sieve. Alternatively, fine chopping with scissors prepares a material from which a representative subsample can be taken (Section 9.3).

Method: digestion and distillation

Weigh 1 g (±0.001) of plant dry matter into a 100 ml Kjeldahl flask or digestion tube. Digest following the procedure of Section 3.6. Unlike the soil digest, no particles remain when digestion is complete, and the whole digest is transferred with washings into a 100 ml volumetric flask and made up to the mark.

Pipette 10 ml of the diluted digest into the distillation flask, distil and titrate against 0.01 M HCl.

Calculation (See also Note 1)

Method summary

plant dry matter ($?$ g N kg^{-1})

   ↓

    1 g → 100 ml digest

      ↓

       10 ml distilled and titrated

             against 0.01 M HCl

Example The titration volume is 10.25 ml. The amount of acid used is

$$0.01 \, \text{mol} \, l^{-1} \times 10.25/1000 \, l = 10.25 \times 10^{-5} \, \text{mol}$$

This is also the number of mol of NH_4OH which reacts in the titration. The N present in 10 ml of digest is

$$14 \, \text{g mol}^{-1} \times 10.25 \times 10^{-5} \, \text{mol} = 1.435 \times 10^{-3} \, \text{g}$$

Therefore in 100 ml of digest there is 1.435×10^{-2} g or 14.4 mg N. This came from 1 g of dry matter, and the result is expressed as 14.4 g N kg^{-1} dry matter. The calculation simplifies to

$$\text{g N kg}^{-1} = 1.4 \times \text{titration volume}$$

The plant materials listed in Table 9.1 contain between 7 and 36 g N kg^{-1}. These would give titration values between 5 and 26 ml.

Note 1 It is advisable to carry out a blank digestion and distillation, and also to check the distillation using a standard ammonium solution. Section 3.6, Note 4 gives the details.

Conversion to field data

The N content of the plant dry matter in g N kg^{-1} is also kg N t^{-1}. This multiplied by the yield of dry matter in t ha^{-1} gives the N content of the crop in kg ha^{-1}. If yield is in terms of fresh material, adjustment is needed because the N content is determined on oven-dried material. For example, potato tubers contain about 22 per cent dry matter (220 kg dry matter t^{-1} tubers). If the dry matter contains 14 g N kg^{-1}, then in 1 t of tubers there is $14 \times 220 = 3080$ g or 3.1 kg N. A yield of 50 t tubers ha^{-1} contains 154 kg N.

Section 11.3 Laboratory methods for studying mineralization

THE PREDICTION OF NITROGEN AVAILABILITY BY LABORATORY INCUBATION

A measurement of mineral-N extractable from a fresh moist soil gives the amount of available N at the time of measurement. No method has proved to be ideal for predicting the likely mineralization of N because of the inevitable differences between the conditions of the incubation experiment and those in the field during the growth of the crop. For this reason incubation measurements have not been used for routine prediction of fertilizer needs. However, they demonstrate the potential for mineralization under standard conditions.

The method of Waring and Bremner (1964) is described here. It is relatively simple and easily used for routine purposes. A soil sample is incubated anaerobically for 7 d at 40 °C. Under these conditions mineralization occurs, but nitrification is prevented by lack of oxygen. The increase in ammonium-N is measured by distillation and titration.

Reagents and equipment

Potassium chloride, 4 M. Dissolve 149 g of KCl in water and make up to 500 ml.
Standard hydrochloric acid, 5 mM.
Reagents and equipment for the determination of mineral-N as in Section 11.1 except that 5 mM HCl is used for titration. Since the volumes of acid used in the titrations are small, a microburette of 5 ml volume

graduated in 0.01 ml intervals is an advantage although not essential.

Soil preparation

The soil should be sampled following the methods of Section 11.1. However, after determining bulk density (if required) the soil is air dried and passed through a 2 mm sieve.

Method

Weigh 5 g ($\pm$0.01) of <2 mm air-dry soil into a 16 $\times$ 150 mm test tube and add 12.5 ml ($\pm$1) of water. Stopper the tube and incubate at 40 °C for 7 d. Shake the tube and transfer the whole soil–water mixture quantitatively into the distillation flask, washing the tube with 12–15 ml of 4 M KCl. Now proceed as in Section 11.1 to distil and determine NH_4^+-N.

Also measure the initial amount of NH_4^+-N in the soil. Follow the above procedure but shake for 1 h instead of incubating for 7 d. Calculate the mineralized N as the difference between the two amounts. Arable soils contain little NH_4^+-N initially, and its determination can be omitted in most cases without seriously affecting the results.

Calculation

Method summary

air-dry soil (? mg N kg^{-1})

↓

5 g extracted, distilled and titrated against 5 mM HCl

Example The titration difference between the incubated and non-incubated samples is 1.45 ml which contains

$$0.005 \text{ mol l}^{-1} \times 1.45/1000 \text{ l} = 7.25 \times 10^{-6} \text{ mol of HCl}$$

This reacts with the same number of mol of NH_4OH which contains

$$14 \text{ g mol}^{-1} \times 7.25 \times 10^{-6} \text{ mol} = 1.015 \times 10^{-4} \text{ g of } NH_4^+\text{-N}$$

which came from 5 g of soil. Thus in 1 kg of soil there is

$$1.015 \times 10^{-4} \times 1000/5 = 0.0203 \text{ g or } 20 \text{ mg N kg}^{-1} \text{ air-dry soil}$$

This calculation simplifies to

$$\text{mg N kg}^{-1} = 14 \times \text{titration volume}$$

Errors Low titration volumes lead to increased errors. However, because the method is only intended to indi-

cate the potential mineralization during cropping, these errors are not important. For the same reason there is no advantage in expressing the result in terms of oven-dry soil.

The use of the measurement

Availability of N to crops will depend on the amount of mineral-N present at the beginning of the growing season and that mineralized during growth. Both need to be measured. In pot experiments good correlation is obtained between N uptake and initial mineral-N plus the NH_4^+-N mineralized in the standard incubation method. Under field conditions correlation is poor, and for this reason computer assisted models of nitrogen mineralization are being developed to predict availability more accurately.

NITROGEN TRANSFORMATIONS

The changes in the forms of N following applications of fertilizer or manure to soil can be studied in laboratory incubation experiments. Applications should be those which simulate field conditions and the release of ammonium- and nitrate-N can be measured. Changes in pH are also important since ammonium fertilizers have a significant acidifying effect.

Quantitative analysis requires the extraction of NH_4^+ and NO_3^- by 2 M KCl, but a simpler semi-quantitative method using spot tests for these ions is also described.

Reagents and equipment

Soil. Obtain a moist sample from the field or from a garden. Pass enough through a 5 mm sieve to give 1.2 kg of moist sample. Air dry the remainder and pass through a 5 mm sieve to give three 1 kg samples.

Manure. A sample of farmyard manure should be air dried and ground to powder. About 10 g of dry material is required. Alternatively any horticultural manure can be used.

Urea solution. Dissolve 1.287 g of urea, $NH_2CO.NH_2$, in water and make up to 1 l.

Reagents and equipment for the determination of NH_4^+ and NO_3^- by the methods of Section 11.1.

pH meter and buffer solution for the measurement of pH (Section 8.1).

Method

Place 1.2 kg of moist soil in a polythene bag. This is the *moist control sample*.

To 1 kg of dry soil add 200 ml of water and thoroughly mix. This is the *dried control* sample.

To 1 kg of dry soil add 200 ml of the urea solution

which contains 120 mg N. This is the *urea treatment* which is equivalent to a field application of 300 kg N ha^{-1} (Section 9.2, 2500 t soil ha^{-1}). Urea has a molar mass of 60 g mol^{-1} which contains 28 g of N.

To 1 kg of dry soil add 6 g of dry manure and thoroughly mix. Add 200 ml of water and thoroughly mix. This is the *manure treatment*. Farmyard manure is very variable in composition. Sampled moist it may contain between 0.3 and 2.2 per cent N. The above application of 6 g kg^{-1} applies N at an equal rate to the urea treatment if the dry manure has 2 per cent N. It is also approximately the same application as that used annually on the farmyard manure plot of the Broadbalk experiment at Rothamsted (Section 13.2, 35 t of moist manure ha^{-1} = 17 t dry matter ha^{-1} if the water content is 50 per cent, or about 7 g dry matter kg^{-1} assuming 2500 t soil ha^{-1}).

Incubation Store the soils in polythene bags with the tops loosely folded to allow aeration. Occasionally shake the bags thoroughly, and add extra water as required to maintain the water content.

Sampling At 3-day intervals or when convenient take 120 g of moist soil from each bag and extract with 100 ml of 2 M KCl and measure NO_3^- and NH_4^+ following the methods of Section 11.1. The pH can be measured in the soil-KCl slurry before filtering. This will give a lower pH than that of a soil-water suspension (the salt effect, Section 8.1) but differences between treatments and changes with time can be determined satisfactorily. If comparison to other data is required, separate soil samples should be taken for pH measurement following the methods of Section 8.1.

The water content of the moist control is not the same as the other treatments. Determine its water content following the methods of Section 3.3, and use this value in the calculations of NH_4^+- and NO_3^--N.

Expected results

The dried control should show the flush of mineralization of soil organic matter resulting from drying and rewetting a soil, and the manure treatment should show a slower release of mineral-N from manure compared to urea. An example of the results for the urea treatment is given in Fig. 11.11 from a similar experiment by M. N. Court using a much larger application rate of 350 mg N kg^{-1} (875 kg ha^{-1}, 2500 t soil ha^{-1}). The transformations of urea are as follows:

- Urease, an enzyme produced by soil micro-organisms, causes hydrolysis which is a simple case of mineralization in that organic N in urea is converted into mineral-N:

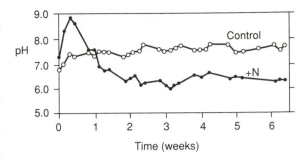

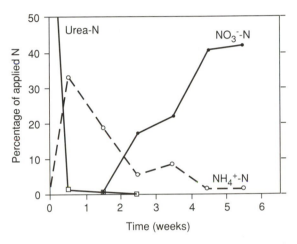

Figure 11.11 Nitrogen transformations in a sandy loam soil following the application of urea fertilizer (350 mg N kg^{-1} soil). From Court *et al.* (1964).

$$NH_2CO.NH_2 + 2H_2O \rightarrow (NH_4)_2CO_3$$

The production of ammonium is seen in the results after less than 1 week.

- Carbonate reacts with water to raise the pH of the soil:

$$CO_3^{2-} + 2H_2O = H_2CO_3 + 2OH^-$$

The carbonic acid forms CO_2 and H_2O with CO_2 being lost to the atmosphere.

- The raised pH causes some NH_4^+ to be lost as NH_3 by volatilization, slightly counteracting the rise in pH due to the carbonate reaction:

$$NH_4^+ = NH_3 + H^+$$

- The next transformation is nitrification:

$$NH_4^+ + 2O_2 \rightarrow NO_3^- + 2H^+$$

The conversion of NH_4^+ to NO_3^- is seen in the results together with a decrease in pH.

Only about 45 per cent of the added N was recovered

after 6 weeks, mostly as NO_3^- (about 150 mg N kg^{-1}). The remainder was probably lost by volatilization, although immobilization and denitrification may have occurred also.

The significance of the data

Soil solution concentrations

The release of NO_3^- must have appreciably raised the soil solution concentration. The change can be calculated. Initially the concentration might have been about 12 mg N l^{-1} (Oxfordshire soils, Table 7.6). A release of 150 mg NO_3^--N into 200 ml of soil solution (kg^{-1} soil) increases the concentration by 750 mg N l^{-1} and raises the NO_3^- concentration to 762 mg l^{-1}. A more normal application rate of 100 kg ha^{-1} (40 mg kg^{-1}) would give a concentration of about 98 mg l^{-1} if the same proportion of the urea-N was converted to NO_3^-.

In order to compare these increases in mineral-N to those which occur under non-fertilized systems the data in Fig. 11.7 can be used. The peak concentration measured after ploughing of grassland was 171 mg N kg^{-1}. Thus the concentrations of mineral-N produced by fertilizers are similar to those produced by the autumn mineralization of organic N. Only around fertilizer granules will very much higher concentrations occur.

Soil pH changes

Figure 11.11 shows a final decrease of about 1.5 pH units resulting from the fertilizer application. It is not possible to calculate the production of acidity from the transformation reactions given above because the relative importance of the various reactions is not known. However, in order to think further about the data assume that of the 350 mg N added, 200 mg has been lost as NH_3 and 150 mg has been converted into NO_3^-. The N lost as NH_3 causes no pH change since

$$(NH_4)_2CO_3 = 2NH_3 + 2H_2O + CO_2$$

The N converted to nitrate releases H^+:

$$(NH_4)_2CO_3 + 4O_2 \rightarrow 2NO_3^- + 2H^+ + 3H_2O + CO_2$$

Thus for each mol of NO_3^- (14 g N) produced, 1 mol of H^+ is released and so 10.7 mmol of H^+ is released with 150 mg of NO_3^--N. If this has caused a pH change of 1.5 units in 1 kg of soil, the buffer capacity is

$$10.7/1.5 = 7.1 \text{ mmol } H^+ \text{ kg}^{-1} \text{ pH}^{-1}$$

Table 8.5 indicates that this would be the buffer capacity of a very sandy soil. The full details for the soil are not available but the sandy loam contained 10.7 per cent clay which means that its sand content was be-

tween 50 and 80 per cent (Section 2.3). Thus the data are consistent with expectations.

Ammonium sulphate could be included as a treatment. Nitrification makes it the most acidifying form of N fertilizer.

$$(NH_4)_2SO_4 + 4O_2 \rightarrow H_2SO_4 + 2HNO_3 + 2H_2O$$

There is no CO_3^{2-} to cause an initial pH rise and little volatilization of ammonia. The nitrification of 150 mg NH_4^+-N kg^{-1} would release 21.4 mmol H^+ kg^{-1} soil which would cause a pH change of 3.0 units if the buffer capacity was 7.1 mmol H^+ kg^{-1} pH^{-1}. If plants were growing in the soil, the acidification effects would also depend on uptake processes and leaching losses (Ch. 8).

ALTERNATIVE MEASUREMENTS OF NITRATE AND AMMONIUM

Simple colorimetric spot tests can be used to demonstrate changes in nitrate and ammonium concentrations in a semi-quantitative manner.

Reagents

Nessler's reagent. This is a poison and should be handled with care. It is purchased as a solution ready for use.
Sulphanilic acid. Add 8 g of $NH_2C_6H_4SO_3H$ to about 700 ml of water. Add 10 ml of concentrated H_2SO_4, 98 per cent *m/m*. Warm if necessary to dissolve the solid. Make up to 1 l.
N-1-Naphthylethylenediamine dihydrochloride (NEDD). Dissolve 1 g in 1 l of 1 M HCl.
Zinc powder.
Standard ammonium nitrate solutions. Dissolve 0.57 g of NH_4NO_3 in water and make up to 1 l. This contains 100 mg NO_3^--N and 100 mg NH_4^+-N l^{-1}. Into two 100 ml volumetric flasks pipette 10 and 1 ml of this solution and make up to the mark. These contain 10 and 1 mg N l^{-1} both as NH_4^+ and NO_3^-.

Method

Only a few drops of soil extract are needed to make the spot tests. Therefore soil extracts can conveniently be prepared by weighing 10 g of moist soil into a 100 ml beaker, adding 25 ml of 2 M KCl and stirring occasionally for 15 min. Measure pH, and then filter through a Whatman No. 41 paper into a test tube.

Spot test for ammonium Into depressions in a spot plate, place three drops of each of the following: distilled water, 1, 10 and 100 mg NH_4^+-N l^{-1} solution. To each add three drops of Nessler's reagent and observe the colour. Table 11.2 gives the expected results.

Table 11.2 Spot test results for ammonium and nitrate

Concentration (mg N l^{-1})	Ammonium	Nitrate
100	Strong brown	Strong purple
10	Yellow	Light purple
1	Trace of yellow	Trace of purple

Develop colour using three drops of soil extract and record its approximate concentration.

Spot test for nitrate (Note 1) Into depressions in a spot plate place three drops of distilled water, 1, 10 and 100 mg NO_3^--N l^{-1} solution. To each add a small quantity of zinc dust. To each add three drops of sulphanilic acid solution followed by three drops of NEDD solution. Observe the colour again using Table 11.2.

Develop colour using three drops of soil extract, observe and record the approximate concentration using the table.

Use of the results

If from the spot test you observe that the nitrate concentration is between 10 and 100 mg l^{-1}, i.e. about 50 mg l^{-1}, then in 25 ml of extract there is 1.25 mg N. This came from 10 g of soil, and so in 1 kg of soil there is 125 mg NO_3^--N kg^{-1} ($\pm$ about 50 mg). The method is obviously only semi-quantitative.

Note 1 The test involves the reduction of nitrate to nitrite by nascent H_2 produced by zinc reacting with acid. The colour test is for nitrite. Thus by carrying out the spot test without zinc, nitrite can be measured.

The nitrate test can be carried out on tap-water to show *very approximately* whether it is above or below the World Health Organization limit of 11.3 mg l^{-1}.

Section 11.4 The ADAS nitrogen availability index

Soils are rated as low, medium or high in relation to the amount of organic N which is expected to mineralize during a growing season. This is a much less exact indication of nutrient availability than that used for potassium, magnesium and phosphorus (Sections 9.1 and 10.3) where a quantity of extractable nutrient is measured.

Tables 11.3 and 11.4 show the system used in Britain. The first table applies to fields where in the last 5 years a permanent pasture, a long ley or lucerne

Table 11.3 The ADAS nitrogen index: following lucerne, long leys and permanent pasture

Crop before ploughing	Crop number (year) after ploughing				
	1	2	3	4	5
Lucerne	2	2	1	0	0
Long leys, cut only	1	1	0	0	0
Long leys, grazed, or cut and grazed, low N†	1	1	0	0	0
Long leys, grazed, or cut and grazed, high N‡	2	2	1	0	0
Permanent pasture, poor quality	0	0	0	0	0
Permanent pasture, low N†	2	2	1	1	0
Permanent pasture, high N‡	2	2	1	1	1

† Received less than 250 kg N ha^{-1} a^{-1} and had a low clover content.
‡ Received more than 250 kg N ha^{-1} a^{-1} or had a high clover content.
From MAFF (1988).

Table 11.4 The ADAS nitrogen index: following arable crops and short leys

Nitrogen index 0	Nitrogen index 1	Nitrogen index 2
Cereals, sugar beet or maize	Peas, beans	Any crop if the field has received large frequent applications of manure or slurry
	Potatoes, oilseed rape	
Vegetables receiving less than 200 kg N ha^{-1}	Vegetables receiving more than 200 kg N ha^{-1}	
Short ley grazed, or cut and grazed, low N†	Short ley grazed, or cut and grazed, high N‡	
Short ley, cut only		
Forage crops removed§	Forage crops grazed	

†‡ See Table 11.3, footnotes.
§ The common forage crops in Britain (other than leys, maize and lucerne which are listed separately) are kale, mustard and rye.
From MAFF (1988).

(*Medicago sativa*) has been ploughed. A *ley* is a grass or grass/clover crop which is grown as part of a rotation with arable crops. A long ley is in the field for 3 or more years before ploughing. Lucerne is a legume which is also grown for several years before the field is ploughed. All the crops in Table 11.3 cause an increase in the organic matter content in the soil. The N content of the organic matter depends primarily on inputs of ferti-lizer-N and N fixation by legumes. The organic matter supplies significant amounts of mineralized N for more than 1 year in all cases except for poor-quality perma-nent pasture where the organic matter has a low N content. Table 11.3 is used as follows:

- Imagine that in the autumn or winter you are con-sidering fertilizer needs for the coming year. Your records tell you for example that a field was ploughed 3 years ago and so you are making plans for the fourth crop after ploughing. The table gives an index based on this crop number and the use of the field before the initial ploughing.
- Arable crops will probably have been grown in the field since the initial ploughing. These may have left residues which will also supply mineralized N. Table 11.4 gives index values based on the crop and its management in the last year. Determine a second index and if the two tables give different indices, use the higher one. For fields which grow continuous arable crops or a rotation of short leys (1–2 years) and arable crops, this table only should be used.

As an example of the use of both tables consider a field which is about to grow its fifth crop after lucerne had been ploughed. In the past year beans had been grown. Because only small amounts of N are likely to be released from the old lucerne residues, Table 11.3 gives an index of zero. In Table 11.4 fresh residues from the crop of beans (a crop which fixes N) will be more im-portant, raising the index to 1. The following general principles should be noted:

- Grass and legumes, together with animal manures and slurry, can contribute large amounts of mineral-izable N. This was the basis of crop production before fertilizers came into common use, and is still an important component of mixed farming systems. Organic farming systems rely almost completely on these sources of N for crops.
- Continuous arable systems cause soil organic matter to decrease (Section 13.3) and leave only small amounts of crop residues.
- Forage crops and short leys do not significantly increase the potential for mineralization unless the N input has been large.
- Grazing animals return N to the soil, leaving larger residues than in fields where all the grass has been cut and removed.
- Legumes (clover, lucerne, peas and beans) leave N-rich plant residues due to N fixation by *Rhizobium* bacteria associated with their roots.
- Following intensively grown vegetables, potatoes and rape, all of which receive large inputs of ferti-lizer, significant residues of mineral-N can remain,

provided rainfall and leaching have not been excess-ive.

Fertilizer recommendations

Based on many field experiments and the experience of farmers, Table 11.5 gives recommended fertilizer-N

Table 11.5 Fertilizer nitrogen recommendations for winter wheat: applications between late March and early May

Soil type	N index		
	0	1	2
	$(kg N ha^{-1})$		
Yield level, up to 7 t grain ha^{-1}			
Sandy and shallow soils over			
chalk or limestone	175	150	75
Deep silty soils	150	50	Nil
Clays	150	75	Nil
Other mineral soils	150	100	50
Organic soils	90	45	Nil
Peaty soils	50	Nil	Nil

Recommendations are increased by 25 kg N for each tonne of expected yield above 7 t ha^{-1}. The full details are in MAFF (1988).

applications for winter wheat in Britain, where the average grain yield is about 7 t ha^{-1}, with a maximum potential yield of about 12 t ha^{-1}. The following prin-ciples should be noted:

- Nitrogen needs depend on the crop being grown and increase as the expected yield increases. The farmer's experience of the potential yield of his site is a basis for decision making.
- Soil type is important. Organic and peaty soils release large amounts of mineral-N. Sandy and shal-low soils have small amounts of organic matter and so release small amounts of mineral-N and also leach easily.
- An increase in the N index indicates an increase in available N and reduced fertilizer requirements.
- Many other factors are considered by farmers and advisers before a decision is made regarding the application rate. For example, the leaching of mineral-N residues during the winter can be assessed from a knowledge of winter rainfall and soil type.

Recommendations and the nitrogen budget

It is difficult to obtain complete and accurate N budgets primarily because of problems in measuring gas losses and net mineralization. Thus typical budgets for soils with different index values cannot be presented. However, much information is accumulating regarding amounts of mineral-N in soils growing winter wheat. In Britain the main flush of mineralization is in the autumn before or shortly after the crop is sown, with another flush in the spring. On clay soils in East Anglia the amounts of mineral-N in late autumn are approximately 80, 130 and 200 kg ha^{-1} where the N index is 0, 1 and 2 respectively. Some is the residue from the previous growing season, but the majority has been mineralized. Much of this N may still be available in the spring: in these soils nitrate is not exhaustively leached in the winter because of low rainfall and retention in soil aggregates (Section 11.5). In wetter regions and on lighter-textured soils potential leaching losses are large from high index soils. Further information on leaching losses in ley-arable rotations is given in Section 13.3. The amounts of soil-derived N likely to be available to wheat on soils of index 0, 1 and 2 are 30–80, 80–130 and 130–200 kg ha^{-1} respectively.

Examples of N budgets have been presented in Table 11.1. Where only fertilizer inputs and crop offtake (or yield) are known approximate budgets can be estimated. Taking Fig. 11.3 as an example, 97 kg fertilizer-N ha^{-1} was applied in May and the whole crop contained 128 kg N ha^{-1}. The input of N from the atmosphere (Colour Plate 17), in the seed and from biological N fixation would have been about 30 kg ha^{-1}. The amount of available soil-derived N can be estimated approximately from the N index. On this site the crop followed potatoes in a ley–arable rotation and so had an index of 1 (Tables 11.3 and 11.4). Thus assuming 80 kg ha^{-1} of soil-derived N, inputs were 207 kg and the offtake about 120 kg. The difference is 87 kg ha^{-1}; some of this has been lost by leaching and denitrification and some remains in the soil at the end of the season. Taking the higher value for soil-derived N (130 kg) raises the difference to 137 kg ha^{-1}. For a subsequent crop the soil would be rated as index 0 with 30–80 kg soil-derived N ha^{-1} likely to be available. Clearly much of the 87–137 kg is not available at the end of the first season.

Responses to fertilizer nitrogen

An index 0 soil in England without fertilizer-N inputs will produce about 4 t of grain ha^{-1}, removing about 80 kg N from the soil (Sylvester-Bradley *et al.*, 1987, Colour Plate 18). For the normal range of inputs the response to fertilizer-N is between 15 and 30 kg of grain for each 1 kg of N applied, becoming smaller as the optimum is approached. The straw increases by between 13 and 26 kg dry matter kg^{-1} N applied with a small increase in the roots. All of this extra plant material contains between 0.4 and 0.8 kg N kg^{-1} fertilizer-N applied. Thus between 20 and 60 per cent of the

added N is not used by the crop, reflecting the large variability in denitrification and leaching losses between sites and between growing seasons.

The typical shape of a response curve (Figs 9.5 and 13.6) is more often seen for N where index 0 soils are common, than for potassium and phosphorus where cultivated soils are commonly held at index 2. The shape of the N response curve depends on the availability of soil-derived N. As this increases, the response curve will become flatter and eventually negative due to the onset of physiological disorders and lodging.

The economic optimum input of N depends on fertilizer costs, grain value and other overheads. In Britain costs are about 36 pence kg^{-1} N and value about 12 pence kg^{-1} grain. Neglecting other costs, a response of more than 3 kg grain kg^{-1} N applied is needed for the application to be worth while. In Nebraska (Fig. 13.2) the respective figures are 24 cents kg^{-1} N and 13 cents kg^{-1} grain, requiring a response of at least 2 kg grain kg^{-1} N applied.

Section 11.5 Methods for studying nitrate leaching

The measurement of mineral-N on soils sampled from the field has been described in Section 11.1 giving the data on nitrate leaching shown in Fig. 11.7. Laboratory studies are also useful as a means of gaining understanding of the mechanisms involved in leaching, and form the basis for computer assisted modelling of leaching in the field.

How fast does nitrate move through the soil?

In soils which have no positive charges (Section 7.5) nitrate moves freely with the water. The distance the water moves depends on both climatic and soil factors. The amount of rainfall over a given period minus the evapotranspiration gives the rate at which water enters the soil. It then moves through water-filled pores and the size, shape and volume of these controls the distance the water and nitrate move.

Cylindrical tubes filled with water

The principles are illustrated in Fig. 11.12(a) and (b). For the same input rate water will move faster through the narrow tube. Similarly the volume required to move a water molecule from the top to the bottom of the tube is smaller for the narrow tube.

The presence of sand or soil particles reduces still further the volume of water in the tube; even less water is required to move a water molecule from top to bottom, this volume being equal to the volume of water-

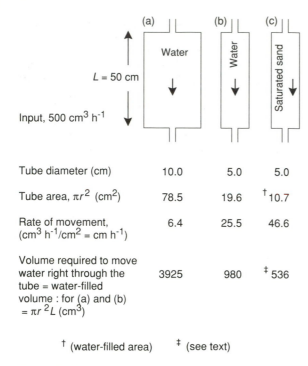

Tube diameter (cm)	(a) 10.0	(b) 5.0	(c) 5.0
Tube area, πr^2 (cm^2)	78.5	19.6	†10.7
Rate of movement, (cm^3 h^{-1}/cm^2 = cm h^{-1})	6.4	25.5	46.6
Volume required to move water right through the tube = water-filled volume : for (a) and (b) = $\pi r^2 L$ (cm^3)	3925	980	‡536

† (water-filled area)   ‡ (see text)

Figure 11.12 The movement of water through cylindrical tubes.

filled pores. Therefore to determine the distance moved by the water, the input and the volume of water-filled pores during leaching have to be known.

In laboratory experiments saturated columns can be used in which case the volume of water-filled pores is the total porosity. In the field the volume of water-filled pores during leaching must be greater than that at field capacity (Ch. 5). In intense storms, the soil may become almost saturated, but for more normal conditions the water content may be only slightly above field capacity.

A column of saturated sand

Imagine the 5 cm diameter column filled with sand (Fig. 11.12(c)). The cross-sectional area through which water can move is reduced by the sand. If the sand is packed at a bulk density of $1.2\,\text{g cm}^{-3}$, the volume of particles (density = $2.65\,\text{g cm}^{-3}$) is $1.2/2.65 = 0.453\,\text{cm}^3\,\text{cm}^{-3}$ of tube volume, and the water volume is $0.547\,\text{cm}^3\,\text{cm}^{-3}$ (Section 4.2). This value also gives a fraction of the cross-sectional area which is water filled ($0.547\,\text{cm}^2\,\text{cm}^{-2}$) and so the area through which water can move is $19.6 \times 0.547 = 10.72\,\text{cm}^2$ and the average rate of movement for an input of $500\,\text{cm}^3\,\text{h}^{-1}$ is $500/10.72 = 46.6\,\text{cm h}^{-1}$. The volume of water required to

move a water molecule through the tube is the water-filled porosity which is $0.547\,\text{cm}^3\,\text{cm}^{-3} \times 980\,\text{cm}^3 = 536\,\text{cm}^3$.

A column of unsaturated sand

Imagine that the column drains so easily that the sand is not saturated but is maintained at a water content near to field capacity which for example might be $0.2\,\text{cm}^3$ of water cm^{-3} of tube volume. The fraction of the cross-section which is now filled with water is 0.2 giving an area of $3.92\,\text{cm}^2$ and a rate of movement of $128\,\text{cm h}^{-1}$. Note that rates of movement and distances moved are averages; not all molecules will move at the same rate.

The above examples show that

rate of movement (cm h^{-1}) = input rate/cross-sectional area filled with water (cm^3 h^{-1}/cm^2).

Similarly for a given input volume

distance moved (cm) = input volume/cross-sectional area filled with water (cm^3 cm^{-2})   [11.1]

Field units

Input of water in the field is normally measured as a depth of rain minus evapotranspiration, normally mm of H_2O. Distances moved can be calculated as follows. Imagine a column of sand with square cross-section, 1 × 1 cm which receives $5\,\text{cm}^3$ of H_2O (Fig. 11.13). Taking again the unsaturated case, the cross-sectional area of water-filled pores is $0.2\,\text{cm}^2$, and the distance moved by Eq. 11.1 is $5/0.2 = 25\,\text{cm}$. However, the depth of water entering the sand is 5 cm and so the distance moved is also given by input depth/water-filled porosity.

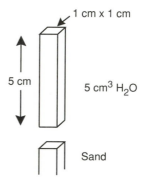

Figure 11.13 A depth of water entering a column of sand.

Mathematically

$$\text{distance moved} = \text{input depth}/\theta \qquad [11.2]$$

where distance moved and input are in the same units (mm or cm) and θ is the volumetric water content in $cm^3 cm^{-3}$. This equation gives the distance moved for both water and nitrate in sand. It can be used for soils if (a) nitrate is not adsorbed on to particle surfaces, and (b) the soil has very little structure. These effects are described below.

Laboratory measurements

Reagents

Potassium nitrate, 20 mM. Dissolve 1.011 g of KNO_3 in water and make up to 500 ml.

Leaching columns

A glass tube 2.5 cm diameter and 20 cm long forms a suitable leaching column. In one end insert a bung holding a glass tube. Place a disc of nylon cloth on the bung. Weigh the whole column. Pour into the column dry sand (or <2 mm air-dry soil), tap to settle the sand and add more until the column of sand is 15 cm long. Place another nylon disc on top of the sand. Reweigh to give the mass of dry sand.

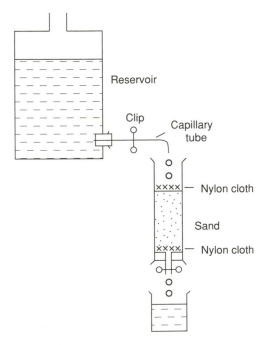

Figure 11.14 Laboratory leaching equipment.

Arrange a reservoir bottle as shown in Fig. 11.14 or a Mariotte bottle (Fig. 5.19) to supply water to the column at about one drop per second. Allow flow to continue until a steady effluent flow rate is maintained. Temporarily stop the flow and pipette 1 ml of 20 mM KNO_3 on to the nylon disc. This is known as adding a *pulse* of nitrate. Start the flow of water again, and collect the effluent in approximately 5 ml samples until a total of 50 ml has been collected. Retain these for measurement. Stop the flow, and immediately weigh the wet sand column to determine the mass of water in the sand.

The measurement of nitrate in the effluent

There are four methods that can be used, the choice depending on the availability of equipment:

1. *By electrical conductivity measurements.* The use of an electrical conductivity (EC) meter is described in Section 14.1. The concentration of KNO_3 in solution is linearly related to the electrical conductivity (approximately) by the equation

$$\text{concentration (mM)} = EC/0.1$$

   where EC is $mS\,cm^{-1}$ (S = siemens). Strictly the units should be $S\,m^{-1}$ in which case the relationship becomes $mM = S\,m^{-1}/0.01$ but electrical conductivity meters normally still give values in the former units. The meter readings can be calibrated if necessary using standard solutions prepared by dilution from the 20 mM KNO_3 solution.

   Measure the EC of each of your 5 ml effluent samples. Measure their volumes using a measuring cylinder or by weighing. Plot EC or the concentration of KNO_3 against cumulative effluent volume.

2. *By distillation and titration.* The method given in Section 11.1 can be used but is time-consuming for 10 samples. Pipette 5 ml of effluent into the distillation flask, distil and titrate against 0.01 M HCl. The calculation simplifies to

$$\text{concentration of } NO_3^- \text{ (mM)} =$$
$$2 \times \text{titration volume (ml)}$$

3. *By the spot test method of Section 11.3.* Although this is only semi-quantitative it will give the peak position approximately.

4. *Using chloride as a tracer.* The experiment could be carried out using 20 mM KCl instead of the nitrate solution, and chloride could be measured using the potentiometric method of Section 7.6. You do not need to titrate each sample. Simply measure the potential in volts, and locate your readings on a calibration graph prepared by measuring the poten-

tial of standard solutions containing 0, 5, 10 and 20 mM KCl. (A 20 mM KCl solution is prepared by dissolving 0.75 g of KCl in water and making up to 500 ml).

Results

The graph obtained (Fig. 11.15) is called a *breakthrough curve*; it shows the water volume required to displace the nitrate through the column. The results should show that the nitrate concentration at the peak is less than 20 mM due to mixing of the pulse of nitrate with water in the column.

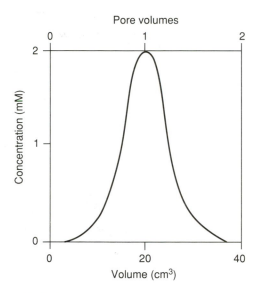

Figure 11.15 A breakthrough curve.

Is the pulse moving at the expected rate?

Take for example the above column of sand (15 cm long × 2.5 cm diameter = 73.6 cm³) packed with 88.3 g of dry sand. The bulk density is

$$88.3/73.6 = 1.21\,g\,cm^{-3}$$

The particle volume is

$$88.3\,g/2.65\,g\,cm^{-3} = 33.3\,cm^3$$

and the pore volume is $73.6 - 33.3 = 40.3\,cm^3$. If the column is saturated, the water-filled pore volume = 40.3 cm³. The breakthrough curve shows that only about 20 cm³ is needed to displace the nitrate peak, showing that the column is unsaturated.

The volume required to displace the nitrate is theoretically the same as the volume of water-filled pores in the column known as *one pore volume*. If the

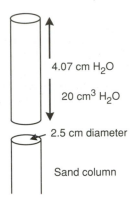

Figure 11.16 An input of water shown as a depth.

nitrate is moving in this way, one pore volume = 20 cm³. This value can be confirmed by the weighings, being simply the volume of water in the sand during the experiment. If for this column there was 20 g of water present, then the breakthrough curve shows that nitrate is moving at the same rate as the water. Breakthrough curves are often plotted with the volume axis labelled in pore volumes rather than cm³ as shown in Fig. 11.15.

Note that for this column, an input of 20 cm³ can be expressed as a depth of water (Fig. 11.16). The column area = πr^2 = 4.91 cm², and so depth = 20/4.91 = 4.07 cm. The water-filled porosity is 20/73.6 = 0.272 cm³ cm⁻³. As expected, using Eq. 11.2 distance moved = 4.07/0.272 = 15 cm which is the column length.

Nitrate leaching in structured soils

Drainage tends to occur predominantly through transmission (macro) pores (Table 4.2). Particularly in heavy-textured soils pores within aggregates are predominantly micropores, and so water flow is mostly around the aggregates, the water inside being almost stationary. This is known as *bypass flow*. It has two important effects:

1. If fertilizer is applied to the soil surface and rain then falls, nitrate will dissolve and be carried through the macropores. It moves further than is expected from Eq. 11.2 because some of the soil water is immobile.
2. If nitrate is present within aggregates as fertilizer residues or mineralized N, rain-water will move past the aggregates leaving much of the nitrate behind. Diffusion of nitrate into the mobile water will occur but leaching may be significantly delayed.

Prediction of movement in either of these situations is possible using computer-assisted models. The effects can be seen in the following laboratory experiments.

Method

Obtain a soil which has a well-developed stable structure (medium to heavy texture but not a clay), air dry and gently push it through a 5 mm sieve. Then shake the <5 mm soil on a 2 mm sieve and retain the aggregates between 2 and 5 mm diameter. Prepare a leaching column of this soil as described above.

Situation 1 After allowing flow of water to become steady (a longer time is needed, perhaps overnight, to allow heavy textured soil to wet thoroughly) add the nitrate pulse and immediately continue the water flow.

Situation 2 Follow the above method but after adding the nitrate pulse leave the column overnight so that the nitrate diffuses into the aggregates. The next day begin the water flow.

Situation 1 should result in nitrate being leached by a small volume of water, whereas in situation 2 nitrate should leach slowly.

Nitrate leaching in positively charged soils

The effects of positive charge in soils can be demonstrated by the delay it causes in the leaching of nitrate or chloride. This is similar to a chromotography experiment and the principles are in Nuffield Advanced Chemistry II, Topic 13.4.

Soil from the humid tropics may not be available, but can be simulated using a mixture of anion exchange resin with either sand or a sandy soil. Use Dowex 1–X8. It has an anion exchange capacity (AEC) of about $250 \, cmol_c \, kg^{-1}$ and is supplied moist. Allow some to partially dry by spreading it on paper in the laboratory until it feels dry (1 h) and then mix thoroughly with a <2 mm air-dry sandy soil, or with dry sand at rates of 0, 0.2, 0.4, 0.6 and 0.8 g resin kg^{-1} soil. This will give mixtures having 0, 0.05, 0.1, 0.15 and 0.2 cmol positive charge kg^{-1}.

Although this resin gives to the mixture a positive charge it differs from that in a real soil because the resin charge is not affected by pH or by the concentration of the solution with which it is in contact. Following the terminology of Ch. 7, it would be called a *permanent positive charge*.

Method

Carry out leaching experiments as for the sand, but with the following modifications. Because the pulse of nitrate is held on anion exchange sites, a second anion is required to displace it. To set up conditions similar to those in the field, initially pass 1 mM KCl through the soil. Add the pulse of KNO_3 and then displace it with

1 mM KCl again, collecting 10 ml samples of effluent until 200 ml has been leached through. Measure nitrate in the effluent samples.

Alternatively if the chloride potentiometric measurement can be made, leach with 1 mM KNO_3 and displace a pulse of 1 ml of 20 mM KCl. Because Cl^- and NO_3^- behave identically they are interchangeable in the experiment. Solutions containing 1 mM KNO_3 or KCl can be prepared by dilution from the 20 mM solutions: using measuring cylinders mix 25 ml of 20 mM solution with 475 ml of water.

Results

Plot the breakthrough curves, and determine the volume at which the peak appears. For each curve express this volume in terms of the pore volume of the column determined by weighing. For example, if one pore volume is $20.0 \, cm^3$ and the peak appears at $45.0 \, cm^3$, then the displacement volume is 2.25 pore volumes. The pulse in this case is moving at $1/2.25 = 0.44$ times the speed of the water which is the *retardation factor* R_f in chromatography theory.

The results can also be considered in terms of the *delay* in the leaching of nitrate. If 1 pore volume is required for the displacement of water through the column, and 2.25 pore volumes for the nitrate, the delay is $2.25 - 1 = 1.25$ pore volumes. A graph of delay against AEC should give an approximately linear relationship. The theoretical relationships are developed further by Wong *et al.* (1990) and in Section 11.6, Project 5.

Can the dilution of the pulse be predicted?

When a pulse of nitrate is added to the soil surface, the boundary between the soil solution and the pulse is initially sharp. In the previous experiment the concentration change was from 1 to 20 mM. The distribution of concentration is called a *nitrate profile*. As in the case of the term 'soil profile' it simply means the distribution of properties in a vertical direction. Addition of water or dilute solution now displaces the pulse downwards. If there were no mixing between pulse and soil solution the boundaries would remain sharp as leaching occurred. However, solution moves rapidly in large pores, but is delayed in small pores. Thus some of the pulse solution moves ahead into the soil solution and some lags behind (Fig. 11.17), causing a spreading of the pulse. Leaching with no mixing is called *piston flow*. The mixing process is known as *dispersion*. The characteristics of the pulse as it moves through the soil reflect the size, shape and continuity of the pores. Soil structure has been shown above to affect the distance moved by nitrate: it also controls the dispersion of the pulse.

Very complex mathematics has been applied to the

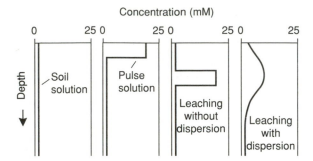

Figure 11.17 Movement of a nitrate pulse.

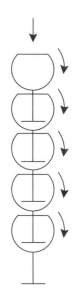

Figure 11.18 The Wineglass model of solute movement.

dispersion process. A simpler and often more useful approach is to set up a model which involves calculating changes resulting from successive small inputs of water, and adjusting the model to fit data obtained from experiments until the best conditions have been chosen for the calculations. Once the model is working successfully it can be used to predict leaching in other soils (Burns, 1974; Addiscott and Wagenet, 1985; Addiscott *et al.*, 1991).

A leaching model

Soil conditions

Consider a column of soil 50 cm deep with a 10×10 cm cross-section. The soil is at field capacity with a volumetric water content of $0.3 \, cm^3 \, cm^{-3}$. Split the column into five 10 cm layers. Each layer contains $1000 \, cm^3$ of soil in which there is $300 \, cm^3$ of soil solution. A $150 \, cm^3$ pulse of nitrate solution containing 100 mg N is added to the soil surface which is equivalent to adding $100 \, kg \, N \, ha^{-1}$. So far, real conditions have been chosen. The model which follows takes unreal conditions as a basis for calculation.

An imagined leaching process Assume that the $150 \, cm^3$ pulse mixes with the soil solution in the top layer before the excess solution moves down into the second layer. There it mixes again before passing to the third layer and so on. After five mixing and displacement steps, $150 \, cm^3$ of soil solution emerges from the bottom of the column. The sequence is repeated with a further 150 ml of water entering the soil and passing down in steps to emerge as effluent. This sequence with water is repeated as many times as necessary to move the pulse through the column.

This imagined process has an analogy in a series of wine glasses stacked inside one another (Fig. 11.18). Initially all the glasses are full of water. Nitrate solution

is poured into the top glass. It mixes with the water and floods over into the second glass where it mixes again and floods into the third glass, and so on.

Although the soil process is imagined, it is closely related to the real situation: it is simply divided into steps to allow calculation.

Calculations

Assuming the initial nitrate concentration to be zero, the first leaching sequence is shown in Fig. 11.19. It can be seen that a decreasing amount of nitrate is carried down in each step, but eventually a small amount of nitrate will appear in the first $150 \, cm^3$ of effluent.

The next leaching sequence is a little more difficult to calculate because nitrate is carried down into layers that already contain some nitrate. The total in each layer is calculated and a third of this is moved down leaving two-thirds behind. However, the calculations are all simple arithmetic and can be carried out using a calculator.

The nitrate in the effluent at the end of each sequence is plotted in Fig. 11.20 (5 layers). It emerges as a dispersed pulse, the peak concentration appearing after about $1200 \, cm^3$ of water has been displaced. This is less than 1 pore volume which is $5 \times 300 = 1500 \, cm^3$. The conditions (parameters) chosen for the calculations therefore predict a more rapid displacement than would probably occur in a real situation.

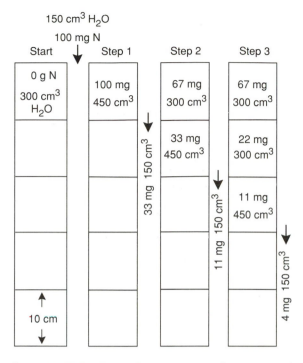

Figure 11.19 Simulation of the movement of nitrate in a soil column.

Tuning the model

Although the calculations have not produced the right answers, the model is in general simulating the process effectively. It can be refined by changing the parameters to see whether agreement can be obtained between predictions and experimental data. The parameters which can be adjusted are the volume of the initial pulse, the volume of displacing water and the number of layers of soil. It is, however impracticable to do the calculations manually. A computer program has been used to produce the second set of data in Fig. 11.20 which shows the peak in its correct position at 1500 cm³. To obtain this fit the column was divided into 20 layers each 2.5 cm deep. Otherwise the conditions were unchanged. The pulse is less dispersed and the peak has a larger concentration. Fitting data in this way is a matter of trial and error; the program takes little time to run and so the input parameters can be varied until a good fit of peak position is obtained. The program named Wineglass is on a computer disk which can be obtained from the address given in Appendix 3. The model is also used in Section 15.4 to simulate the leaching of pesticides.

A comparison of simulated and experimental data is shown in Fig. 11.21. An undisturbed column of soil (Section 11.1) 16 cm diameter and 65 cm long in a PVC

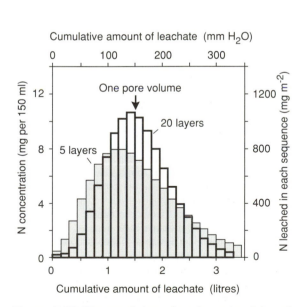

Figure 11.20 The prediction of a nitrate breakthrough curve using the Wineglass model.

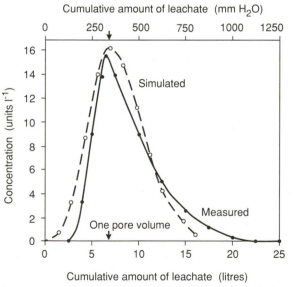

Figure 11.21 The leaching of a non-adsorbed solute through a column of soil in the laboratory and its simulation using the Wineglass model. Measurements by M. Wong, computing by L. Simmonds, Reading University. Amounts of radioactive tritium are given in units of 10^6 disintegrations min^{-1}. The activity of the tritium added was 119 units.

tube was installed in the laboratory. Water added uniformly to the soil surface flowed through the column and was removed at the base at a suction of 10 kPa, so maintaining conditions similar to those in the field. A pulse of water labelled with tritium was added uniformly over the surface and leaching continued. The effluent was collected and its volume and tritium concentration measured. Tritium, ^3H, is an isotope of hydrogen and the ^3H_2O molecules move in the soil like nitrate. Thus Fig. 11.21 shows the way nitrate or any other non-adsorbed solute would move through the soil. The peak appeared after 6.5 l of water had passed which is close to the pore volume determined by measuring the water content of the soil (7.03 l). The curve is not symmetrical with some indication of delay in finally removing all the tritium. This, and the slightly early appearance of the peak may result from the presence of very small pores through which water flow is very slow. The position and dispersion of the pulse are well simulated by the Wineglass model if the column is divided into 20 layers with 45 mm of water passing down in each sequence. The curve is now almost symmetrical with the peak at one pore volume. The data used to run the program is included in the file named Tritium-dat on the disk.

The model can in addition simulate the effects of bypass flow and delay caused by adsorption (Section 15.4). When extended to deal with mineralization, water availability and root growth, computer-assisted modelling has proved itself to be a powerful technique in the prediction of leaching and N availability.

Section 11.6 Projects

1. Using the methods of Section 11.3, compare the effects of the addition of N to a soil in the form of urea, ammonium sulphate and ammonium nitrate. For the same amount of N (e.g. 120 mg N kg^{-1} added as $257 \text{ mg NH}_2CO.NH_2$, $566 \text{ mg (NH}_4)_2SO_4$ or $343 \text{ mg NH}_4NO_3 \text{ kg}^{-1}$) you should see a difference in the acidification caused by the three compounds. Their reactions can be summarized as

   $$NH_2CO.NH_2 \rightarrow (NH_4)_2CO_3 \rightarrow 2HNO_3 + H_2O + CO_2$$
   $$(NH_4)_2SO_4 \rightarrow H_2SO_4 + 2HNO_3$$
   $$NH_4NO_3 \rightarrow 2HNO_3$$

   Thus urea and NH_4NO_3 produce half the acidity of $(NH_4)_2SO_4$.

   If an acidic soil is used, a reduced rate of nitrification should be measured compared to a neutral soil.
2. Set up an incubation study using a flooded soil. Add urea and ammonium nitrate at the above rate to 5 kg of soil in two buckets. Flood with water and stir to wet the soil thoroughly and remove air bubbles. At intervals of 1 week, stir thoroughly and measure pH (Section 8.1), E_h (Section 6.7) and NH_4^+ and NO_3^- concentrations (Section 11.1). Your measurements should show mineralization of urea, inhibition of nitrification, denitrification and NH_3 volatilization.
3. If you have access to a farm's cropping records and can obtain soil samples from its fields compare the amounts of mineral-N released in the standard Waring and Bremner method (Section 11.3) for soils with different N indices. Tables 11.3 and 11.4 give the relationships between management and index.
4. Leaching studies in field plots are possible if you have access to a non-stony medium to light-textured soil. Isolate two plots of soil by sinking $50 \times 50 \text{ cm}$ wooden frames a few inches into the ground, with a few inches projecting above the soil surface. This allows known volumes of water to be applied only to these areas. Apply water over a few days to ensure that both plots have come to field capacity; a very dry soil may need an application of up to one-third of its volume of water for this purpose (Table 4.3). Thus to 1 m depth, the plot volume is

   $$0.5 \times 0.5 \times 1 = 0.25 \text{ m}^3$$

   and 0.075 m^3 of water may be needed. This is 75 l, or about eight 2 gallon watering cans. Apply the same volume to both plots. Cover the plots with a polythene sheet to prevent evaporation, and leave for 2 days to drain.

   Apply nitrate to the surface of one plot. An application of 100 kg N ha^{-1} would be 2.5 g N on the plot surface. Dissolve 18 g KNO_3 in 2 l of water and sprinkle from a watering can uniformly over the surface. This should produce a nitrate pulse in the top 2.5 cm of soil. Add the same volume of water to the other plot.

   A sandy soil with a volumetric water content at field capacity of $0.3 \text{ cm}^3 \text{ cm}^{-3}$ requires about 3 cm of H_2O to displace the pulse 10 cm through the soil (Eq. 11.2). Over the area of the plot (2500 cm^2) this is 7.5 l. Apply water uniformly at a rate of 11 d^{-1} until you have displaced (theoretically) the pulse to the depth required. Cover between applications to prevent evaporation.

   Take a core of soil from the centre of each plot (Section 11.1) and measure bulk density, water content and nitrate in suitable (5 cm) sections of the core. Compare the theoretical (Eq. 11.2) and measured distance of movement of the nitrate pulse. Did you recover the expected amount of nitrate? Note that nitrate can be measured semi-quantitatively by the spot test method of Section 11.3.

   If a longer project is possible natural leaching

could be studied. Prepare the plots as above and then leave them exposed to rain and evaporation. Section 12.2 gives information on the calculation of net water input.

5. Section 11.5 describes a leaching experiment for a positively charged soil. This can be modified to allow measurement of delay to be compared to prediction using chromatography theory, which shows that

$$R_f \text{ (the retardation factor)} = 1/[1 + (b\rho/\theta)]$$

where b is the adsorption coefficient given below, ρ is the bulk density of the soil, $(g\,cm^{-3})$, θ the volumetric water content during leaching $(cm^3\,cm^{-3})$, and $b\rho/\theta$ is the delay in pore volumes (Section 15.4).

If the experiment is carried out in a soil with 0.2 cmol positive charge kg^{-1} and using 10 mM KNO_3 initially followed by a pulse of 10 mM KCl followed by 10 mM KNO_3 again, then the value of b is a constant and is shown in Fig. 11.22. This is a $Cl^- - NO_3^-$ exchange isotherm (Section 9.4).

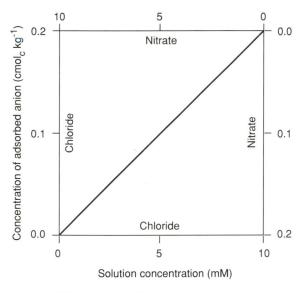

Figure 11.22 A chloride–nitrate exchange isotherm.

The amount of adsorbed Cl^- is linearly related to its concentration in solution (the slope is constant) because both ions are equally attracted to the exchange sites.

The adsorption coefficient is related to the slope of the graph but has to be in the correct units of $\mu mol\,g^{-1}/\mu mol\,cm^{-3} = cm^3\,g^{-1}$. The slope is $0.2\,cmol_c\,kg^{-1}/10\,mM$ and so the adsorption coefficient is

$$2\,\mu mol\,g^{-1}/10\,\mu mol\,cm^{-3} = 0.2\,cm^3\,g^{-1}$$

For the above conditions and for a column with $\rho = 1.2$ and $\theta = 0.27$ as in the experiment in Section 11.5, then

$$R_f = 1/[1 + (0.2 \times 1.2/0.27)] = 1/(1 + 0.89) = 0.53$$

and the pulse is expected to be delayed by 0.9 pore volumes.

The linear relationship between charge and delay discussed in Section 11.5 is seen mathematically in the above equation because b is linearly related to charge.

6. Use the Wineglass model to check the simulations shown in Figs 11.20 and 11.21.

Block 11.7 Calculations

1. A moist topsoil was sampled and 40 g was extracted with 200 ml of 2 M KCl. The measured concentration of NO_3^--N was 5 mg l^{-1}. The water content of the moist soil was 26 g water per 100 g oven-dry soil. Calculate the amount of NO_3^--N in the soil as mg kg^{-1} oven-dry soil. (*Ans.* 33) What was the concentration of nitrate in the soil solution? (*Ans.* 126 mg l^{-1}) Calculate the amount of NO_3^--N in the topsoil (2500 t ha^{-1}) in terms of kg ha^{-1}. (*Ans.* 82)

2. An application of 100 kg N ha^{-1} is made using NH_4NO_3 fertilizer. What mass of fertilizer is used if it is 98 per cent pure? (Molar masses are in Appendix 2.) (*Ans.* 292 kg ha^{-1})

   If all the NH_4^+ is nitrified, calculate the increase in concentration of nitrate-N in the soil solution if the fertilizer is uniformly mixed with 2500 t of soil at a water content of 30 g H_2O per 100 g of dry soil. (*Ans.* 133 mg l^{-1} or 9.5 mmol l^{-1})

   If the nitrification process releases 2 mol H^+ for each mol of NH_4^+ mineralized calculate the pH change in the soil if its buffer capacity is 50 mmol H^+ kg^{-1} pH^{-1}. (*Ans.* 0.06)

3. A sample of dried grass is taken from a crop yielding 10 t dry matter ha^{-1}. The sample contains 15 g N kg^{-1} dry matter. What mass of fertilizer-N per hectare would be needed to supply that removed in the crop? (*Ans.* 150 kg)

4. In a laboratory incubation experiment, an input of 40 mg N kg^{-1} dry soil is required. Calculate the mass of each of the following which would give the above input: NH_4NO_3, $(NH_4)_2SO_4$, urea, dry manure containing 2 per cent N. (Molar masses are in Appendix 2.) (*Ans.* 114, 189, 86 and 2000 mg kg^{-1}).

CHAPTER 12

The Availability of Water in Soils

The growth of vegetation and crops is broadly controlled by climate primarily through constraints due to temperature, rainfall and exposure to wind. In a given climatic region, soil adds further constraints through (a) depth, stoniness, texture and structure which together influence both the amount of water which can be held in the profile and the ease with which roots can penetrate and (b) the availability of nutrients. These soil factors together determine *soil fertility*, a topic developed further in Chapter 13.

The soil profile lies at the heart of the hydrological cycle shown in Fig. 12.1, and can be regarded as a bank account with additions and removals controlling the balance which in this case is the amount of water stored in the profile. The balance sheet has the following components:

- Rain, melted snow and overland flow deliver water to the land surface. However, not all the water will necessarily enter the soil. Some remains on vegetation and is lost by evaporation, and if the rate of supply exceeds the ability of the soil to absorb water, it will run off the soil surface possibly causing erosion and flash floods.

- After periods of heavy rain the ability of the soil to hold water may be exceeded and drainage will occur with water moving into streams or aquifers. Throughput of water will leach solutes from the soil.

- The water held in the profile is the reservoir supplying the water which evaporates from the soil surface and maintains the water supply to plants between rain events. In dry conditions, the extraction of water by plants becomes limited by the supply from the soil, with resultant water stress. The rate of evaporation from the soil surface is also controlled by the dryness of the soil.

- Upward movement of water into a profile will occur if moist subsoil is below a dry soil. Sections 5.4 and 5.5 give the principles underlying water movement.

Short-term imbalance between gains and losses of water by a profile results in a net change in water content. However, unlike a bank balance the soil has an upper limit to the amount of water it can hold known as its *field capacity*. Because this capacity is small compared with annual rainfall (except in very dry regions), the gains and losses over a year or more will be approximately balanced, with rainfall inputs equal to losses by evaporation and drainage. Again, unlike a bank balance, there tends to be a lower limit to the amount of water held in the profile known as the *permanent wilting point*, which is the limit in terms of plant uptake although evaporation from the soil surface can reduce the water content further. Thus, in soils covered by vegetation or crops, the water content fluctuates in the short term between field capacity and the permanent wilting point. Table 4.3 shows that storage pores occupy between 0.05 and 0.25 $cm^3 cm^{-3}$ of soil, and so in a 1 m deep profile this fluctuation is equivalent to between 50 and 250 mm of rain (Section 5.1).

This chapter considers the various components of the *soil water balance*, and the ways in which they can be measured or predicted in field situations.

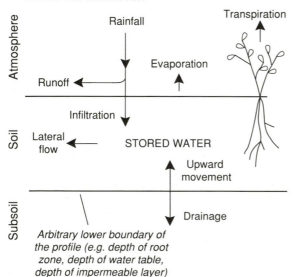

Figure 12.1 Components of the soil water budget.

INFILTRATION AND SURFACE RUNOFF

During gentle rain, water infiltrates the soil surface as rapidly as the rain falls with no accumulation of water on the surface. With increased intensity the rate at which the soil can accept water, known as its *infiltration*

244

capacity, may be exceeded, causing water to accumulate in puddles. This condition is known as *ponding* which is the precursor of surface runoff and erosion.

The factors which control the infiltration capacity are those controlling water movement discussed in Chapter 5 and Section 5.5. The force moving water down through soils is the result of a combination of gravity and a soil water suction gradient driving water from soil with a small suction (wet) to soil with a large suction (dry). During the early stages of infiltration, the wetted soil surface tends to be at a smaller suction than the soil beneath: there is a large suction gradient acting with gravity to drive the water downwards. As rainfall continues and the upper part of the profile becomes wetter, the suction gradient and the downward driving force decrease. If the soil becomes saturated, then gravity remains as the only force responsible for infiltration. Figure 5.17 shows this sequence in a special case where a dry, deep, freely draining soil is suddenly flooded with the floodwater maintained on the surface during infiltration. The more normal case resulting from occasional rainfall is shown in Fig. 12.5. Note that the infiltration capacity of a soil is not a fixed value but varies depending on the distribution of water in the profile.

One of the consequences of these changes is that during steady rainfall, puddles do not form immediately, but eventually will form when the decreasing infiltration capacity becomes less than the rainfall intensity. Also, puddles are likely to form more quickly when rain falls on to wet rather than dry soil.

Section 5.5 shows that the rate of water movement depends both on the driving force and the hydraulic conductivity of the soil. Infiltration rate depends on the hydraulic conductivity of the wet soil, which in turn depends on the size and continuity of the transmission pores (Table 4.2). When the profile is eventually saturated with no suction gradient, the gradient of water potential is equal to the gradient of gravitational potential and the rate of water flow (Eqn. 5.7) is numerically equal to the saturated hydraulic conductivity when expressed as velocities, normally $mm\,d^{-1}$. This provides the basis of the field method for measuring the hydraulic conductivity of a saturated soil described in Section 5.5.

One further aspect has to be taken into account when considering the factors controlling infiltration rates. The surface of the soil is prone to structural damage by traffic, raindrop impact and runoff leading to the saturated hydraulic conductivity of the thin surface layer being less than that of the soil beneath. Many cultivation treatments aim to maintain a network of large continuous pores to the soil surface. This is particularly important in semi-arid tropical areas where the limited rainfall usually arrives in intense storms with a twofold hazard. Firstly, they tend to damage the structure of the soil surface resulting in blocking of the large pores. The problem is exacerbated in these regions because structure is often weak because of small organic matter contents (Ch. 13). Secondly, the storms are likely to cause runoff which adds to the stress already imposed by raindrop impact, reducing further the infiltration capacity which in turn leads to more runoff and further structural damage.

To break into this 'vicious circle' the soil surface should be protected from raindrop impact by vegetation or a mulch, cultivations should leave large aggregates at the surface with a stable network of large pores, and overall management should aim to maintain or increase the organic matter content of the soil.

EVAPORATION FROM SOIL

Water is lost more or less continuously from land surfaces by direct evaporation from the soil surface or by evaporation from leaves which is termed *transpiration*. The combined loss is often termed *evapotranspiration* but in this chapter evaporation will be used as a general term for both losses. The prevailing weather is a major factor controlling the rate of evaporation: a hot, sunny day with a dry wind is ideal 'drying weather'. The *evaporative demand* of the atmosphere can be assessed in terms of the rate of evaporation that would be achieved from an extensive open water surface such as a lake, expressed as the depth of water removed each day (e.g. $mm\,d^{-1}$) and known as the *potential evaporation*. It can be estimated by measuring the rate of loss of water from an open pan. Because of the limited pan size, the measurements have to be corrected if the rate of evaporation is enhanced (often doubled or even trebled) by the surrounding area being dry. In this case the pan is acting as an 'oasis' over which air is passing which has not been cooled and moistened by surrounding soil and vegetation.

Alternatively, potential evaporation can be estimated from measurements of the controlling weather variables (solar radiation, the temperature and humidity of the air and windspeed), based on an understanding of the physics of evaporation. In order for evaporation to occur, an input of energy is needed to break the bonds between the water molecules. The energy comes from the absorption of solar radiation by the evaporating surface. The removal of the water vapour which is produced depends on the dryness of the overlying air and the windspeed.

If the soil surface is wet, the rate of evaporation is similar to that from a lake. When the soil surface dries the *actual evaporation* becomes less than the potential evaporation for two reasons which are interrelated:

1. As soil water suction increases, the vapour

concentration in the soil air decreases, reducing the concentration difference between the soil air and the drier overlying air. This concentration difference is one of the driving forces for evaporation and so the rate of evaporation is reduced.

2. The hydraulic conductivity of the soil is reduced, so decreasing the rate at which the surface can be resupplied with water to meet the losses caused by evaporation.

Figure 12.2 shows the progress of potential and actual evaporation from a sandy soil in Niger following rainfall. The soil and site are described in Sections 5.4 and 5.5. The rain (8.4 mm) fell during the evening before the first day of measurement, with no subsequent rain for 2 days. The potential evaporation reflected the diurnal changes in atmospheric conditions. The first day was cloudy, and potential evaporation rose to about $0.5 \, \mathrm{mm \, h^{-1}}$ by noon. The actual evaporation equalled the potential evaporation except for a short period in the afternoon when the upward movement of water was being limited by the small hydraulic conductivity of the soil. Fig. 5.3(b) shows that the hydraulic conductivity of the soil decreases sharply as suction increases around the field capacity condition (10 kPa). Overnight, the slow upward flow of water continued, partially rewetting the surface. The next day was clear and hot and the potential evaporation rose to $0.8 \, \mathrm{mm \, h^{-1}}$. The evaporative demand was met for a short period in the morning before the dry surface layer was re-established and evaporation was restricted. A well-defined layer of dry soil (about 1 cm thick) was

seen to develop which effectively acted as a barrier to water flow. The potential evaporation was 3.1 and 5.5 mm and the actual evaporation was 2.7 and 1.6 mm during the first and second days respectively.

EVAPORATION FROM VEGETATION

For a low crop such as mown grass growing in a moist soil, the cells within the leaves effectively act as water surfaces because the relative humidity in the intercellular spaces of the leaves is virtually 100 per cent. However, there are important differences between evaporation from leaves and from a water surface:

- The surface layer of cells in the leaves is more or less impermeable to water vapour. The vapour flows from the leaves via the stomata which are apertures which open during the day to facilitate the uptake of CO_2 for use in photosynthesis. The stomata present a resistance to flow of water vapour which restricts transpiration, although the resistance is small when the stomata are fully open.
- The effect of stomata as a barrier to flow is partly offset by the canopy of leaves having a much larger surface area than the area of land on which they are growing. For example, a good wheat crop at the time of maximum leaf area may have more than $6 \, \mathrm{m}^2$ of leaf m^{-2} of land surface and evaporation can occur from both sides of the leaves.

The combined effect of these differences is that during the summer, the rate of evaporation from dense vegetation such a grassland or a wheat field well supplied with water is typically about 80 per cent of that from an open water surface. Although the leaf area can be very large, the ultimate limitation to evaporation is usually the energy available from solar radiation.

As soils dry, the availability of water to plants is reduced. To avoid excessive dehydration, plants respond by closing their stomata or shedding leaves, which reduces both the loss of water and the rate of uptake of CO_2. Hence one of the consequences of water shortage is reduced uptake of CO_2 which results in a reduced rate of growth and leads to the need for irrigation to maximize crop productivity. To understand the way in which plants and soils interact to control water loss to the atmosphere, a basic understanding of the processes involved in water flow from the soil to the site of evaporation in the leaf is needed.

LEAF WATER POTENTIAL

Water moves from soil to leaves along a gradient of decreasing water potential, as it does within the soil (Section 5.5). There is a large resistance to flow through the plant tissue and so a large difference in water poten-

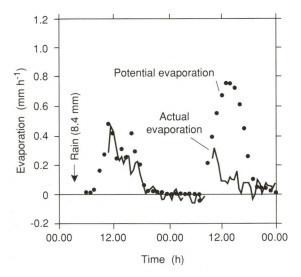

Figure 12.2 Potential and actual evaporation from a bare soil surface at Sadoré, Niger. Data by C. Daamen, Reading University.

tial between soil and leaf is needed to cause water to flow fast enough to meet the evaporative demand. The water in soil normally has a negative water potential (Section 5.4) because it is below the soil surface and because it is held by the soil matrix, i.e. both the gravitational and matric components of soil water potential are negative. Thus the water in the leaf must be at an even lower potential (more negative) for water to move up through the plant. Leaves maintain a low *leaf water potential* by accumulating a large concentration of salts in the cell sap which results in a large negative osmotic potential. However, the osmotic effect on leaf water potential is partially offset because the cells have a positive internal pressure known as *turgor pressure*, which provides a positive contribution to leaf water potential known as the *turgor potential*.

An alternative way of viewing these concepts is that during growth the plant uses energy to maintain a large salt concentration in the leaf cells. By osmosis, water is drawn into the cells and because their volume is limited, the hydrostatic pressure in the cell increases. A state of dynamic equilibrium is established in which the rate of water uptake from the soil equals the rate of loss by evaporation.

Figure 12.3 shows how soil and leaf water potentials might change during a day of high evaporative demand. At night, assuming zero water flow, the leaf water potential is the same as the soil water potential ($-0.05\,MJ\,m^{-3}$ in this case). The leaf water potential is made up of the osmotic potential ($-1\,MJ\,m^{-3}$) and the turgor potential ($+0.95\,MJ\,m^{-3}$). The latter is the result of the hydrostatic pressure in the cell being 0.95 MPa above the ambient air pressure, and the plant tissue is therefore turgid. After dawn the stomata open and evaporation begins. The leaf cells partially dehydrate decreasing the turgor potential. Also the cell volumes decrease causing the cell sap to become slightly more concentrated and the osmotic potential to become more negative. The combined effect is to reduce the leaf water potential, so giving the water potential difference required to cause water to flow from the soil to the leaves. Thus, the leaf water potential has a 'floating value' which responds to both the evaporative demand and the availability of soil water in order to maintain a balance between water supply and demand.

To prevent excessive dehydration and physiological damage plants limit water loss by reducing the stomatal aperture. The arrow in Fig. 12.3(a) shows the time at which the stomata begin to respond in this way. Thus during times of high evaporative demand the actual rate of transpiration may be reduced below the potential rate. At times of very high demand, turgor pressure may be reduced to zero, causing the leaves to become flaccid: this condition is termed *wilting*. If plant turgor recovers when the evaporative demand falls, i.e. during

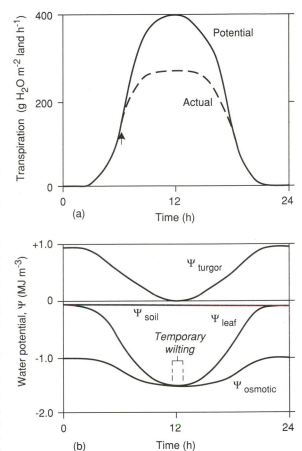

Figure 12.3 Changes in (a) transpiration, and (b) water potentials during a day of large evaporative demand.

the following night, the condition is termed *temporary wilting*. If the soil has become so dry that plant turgor does not recover overnight, the plant's condition is termed *permanent wilting*.

WATER AVAILABILITY TO PLANTS

How dry does a soil have to be before water availability to plants is restricted? The classical approach to this question is to grow plants in pots of watered soil until they are well established, and then to seal the pots to prevent evaporation, allowing the plants to extract water until they are permanently wilted. At this stage it is assumed that the plants have extracted all the water that is available to them. This lower limit of water availability occurs when the soil water suction is about 1.5 MPa and is termed the permanent wilting point. The suction value is little influenced by soil type although the water content at this suction will be very

Table 12.1 Typical values for the volumetric water contents of soils of varying texture at field capacity and the permanent wilting point together with their available water capacities. Derived from Campbell (1985)

Texture	Water content ($cm^3 cm^{-3}$)		
	Permanent wilting point	Field capacity	Available water capacity
Clay	0.28	0.44	0.16
Silty clay	0.28	0.44	0.16
Clay loam	0.23	0.44	0.21
Silty clay loam	0.20	0.42	0.22
Sandy clay loam	0.16	0.36	0.20
Loam	0.14	0.36	0.22
Silt loam	0.14	0.36	0.22
Sandy loam	0.08	0.22	0.14
Loamy sand	0.06	0.18	0.12
Sand	0.05	0.15	0.10

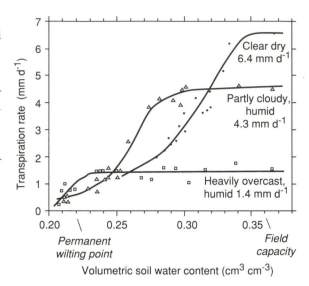

Figure 12.4 The effect of evaporative demand and soil water content on the transpiration rate of maize. The potential transpiration rates ($mm\,d^{-1}$) are marked on the graph. From Denmead and Shaw (1962).

dependent on soil texture and organic matter content. Table 12.1 gives typical values. Permanent wilting occurs when there is no flow of water during the night to rehydrate the leaves, i.e. there is no water potential difference between soil and leaves. Hence, if wilting is permanent the soil water potential must be equal to the leaf water potential at which wilting occurs. For the plant illustrated in Fig. 12.3(b), permanent wilting would not be expected until the soil dried to a water potential of $-1.5\,MJ\,m^{-3}$.

Temporary wilting can occur in soils much wetter than the permanent wilting point. For example, Fig. 12.3(b) shows that the leaf water potential can fall to the point where wilting occurs even though the soil is relatively wet. Thus, shading can be at least as effective in reducing temporary wilting as applying water to the soil.

The permanent wilting point is therefore a simple approach to determining the absolute lower limit of water availability to plants, but in practice plants can be affected by the availability of water long before soils have dried to this point. Figure 12.4 gives the results of one of the classical experiments in this area of water relations. Water loss from maize grown in containers was monitored and the figure shows the transpiration rates on 3 days with contrasting weather. On a heavily overcast, humid day the evaporative demand was $1.4\,mm\,d^{-1}$ and this rate of transpiration was maintained except in soils close to the permanent wilting point. On a clear dry day, the potential evaporation rate of $6.4\,mm\,d^{-1}$ could only be sustained in soils close to field capacity. In all cases the rate of water loss was restricted by supply before the permanent wilting point was reached.

The amount of water held by a soil between field capacity and the permanent wilting point is termed the *available water capacity*. Typical values are given in Table 12.1 and are expressed as $cm^3 cm^{-3}$ which is equivalent to a depth of water per unit depth of soil (Section 5.1). The amount of water available to a crop can be found from the product of rooting depth and available water capacity. However, not all of this water is equally available. The effects of spatial variability of soil properties on available water capacity and rooting depth and thus on the growth of a crop are shown in Plate 1.2.

SOIL WATER DEFICITS AND IRRIGATION NEEDS

Ideally irrigation should be applied when crop growth is just beginning to be limited by the deficit of water in a soil. The amount of water then required to restore the profile to field capacity is called the *soil water deficit*, often expressed as mm of water. In practice irrigation planning can be based on the simple idea that there is a critical soil water deficit (the *limiting deficit*) beyond which crop growth is restricted by water shortage. The water held between field capacity and the limiting deficit is considered to be *readily available*. Table 12.2 gives typical values. The fraction of the available water in the root zone which is readily available depends on the

Table 12.2 Typical values of rooting depth and the fraction of the available water capacity of the root zone which is readily available (Doorenboss and Pruit, 1977)

Crop	Rooting depth (m)	Readily available fraction
Alfalfa	1.0–2.0	0.55
Barley	1.0–1.5	0.55
Beans	0.5–0.7	0.45
Grass	0.5–0.7	0.45
Maize	1.0–1.7	0.60
Potatoes	0.4–0.6	0.25
Vegetables (various)	0.3–0.6	0.20
Wheat	1.0–1.5	0.55
Groundnuts	0.5–1.0	0.40

density of the root system and the responsiveness of the stomata to water deficits.

Section 12.1 shows how measurements of changes in soil water content can be used to calculate the rate of water uptake by crops during a growing season, and Section 12.2 shows how an irrigation manager might use weather data and crop and soil factors to determine when the limiting deficit is reached. Section 12.3 extends this approach to include drier soils and presents a model which can be used to follow changes in the soil water balance using weekly values of rainfall and potential evaporation to determine changes in the soil water deficit and the amounts of water draining from the profile. The model can be used manually or with a computer program.

FURTHER STUDIES

Ideas for projects are in Section 12.4 and calculations in Section 12.5.

Section 12.1 The measurement of water loss from soils

Water is removed from soils by evaporation from bare soil, by transpiration from plants and by drainage. The resulting changes in soil water content can be measured in layers of soil down through the profile in volumetric terms. Three methods are commonly used:

1. Using a Jarret auger (Section 1.3) or a root sampling auger (Section 3.1) samples of soil from known depths are taken and gravimetric water content measured (Section 3.3). Bulk density can be measured approximately by recording the mass of moist soil removed by the auger from known depth

increments and calculating its oven-dry mass, or by the methods of Section 4.2.

2. An undisturbed core is taken (Section 11.1) and sectioned. The mass of each moist section is recorded and its gravimetric water content measured. Knowing the volume of each section its volumetric water content is calculated.

3. Both the above methods are time-consuming and do not allow subsequent measurement at the same position. Thus, to follow changes over time cores have to be taken at different positions in a plot. Also the hydraulic properties of the plot may be altered even though the holes are refilled after sampling. A *neutron moisture meter* avoids these problems. The method is beyond the scope of this book (Smith and Mullins, 1991) but involves auguring a hole through the profile, into which a metal access tube is inserted. The meter is lowered down the tube and with suitable calibration gives a direct reading of volumetric water content in the soil around the depth of measurement. Measurements can be repeated at a later date using the same tube so allowing changes with time to be monitored at the same site.

Changes in water content of soils under crop and fallow in Niger

An example of the use of measurements of soil water content is shown in Fig. 12.5. The site was at N'Dounga, 20 km north of Niamey in Niger. The soil was a sandy loam, and access tubes were inserted to allow measurements using a neutron moisture meter on a plot that was cropped with millet and on an adjacent plot that remained bare. Rainfall was also measured at the site. Figure 12.8 shows the average weather conditions.

When the crop was sown the early rains had wetted the upper few cm of soil, but the soil below was dry. In the bare soil rain penetrated to increasing depths and by August 10, some water had drained below 150 cm with water contents in the profile being close to field capacity. In the cropped site uptake prevented water from penetrating beyond 100 cm, and the soil was wettest at around 50 cm. The measurements at 5 and 13.5 cm depth are unreliable because of difficulties in calibrating the meter in surface soils.

Table 12.3 shows how the measurements can be used. The data for the cropped site have been chosen because the water profiles show that no drainage occurred. Thus, because soil at the bottom of the profile was never wetted, any decrease in the amount of water stored in the profile must be equal to the excess of evaporation over rainfall. The water content at each depth was assumed to be the water content of the

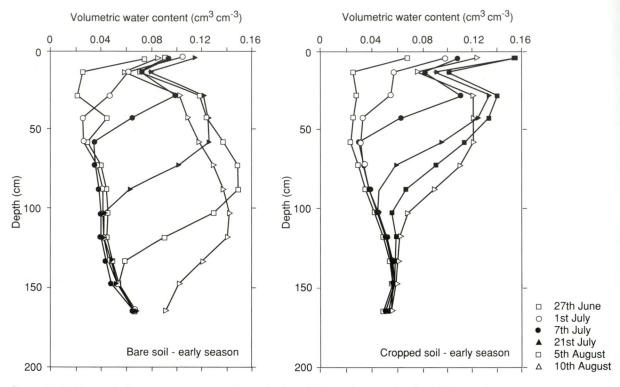

Figure 12.5 Changes in the water content profiles under bare fallow and cropped soil at N'Dounga, Niger. Data by W. A. Payne, Texas A & M.

Table 12.3 Calculations of changes in stored water for a soil cropped with millet

Depth (cm)	Volumetric water content at date of measurement:		Layer thickness (mm)	Increase in stored water in each layer (mm)
	27 June	1 July		
5.0	0.068	0.100	92.5	2.96
13.5	0.025	0.058	117.5	3.88
28.5	0.028	0.054	150	3.90
43.5	0.024	0.034	150	1.50
58.5	0.023	0.031	150	1.20
73.5	0.030	0.026	150	−0.60
88.5	0.035	0.038	150	0.45
103.5	0.042	0.044	150	0.30
118.5	0.049	0.052	150	0.45
133.5	0.055	0.057	150	0.30
148.5	0.056	0.059	150	0.45
163.5	0.049	0.054	150	0.75
	Total increase in stored water (mm)			15.54
	Rain (mm)			24.5
	Actual evaporation (mm)			8.96
	Rate of evaporation (mm d^{-1})			2.24

surrounding layer, so allowing the change in water stored in the layer to be calculated. Summing the change in the layers gives the total increase in stored water. A similar procedure is described in Section 3.7 to determine amounts of carbon in a soil profile. The difference between the amount of rain and the increase in stored water gives the actual evaporation for the period with an average value of $2.2 \, \text{mm d}^{-1}$.

The complete data for the cropped site of this experiment are in a file named Water.dat on the computer disk described in Section 12.3 and Appendix 3. Figure 12.6 shows the changes in evaporation for the whole growing season which lasted until 25 September, and can be compared to Fig. 12.8 which shows similar data estimated from meteorological information. Figure 12.6 shows that initially the actual evaporation rate was only about one-third of the potential rate due to the small leaf area of the young plants, and because in this sandy soil evaporation is soon restricted after rain by the formation of a dry surface layer (Fig. 12.2). Later in the season, the large plants transpire water at rates approaching the potential rates until lack of rainfall and the senescing crop reduce the water losses.

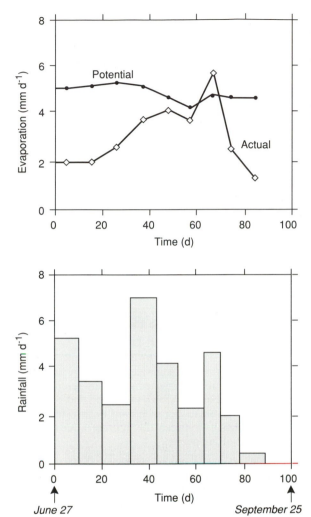

Figure 12.6 The evaporation rate from a millet crop and the rainfall at N'Dounga, Niger. The plotted values are the means for the periods indicated (mostly 10 d). Data derived from measurements by W. A. Payne, Texas A & M.

Section 12.2 The estimation of water use by crops

Because water is lost through the stomata at the same time as CO_2 is taken in for use in photosynthesis, water loss through transpiration is an inevitable cost of dry matter production. Between about 200 and 800 g of water is used to produce 1 g of dry matter for a range of crops and climates. Thus the volume of water required to produce a crop is immense. For example, $20\,000\,m^3$ of water may be used by 1 ha of sugar cane giving a yield of 40 t of dry matter ha^{-1}.

The actual evaporation from a crop normally falls well short of the potential evaporation as a result of insufficient rainfall during the growing season. For example, the evaporation from a wheat crop in Britain in 1975 (a dry year) during 3 summer months was 188 mm compared to potential evaporation of 270 mm and rainfall during the growing season of 68 mm. The difference between actual evaporation and rainfall (120 mm) is the extent to which the available water in the soil has been depleted. However, the amount of water stored in the soil was insufficient to maintain the potential evaporation and irrigation would probably have increased crop growth. More extreme cases occur in the dry regions of the world. For example, the actual evaporation from a groundnut crop in India during 2 months in 1982 was 102 mm, compared to the potential evaporation of 554 mm. During the growing season there was no rainfall and so all the water taken up by the crop came from stored water.

In both the above cases, the soil was at field capacity at the beginning of the growing season and was almost at the permanent wilting point when the crop was harvested. Hence the available water capacities of the root zones were 120 and 102 mm respectively.

Available water capacities are given in Table 12.1. When multiplied by the depth of rooting (Table 12.2), they give the *available water capacity of the root zone*. For example, a sandy loam soil with an available water capacity of $0.14\,cm^3\,cm^{-3}$ exploited by groundnut roots to a depth of 50 cm (Table 12.2) has 70 mm of available water which is equivalent to 12 days' supply at a potential evaporation rate of $6\,mm\,d^{-1}$, typical of central India in the dry season. In practice the actual evaporation soon becomes less than the potential evaporation and crop growth is retarded. This has led to the concept of readily available water which is the amount of water in a soil at field capacity which can be taken up before the actual evaporation rate becomes less than the potential rate and growth becomes restricted. In the light of Fig. 12.4 it is clearly a rather arbitrary concept. It has to be determined by experiment for a given crop and site, but as an approximation about half the available water capacity is freely available water (Table 12.2).

For the purposes of planning irrigation, the limiting soil water deficit (p. 248) is equal to the amount of readily available water. A soil water balance sheet is given below in which the deficit is not allowed to increase beyond the limiting value, so allowing the atmospheric demand to be met and growth to be unrestricted. Section 12.3 shows how actual evaporation can be predicted from potential evaporation in drier soils.

A soil water balance sheet

Table 12.4 is the balance sheet for a 10 d period during

Table 12.4 A soil water balance sheet for a crop of groundnuts grown in central India

Day	1	2	3	4	5	6	7	8	9	10
Potential evaporation (mm)	4.5	6.0	7.2	5.4	6.6	3.2	6.6	7.2	6.5	5.5
Crop factor	0.8	0.8	0.8	0.8	0.9	0.9	0.9	0.9	0.9	0.9
Estimated evaporation (mm)	3.6	4.8	5.8	4.3	5.9	2.9	5.9	6.5	5.9	5.0
Rainfall (mm)	0	0	5.2	0	0	0	0	0	25.0	0
Soil water deficit at the beginning of the day (mm)	23.0	26.6	31.4	32.0	36.3	0	2.9	8.8	15.3	0
Soil water deficit at the end of the day (mm)	26.6	31.4	32.0	36.3	42.2	2.9	8.8	15.3	0	5.0
Irrigation (mm)	0	0	0	0	50.0	0	0	0	0	0
Drainage (mm)	0	0	0	0	7.8	0	0	0	3.8	0

the growth of a crop of irrigated groundnuts in India. The data were obtained as follows:

- *Potential evaporation* was calculated from meteorological information (Doorenbos and Pruitt, 1977). Alternatively these data are often available from the local meteorological service. For example, in the UK the Meteorological Office maintains a data base from which records of daily or weekly potential evaporation are available: contact The MORECS Unit, The Meteorological Office, Johnson House, Room JG2, London Road, Bracknell, Berkshire RG12 2SY.
- *The crop factor* takes into account the limited evaporation from bare soil between plants and is based on the extent to which the crop covers the ground. The value rises to approach 1.0 for full ground cover; values can be gauged from the extent of ground cover or published tables (Doorenbos and Pruitt, 1977). Factors influencing the crop factor include species, amount of ground cover, how frequently the soil surface is wetted and degree of senescence.
- Estimated evaporation was obtained by multiplying the potential evaporation by the crop factor. In this case it was assumed that water supply was not limiting, i.e. the limiting soil water deficit had not been reached, and so the estimated value is the *potential evaporation* of the crop.
- *Rainfall* was measured on site.
- *Soil water deficit*. The previous records showed that the deficit had reached 23 mm by the beginning of day 1. On day 1, 3.6 mm of water was lost and the deficit rose to 26.6 mm. On day 2 the deficit rose to 31.4 mm. On day 3, evaporation was only 0.6 mm more than the rainfall and so the deficit rose to 32.0. On day 5, the limiting deficit of 40 mm was exceeded, and so during the evening irrigation was applied which brought the soil to field capacity and added extra water for controlling salinity by leaching (Section 14.6). On day 9, rainfall was in excess of that required to bring the soil back to field capacity and 3.8 mm of water must have drained through the root zone.

The data illustrate the use of a soil water balance sheet as a basis for planning irrigation. Water is applied when the deficit reaches the limiting value, although not necessarily in amounts required to bring the soil back to field capacity. A further example for a crop in the UK is in Section 12.4 Project 6.

Section 12.3 The Bucket model for the soil water budget

The components of the soil water budget are discussed in Section 12.2 for an irrigated groundnut crop. To be of more general use for crops, vegetation and bare soil, estimates of actual evaporation need to be extended to soils with deficits between the limiting value and the permanent wilting point.

A simple procedure is illustrated in Fig. 12.7 where, for a soil fully covered by a crop, the ratio of actual evaporation (EVAP) to potential evaporation (EPOT) termed ERATIO is assumed to have a value of 1.0 between field capacity and the limiting soil water deficit. In drier soils, its value decreases linearly to become zero at the permanent wilting point. To use this model, a value is needed for the available water capacity of the root zone (RZAWC) and the limiting soil water deficit (LIMDEF). The former can be found from the available water capacity (Table 12.1) and rooting depth (Table 12.2). The limiting deficit can be found from RZAWC and the readily available fraction (Table 12.2). Then if the soil water deficit (SWD) at the beginning of a time period is known the value of ERATIO can be calculated. In the example shown in Fig. 12.7 the soil water deficit is 50 mm which is more than the limiting value of 35 mm. Using similar triangles the value of ERATIO is 0.57. Thus, when SWD < LIMDEF then ERATIO = 1 and when SWD > LIMDEF, then

$$\text{ERATIO} = (\text{RZAWC} - \text{SWD})/(\text{RZAWC} - \text{LIMDEF})$$

The actual evaporation for full crop cover is then given by EPOT × ERATIO. If the crop factor

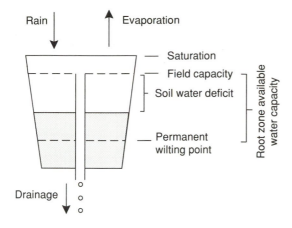

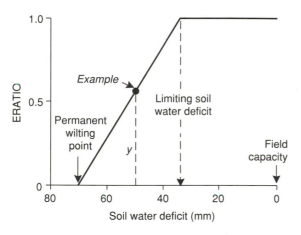

RZAWC = 70 mm
LIMDEF = 35 mm
Example:
$$\frac{y}{20} = \frac{1}{35}$$
$$y = 0.57$$

Figure 12.7 The Bucket soil water budget model.

(CROPFACT) is known (Section 12.2) then the actual evaporation is

$$CROPFACT \times EPOT \times ERATIO.$$

A bare soil is a special case. Evaporation will become restricted when only small amounts of water have been lost from the profile. A simple model is to assume that water is only lost from the upper 40 cm of soil, and that the limiting deficit is reached when only a quarter of the available water in this zone has been used. However, variations in hydraulic conductivity of soils (Section 5.5) make this an arbitrary assumption. Values between

0.3 and 1.0 are used as the crop factor for bare soils, high values being appropriate for cold wet conditions and low values for hot dry conditions (Doorenbos and Pruitt, 1992, Fig. 6).

The soil water budget at Niamey, Niger

A data set for the potential evaporation and the rainfall on a weekly basis is given in Table 12.5. It is possible to calculate manually the water budget for this site following the approach in Table 12.4 combined with the model to estimate the actual evapotranspiration. A BASIC computer program to aid these calculations is given in Appendix 3. It is available on a disk from the address given in Appendix 3. One further aspect has to be dealt with in the model. Whereas in Table 12.4 a starting value for the soil water deficit was known, Table 12.5 gives no such value. For this data set, the soil can be assumed to be at the permanent wilting point at the beginning of January because of the preceding dry months (i.e. SWD = RZAWC). This is not quite correct, but the program then runs through an annual cycle of calculations 10 times (average weather is assumed each year), so removing the influence of the initial choice of deficit. Even for sites where the soil is wet at the beginning of January, this procedure is satisfactory.

The balance sheet calculations in Section 12.2 are the basis of the program and can be written

> SWD this week =
> SWD last week + actual evapotranspiration − rainfall

When SWD becomes negative (i.e. the soil is wetter than field capacity), then water drains and field capacity is restored.

Using this program the data in Fig. 12.8 were calculated. At Niamey the dry period extends from November to April with soils close to the permanent wilting point. The model predicts that soil only reaches field capacity in late June and 37 mm of drainage pass beyond the 100 cm root zone. Not surprisingly, in this climate the actual evaporation follows closely the rainfall distribution.

Also included on the computer disk are files containing data sets for other sites around the world. The program can be used to compute data similar to Fig. 12.8 for each of these sites.

Applications of the Bucket model

Length of the growing season

The climate at Niamey imposes limitations on crop growth because of the limited rainfall. More specifically, Fig. 12.8 shows that the soil is above the limiting

Table 12.5 Average weekly rainfall and potential evaporation at Niamey, Niger (mm)

Week	EPOT	Rain	Week	EPOT	Rain	Week	EPOT	Rain	Week	EPOT	Rain
1	39	0	14	49	0	27	35	47	40	30	2
2	41	0	15	53	2	28	38	35	41	35	0
3	35	0	16	55	0	29	41	80	42	37	15
4	42	0	17	53	0	30	36	29	43	38	4
5	40	0	18	58	5	31	31	95	44	39	0
6	38	0	19	49	1	32	29	42	45	38	0
7	41	0	20	52	15	33	29	35	46	36	0
8	45	0	21	51	25	34	31	10	47	39	0
9	48	0	22	48	18	35	29	25	48	35	0
10	49	1	23	49	12	36	29	40	49	38	0
11	53	0	24	52	28	37	31	5	50	34	0
12	55	0	25	47	36	38	32	25	51	35	0
13	50	1	26	44	15	39	28	31	52	36	0

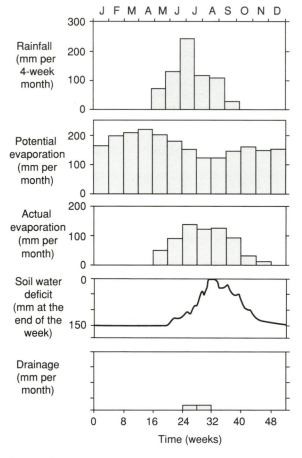

Figure 12.8 The soil water budget at Niamey, Niger, for a cropped area calculated using the Bucket model. Depth of root zone = 100 cm; available water capacity of the root zone = 150 mm; limiting soil water deficit = 60 mm; crop factor = 1.0.

deficit for a 12-week period. The length of the growing season is, however, longer than this and is normally considered to be the period during which the actual evaporation is greater than 50 per cent of the potential evaporation. The figure shows that this period is about 20 weeks, and running the computer program on a weekly basis tells us that the period is in fact 18 weeks. Thus the model can be used to predict the length of the growing season and help in the choice of suitable crops. This is one aspect of agro-ecology which looks at the matching of agricultural systems to ecological zones. Figure 12.9 shows the effect of length of growing sea-

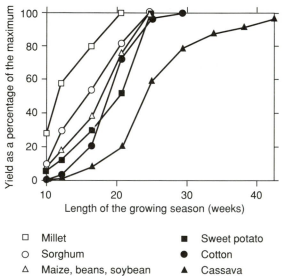

Figure 12.9 The effect of the length of the growing season on the yield of crops in tropical and subtropical regions. From FAO (1978).

son on the yield of crops commonly grown in tropical regions. In most years, conditions at Niamey are unsuitable for the growth of cassava and sweet potato. Only millet can be expected to give reliable yields in most years, and not surprisingly is the staple food crop of the area. Section 12.4, Project 2 uses the file data to examine further this problem.

Nitrate leaching

The model gives predictions of the times of the year when drainage occurs. If these coincide with the release of nitrate in the soil by mineralization or with fertilizer application, there is a likelihood that nitrate will be leached into groundwater. The model is unsophisticated in that it does not take account of where in the soil the nitrate is concentrated. Section 12.4, Project 3 looks at this problem, and in Sections 11.5 and 15.4 a more detailed model to predict leaching is presented.

Section 12.4 Projects

1. In pots of soil (Section 9.2) grow sunflowers with adequate water until the plants are well established. Then cover the pots to prevent evaporation and continue growing the plants until they wilt permanently (no overnight recovery). Measure the gravimetric water content (Section 3.3) and the soil water suction by the filter paper method (Section 5.2). Compare your results to the standard suction value of 1.5 MPa, and the water contents given in Tables 5.2 and 12.1.
2. Use the computer file data for Aleppo, Niamey, Kaduna and Port Harcourt to determine the length of the growing season making the same assumptions that were used for Fig. 12.8. Repeat the calculations for a crop which roots only to 50 cm.

3. Use the computer file data to predict the times of the year when drainage occurs under various climatic conditions. Make suitable assumptions regarding root zone depth and the limiting deficit. Compare the dry and wet years at Reading and Cairngorm. Using the information in Ch. 11, when should farmyard manure or fertilizer be applied and when should grassland be ploughed?
4. Use the data in Figs 12.5 and 5.3 to plot the soil water potential profile for 10 August in the bare soil as shown in Fig. 5.14. At what depth is the zero flux plane? Using Fig. 5.3 calculate the rate of water movement through the 150-165 cm layer. Using the same method determine whether the water potential gradient between 50 and 165 cm in the bare soil on 1 July is tending to move water up or down in the profile.
5. Figure 12.2 shows diurnal changes in evaporation rate. The values for the amount of evaporation per day given on p. 246 were calculated from the figure. Check the values by determining the areas under the curves. The area represents the integrated value of $mm\ h^{-1} \times h$ and is thus the evaporation in mm. The area can be determined by redrawing the graphs on squared paper and counting the squares under the curve, or more simply by preparing enlarged photocopies of the figure and cutting out with scissors the area under the potential evaporation curves, and a rectangle of paper $1\ m\ h^{-1} \times 48\ h$ (= 48 mm). Weigh each separately. The mass of paper is proportional to its area, with the mass of the rectangle representing 48 mm of evaporation.
6. Table 12.4 shows how irrigation planning can be based on estimates of evaporation from meteorological data. The computer disk (Appendix 3) gives a further set of data for a potato crop grown near Reading, UK (file Sonning.dat). Following the approach used in Table 12.4 calculate the water deficits and drainage during the 27 d period.

CHAPTER 13

Soil Fertility

The quest for increased and sustainable productivity to match population growth has been a central issue in agriculture for as long as crops have been grown. Its stark significance is seen today in areas of Africa suffering from drought. In countries with capacity for excess food production maintenance of soil fertility is a requirement for both economic and environmental viability of the farming system, with production matched to national needs and export demands. Scientists have attempted to develop the understanding and supply the information required to allow the production of sustainable increased yields through the breeding of new crop varieties, the use of fertilizers and lime, the development of chemical control of pests and diseases and the application of suitable cultivation and water management techniques.

The ability of soils to sustain increased production depends largely on the way in which their properties are altered as a result of cultivation. Although the pace of change in farming methods has quickened in the last 50 years and agricultural chemicals have added a new dimension, the basic problems listed below are the same as when man first began to cultivate soils:

- The removal of vegetation exposes soil to rain and wind and erosion may result.
- With time organic matter levels decrease, reducing soil stability and increasing sensitivity to erosion.
- Reduced organic matter contents and the associated reduction in microbial biomass cause a decrease in the amounts of nutrients mineralized each year.
- Reduced organic matter contents decrease the cation exchange capacity (CEC) of soils, so reducing their ability to hold nutrient cations against leaching.
- Removals of nutrients from soils by crops have to be balanced by inputs, either by natural processes or through the use of manures and fertilizers.
- Irrigation in dry regions may increase the salinity of soils.
- In wet regions the removal of vegetation increases leaching which contributes to the development of acidity.

The significance of agricultural and other chemicals in relation to sustainable production has three main aspects:

1. Fertilizers add nutrients in forms which are already present naturally. They change nutrient concentrations in the soil and may increase leaching losses but increase production and allow much greater removals without impoverishing the soil.
2. Pesticides and growth regulators add organic chemicals in forms which are not naturally present in soils. Their effects must be understood.
3. Disposal of waste materials on land adds domestic and industrial chemicals to soils. Of particular importance is sewage sludge which contains metals which are present in soils naturally in trace amounts but may be toxic to plants or animals at increased concentrations.

Apart from the need for agricultural production to be sustainable, it also has to supply food of satisfactory quality and must be acceptable in an environmental context. Although there is much discussion regarding intensive agricultural systems and food quality there is little evidence to suggest that this should be a matter of concern. However, any form of agricultural production must result in environmental change. Understanding the cause and degree of change is important. The salinization of large areas of the Middle East, the effects on soil fertility of extended periods of cropping in parts of the tropics traditionally used for shifting cultivation and erosion of the loessial soils of China are examples of major environmental changes which can lead to disasters. They are environmentally unacceptable because of damage to the land and because agricultural production is not sustainable. In comparison, the use of agricultural chemicals seems very safe although from the viewpoint of maintaining soil fertility little is yet known about their effects on constituents of the soil microbial biomass.

Changes in soil properties resulting from man's use of soil fall into two broad categories. Irreversible changes have a different significance to those which can be reversed. For example, erosion and metal pollution are irreversible, whereas decreases in organic matter content due to intensive cropping are reversible.

POTENTIAL PRODUCTIVITY

The potential yield of food crops depends first of all on the genetic capacity of the plant. For example, both wheat and rice can produce 14.5 t grain ha^{-1} under ideal conditions but maize can produce 22 t ha^{-1}. To produce these yields optimum climatic and soil conditions are required and the crop must be free from pests and diseases. However the average world production of these crops in 1977–79 was 1.78, 2.61 and 3.09 t ha^{-1} respectively. Identifying the factors which cause low average yields is essential if forecasts of total agricultural production and the world population that this could support are to be realistic.

Figure 13.1 shows the range of factors that determine yield. They can be broadly grouped under the headings of climate, soil fertility, pests and diseases, and management.

- In the absence of other limitations, climate will impose a limit directly through sunlight energy input or air temperature. As an example, potential maximum yields of cereals are higher in Scotland than in southern England; the increased day length at higher latitudes during the summer growth period supplies a larger input of energy. Even greater yields are possible at higher altitudes in Turkey because of high light intensities.

- Climate also controls soil temperature which affects growth rates.

- Rainfall inputs and potential evaporation rates control the water supply to crops which is a major yield-limiting factor.

- Physical characteristics of soils impose limitations on growth through water supply, aeration and root distribution. These properties are interrelated.

- Chemical characteristics of soils impose limitations through the availability of nutrients and toxic elements.

- Biological processes are involved in making nutrients available, for example by mineralization of organic matter and N fixation.

- Limitations are imposed by pests and diseases both in the soil and on the above-ground parts of the crop. Soil-borne pests and diseases can be as important as nutrient availability and soil physical properties in controlling yield.

- Management can have direct effects on soil fertility and pests and diseases, but also influences yield through timeliness of cultivations, sowing, fertilizer and pesticide applications and irrigation.

RESPONSES AND INTERACTIONS

Responses

The presentation of yield data in the form of a response curve has been shown in Fig. 9.5. They are either exponential with response decreasing as yields approach a maximum, or have two approximately straight line sections again with decreasing slope as the maximum is approached (Cooke, 1982). The optimum economic input of a nutrient is normally less than the input required for maximum yield (Ch. 9 and Section 11.4).

Interactions

In Fig. 13.1 vertical bars have been marked on the right-hand side to indicate links between various factors. For example, climate links with soil physical factors in affecting soil temperature and water supply, and soil physical properties link with chemical properties in controlling nutrient availability. Thus it is not surprising that when one constraint to growth is removed, the crop responds by making better use of light, water and nutrients even though these factors may not have been changed. A simple example would be differences in

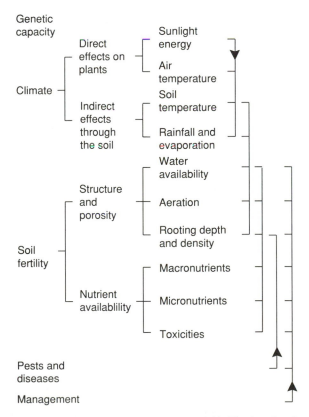

Figure 13.1 Factors influencing crop yield. The bracket [indicates component factors and] indicates direct links between factors. The arrows indicate a one-way effect.

growth between a cold and a warm climatic region. The crop will take up more nutrients when the constraints of low temperature are removed.

Although links between factors may be termed interactions within the general meaning of the word, it is used in a special way in relation to the response of crops to increased supply of nutrients and water, or to the removal of constraints resulting for example from soil acidity or compaction. *An interaction occurs when the gain in yield resulting from changing two factors together is greater (or less) than the sum of the gains (or losses) caused by changing them separately.* The term is also used in statistics with an even more specific meaning.

Probably the most commonly occurring interaction is between water and N. An example is shown in Fig. 13.2. Winter wheat was grown in Nebraska with the amount of available water adjusted before planting, and with four rates of applied N. When no available water was present at the time of planting the crop relied on subsequent winter rainfall and the response to N was small with a depression in yield at high N applications.

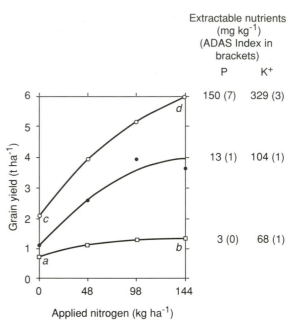

Figure 13.3 Interaction between nitrogen and residual phosphorus and potassium. Hoosfield spring barley experiment, Rothamsted.

Without N there was a moderate response to water. The response to N and water together is much larger than the sum of the separate responses: this is a *positive interaction*.

The Hoosfield spring barley experiment at Rothamsted (Colour Plate 19) provides an example of interaction between nutrients. Applications of P and K^+ over many years have produced plots with low, medium and high fertility in terms of available P and K^+. In 1968 these plots were split and treated with four levels of N and the yields obtained are shown in Fig. 13.3. The crop responds to N and also responds to the *residual* P and K^+ in the soil; any reponse to a nutrient applied in a previous year is known as a *residual effect*. The graph shows a positive interaction between N and residual P and K^+. For example, the response to $144\,kg\,N\,ha^{-1}$ in the low fertility soil is $0.60\,t\,ha^{-1}$ ($b - a$) and the response to residual P and K^+ in the soils with no applied N is $1.34\,t\,ha^{-1}$ ($c - a$). Together the response is $5.27\,t\,ha^{-1}$ ($d - a$) which is $3.33\,t\,ha^{-1}$ more than the sum of the separate responses. In practical terms, a worthwhile response (Section 11.4) is only obtained with $48\,kg\,N\,ha^{-1}$ on the low fertility soil. On the medium fertility soil $96\,kg\,ha^{-1}$ could be applied, and on the high fertility soil $144\,kg\,ha^{-1}$ would be worth while.

Two important implications follow from the occurrence of interactions:

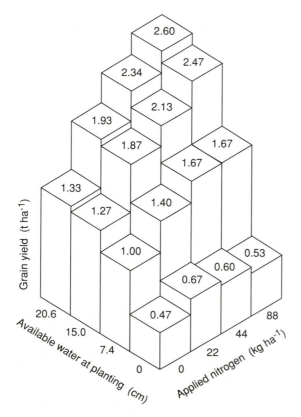

Figure 13.2 Interaction between water and nitrogen. Winter wheat grown in the central Great Plains of Nebraska. Mean values for 3 years. From Ramig and Rhoades (1962).

1. To obtain high yields, all the growth requirements of a crop have to be considered. It is not efficient to use large inputs of one nutrient if the ablility of the crop to respond is limited by other factors. An example occurs in the production of cereal crops on thin soils over chalk in southern England. It is difficult to maintain adequate potassium availability probably because of leaching losses (Section 9.5), and the ability of the crop to respond to N may be reduced.

2. Where there is overproduction a reduction in yield may be necesary. Similarly an environmental hazard associated with inputs of agricultural chemicals may require a reduction in inputs. Taking again the case of cereals on chalk soils, a reduction in yield could be obtained by reducing inputs of either K^+ or N. If leaching of N is a hazard, there are advantages in maintaining inputs of K^+ so that the minimum N inputs can be used for a given yield.

Table 10.2 shows another experiment in which responses were measured. In this case there was a *negative interaction* resulting from the application of lime and P. The response to the treatments together was less than the sum of the separate responses because liming simply removed the constraints caused by acidity and allowed the crop to use the P that was already present in the soil. The effects of weeds on the response of a crop to N is another example of an interaction (Table 15.3). There is clearly a need for a combined approach to the management of the factors controlling yield.

Section 13.1 describes pot and plot techniques to study responses and interactions.

Conclusions

Crops respond to the addition of a nutrient when its availability is low. The characteristics of the response curve will depend on the crop, the climate and on the many soil properties which together control soil fertility. Interactions normally occur as constraints are removed. The optimum application of a nutrient can be considered from three viewpoints:

1. In a scientific experiment it may be important to determine the maximum potential yield. In this case the optimum input is that required to give maximum yield.

2. Farming is a business and economics are important. From this viewpoint the optimum input is that required to satisfy market demand and to give maximum profit and is usually less than that required to give maximum yield.

3. Because soil and land are arguably the most important natural resource of any country, a production system must maintain them in good condition. It

must also produce food of acceptable quality. Thus there may be an optimum input to meet environmental criteria which may be less than the economic optimum.

SOIL FERTILITY AND ITS MAINTENANCE

The meaning of the term *soil fertility* is seen in Fig. 13.1. It includes all those soil factors which influence growth and impose limitations on yield. An area of land under natural vegetation has an inherent fertility. Its use for crop production will change its fertility, but the various factors influencing growth will be changed in different ways. For example, nutrient availability can be improved relatively easily by applying fertilizers, water availability which depends on the volume of storage pores (Table 4.2) cannot easily be changed, and root penetration and aeration which depend on transmission pores can be decreased by compaction or improved by careful cultivation (Section 4.4).

Organic matter

Soil organic matter is central to the maintenance of soil fertility: mineralization of N, P and S, the soil's ability to hold nutrient cations, structural stability and water-holding capacity are all affected by organic matter content. Because nutrients can easily be supplied by the use of fertilizers, the extent to which the natural fertility of a soil should be allowed to fall is a question that has been of much interest since fertilizers first became widely available in the middle of the nineteenth century. Is there a natural fertility level and an associated critical organic matter content for a given site below which problems of soil management and reduced yield become unacceptable?

Work at Rothamsted Experimental Station, begun by J. B. Lawes in 1843, has made a major contribution to understanding the way soils respond to reduced organic matter contents. The classical Broadbalk Experiment (Colour Plate 19) has compared the effects of organic and inorganic inputs for winter wheat grown each year since 1843, and details are given in Section 13.2. With fertilizer inputs, soil organic C contents are now between 0.84 and 1.04 per cent compared to 2.59 per cent where farmyard manure has been applied each year. However, fertilizers maintain yields close to those of the farmyard manure plots. Although the natural fertility of the soil diminished initially in terms of its ability to supply N by mineralization, fertility (and organic matter content) has stabilized over the last 100 years. There has been some deterioration in soil structure. Both these changes are reversible when sites are returned to grass crops, with soil organic matter increasing slowly towards its original content

(Johnston, 1992). Continuous arable cropping causes more serious structural deterioration on silty and poorly drained soils (MAFF, 1970). An example of soil structural changes due to cultivation is shown in Colour Plate 5.

Organic farming systems

Systems which rely on organic inputs for crop production normally use a combination of crop rotations, animal manures and certain inorganic compounds with low solubility which are considered to be 'natural'. Fertilizers with higher solubility are considered to be 'artificial' and are not used (Section 13.3). In these systems organic matter has a central role in maintaining soil structure and supplying plant nutrients.

The yield-limiting factor in organic systems is normally N supply, and legume–grass leys are grown to make use of N fixation by *Rhizobium* bacteria (Section 13.4). When the ley is ploughed mineralization releases more nitrate than crops need in the autumn and large leaching losses may occur in the winter unless cover crops are used. However, in these mixed farming systems leys are ploughed each year on only part of a farm area. Thus the concentration of nitrate in the water draining from a catchment area will be smaller than from the fields where the ley has been ploughed. In the following spring as soil temperature rises, mineralization rates increase approximately in phase with crop demand. However, release may not occur early enough in the season to produce maximum cereal yields.

Discussions about the relative merits and disadvantages of so-called organic and inorganic farming systems include many questions which are outside the realm of soil science. There is little doubt that the natural fertility of the soil is normally better in organic systems; this does not, however, mean that the use of fertilizers is unacceptable. From a scientific viewpoint it is sensible to use systems which maintain soils in good condition and also take advantage of fertilizers; the two are completely compatible. On stable soils such as those at Broadbalk, well managed continuous arable cropping with fertilizers is an acceptable production system. On soils which develop poor physical conditions, crop rotations are needed combined with fertilizer use to maintain good soil conditions and high yields.

Shifting cultivations

The agricultural system known as *shifting cultivation* is perhaps the most important example of an organic farming system. It was practised for centuries over large parts of the world and is still used in parts of the tropics. It involves clearing an area of forest or savanna by burning and then growing crops for a few years before allowing natural vegetation to regenerate. Soil fertility decreases during the cropping period and increases during the much longer period under natural vegetation. Section 13.5 gives details of the changes in soil properties during this cropping cycle. Increased population pressures and an extended cropping period have led to a breakdown of this system in many areas with the formation of infertile and eroded soils.

The rain forests

The consequences of clearing rain forest for agricultural use have implications well beyond the soil. However, within the soil fertility gradually decreases, but unlike shifting cultivation there is often no intention of regenerating the forest and future agricultural production depends on maintaining fertility through the use of grassland and fertilizers. Particularly at risk are the light-textured soils in which CEC is primarily associated with organic matter (Section 13.5). As this decomposes (rapidly under humid tropical conditions) the soil's ability to hold nutrient cations decreases. Leaching losses are large, and the soil may become more susceptible to erosion. If soils from these regions must be used for agriculture, successful management will have to take account of differences in soil properties with those soils most at risk being maintained in forest or used for grass crops.

Soil fertility classification

Soil fertility comprises so many soil factors that classification is difficult. Assessments of the suitability of land for particular uses include simple classifications (Dent and Young, 1981).

FURTHER STUDIES

Calculations are in Section 13.6.

Section 13.1 Pot and plot techniques to study responses and interactions

Techniques for pot experiments have been discussed in Section 9.2. These may be used to study responses and interactions provided soils with low availability of nutrients can be obtained. Note the following:

- Most agricultural soils are well supplied with P and K^+. Responses in pot experiments are unlikely. Woodland soils may be suitable, but an availability index should be measured before time is spent setting up an experiment (Sections 9.1 and 10.3).

- Sampling, air drying and sieving a soil normally result in a flush of mineralization when the soil is rewetted. Nitrogen supply may therefore be adequate at least initially. For a study of response to N use a light-textured soil with small organic matter content (a continuous arable site, nitrogen index 0, Section 11.4) and sample at the end of the growing season.
- Overwatering can easily cause denitrification resulting in poor responses to applications of N.

A pot experiment to determine a response curve for nitrogen

Following the techniques of Section 9.2, choose levels of N up to $0.12\,g\,N\,kg^{-1}$ ($100\,kg\,ha^{-1}$ pot area basis) and one higher application ($0.25\,g\,N\,kg^{-1}$) together with lime, P and K^+ as required. Ryegrass is an easy crop to use. Replication is less important here than in experiments where the effect of a single treatment is being determined. For a given number of pots it may be more useful to increase the number of levels of N rather than replicating each treatment, because in fitting a curve to the data, replication is in effect built into the experiment.

The minus-one technique

This type of experiment is used to identify which nutrients are likely to limit growth in a soil. It provides preliminary information before setting up more detailed pot or plot experiments, and has proved particularly useful when areas of land are first brought into agricultural use. Pots of soil are treated either with all nutrients (the control) or with each one omitted in turn. If the omitted nutrient is naturally deficient, growth will be depressed compared to the control. This is a more effective technique than adding one at a time because if several nutrients are deficient the response to one is limited by the others.

There are many ways in which minus-one experiments can be set up. That described here examines N, P and K^+ separately and the other nutrients together, but any nutrient can be isolated by modifying the inputs.

Nutrient solutions

Nutrients are added as mixtures in solution. If each nutrient were added separately another ion would be required in each case, e.g. to add potassium, KCl might be used and an unnecessary excess of chloride would be added.

The following solutions should be prepared. In each case the masses of reagents are followed by the volume of distilled water in which the reagents are dissolved.

Analytical quality reagents should be used as less pure substances will supply micronutrients.

Solution 1: $21.1\,g\,Ca(NO_3)_2.4H_2O$, $18.0\,g\,KNO_3$, $6.3\,g\,NaH_2PO_4.2H_2O$, 500 ml

Solution 2: $15.5\,g\,K_2SO_4$, $6.3\,g\,NaH_2PO_4.2H_2O$, 500 ml

Solution 3: $21.1\,g\,Ca(NO_3)_2.4H_2O$, $18.0\,g\,KNO_3$, 500 ml

Solution 4: $21.1\,g\,Ca(NO_3)_2.4H_2O$, $15.1\,g\,NaNO_3$, $6.3\,g\,NaH_2PO_4.2H_2O$, 500 ml

Solution 5: $5.1\,g\,MgSO_4.7H_2O$, 250 ml

Solution 6: $9.9\,g\,CaCl_2.6H_2O$, 250 ml

Solution 7: $300\,mg\,H_3BO_4$, $200\,mg\,MnSO_4.4H_2O$, $20\,mg\,CuSO_4.5H_2O$, $20\,mg\,ZnSO_4.7H_2O$, $3\,mg\,(NH_4)_6Mo7O_{24}.2H_2O$, 1000 ml

Soil preparation

The pot size can be varied depending on how much soil is available. The details which follow apply to 0.5 kg of soil. The solution volumes should be adjusted if the mass is different. However, because an excess of each nutrient is used simple pipette volumes can be retained, i.e. 5, 10, 20, 50 ml, even though the inputs are altered slightly.

Spread 0.5 kg batches of soil on to polythene sheets. On to each 0.5 kg of soil distribute the appropriate volumes of the required nutrient solutions (Table 13.1) and mix well. Solution 7 should be added to the saucers in which the pots stand. In each case the solution number and volume are given.

These solutions supply the following amounts of nutrients (mg per pot): N 200, P 50, K^+ 280, Ca^{2+} 140, Mg^{2+} 20, S 30, B 1, Mn^{2+} 1, Cu^{2+} 0.1, Zn^{2+} 0.1, Mo 0.05. These applications have been chosen to be adequate for plant needs and for convenience in making mixed solutions.

Place a glass fibre filter disc in the base of each pot and fill with the treated soil. Place in a saucer. Prepare three replicates per treatment.

Table 13.1 Volumes of solution (ml) needed to achieve the required nutrient additions

Treatment	Solution number						
	1	2	3	4	5	6	7
All nutrients	20	0	0	0	10	0	20
Minus N	0	20	0	0	10	10	20
Minus P	0	0	20	0	10	0	20
Minus K^+	0	0	0	20	10	0	20
Minus Mg^{2+}, S and micronutrients	20	0	0	0	0	0	0

Plant growth

Ryegrass is a suitable test crop. Plant, water and harvest as directed in Section 9.2. If small amounts of soil are used, e.g. 250 g, grow the plants for about 3 weeks and then repeat the addition of nutrients, this time directly on to the soil in the pots.

Express the mean yield for each treatment as a percentage of the yield of the treatment with all nutrients added.

Field plot techniques

Field scale trials are described by Dyke (1974). However, details are given here of a simple plot experiment that can be set up on any area of grass provided that it can be fenced to avoid treading. The aim of the experiment is to show for a dry part of the country the response of grass to both N and water, and to determine whether an interaction occurs. Either treatment could be replaced by another; for example, in a wet area, the water treatment should be replaced by phosphorus and potassium applied together (see the alternative treatments below).

The plots

An area approximately 8×5 m is required (Fig. 13.4). The mixture of grass and other plants should be reasonably uniform over the area. Cut the grass. Mark out a block containing 12 plots (Note 1), each 1×1 m, with a path between each plot which is the width of the mower available for grass cutting; the mower should have a box to catch the grass cuttings. The corner of each plot can be marked with a peg pushed through a white disc. They should be driven into the ground so that the mower will pass over them. The area should be fenced

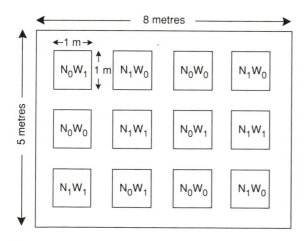

Figure 13.4 A randomized block of 12 plots.

to prevent treading, and the paths should be used when working on the plots. Mow the paths as necessary.

Treatments

With and without added nitrogen: N_1 and N_0
With and without added water: W_1 and W_0
All plots receive P and K^+

The treatments are N_0W_0, N_1W_0, N_0W_1, N_1W_1, with three replicates per treatment requiring 12 plots.

Nitrogen Using applications which might be given to productive grassland, $100 \, \text{kg N ha}^{-1}$ should be applied at the beginning of April, and repeated after each cutting at 4–6 week intervals. The application is $100 \times 1/10^4$ kg per plot which is 10 g per plot and can be supplied as 29 g NH_4NO_3. This should be dissolved in 5 l of water and applied uniformly over the plot using a watering can and a fine rose. Wash in the solution with another 5 l of water.

Water When the nutrients are applied in solution, all plots should receive an equal amount of water. The water treatment can be based on the information in Table 12.1. Alternatively, they can be estimated. For example in southern England in dry weather in the summer, grass will transpire up to 5 mm of water d^{-1} from a moist soil. Thus the volume of water that could be lost from a plot is

$$5 \times 10^{-3} \, \text{m} \times 1 \, \text{m}^2 = 5 \times 10^{-3} \, \text{m}^3$$

or

$$5 \, \text{l d}^{-1} \, (1 \, \text{m}^3 = 1000 \, \text{l})$$

or about 35 l per week. (A 2 gallon watering can holds about 9 l.) The difference between estimated losses and rainfall gives the approximate soil water deficit (Section 12.2).

Apply water uniformly over the W_1 plots at 2 weekly intervals and after cutting to reduce the soil water deficit to between 25 and 50 mm.

Phosphorus and potassium The amounts of P and K^+ applied to grassland depend on how much is returned by grazing animals. In this plot experiment all the grass is being removed and your soil may initially have low P and K^+ availability. Grass takes up about the same amounts of K^+ and N. All plots should receive $100 \, \text{kg K}^+ \, \text{ha}^{-1}$ when N is applied. If K^+ is applied as K_2HPO_4, 40 kg P is also added; this is probably more than is needed but is convenient. Apply to each plot 5 l of solution containing 22 g KH_2PO_4. This can be dissolved with the NH_4NO_3 for application to the N_1, plots or dissolved alone for the N_0 plots.

Harvesting

At approximately 1 month intervals, cut the grass to about 2 cm above the ground from a 0.75×0.75 m square in the centre of each plot (Note 2). Dry at 100 °C and weigh the dry matter. When all plots have been harvested, mow the whole block removing the grass and reapply the nutrients.

Data

After each harvest draw a graph of mean yields (g dry matter per plot) against N application (Fig. 13.5). Around each mean value mark the replicate yields, and the 95 per cent confidence limits of the means (Section 3.8). At the end of the growing season draw a graph of cumulative (total) yield for each plot in the same way.

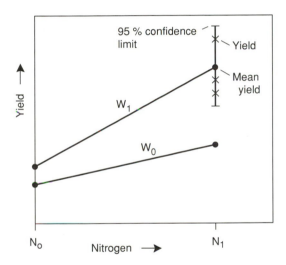

Figure 13.5 Nitrogen–water interactions.

Note 1 Replication is important even though it increases the work required to carry out the experiment. Unlike a pot experiment where soil is homogenized to ensure uniform pot conditions, field soils are variable (Section 1.3). Also in this experiment the natural variability of plants over the block must be accepted. Variability is reduced if plot size is increased. The replicate plots for the treatments are randomly distributed over the block: hence the name *randomized block*.

Note 2 The edges of each plot may be influenced by the different soil conditions of the adjacent paths. This is known as an *edge effect*. Hence the area used to estimate the yield is around the centre of each plot.

Alternative treatments

N, PK interaction In wet areas replace the water treatment with a potassium and phosphorus treatment. Instead of W_1 and W_0 in the design use P_1K_1 and P_0K_0. The treatments are $N_0 P_0K_0$, $N_1 P_0K_0$, $N_0 P_1K_1$ and $N_1 P_1K_1$ and apart from the water needed to apply the nutrients, no further water should be added.

The effects of clover If the plots can be maintained for 2 or 3 years, the role of clover in supplying N can be studied. Choose a clover-free area or use a selective lawn weedkiller to clear clover and broad-leaved weeds. The following treatments can be used, with three replicates of each.

N_0: no fertilizer, no clover
N_1: 25 kg N ha^{-1} initially and after each cut
N_2: 100 kg N ha^{-1} initially and after each cut
C: no fertilizer, with clover

After the initial mowing of the block, rake each plot to loosen the soil surface. On to each C plot sprinkle 5 g of clover seed and rake again. Tread the surface of all the plots. Add P and K^+ each spring by sprinkling 40 g of KH_2PO_4 over the surface of each plot and water. Each summer use a selective herbicide on the plots without clover. Draw a graph of annual yield against added N. From the graph determine the fertilizer-N input required to give the same yield as the grass–clover plots. This is a rough estimate of the amount of N fixed by the clover (see also Section 13.4).

Sward composition Again if the experiment can be maintained for a few years, the influence of nutrient inputs on sward composition can be studied. This is similar to the experiment set up by Lawes and Gilbert in 1856 at Rothamsted which is known as Park Grass. Ideally the plot size should be increased to several m^2. Replication is not needed if the experiment is primarily for observation purposes.
The following treatments are suggested:

1. Ammonium sulphate without lime: 47 g $(NH_4)_2SO_4$ m^{-2} applies 100 kg N ha^{-1}. This should be applied each year in the spring.
2. Ammonium sulphate with lime: as above plus 0.5 kg $CaCO_3$ m^{-2} which applies 5 t ha^{-1}. The $CaCO_3$ should be applied at the beginning of the experiment.
3. Sodium nitrate: 61 g $NaNO_3$ m^{-2} applies 100 kg N ha^{-1}. This should be applied each spring.
4. Phosphorus and potassium: 22 g K_2HPO_4 m^{-2} applies 100 kg K ha^{-1} and 40 kg P ha^{-1}. This should be applied each spring.
5. Treatments 1 and 4 combined.
6. Treatments 2 and 4 combined.

Treatment 1 has a strong acidifying effect (due to the nitrification of ammonium and the leaching of sulphate) which is neutralized by $CaCO_3$ in treatment 2. Treatment 3 causes no acidification provided all the nitrate is taken up and should give similar yields to treatment 2.

The plots should be cut after flowering and seeding, and the grass allowed to dry on the plot so that seeds are left when the grass is removed. For details of the Park Grass experiment see Rothamsted Experimental Station (1970, 1977b). A slide set and accompanying guide is available from the Librarian, Rothamsted Experimental Station, Harpenden, AL5 2JQ.

Section 13.2 The Broadbalk Experiment

When J. B. Lawes and J. H. Gilbert began their experiments 150 years ago arable crops were produced in the Rothamsted area following a four (or five) course rotation of turnips, barley, clover, wheat, (barley) with the clover and farmyard manure maintaining soil fertility. The yield of wheat was about $1.5\,t\,ha^{-1}$.

The soil is a non-calcareous silty clay loam formed in a clay-with-flints parent material over Upper Chalk. Plots were laid out and treatments begun in 1843 and wheat has been grown every year since then on part of each plot (Colour Plate 19). Over the years the main nutrient treatments have been unchanged, but husbandry practices have altered and new varieties have been introduced. The last major modifications were made in 1968 and a high N plot was introduced in 1985. Table 13.2 summarizes those treatments which will be discussed here. Fig. 13.6 shows the average yields produced between 1979 and 1984 when the yield potential

Table 13.2 Treatments in the Broadbalk Experiment

Cropping:	wheat grown continuously or after a 2-year break to control soil borne pathogens. Weeds, pests and diseases controlled by agrochemicals
Manure plot:	$35\,t\,ha^{-1}$ of farmyard manure applied in the autumn. This contains about 225 kg N, 45 kg P and 145 kg K^+
Fertilizer plots:	P, K^+, Mg^{2+} and lime applied in the autumn, N in the spring
Phosphorus:	35 kg P ha^{-1} as triple superphosphate (monocalcium phosphate)
Potassium:	90 kg K^+ ha^{-1} as potassium sulphate
Magnesium:	30 kg Mg^{2+} ha^{-1} as magnesium sulphate
Nitrogen:	48, 96, 144, 192, 240 and 288 kg N ha^{-1} as ammonium nitrate
Lime, as required to maintain pH 7.0	

From Rothamsted Experimental Station (1969 and 1991).

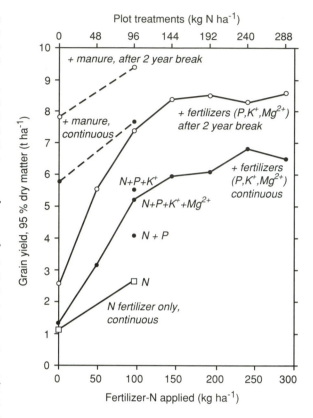

Figure 13.6 Responses of winter wheat (variety Brimstone) to manure and fertilizers in the Broadbalk Experiment, Rothamsted, 1985–90. From Rothamsted (1991).

of the crop was protected by the use of fungicides against foliar pathogens. The main findings are as follows:

- Continuous wheat with no inputs for 140 years produced $1.2\,t\,ha^{-1}$, being about the same as the yield in 1843. The crop has drawn on reserves of P, K^+, Ca^{2+} and Mg^{2+} in the soil, on atmospheric inputs of S, and on N fixed by soil bacteria and supplied from the atmosphere.
- With adequate P, K^+ and Mg^{2+}, application of fertilizer-N raised the yield from 1.3 to $6.5\,t\,ha^{-1}$ and the manure plot gave $5.8\,t\,ha^{-1}$. Where extra fertilizer-N was applied with the manure, yield rose to $7.7\,t\,ha^{-1}$. It is not easy to identify the cause of the extra yield from manure and fertilizer together compared to fertilizer alone. Possible reasons include better soil physical conditions which enhanced the supply of water in dry years and maybe a more uniform supply of N through the growing season.
- The figure also shows that higher yields were produced following a 2-year break from cereals which

Table 13.3 The effects of management on the properties of Broadbalk soils

Plot	N in the 0–23 cm layer† (%)			
	1865	1944	1966	1987
Zero inputs	0.105	0.106	0.099	0.102
Zero N, + P K$^+$ and Mg^{2+}	0.107	0.105	0.107	0.104
96 N, + P K$^+$ and Mg^{2+}	0.117	0.121	0.115	0.124
Manure	0.175	0.236	0.251	0.270

	C (1966)† 0–23 cm (%)	Biomass −C‡ 0–23 cm (µg g^{-1})
Zero inputs	0.84	158
Zero N, + P K$^+$ and Mg^{2+}	0.91	—
96 N, + P K$^+$ and Mg^{2+}	1.00	—
144 N, + P K$^+$ and Mg^{2+}	1.04	190
Manure	2.59	342
	C% × 1.72 ≃organic matter %	

	Bulk density§ 0–23 cm (g cm^{-3})	Available water capacity§ 0–25 cm (mm H$_2$O)
Zero inputs	1.44	30
Zero N, + P K$^+$ and Mg^{2+}	1.44	30
144 N, + P K$^+$ and Mg^{2+}	1.44	30
Manure	1.32	45

† Rothamsted Experimental Station (1969 and 1991).
‡ Brookes *et al.* (1984).
§ A.E. Johnston, personal communication. The available water capacity was determined between 5 kPa and 1.5 MPa soil water suction (Section 5.2). See also Salter and Williams (1969). The bulk density is the mass of <2 mm soil cm^{-3}.

reduced the levels of soil-borne pests and diseases. Farmyard manure plus fertilizer-N again outyielded the fertilizer plots.

- The omission of Mg^{2+} from the continuous wheat plots has no significant effect on yield. At N = 96 kg ha^{-1} the omission of Mg^{2+} and K$^+$ reduces the yield from 5.3 to 4.1 t ha^{-1}, and the omission of Mg^{2+}, K$^+$ and P reduces the yield to 2.7 t ha^{-1}. Soil reserves of these nutrients are still adequate to maintain a moderate yield. Note that there is a clear interaction between N and these nutrients.

The nitrogen budget

The Broadbalk Experiment is unique and is still supplying new information of considerable interest, for example in relation to the efficiency of N use, where the well-documented history of the site is of great importance. Constant management since 1843 has produced plots in which total soil-N has now reached stable levels (Table 13.3). This means that in the annual N budget there is no net mineralization or immobilization. Uncertainty in measuring changes in total soil-N which hinders the interpretation of modern experiments is therefore of no consequence in considering the Broadbalk data. The N budget for the crops grown in 1980 and 1981 was analysed in detail by Powlson and others (1986) using isotopically labelled fertilizer-N. Their findings are shown in Table 13.4. The following conclusions can be drawn:

- With low inputs of N the crop removes more N than that applied as fertilizer. However, non-fertilizer inputs (from the atmosphere, in the seed and by biological N fixation) are about 45 kg ha^{-1}a^{-1}, and losses by leaching and denitrification occur from all plots, ranging from 9 to 70 kg ha^{-1}a^{-1}. The non-

Table 13.4 The nitrogen budget for the Broadbalk Continuous Wheat Experiment. Mean values for the 1980 and 1981 seasons (kg N ha^{-1}a^{-1})

(a) Fertilizer plots

Fertilizer−N applied	N removed in the grain and straw	Calculated input of non-fertilizer N	Calculated loss of N
0	26	35	9
48	62	36	22
96	114	51	33
144	145	55	54
189	169	50	70
		Mean = 45	

(b) Manured plot

Manure−N applied	N removed in the grain and straw	Assumed input of non-manure N	Estimated loss
225	145	45	125

fertilizer inputs calculated from ^{15}N measurements are greater than estimates using other methods (Table 11.1).

- Input of N in the seed is $4\,kg\,ha^{-1}a^{-1}$. There are still uncertainties regarding the amount of N received from the atmosphere. Colour Plate 17 gives a value of $20{-}24\,kg\,ha^{-1}a^{-1}$, but these estimates are being revised upwards, and Rothamsted measurements indicate a value of $30{-}35\,kg\,ha^{-1}a^{-1}$, leaving $6{-}11\,kg\,ha^{-1}a^{-1}$ added by biological fixation.
- The apparent increase in non-fertilizer input on the high N plots may not be significant.
- The manure plot was not included in Powlson's study. Table 13.4 shows an estimated loss of $125\,kg\,ha^{-1}a^{-1}$, even greater than the loss from the fertilizer plots. To supply the N needs of the crop by mineralization of organic N, there will inevitably be a large mineralization flush in the autumn which, combined with mineral-N in the fresh manure, gives the nitrate which is lost by winter leaching. Thus an organically produced crop which relies on manure may be less efficient than a fertilized crop in the use of N.

Where autumn sown crops are not grown, a catch crop sown in September and ploughed in before planting spring cereals may prevent some of the leaching loss.

Soil properties

Table 13.3 shows the total N content of the Broadbalk soils. The C:N ratio of all plots is about 10:1 and so these figures also indicate the changes in C content. The C contents measured in 1966 are also shown. The manure plot has about three times as much organic matter as the zero input plot and the microbial biomass is approximately doubled. Note that fertilizer additions increase both yield and crop residues, so increasing the C content and biomass above the values for the zero plot.

Soil structure is better in the manure plot as indicated by a smaller bulk density and larger available water capacity. The soil is darker in colour which means that it could warm up slightly more quickly in the spring. There is a surprising lack of information on changes in soil physical properties for this site, reflecting the difficulties in making measurements of soil structure which are relevant in terms of crop growth. However, there is no evidence that structural deterioration has influenced crop yields significantly.

For further details of the Broadbalk Experiment see Rothamsted Experimental Station (1969, 1970, 1977a, 1977b, 1983, 1991). A slide set and accompanying guide is available from the Librarian, Rothamsted Experimental Station, Harpenden AL5 2JQ.

Section 13.3 The effects of management on soil organic matter and issues raised by organic farming systems

The effect of a farming system on soil organic matter content is a good indication of the change in soil fertility produced by that system, particularly in relation to soil structure and potential N supply by mineralization. However, sites under the same management will develop different organic matter contents because of differences in soil texture, drainage and climate.

The ADAS representative soil sampling scheme, 1974–83

Under this scheme fields representative of different farming systems are sampled each year. The soils are allocated to classes depending on their organic matter content and Table 13.5 shows the percentage of the

Table 13.5 Organic matter in soils under different cropping systems in England and Wales. The values given are the percentage of the fields in each organic matter class

Cropping system	Organic matter (%)					
	<2	2–5	5–8	8–10	10–13	>13
Arable	11	67	15	3	2	2
Ley–arable	4	56	31	6	2	1
Continuous grass	—	18	40	23	13	6

From Church and Skinner (1986).

fields in each class. The management of the fields was unchanged for 10 years, and the organic matter contents were almost constant over this period. Most arable soils had organic matter contents between 2 and 5 per cent. An increased proportion of the soils in ley–arable rotations had between 5 and 8 per cent. Soils in continuous grass had even higher contents. The spread of values within any system primarily reflects differences in texture, the heavy-textured soils tending to protect the organic matter (Table 3.4 and Section 3.5). Organic matter contents are also dependent on climate: the data are for the cool temperate conditions of England and Wales.

The Rothamsted and Woburn classical experiments

The Norfolk four-course rotation was the basis for

maintaining fertility in arable soils in the middle of the nineteenth century when the Rothamsted and Woburn experiments on arable crops started (1843 and 1876 respectively). Organic matter contents would tend to be low and were maintained by occasional dressings of manure (once per rotation) and by the weeds ploughed into the soil in late autumn. Other fields on the farm would be in permanent grass and would have high organic matter contents. Sir George Stapledon was one of the first to suggest that soil fertility could be enhanced if arable and grass followed one another on the same piece of land. Such systems became known as ley–arable farming, and the first experiments were started at Woburn in 1937. This was followed after 1945 by further experiments at Rothamsted and on the Ministry of Agriculture experimental husbandry farms.

Figure 13.7(a) shows the changes in organic matter in the Hoosfield continuous barley experiment at Rothamsted, which had been used for arable crops prior to 1850 and so initially had a small organic matter content. On the unmanured and fertilized plots the C content has remained almost constant for more than 100 years. The amount in the fertilized soils is larger than in the unmanured soil because the former has grown larger crops and inputs from roots and stubble have been larger. Farmyard manure has trebled the C content. Residues from applications of farmyard manure made between 1852 and 1871 can still be measured.

The Woburn soil (Fig. 13.7(b)) contained more organic matter initially. This has decreased to a lower level than at Rothamsted under similar management, and a manured four-course rotation has not prevented the decrease. The light-textured soil appears to have a poor ability to protect its organic matter from decomposition: the eventual equilibrium value for sandy soils under similar management and climate will depend on their clay content.

Other Rothamsted experiments were on an old grassland site and an old arable site (Fig. 13.8(a)). On the old grassland site, continuous arable cropping caused organic matter to decrease towards a new equilibrium value of about 2 per cent C, whereas when kept in grass organic matter increased as a result of improved management. On the old arable site soil organic matter has remained essentially constant with continued arable cropping, whereas permanent grass has caused a slow increase. On this soil it would take about 100 years for the C content characteristic of an old arable site to increase to that of an old grassland site and about 25 years to reach the half-way stage (Johnston, 1992).

The benefits in this experiment of a rotation of 3 years arable and 3 years grass–clover ley compared to continuous arable cropping has been small in terms of organic matter content (not shown in the figure). The mean increase between 1949 and 1975 was only about

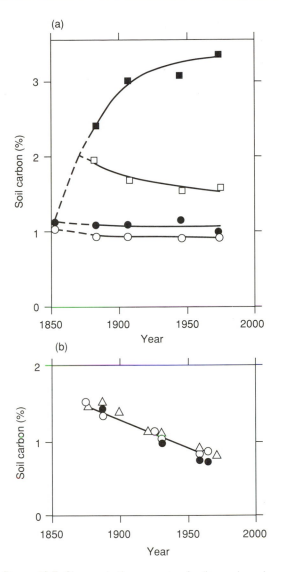

Figure 13.7 Changes in the amounts of soil organic carbon in classical experiments at Rothamsted and Woburn. (a) In soil from Hoosfield, Rothamsted. Barley each year, with the following annual treatments since 1852: (O) unmanured; (●) NPK fertilizers (48 kg N ha^{-1}); (■) manured (35 t ha^{-1}); (□) manured 1852–71 with none subsequently. (b) In soil at Woburn. Cereals each year with the following treatments: (O) unmanured; (●) NPK fertilizers; (△) manured four-course rotation. From Johnston (1986).

0.25 per cent C with the ley rotation. However, there was a significant benefit in terms of N supply to the crops following the ley.

At Woburn (Fig. 13.8(b)) a combination of a grass-arable rotation and farmyard manure raised the C content on an old arable soil during 32 years by 0.6 per cent

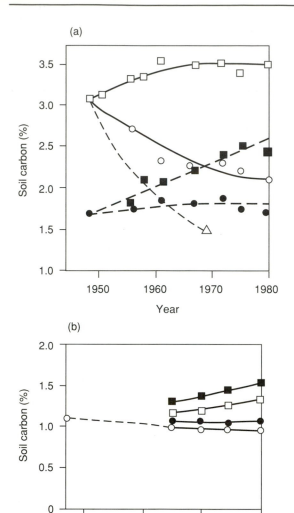

(a)

(b)

Figure 13.8 Changes in the amounts of soil organic carbon in ley–arable experiments at Rothamsted and Woburn. (a) At Rothamsted. Old grassland soil: (□) kept in grass; (○) ploughed and kept in arable crops; (△) ploughed and fallowed each year. Old arable soil: (●) kept in arable crops; (■) sown to grass which remained unploughed.
(b) At Woburn. All arable rotation: (○) no manure; (●) manure (37.5 t ha^{-1} once in 5 years). Rotation of 3 years' grass, 2 years' arable crops: (□) no manure; (■) manure (37.5 t ha^{-1} once in 5 years). From Johnston (1986).

compared to the fertilized plot, with significant benefits to the arable crops. The data show that many years are required to build up organic matter using a ley–arable rotation, although there are immediate short term benefits.

The effect of fallows

Fig. 13.8(a) shows the effects of a continuous fallow on the old grassland soil. The very rapid decrease in organic matter was the result of cultivation to control weeds with no crop residues returned to the soil. Fallows are rarely used in British agriculture. However, as a result of Government proposals in 1987 fields have been 'set aside' to reduce cereal production. One of the options is simply to cultivate the land to keep it free from weeds. This will cause a decrease in fertility as organic matter decomposes, and nitrate mineralized from soil organic matter will be leached. Vegetation of some sort is needed to prevent such losses.

For further details of the long-term experiments, see Rothamsted Experimental Station (1970, 1977b).

Organic farming and sustainable agriculture

Many issues are raised by those who believe in organic systems of food production (Hodges, 1992). From a soil science viewpoint certain points need to be emphasized:

- Truly sustainable systems require that all nutrients removed are replaced. If products leave the farm, equivalent amounts of nutrients must be returned to the soil. Nitrogen is replenished by N fixation and atmospheric deposition. Weathering of soil minerals may provide some potassium, phosphorus, calcium and magnesium with small amounts from atmospheric deposition. Sulphur is supplied by atmospheric deposition in amounts which depend on emissions from fuel burning. Alongside the use of organic matter to maintain good soil structure and to supply mineralizable N there must be adequate inputs of other nutrients.

- Urea and ammonium nitrate fertilizers do not add unnatural compounds to the soil. For example, the total concentration of soluble N in animal urine is about $8 \, \text{g N l}^{-1}$ of which about $6 \, \text{g l}^{-1}$ is urea-N. This is converted to ammonium and nitrate in the soil (Haynes and Williams, 1992). When urea or ammonium nitrate fertilizers are added at a rate of $100 \, \text{kg N ha}^{-1}$ and washed into the top 20 cm of the soil, the concentration is about $38 \, \text{mg N kg}^{-1}$ soil or $167 \, \text{mg l}^{-1}$ in soil solution (assuming a bulk density of $1.3 \, \text{t m}^{-3}$ and a volumetric water content of $0.3 \, \text{m}^3 \, \text{m}^{-3}$). Similarly when grassland is ploughed ammonium and nitrate are released: Fig. 11.7 shows a peak concentration of $171 \, \text{mg mineral-N l}^{-1}$ in soil solution.

- Animal urine contains about $10 \, \text{g K}^+ \, \text{l}^{-1}$ and $4 \, \text{g Cl}^- \, \text{l}^{-1}$. Good quality irrigation water can have up to $210 \, \text{mg Cl}^- \, \text{l}^{-1}$ (Table 14.4). An application of $60 \, \text{kg K}^+ \, \text{ha}^{-1}$ and $54 \, \text{kg Cl}^- \, \text{ha}^{-1}$ (Table 9.3) in the form of KCl fertilizer would give, using the above calculations, $23\,mg\,K^+\,kg^{-1}$ or $100\,mg\,K^+\,l^{-1}$ and $21\,mg\,Cl^-\,kg^{-1}$ or $90\,mg\,Cl^-\,l^{-1}$. However, K^+ is immediately adsorbed on to cation exchange sites giving a much smaller increase in solution concentration: Section 9.4 contains an example which shows that the concentration in soil solution might be raised from about 40 to $50\,mg\,K^+\,l^{-1}$ (1.1–$1.3\,mM$). Chloride is not adsorbed and so, like nitrate, does not accumulate but is leached away each winter. The concentrations from fertilizer can also be compared to those entering soil in rain. Warren Spring Laboratory (Campbell *et al.*, 1987) reports a range of concentration from 1 to $17\,mg\,Cl^-\,l^{-1}$, and a range of annual inputs from 9 to $200\,kg\,Cl^-\,ha^{-1}$. The ranges for K^+ are 0.04–$0.4\,mg\,l^{-1}$ and 0.3–$5\,kg\,ha^{-1}$.

- The leaching of fertilizer-chloride will carry away mainly calcium ions and so causes acidification. If $54\,kg\,Cl^-\,ha^{-1}$ is lost as $CaCl_2$ then $30\,kg\,Ca^+$ is leached which could be replaced by 76 kg of $CaCO_3$. This is equivalent to an input of $1.5\,kg\,H^+$ and is about one tenth of the annual acidity production in an agricultural soil (Table 8.1). There is no advantage in the use of potassium sulphate if sufficient sulphate for crop needs enters from the atmosphere. Little sulphate is adsorbed in neutral soils and so like chloride is leached away and causes acidification. As atmospheric inputs decline gypsum and potassium sulphate become obvious choices as fertilizers.
- Animal urine contains a small amount of P (about $44\,mg\,P\,l^{-1}$). An application of $21\,kg\,P\,ha^{-1}$ (Table 10.1) in the form of ammonium phosphate fertilizer would produce a concentration of $8\,mg\,kg^{-1}$ or $35\,mg\,l^{-1}$ if not adsorbed. However, P is strongly adsorbed: using the Great Field sorption isotherm of Fig. 10.10 the rise in solution concentration might be from 0.2 to $0.8\,mg\,l^{-1}$. Because of its strong adsorption, P does not wash very far into the soil. If it is retained in the top 2 cm of soil its concentration might rise to about $10\,mg\,l^{-1}$. In soils which have a low availability of P it is necessary to increase the solution concentration in order to supply crop demands regardless of the farming system.
- To supply N from organic sources may cause a similar acid input to the application of an equivalent amount of ammonium nitrate depending on the amount of nitrate leached. However, the increased buffer capacity of a soil with larger amounts of organic matter may reduce the effects of acidification. There is more acidification during the growth of a legume than during the growth of crops which do not fix N (Jarvis and Robson, 1983). Thus control of acidity is especially important in organic systems because clover is an acid-sensitive crop (Table 8.2).

Section 13.4 Measurements of legume growth and nitrogen fixation

In soil *Rhizobium* bacteria exist in a free-living state and use a wide range of C compounds as food. In this condition they do not fix N. They also live in symbiotic association with legumes where they infect the root by entering cells which then proliferate into a nodule. The bacteria fill these cells where they convert nitrogen, N_2, into ammonium ions which are released into the cells of the legume and synthesized into organic N. Legumes normally have about twice the N content per unit dry matter compared to non-legumes.

Rates of fixation vary greatly in the field. Values between 50 and $450\,kg\,N\,ha^{-1}\,a^{-1}$ have been reported for clover, with similar amounts for lucerne and broad beans and rather less for tropical legumes. The presence of mineral-N in the soil reduces fixation, the plant taking up ammonium and nitrate instead. There has been considerable research effort to develop associations of clover and *Rhizobium* which will continue to fix N in the presence of applied fertilizer-N as a means of increasing the productivity of ryegrass–clover leys. Progress to date has been slow and for temperate crops relatively unsuccessful.

Little if any of the fixed N is lost from the plant until the root dies. Thus in a ryegrass-clover ley there is little transfer of N to ryegrass in the first year of growth. In subsequent years considerable amounts of mineral-N are released as the nodules and roots decompose. The presence of ryegrass may encourage an efficient system because clover on its own would fix less N once this release begins. When a ryegrass–clover ley is ploughed the N-rich plant material together with soil organic matter produce a large flush of mineralized nitrate in the first year, with decreasing amounts in subsequent years (Table 11.3).

A nitrogen fixation experiment

The N in clover plants is derived partly from soil mineral-N and partly from fixed N. If ryegrass is grown on its own it can only take up mineral-N. If each is grown separately in a pot of the same soil and it is assumed that in both cases the same amount of mineral-N is taken up, then by difference the fixed N can be determined. A pot experiment is described below to measure N fixation in monocultures and mixtures.

Method

Soil A neutral soil with little organic matter and

mineral-N is needed. Use a light-textured soil from a site which has been in arable crops for a number of years and sample at the end of a growing season if possible. Allow the soil to dry until it readily passes through a 5 mm sieve. To 9×1 kg samples add 60 mg N kg^{-1} (50 kg ha^{-1}) following the methods of Section 9.2. To these and 9 further 1 kg samples add P and K$^+$. Place the soils in 5-inch pots in saucers. The above treatments are termed N_1 and N_0.

Plants Sow ryegrass and white clover (*Trifolium repens*) in trays of soil, compost or Perlite. Cover to maintain moist conditions until germination begins. Uncover and grow on until the seedlings are large enough to handle (2–3 weeks).

Ryegrass: transplant 8 seedlings per pot approximately equally spaced ($3 \times N_0$ pots and $3 \times N_1$ pots).

Clover: plant as for ryegrass (6 pots).

Ryegrass and clover: transplant 4 seedlings of each mixed together and equally spaced in each pot (6 pots).

Maintenance Water as directed in Section 9.2 and grow on for 6–8 weeks. Remember that denitrification will occur if the soils are overwatered .

Harvest and analysis Cut the crops at soil level. Separate the ryegrass and clover from the mixed crop and dry at 100 °C. Weigh the dry matter.

Determine the N content of the dry matter following the methods of Section 11.2. For each determination 1 g of dry matter is needed. Calculate the N taken up per pot.

Data

The experiment as described neglects the N in plant roots. Ideally roots should also be harvested (Section 3.1) and analysed but separation of ryegrass and clover roots is difficult. The N content of the roots is small compared to the shoots.

The difference between the N content of the clover and ryegrass grown separately is a measure of the N fixed by the clover. The assumption is made that both crops take up the same amount of mineral-N: this will not be correct for young crops, but is correct when both crops have fully depleted the soil. Table 13.6(a) (7 weeks) gives an example of data obtained in a similar experiment (N_0 treatments) in which the pots contained 350 g soil and roots were also analysed to give data for the whole plant. The N content of the plants has been calculated (Table 13.6(b)) by multiplying mg N g^{-1} dry matter by g dry matter per pot to give mg N per pot.

The following conclusions can be drawn:

- In the monoculture, grass takes up 1.8 mg mineral-

Table 13.6 (a) Data from a ryegrass–clover experiment

| | After 7 weeks' growth | | | | After 9 weeks' growth | | | |
| | Monoculture | | Mixture | | Monoculture | | Mixture | |
	DM	N	DM	N	DM	N	DM	N
Grass	0.15	12.1	0.11	15.6	0.30	19.7	0.21	16.2
Clover	0.69	23.8	0.25	22.4	0.94	24.6	0.44	20.9

DM = g dry matter per pot.
N = mg N g^{-1} dry matter.

(b) Plant–N content after 7 weeks' growth (mg N per pot)

	Monoculture	Mixture
Grass	1.8	1.7
Clover	16.4	5.6

(c) Increase in plant–N content between 7 and 9 weeks (mg N per pot)

	Monoculture	Mixture
Grass	4.1	1.7
Clover	6.7	3.6

From McNeill and Wood (1990).

N. If the clover also takes up this amount, fixation of N is $16.4 - 1.8 = 14.6$ mg per pot or 1.8 mg per plant.

- In the mixture, total plant-N is $1.7 + 5.6 = 7.3$ mg. The amount of fixed N cannot be calculated because the amount of mineral-N taken up by the clover is not known. However, if the ryegrass and clover take up equal amounts of mineral-N, fixed N = $5.6 - 1.7 = 3.9$ mg per pot or 1.0 mg per plant. The dry matter per clover plant is reduced from 0.08 to 0.06 mg by competition with the ryegrass.

An alternative experiment

Although a single harvest provides data which can be used to calculate N fixation it is better to establish the crops, and then measure uptake over a period of time. An example of the data obtained in this way is shown in Table 13.6(a) and (c). Following the above procedures, two sets of N_0 pots were set up. The first set of pots was harvested after 7 weeks. The second set was harvested after 9 weeks having received 3.3 mg of nitrate-N per pot when the first set of pots was harvested.

The data can be analysed as follows:

- When grown in monoculture grass took up 4.1 mg N. This came from soil-mineral N plus fertilizer-N. Grass takes up N very efficiently and it is likely therefore that the amount of soil-mineral N available during the 2-week period was $4.1 - 3.3 = 0.8$ mg (3.3 mg of fertilizer-N was added).
- The N in the clover (monoculture) increased by 6.7 mg N. If clover took up the same amount of mineral-N as the ryegrass, then fixed-N = $6.7 - 4.1 = 2.6$ mg.
- In the mixture, grass took up 1.7 mg of mineral-N, about half of that available (4.1 mg). There are equal numbers of grass and clover plants growing together and so clover may also have taken up 1.7 mg of mineral-N. If this was so, the clover fixed $3.6 - 1.7 = 1.9$ mg N.
- Assuming therefore that clover fixed 2.6 mg N per pot (8 plants) in monoculture and 1.9 mg (4 plants) in the mixture, the amounts fixed per plant were 0.33 and 0.48 mg N respectively, the plants being of almost equal size (0.12 and 0.11 g dry matter per plant respectively). Interpretation of the data depends on the assumptions about the uptake of mineral-N by clover: McNeill and Wood (1990) show how isotopically labelled N can be used to partition plant-N into fixed and soil-derived N.

Clover–grass competition

When ryegrass and clover grow together they are competing for light, nutrients and water. The survival of both plants in the field requires that soil conditions and management do not depress the growth of one so that the other becomes dominant. A ready supply of mineral-N, low pH, low phosphate and potassium availability and overgrazing all tend to cause clover to die out. The following pot experiment examines the effects of acidity and phosphate supply on the survival of clover.

Soil

An acidic soil with low phosphate availability is needed. A woodland soil is most likely to have these properties. Dry as far as is necessary to pass the soil through a 5 mm sieve.

Treatments

Lime Table 8.5 gives buffer capacities. Assuming that the soil is medium textured and has a pH of 4.5, then about $6 \text{ g CaCO}_3 \text{ kg}^{-1}$ is needed to raise the pH to 6.5. Apply as a powder and mix well (with $CaCO_3$, L_1; without $CaCO_3$, L_0).

Phosphorus Apply as a solution. Section 9.2 gives details for an application of 20 kg P ha^{-1}, but for a low P availability soil, more is needed. Dissolve 11 g of $CaHPO_4$ in 500 ml of water and apply 25 ml to 1 kg of soil. This is equivalent (pot area basis) to a field application of 100 kg P ha^{-1} (with phosphorus, P_1; without phosphorus, P_0). The following treatments are needed with 3 replicates of each: L_0P_0, L_1P_0, L_0P_1, L_1P_1 so requiring 12 pots.

Nitrogen Mineralized nitrogen is likely to be adequate.

Potassium Apply to all pots as indicated in Section 9.2.

Plants

Follow the instructions for the N fixation experiment planting four clover and four ryegrass seedlings together in each pot. Water as required. Harvest and separate the clover and ryegrass. Weigh the dry matter. Allow regrowth to occur and take a second harvest to determine the effect of cutting on plant survival and growth. Use a statistical analysis (Section 3.8) when examining differences between treatments.

Alternative treatments

Apply the P and lime together as one treatment and examine the effects of fertilizer-N as the second treatment. Use the same N application as in the fixation experiment and after cutting apply again to stimulate regrowth.

Section 13.5 Forest clearance and shifting cultivations

In 1960, P. H. Nye and D. J. Greenland wrote

> Over 200 million people, thinly scattered over 14 million square miles of the tropics obtain the bulk of their food by the system of shifting cultivation. They form a little under 10% of the world's population and are spread over more than 30% of its exploitable soils.

In describing the system they added

> People ... burned the forest, planted their crops with a simple digging stick, and after taking one or two crops they abandoned their plot to the invading forest. Each year a new plot of land was cleared, hence the term – shifting cultivation. Later some of them developed the hoe, and with this implement they were able to clear tenacious grass roots and

cultivate the savannas, abandoning their plots as in the forest after a few years of cropping.

Of recent changes they commented

Up to a century or so ago shifting cultivation had no very serious effect on the farmland of the tropics, since the soil and vegetation were given adequate time to regenerate after a period of cropping. The recent ... rapid increase in the population, and ... a demand for export crops like cocoa, coffee and rubber ... has increased the area under cultivation and reduced that available for subsistence crops. In consequence, vast areas of tropical forest and woodland have been transformed into less productive savanna grassland.

To this could be added the problems of drought years which have led to recent disasters in Africa.

CHANGES IN SOIL PROPERTIES FOLLOWING FOREST CLEARANCE

Burning

Soil properties after burning a 40-year-old forest in Ghana are shown in Table 13.7. There are three main effects from burning:

1. Large quantities of nutrient ions are spread in the ash. The amount of ash is much greater when virgin forest is burnt compared to a regenerated forest (between 4 and 50 t ha^{-1}). Much of the N in the vegetation is lost to the atmosphere as ammonia and oxides of N. Some S is lost as sulphur dioxide gas. Both vegetation and litter are burnt, but not the organic matter in the soil. The composition of wood ash is very variable depending on the vegetation and the temperature of burning. Typically ash may contain (% of dry ash): 0.07–1.7 N, 0.1–0.4 P, 1–19 Ca^{2+}, 0.5–4 Mg^{2+}, 0.1–5 K^+, about 0.3 Cl^- and about 1 S (as SO_4^{2-}) together with variable amounts of oxide, hydroxide, carbonate, bicarbonate and silicate ions and charcoal.
2. The soil surface is heated, killing weed seeds and initially improving structure in heavy-textured soils.
3. Soil pH is increased by the carbonates (initially the ash contains oxides but CO_2 is absorbed to produce carbonates).

The table shows marked increases in pH, and in extractable P, K^+, Ca^{2+} and Mg^{2+}. The measured increases in the amounts of extractable nutrients have been converted into additions to the whole soil. Some N has apparently been lost but the change (and that of the C content) may not be statistically significant. The rise in pH stimulates mineralization of soil organic matter.

Table 13.7 (a) Changes in the chemical properties of a soil caused by the burning of forest (Kade, Ghana)

Depth (cm)	pH	C (%)	N (%)	Extractable P† (μg g^{-1})	K^+	Ca^{2+} (cmol$_c$ kg$^-$)	Mg^{2+}	CEC
Before clearing								
0–5	5.21	2.22	0.21	9.8	0.41	5.7	1.2	10.1
5–15	4.73	1.11	0.11	3.6	0.33	3.6	1.0	7.0
15–30	4.63	0.87	0.09	1.9	0.32	3.0	1.0	6.7
After clearing and burning								
0–5	7.9	2.26	0.20	30.0	2.01	17.9	2.7	9.9
5–15	6.4	1.26	0.12	8.0	0.81	4.6	1.3	5.7
15–30	5.7	0.94	0.08	5.0	0.42	3.0	1.1	6.4

(b) Additions to the 0–30 cm layer resulting from clearing and burning (kg ha^{-1})

N	P	K^+	Ca^{2+}	Mg^{2+}
−110	25.3	737	1551	187

(c) Bulk density before clearing

Depth (cm)	Bulk density (t m^{-3})
0–5	1.00
5–15	1.54
15–30	1.68

† P was extracted using 0.03 M NH$_4$F + 0.025 M HCl solution (Bray solution). It extracts rather more than NaHCO$_3$ (Section 10.3).
From Nye and Greenland (1960).

Cropping

Good yields are normally obtained in the first year after clearing, but then decline rapidly. Inputs of manure or fertilizer are beneficial but may not prevent falling yields. The reasons for the decline are:

- Multiplication of weeds, pests and diseases.
- Deterioration in soil structure, surface capping and erosion of topsoil.
- Reacidification and deterioration in the nutrient status of the soil.

Table 13.8 shows the changes in soil properties which occurred during 8 years of continuous cropping following forest clearance in Ghana. The site is not the same as that in Table 13.7. The soil became more acidic and organic matter decreased, as did the available P. Surprisingly the exchangeable K^+ increased although leaching would be expected to cause a reduction as it did for Ca^{2+}. On this site the most likely nutrient to be limiting yield was P because large amounts of N were mineralized.

Changes in soil structure have proved difficult to measure and to interpret in relation to yield. There is a deterioration in structural stability in the surface soil associated with the rapid decomposition of organic matter. Heavy rainfall may lead to the formation of a surface cap and may reduce porosity. The reduced infiltration capacity and poor stability can lead to erosion.

Questioning the data

A set of data is often accepted at face value. It is important to question data both in terms of their reliability and meaning. Tables 13.7 and 13.8 can be approached in this way.

Table 13.8 (a) Changes in the chemical properties of a soil under continuous cropping (Aiyinasi, Ghana)

Depth (cm)	pH	C (%)	N (%)	Extractable P† ($\mu g\,g^{-1}$)	K^+	Ca^{2+}	CEC
					($cmol_c\,kg^-$)		
Immediately after clearing							
0–15	6.0	2.19	0.164	16.0	0.22	4.62	10.0
15–30	5.0	1.28	0.103	6.0	0.16	1.37	7.2
Eight years later							
0–15	5.0	1.50	0.128	3.7	0.26	2.05	9.3
15–30	4.8	1.02	0.089	2.6	0.25	1.18	8.6

(b) Losses from the 0–30 cm layer in 8 years ($kg\,ha^{-1}$)

N	P	K^+	Ca^{2+}
1100	34	Increase	1210

† See Table 13.7.
From Nye and Greenland (1960).

Are the quoted total additions of nutrients from clearing and burning in Table 13.7 in agreement with the measurements?

Taking calcium as an example, the table records an addition of $1551\,kg\,ha^{-1}$ which has presumably been calculated from the measured changes in exchangeable Ca^{2+}. It entered the soil as CaO in the ash. In the three layers, the increases are 12.2, 1.0 and $0\,cmol_c\,kg^{-1}$. The molar mass of Ca^{2+} is $40\,g\,mol^{-1}$, and so $1\,mol_c\,Ca^{2+} = 20\,g$.

Taking the 0–5 cm layer, the increase in Ca^{2+} is

$$12.2 \times 10^{-2}\,mol_c\,kg^{-1} \times 20\,g\,mol_c^{-1} = 2.44\,g\,kg^{-1}$$
$$(= kg\,t^{-1})$$

The layer contains $0.05 \times 10^4\,m^3$ soil ha^{-1}, and with a bulk density of $1.0\,t\,m^{-3}$, the mass of soil is $500\,t\,ha^{-1}$. Therefore the increase in Ca^{2+} is $2.44 \times 500 = 1220\,kg\,ha^{-1}$. Similarly in the second layer ($1540\,t\,ha^{-1}$) there was an increase of $308\,kg\,ha^{-1}$. There is no increase in the third layer and so the total increase is $1220 + 308 = 1528\,kg\,ha^{-1}$. Allowing for some 'rounding off' in the presentation of the data the values are consistent.

Are the pH changes sensible in relation to the input of ash?

We do not know the amount or composition of the ash added to the soil and so the changes in soil properties cannot be related quantitatively to ash input. However, assuming that the changes in exchangeable cations result from an input of K_2O, CaO and MgO in the ash, and that these liming materials cause the pH change, then a buffer capacity can be calculated and compared to expected values.

The changes in exchangeable cations in the 0–5 cm layer are 1.6, 12.2 and $1.5\,cmol_c\,kg^{-1}$ of K^+, Ca^{2+} and Mg^{2+} respectively. In the form of its oxide, $1\,cmol_c$ of each cation would contribute 1 cmol of OH^-. Thus the total input of OH^- is

$$1.6 + 12.2 + 1.5 = 15.3\,cmol\,kg^{-1}$$

The pH change in this layer is $7.9 - 5.21 = 2.69$. Therefore the buffer capacity is apparently

$$15.3/2.69 = 5.69\,cmol\,OH^-\,kg^{-1}\,pH^{-1}$$

or

$$57\,mmol\,OH^{-1}\,kg^{-1}\,pH^{-1}$$

Table 8.5 shows that this is the expected value for a medium textured soil. A pH value of 7.9 may indicate that unreacted ash (carbonate) was present when the measurements were made. If so, apparent buffer capacities would be larger than in Table 8.5 which relate to acidic and neutral soils. The apparent buffer capacity of the second layer is $11\,mmol\,kg^{-1}\,pH^{-1}$. The data therefore appear to be consistent, the surface layer having a larger C content and CEC than the second layer. In the third layer the pH value has risen by 1.07 with almost no change in exchangeable cations which is a surprising result. The buffer capacity of the combined first and second layers (used below) is $22\,mmol\,kg^{-1}\,pH^{-1}$; this is found by considering the amount of soil in the two layers under unit area of soil surface (50 and $154\,kg\,m^{-2}$ for layers 1 and 2).

Is the quoted total loss of nitrogen during cropping consistent with the recorded change in soil nitrogen?

The procedure adopted above for Ca^{2+} can be used for

the data in Table 13.8. The bulk densities are not known, but the values in Table 13.7 can be used for the calculations. In the 0–15 cm layer the change in N (0.036% m/m, 2040 t soil ha^{-1}) indicates a release of 734 kg ha^{-1}. In the 15–30 cm layer 353 kg ha^{-1} is released (0.014% N, 2520 t ha^{-1}), giving a total of 1087 kg ha^{-1} in agreement with the quoted value of 1100 kg ha^{-1}. We can conclude that the bulk density at this site must have been similar to the previous one.

Is nitrification the main source of acidity during cropping?

Mineralization followed by nitrification produces acidity. If the nitrate is taken up by crops the acidity is probably neutralized, but 1 mol H$^+$ remains mol^{-1} of NO$_3^-$ leached (Reuss and Johnson, 1986). Denitrification may also occur. The estimate below shows that about 10 per cent of the mineralized N may be taken up and 90 per cent leached. In the 0–15 cm layer (Table 13.8) 734 kg N ha^{-1} is mineralized, and so 73 kg may be taken up and 661 kg leached. Acid production can therefore be estimated as follows: the molar mass of N is 14 g mol^{-1} and so 661 kg N ha^{-1} is 47.2 kmol N ha^{-1}. Leached as nitrate, this leaves 47.2 kmol H$^+$ ha^{-1} in the soil. In this layer there is 2040 t soil ha^{-1}, giving an acidity input of

$$47.2/2040 = 0.023 \, \text{kmol H}^+ \, \text{t}^{-1}$$

or

$$23 \, \text{mmol kg}^{-1}$$

There must have been other sources of acidity or alkalinity, but for this input alone the measured pH change of 1 unit would have been produced if the soil had a buffer capacity of 23 mmol H$^+$ kg^{-1} pH^{-1}. This is similar to the value for the combined 0–5 and 5–15 cm layers which has been calculated above using ash inputs (22 mmol kg^{-1} pH^{-1}). Following similar calculations, the 15–30 cm layer has an acid input of 9 mmol kg^{-1}, the pH change is 0.2 units and the apparent buffer capacity is 45 mmol H$^+$ kg^{-1} pH^{-1}.

In recent experiments in the United States, deforestation of a watershed in the Hubbard Brook Experimental Forest (Likens *et al.*, 1970 and Ch. 8) caused a release of 340 kg N ha^{-1} during 3 years and the pH of the stream water decreased by 1 unit, indicating similar changes in both temperate and tropical regions following forest clearance. Nitrification is clearly a major source of acidity.

What happens to the lost nutrients?

Leaching and crop removals are likely to be the major losses together with denitrification. The amounts in crops can only be guessed but assuming an annual yield of about 1 t grain ha^{-1} they may have been 17 N, 3 P, 5 K$^+$, 0.5 Ca^{2+} and 1 Mg^{2+} (Table 9.1) expressed as kg ha^{-1} a^{-1}.

Taking N as an example, crop removals over 8 years may have been 136 kg ha^{-1} out of a total loss of 1100 kg ha^{-1}. If inputs are negligible, estimated denitrification and leaching losses are 964 kg ha^{-1} or about 121 kg ha^{-1} a^{-1}. The system appears to lose very large amounts of nitrate, reflecting both the large amounts of mineralized N and the heavy rainfall of the area.

The P removals in crops using the above figures might be 24 kg ha^{-1} over the 8 years compared to the measured change of 34 kg ha^{-1}. Phosphorus is tightly held in soils.

The Ca^{2+} removed in crops over 8 years might be 4 kg ha^{-1} compared to a total loss of 1210. The difference indicates a large leaching loss. This is to be expected if NO$_3^-$ is being leached because there must always be a charge balance between the cations and anions in the leaching water. If Ca(NO$_3$)$_2$ is the major component of this water each mol of Ca^{2+} will be balanced by 2 mol of NO$_3^-$. This can be checked: if the amount of leached Ca^{2+} is about 1200 kg ha^{-1} this can be expressed as

$$1200 \, \text{kg ha}^{-1}/40 \, \text{g mol}^{-1} = 30 \, \text{kmol ha}^{-1}$$

If 960 kg N ha^{-1} is leached with no denitrification, then a similar calculation gives

$$960 \, \text{kg N ha}^{-1}/14 \, \text{g mol}^{-1} = 69 \, \text{kmol ha}^{-1}$$

which is approximately twice the Ca^{2+} value. Thus the data suggest that most of the lost Ca^{2+} is leached with NO$_3^-$.

Thus NO$_3^-$ formed by mineralization is both the source of acidity and the means of carrying away large amounts of Ca^{2+} and other cations. The loss of K$^+$, Ca^{2+} and Mg^{2+} may be the major cause of the decreases in crop yield on some sites. This points to the need for improved cropping systems to trap the nitrate in the crop.

Sensitive soils

Fertility

The soils most likely to show rapid decreases in fertility after forest clearance are those in which the cation exchange sites are primarily on the organic fraction. As organic matter decomposes the soil's ability to hold nutrient cations decreases. Soil B in Table 7.4 is an example from Brazil: despite its large clay content, its exchange capacity is small. Charge varies in proportion to the organic matter content which may therefore contribute much of the charge. This conclusion can be evaluated as follows. The 0–17 cm layer has an ECEC of $0.94\,cmol_c\,kg^{-1}$ soil and a C content of 2.6 or 4.5 per cent organic matter. Fig. 7.2 indicates that organic matter (humus) may carry between 5 and 20 $cmol_c\,kg^{-1}$ organic matter at pH 5 depending on the amount of aluminium present. Assuming $10\,cmol_c\,kg^{-1}$, the organic matter contributes $0.45\,cmol_c\,kg^{-1}$ soil. Similarly for the 51–106 cm layer, the calculated contribution is $0.12\,cmol_c\,kg^{-1}$ soil compared to a measured ECEC of $0.10\,cmol_c\,kg^{-1}$. Thus a large fraction of the charge may be on the organic matter and cultivation will put at risk the soil's ability to hold cations.

Erodibility

The suitability of land for clearing and cultivation depends on its likely sensitivity to erosion. Assessment is difficult (Lal *et al.*, 1986).

FURTHER STUDIES

Calculations are in Section 13.6.

Section 13.6 Calculations

1. Convert the N data in Table 13.3 into percentage C, and plot changes in percentage C for the zero input and manure plots against time (abscissa, 1840–1980). Plot on the same graph data from Fig. 13.8(a). Draw on the graph the changes that might have occurred between 1843 and 1865 when N contents were first measured on the Broadbalk site.

2. In experiments at Rothamsted (Fig. 13.8) a permanent pasture was ploughed and used for arable crops for 32 years. Over that period the organic C content decreased from 3.1 to 2.0 per cent. If the C:N ratio remained at 10:1, what was the average release of mineral-N per year? Assume 2500 t soil ha^{-1}. (*Ans.* 86 kg)

   In another field which had grown arable crops for many years, grass was sown and was not ploughed for 32 years. Over this period the soil C content rose from 1.7 to 2.4 per cent. Again assuming a C:N ratio of 10:1 what input of N was required on average each year to supply the stored N? (*Ans.* 55 kg)

   For the Broadbalk manure plot (Table 13.3) calculate the average amount of N stored in the organic matter each year for the periods 1865–1944 and 1944–66. Assume 2840 t soil ha^{-1}. (*Ans.* 22 and 19 kg ha^{-1})

3. In field plots sown in 1985 (McNeill and Wood, 1990), ryegrass and white clover grown together were compared to ryegrass grown alone . No fertilizer was applied. Over the period 29 April–19 October 1987 the following data were obtained for the grass grown alone, grass in the mixture and clover in the mixture respectively: shoot dry matter yield 431, 492 and 476 g m^{-2}, nitrogen percentage in the dry matter 1.45, 2.14 and 3.97. Calculate the uptake of N by each crop. (*Ans.* 6.25, 10.53 and 18.9 g N m^{-2}) Calculate the total N taken up by the ryegrass–clover mixture. (*Ans.* 29.43 g m^{-2}) How much more N is present in the mixture than in the ryegrass grown alone? (*Ans.* 23.18 g m^{-2}) This extra N has either been fixed, or is mineralized from the N-rich residues of clover roots from the previous 2 years' growth. Calculate the amount fixed by assuming that clover and ryegrass in the mixture take up equal amounts of mineral-N. (*Ans.* 8.37 g m^{-2}). What is the total mineral-N taken up by the mixture? (*Ans.* 21.06 g m^{-2}) How much extra N is mineralized in the soil supporting the mixture compared to the soil supporting ryegrass alone? (*Ans.* 14.81 g m^{-2})

   Compare the N percentage of the ryegrass grown alone (which is unfertilized) and in the mixture to the typical value for fertilized ryegrass in Table 9.1.

   Calculate the rate of fixation by clover and the rate of uptake of N by the mixture as kg N $ha^{-1}\,d^{-1}$. (*Ans.* 0.48 and 1.7) Compare to the rates of uptake given for fertilized ryegrass in Fig. 11.5

4. Prepare an estimated N budget for the wheat grown in the Nebraska experiment shown in Fig. 13.2. Assume that the grain contains 17 kg N t^{-1}, that the straw yield is 0.6 × the grain yield and contains 7 kg N t^{-1} (Table 9.2). Calculate the N taken up in the crop. Tabulate your estimates as in Table 13.4. The non-fertilizer inputs can only be guessed: assume an input of 5 kg N ha^{-1} in the seed, a relatively clean atmosphere giving an input of 5 kg N ha^{-1} (Colour Plate 17), and biological N-fixation equal to 5 kg N ha^{-1}. The amount of net mineralized N is not known. However, an indication of the amount can be obtained from the N removed in the plot giving 1.33 t ha^{-1} yield where no N was applied: you have calculated this value. If there were no losses on this plot, this value is the total input of non-fertilizer N, and so by difference the net mineralized N can be calculated. (*Ans.* 13 kg ha^{-1})

   Plot the response curves for Nebraska wheat (Fig. 13.2, +20.6 cm available water) and for Rothamsted continuous wheat (Fig. 13.6) on the same graph. Calculate and mark on the graph the increase in yield per kilogram N applied for each N increment. If the cost of fertilizer-N and the price of wheat to the farmer is 36 pence kg^{-1} and £120 t^{-1} respectively in the UK and 25 cents kg^{-1} and $130 t^{-1} in the USA, calculate the cost of the fertilizer and the value of the extra yield for each increment (Section

11.4). Because of other production costs these values do not give a direct indication of the economic optimum input of fertilizer-N: they simply show for which increments the value of the extra yield is greater than the cost of the fertilizer.

5. Burning a forest in Africa produced 20 t ash ha^{-1}. Its composition was 8.2 per cent Ca^{2+}, 1.7 per cent Mg^{2+} and 2.8 per cent K^+. Calculate the input of the three nutrients as kg ha^{-1}. (*Ans.* 1640, 340 and 560). The ash was cultivated into the 0–15 cm soil layer which contained 2000 t soil ha^{-1}. If all the nutrients dissolved and became exchangeable, calculate their input in terms of cmol$_c$ kg^{-1}. (*Ans.* 4.10, 1.40 and 0.72) If the buffer capacity of the soil was 5 cmol OH$^-$ kg^{-1} pH^{-1}, calculate the change in pH produced by the ash in this layer. Assume that the cations in the ash are in the form of oxides, hydroxides and carbonates and so their liming effect (cmol OH$^-$) is equal to their amount expressed as cmol$_c$ cation. (*Ans.* 1.24) Compare your results to those in Table 13.7.

CHAPTER 14

Salinity and Sodicity

SOIL SALINITY

Soil salinity is the oldest soil pollution problem. The collapse of the Babylonian Empire is considered to be partly the result of failure of irrigated crops due to the accumulation of salts (Hillel, 1992). Despite present-day understanding of the problem and its management, about one-third of the world's irrigated land is subject to degradation and loss of production.

The problem is primarily associated with the arid and semi-arid regions of the world, where there is insufficient rain to leach away soluble salts. These occur naturally in the soil and are added in irrigation water, rain and wind-blown dust (loess) and by upward movement of groundwater. Evaporation from the soil surface and transpiration from crops removes water but leaves salts in the soil (Colour Plate 21(b)). Salts are removed in crops and in percolating water.

All irrigation waters contain dissolved salts: rivers acquire them as water flows over the land surface or passes through the ground before emerging again in springs, and groundwater contains salts dissolved during drainage through porous rocks and sediments. The presence of even a small concentration of salts in good quality irrigation water leads to salt accumulation in soils unless leached away by rain or irrigation water. The displaced salts should ideally be removed in a drainage system, but often are simply leached to accumulate below the rooting zone of the crop.

Soil salinity also occurs as a result of flooding by sea water (Colour Plate 21(a)). In 1953, extensive flooding occurred in parts of eastern England, Belgium and the Netherlands leaving large quantities of salt in the soil. In these temperate climatic regions rain leaches the salts away, but careful management is required to prevent damage to the physical properties of the soil.

The reclamation of land from the sea also involves salinity management. The Dutch polders were originally sediments under the sea. Drainage left a saline sediment from which the salts had to be leached away during the process of soil development.

Intensive cropping in glasshouses may also lead to salinity due to high input of fertilizers.

Salt inputs in irrigated agriculture

Inputs depend primarily on the amount of irrigation water used and its concentration of dissolved salts or in the case of coastal flooding on the composition of sea water. Table 14.1 shows the composition of good and poor quality irrigation waters and sea water as compared to rain-water. The total salt content (the salinity) of the water is normally determined by measuring its *electrical conductivity* or EC_w value (Section 14.1). Individual ions, however, have special effects, and analysis of irrigation water quality normally involves the measurement of sodium, calcium, magnesium, and carbonate plus bicarbonate (Section 14.1). Boron and chloride may also be measured as these can have toxic effects on plants.

Significant amounts of nutrient ions may be present in irrigation water, particularly nitrate, reducing the requirement for nutrients from fertilizers and other sources. A high concentration may supply an excess and cause problems to susceptible crops.

The salt content of soils

The accumulated salts are present either dissolved in the soil solution in a moist soil, or as crystals in dry soil. Colour Plate 21(b) shows massive crystals of mixed salts in a surface soil close to the Dead Sea. This is an extreme case of salinity resulting from water flowing via the River Jordan into what is really a lake from which there is no outflow. Evaporation leaves the salts behind. The salt accumulation is so large that it is extracted for commercial use.

The measurement of salts dissolved in soil solution is normally made by preparing a *saturation extract*. A saturated paste of soil and water is prepared from which the solution is extracted. The total salinity of this solution is determined by measuring the electrical conductivity (EC_e) and individual ions are measured as required (Section 14.2).

The simplest classification of salt-affected soils is based on EC_e (p. 278). The unit $dS\,m^{-1}$ (decisiemens per metre) is explained in Section 14.1.

Table 14.1 The composition of irrigation waters, sea water and rain water

	River Colombia, Washington	River Pecos, New Mexico	Lake Galilee, Israel	Groundwater, Nahal Oz, Israel	Sea water	Rain, Malham, UK
Electrical conductivity, at 25 °C, EC_w (dS m^{-1})	0.15	3.21	1.00	4.6	51	0.009
Ion concentrations (mmol l^{-1})						
Ca^{2+}†	0.45	8.65	0.10	0.65	10	0.007
Mg^{2+}	0.2	4.6	1.29	1.85	54	0.008
Na^+	0.2	11.5	5.1	41.3	470	0.06
$2CO_3^{2-} + HCO_3^-$‡	1.2	3.2	1.9	9.6	2.3	—
SO_4^{2-}	0.1	11.6	0.53	0.75	38	0.02
Cl^-	0.1	12.0	6.0	35.2	550	0.08
Sodium adsorption ratio§	0.2	3.2	4.3	26.1	59	0.5

† Ion concentrations are mmol l^{-1}. For Na^+ and Cl^-, mmol l^{-1} = mmol$_c$ l^{-1}. For the divalent ions (Ca^{2+}, Mg^{2+}, SO_4^{2-}) multiply mmol l^{-1} by 2 to give mmol$_c$ l^{-1}.
‡ HCO_3 and CO_3 are given together as the *alkalinity*. The values are mmol $[2CO_3^{2-} + HCO_3^-]$ l^{-1} which is equal to mmol$_c$ $[CO_3^{2-} + HCO_3^-]$ l^{-1} (Section 14.1).
§ Sodium adsorption ratio = $[Na^+]/([Ca^{2+}] + [Mg^{2+}])^{1/2}$ where [ ] indicates concentration in mmol l^{-1}.
Data from Richards (1954), R. Keren and T. Shainberg (personal communication) and Campbell *et al.* (1987).

saline soils $EC_e > 4\,dS\,m^{-1}$
non-saline soils $EC_e < 4\,dS\,m^{-1}$

Examples of the chemical properties of these soils are shown in Table 14.2. The use of either the River Pecos or the Nahal Oz water (Table 14.1) to irrigate the non-saline, non-sodic soil in Table 14.2 would increase the salinity: evaporation of water would further increase it unless the salts were leached away.

The effects of salinity on plant growth

There are three main effects (Fitter and Hay, 1987):

1. Direct toxicities e.g. sodium, chloride, boron.
2. Ionic imbalance in the plant.
3. A reduction in the availability of water by lowering the osmotic potential. This has been termed *physiological drought* because plants are affected by lack of water even though the water content of the soil is apparently adequate for crop needs. Section 5.4 explains these effects in more detail. Thus the effects on plants are complex, and plant, soil, water and climate interact to influence the *salt tolerance* of plants.

Plant tolerance to salinity is assessed in various ways:

- Effects on germination are normally given as the proportion of plants emerging from a saline soil

expressed as a percentage of those emerging from a non-saline soil. Usually crops are at least as tolerant during germination as they are during subsequent growth, sugar beet being an exception. Maize, rice and wheat are most sensitive at the early seedling stage and become increasingly tolerant as they mature.

- The effects on growth after emergence are normally given as a relative yield, which is the yield of the crop grown in a saline soil expressed as a fraction of its yield in non-saline conditions. For convenience data are often presented as the EC of the saturation extract causing a 50 per cent reduction in yield. Sensitivity can then be rated according to Fig. 14.1 which shows that there is a salinity threshold below which yield is not affected followed by an approximately linear decrease in yield with increase in salinity.

Table 14.3 gives data for common crops. Maas (1986) has reviewed the extensive data and information on other plants can be found in his tables and in Ayers and Westcot (1985). Because data are given in various ways in the literature, Fig. 14.1 is a useful way of standardizing information. For example, if the EC giving 50 per cent reduction in yield is known for a crop, this value can be marked on the figure and an approximate threshold and slope can be predicted. The tolerance of ryegrass has been shown in this way.

Table 14.2 Chemical properties of four soils from arid regions

	Non-saline, non-sodic	Saline	Sodic	Saline, sodic
Soil				
Saturation percentage (g H_2O per 100 g oven-dry soil)†	40.4	46.4	38.7	59.7
pH (saturated soil)	7.9	8.0	9.6	7.8
Cation exchange properties (cmol$_c$ kg^{-1})				
Cation exchange capacity	17.4	17.0	21.9	40.3
Exchangeable Na^+	0.5	1.4	10.1	10.5
Exchangeable Ca^{2+} + Mg^{2+}	16.7	15.3	4.8	29.0
Exchangeable Na^+ percentage‡	3	8	46	26
Saturation extract				
Electrical conductivity at 25 °C (dS m^{-1})	0.84	12.0	3.16	16.7
Ion concentrations (mmol l^{-1})				
Ca^{2+}	1.4	18.5	0.6	16.2
Mg^{2+}	0.9	17.0	0.2	19.2
Na^+	5.2	79.0	29.2	145.0
HCO_3^-	6.6	7.2	27.1	3.3
SO_4^{2-}	1.4	31.1	2.3	53.0
Cl^-	0.4	47.0	7.5	105.0
Sodium adsorption ratio	3.5	13.3	32.6	24.4

† Saturation percentage is the water content of the saturated paste (Section 14.2).
‡ Exchangeable Na^+ percentage = 100(exchangeable Na^+/cation exchange capacity) (Section 14.2).
Data from Richards (1954).

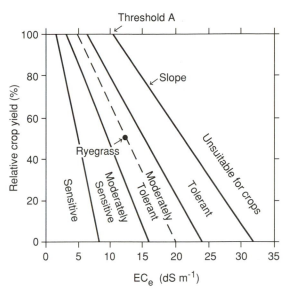

Figure 14.1 Crop sensitivity to salinity. Yield = 100 − slope × (EC_e − A) when EC_e > A, where yield is a percentage of the maximum, slope is per cent/(dS m^{-1}), EC_e is the electrical conductivity of the saturation extract at 25 °C (dS m^{-1}) and A is the threshold value of EC_e. From Maas (1986).

Table 14.3 The tolerance of common crops to soil salinity

Crop	Tolerance†	Electrical conductivity of the saturation extract at 25 °C (dS m^{-1})	
		50% emergence	50% yield
Barley	T	16–24	18
Cotton	T	15	17
Sugar beet	T	6–12	15
Sorghum	MT	13	15
Wheat	MT	14–16	13
Ryegrass	MT	—	12.2
Lucerne	MS	8–13	8.9
Tomato	MS	7.6	7.6
Maize	MS	21–24	5.9
Lettuce	MS	11	5.2
White clover	MS	—	—
Rice	S	18	3.6
Beans	S	8.0	3.6

† T = tolerant, MT = moderately tolerant, MS = moderately sensitive, S = sensitive.
From Maas (1986).

The significance of the simple classification limit set for saline soils (EC_e = 4 dS m^{-1}) is seen in Table 14.3. Sensitive crops begin to be affected at about this level of salinity.

Methods for measuring plant tolerance to salinity are described in Section 14.3.

THE SODICITY PROBLEM

The management of salt-affected soils would be a reasonably simple matter if the only consideration was the need to leach away salts. A more serious problem occurs, however, due to sodium salts, known as *sodicity*. Their accumulation increases the amount of exchangeable sodium which is normally expressed as an *exchangeable sodium percentage* (ESP) which is the amount of exchangeable sodium expressed as a percentage of the cation exchange capacity (CEC). Section 14.2

describes its measurement. At an ESP of between 10 and 15 soil clays are liable to swell and disperse causing a deterioration of soil structure (Section 14.4), particularly when the soil solution is diluted by rain-water or when good quality (low salinity) irrigation water is applied (Shainberg and Letey, 1984). There is a dilemma here: soluble salts need to be removed to avoid salinity damage, but the same process causes physical deterioration if the soil is sodic. Calcium is the ion which protects the soil from deterioration, and also defends plants against the toxic effects of sodium.

The effects of swelling and dispersion can be illustrated using Fig. 4.3: the swelling of clay crystals has to be accommodated in adjacent spaces, and transmission pores are closed up with an increasing volume of storage and residual pores being formed. The deterioration of soil structure has several important effects:

- Heavy-textured soils become more sticky and plastic when wet and hard when dry, leading to cultivation problems.
- Hydraulic conductivity is decreased. Irrigation water moves more slowly through the soil leading to ponding on the soil surface. It becomes more difficult to leach salts out of the profile.
- If the soil surface becomes saturated during irrigation, air entry is restricted and anaerobic conditions may develop, causing denitrification and the production of plant toxins. Plants are also more sensitive to salt damage in anaerobic soils.
- The soil surface becomes particularly sensitive to the mechanical effects of rain or irrigation water especially if applied by sprinklers. This leads to capping and reduced infiltration rates and exacerbates the above problems. In soils exposed to rain even 3–5 per cent of exchangeable sodium is enough to cause problems (Shainberg, 1985).

The simplest classification of soil sodicity is based on the ESP:

$$\text{sodic soils} > 15 \text{ per cent ESP}$$
$$\text{non-sodic soils} < 15 \text{ per cent ESP}$$

Examples are given in Table 14.2.

The alkalinity problem

Sodic soils have pH values up to 10.5 resulting from the high bicarbonate concentration of the irrigation water or groundwater entering the soil, and from the hydrolysis of the sodium-rich clays (Section 14.4). Thus sodic soils are alkaline (more so than calcareous soils, Section 8.3), which contributes to their problems primarily through causing soil organic matter to disperse, further weakening soil structure.

Examples of changes caused by sodicity

Clay swelling

Smectite clays have the greatest capacity for swelling and dispersion resulting from their high negative charge and small particle size (Section 14.4). Fig. 14.2 gives data on the swelling of flakes of oriented clay particles at a range of ESP values and soil solution concentrations. The technique is discussed in Section 14.5. Below ESP 15 little swelling occurs. Above this, swelling increases both with an increase in ESP and a decrease in solution concentration.

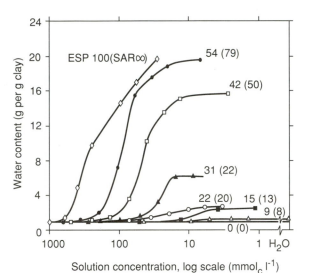

Figure 14.2 The swelling of smectite in mixed sodium–calcium solutions:
$$mmol_c \, l^{-1} = [Na^+] + 2[Ca^{2+}] \text{ where } [\ ] = mmol \, l^{-1}$$
$$= \{Na^+\} + \{Ca^{2+}\} \text{ where } \{\ \} = mmol_c \, l^{-1}.$$
The numbers against the lines are the approximate ESP values with the experimental SARs (Sodium Adsorption Rates) in brackets (see Section 14.5). The ESP values have been calculated using the data in Fig. 14.7. Data from Rowell (1963).

Hydraulic conductivity changes

The effects of swelling and dispersion on hydraulic conductivity are shown in Fig. 14.3 using sieved soil packed in leaching tubes. Serious reductions in hydraulic conductivity occur at ESP values above 15 when the solution concentration is reduced. The method is given in Section 14.5.

In the field critical conditions occur when winter rain or good quality irrigation water enters the soil reducing the soil solution concentration to low values. Except at the soil surface, total ionic concentrations are main-

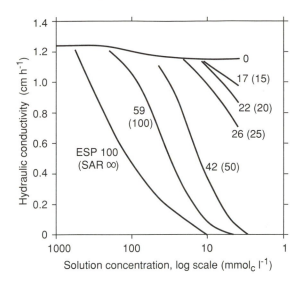

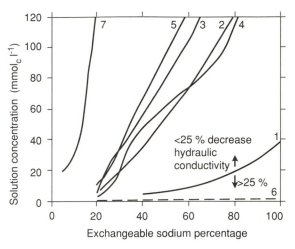

Figure 14.3 The effect of exchangeable sodium percentage on the hydraulic conductivity of a smectite-rich soil. The data are for soil 2 in Figure 14.4. The numbers against the lines are the approximate ESP values with the experimental SAR values in brackets (see Section 14.5). The ESP values have been calculated using the data in Fig. 14.7. Data from McNeal and Coleman (1966).

Soil	Clay %	Mineralogy
1	11	Mica / smectite / sesquioxide
2	13	Mica / smectite
3	14	Mica / smectite / vermiculite
4	23	Mica / smectite
5	30	Mica / smectite / vermiculite
6	46	Kaolinite
7	60	Mica / smectite

Figure 14.4 The effect of clay content and mineralogy on hydraulic conductivity changes in soil. The graph shows the solution concentration for each ESP value which gives a 25 per cent decrease in hydraulic conductivity. Lower concentrations give larger decreases. Data from McNeal and Coleman (1966).

tained at about $3\,mmol_c\,l^{-1}$ (see Section 14.1, Note 1) even during rain as a result of buffering by exchangeable cations and weathering of minerals. At this concentration, the hydraulic conductivity of soils with ESP values as low as 5 may be affected. At the surface small solution concentration combined with the mechanical effects of raindrops can cause capping and reduce infiltration at even lower ESP values.

The data in Figs 14.2 and 14.3 have important implications in terms of soil management: a sodic soil requires a poor quality irrigation water (large salt concentration) to protect it from structural damage, and the damaging effects of rain can be prevented by applying to the soil a soluble salt such as gypsum or calcium chloride which maintains the soil solution concentration and supplies exchangeable calcium. Thus although ESP 15 is used in the simplest classification of sodicity, critical ESP values depend on solution concentration and both are taken into account in considering water quality for irrigation.

Differences between soils

There is a further complication: the damage caused by sodicity varies between soils. Fig. 14.4 shows the differences in terms of the *threshold concentration* which is the concentration in soil solution below which significant

changes in hydraulic conductivity occur. For these data it is the concentration giving a 25 per cent decrease in hydraulic conductivity. Structural damage is prevented by maintaining the soil solution above the threshold value. The data were obtained using the experimental procedure described for Fig. 14.3.

The kaolinitic soil No. 6 is hardly affected by sodicity. Soil No. 1, with a small clay content and the stabilizing influence of sesquioxides associated with the mica and smectite, is also resistant to damage. The sensitivity of the other soils all dominated by 2:1 clays depends primarily on clay content.

Summary

The sensitivity of soils to sodicity depends on clay content, clay mineralogy, sesquioxide content, organic matter content and bulk density. The extent to which damage occurs depends on these soil properties, the amounts of exchangeable cations particularly sodium, the soil solution concentration and pH. Weatherable minerals, e.g. feldspars, calcite and gypsum, confer protection: their solubility, although limited, is sufficient to prevent the occurrence of small soil solution concentrations, and they are a source of calcium ions.

THE MANAGEMENT OF SALINITY AND SODICITY

Reclamation

The introduction of irrigated crop production on a previously uncultivated soil should be preceded by soil analysis to determine whether the soil is saline or sodic. If either problem exists, reclamation is required before using the soil. The details are beyond the scope of this book (see van Schilfgaarde, 1974) but the following general principles apply:

- *Saline–non-sodic soils*. Good quality water can be used to leach salts out of the profile. There is little danger of structural damage.
- *Saline–sodic soils*. Good quality irrigation water will cause structural damage. Gypsum applied to the soil surface will maintain the solution concentration and supply calcium ions to displace exchangeable sodium.
- *Sodic–non-saline soils*. Structure will already have been damaged and infiltration rate and hydraulic conductivity may be low. Surface applications of gypsum will slowly introduce exchangeable calcium. The formation of soil structure is a gradual process: grass or other forage crops (for grazing) will develop and deepen a layer of soil with good structure. The cost of reclamation is high.

Water quality

Soils may be maintained in good condition (non-saline, non-sodic) by the use of good quality irrigation water and adequate leaching. The criteria of quality are low salinity, a low ratio of Na^+ to Ca^{2+} + Mg^{2+} to prevent the development of sodicity, and small concentrations of those ions which may have specific toxic effects. The most widely used classification of irrigation water quality is shown in Table 14.4 (Ayers and Westcot, 1985). Interpretation of the table is as follows:

- *Restrictions on water use*. These are broad guidelines. Problems develop over a period of time. Suitable management is required even when no problems are indicated.
- *Salinity*. The high salinity water may have a direct effect on sensitive crops (Table 14.3). At lower salinities, salts may accumulate eventually leading to

Table 14.4 FAO guidelines for the interpretation of water quality for irrigation

Irrigation problem	Restrictions on use		
	None	Slight to moderate	Severe
Salinity			
EC_w at 25 °C (dS m^{-1})	<0.7	0.7–3.0	>3.0
Infiltration (sodicity)			
SAR† = 0–3 and EC_w =	>0.7	0.7–0.2	<0.2
= 3–6 =	>1.2	1.2–0.3	<0.3
= 6–12 =	>1.9	1.9–0.5	<0.5
= 12–20 =	>2.9	2.9–1.3	<1.3
= 20–40 =	>5.0	5.0–2.9	<2.9
Specific ion toxicity			
Sodium‡ (SAR)	<3	3–9	>9
Choride‡ (mmol l^{-1})	<4	4–10	>10
Boron (mg l^{-1})	<0.7	0.7–3.0	>3.0

† Read across each line to obtain the EC_w values giving restrictions for each range of SAR values.
‡ The values apply to surface (drip) irrigation.
From Ayers and Westcot (1985).

crop damage unless leached away by irrigation water.
- *Sodicity*. The major hazard is a reduction in infiltration rate due to structural damage caused by exchangeable Na^+. The tendency of the water to increase the ESP of the soil is rated using the *sodium adsorption ratio* (SAR) which is $[Na^+]/([Ca^{2+}] + [Mg^{2+}])^{1/2}$ where [ ] is the concentration in mmol l^{-1} (Section 14.2). If the soil and water come to equilibrium, ESP is approximately equal to SAR. The classification limit for sodic soils at ESP (=SAR) 15 (p. 280) is seen in the table to be at the more severe end of the range of restrictions. The extent of structural damage at a given sodicity depends on the salinity of the water. Surprisingly in the light of Fig. 14.4, no account is taken of variations in clay content or mineralogy of the soil, but the table is intended to cover the wide range of conditions found in irrigated agriculture and is modified to fit local conditions more closely where information is available.
- *Specific ion toxicity*. These are broad guidelines, the effects depending on crop sensitivity. The toxicity of sodium depends on the ratio of Na^+ to Ca^{2+} + Mg^{2+} in the soil and so SAR is a useful measure of likely problems.

A further problem, not considered in Table 14.4, arises if the water contains high concentrations of Ca^{2+}

and $CO_3^{2-} + HCO_3^-$. When the water enters the soil, $CaCO_3$ may precipitate blocking the conducting pores, and causing an increase in the SAR through the reduced Ca^{2+} concentration (Ayers and Westcot, 1985).

The need for leaching

Salt accumulation is an obvious result of using irrigation water in dry regions. However, a less obvious effect is that as salts accumulate the SAR of the soil solution increases above that in the irrigation water. Leaching is required both to prevent the accumulation of salts, and to prevent the SAR (and ESP) rising to unacceptable levels.

An example The River Pecos water (Table 14.1), has $11.5 \, \text{mmol Na}^+ \, l^{-1}$, $13.25 \, \text{mmol Ca}^{2+} + Mg^{2+} l^{-1}$ and an SAR of 3.2. If accumulation of salts doubles their concentrations to 23 and $26.5 \, \text{mmol} \, l^{-1}$ respectively, the SAR rises to 4.5.

The leaching fraction

Irrigation water is needed to supply the evapotranspiration needs of the crop. To leach salts out of the root zone to maintain acceptable salinity and sodicity levels extra water is needed which, expressed as a fraction of the total water use, is known as the *leaching fraction*, abbreviated LF.

Conditions within a soil profile can be predicted if it is assumed that water enters the surface at a constant rate, is removed at a constant rate by roots, and drainage water leaves at a constant rate. These are known as *steady state* conditions. They do not occur in the field because irrigation is applied when water and equipment are available and because evapotranspiration rates vary. However, such predictions are a useful guide to management.

Fig. 14.5 gives examples of steady state conditions for a 1 m deep profile with uniform removal of water by roots. The shape of the curves results from the fact that water is moving steadily down the profile with root uptake of water progressively increasing the concentration. At the surface the soil solution concentration is equal to that of the irrigation water. When half the water is used by roots (LF = 0.5), the EC rises from 4 to $8 \, \text{dS} \, \text{m}^{-1}$ through the profile, but if 9/10 of the water is used (LF = 0.1), the EC rises to $40 \, \text{dS} \, \text{m}^{-1}$. Section 14.6 describes how these data were obtained. Crop tolerance (Table 14.3) is used to decide the maximum acceptable salt concentration in the root zone, and the LF which maintains this condition is termed the *leaching requirement for salinity control*. Ayers and Westcot (1985) present similar data (their Fig. 2) which takes

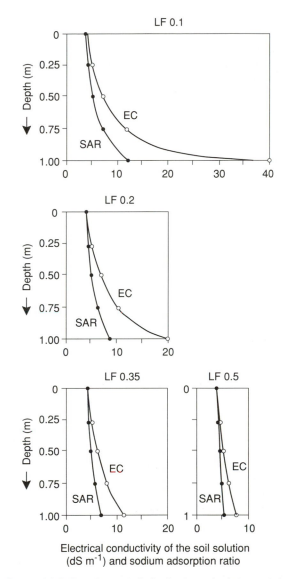

Electrical conductivity of the soil solution (dS m^{-1}) and sodium adsorption ratio

Figure 14.5 The theoretical distribution of salt in an irrigated soil profile. The irrigation water entering the profile has an EC of $4 \, \text{dS m}^{-1}$. It also has an SAR of 4. The data have been calculated from Eqs 14.3 and 14.4.

account of the non-uniform uptake of water through the root zone.

The changes in SAR are also shown in Fig. 14.5 calculated assuming that sufficient water has passed through the soil for equilibrium to be established between exchangeable cations and the solution. Initially cation exchange will buffer the solution, with SAR rising slowly to the equilibrium condition. The susceptibility of the soil to the effects of sodicity is used to decide the maximum acceptable SAR at the base of the profile. From this a *leaching requirement for sodicity control* can be found. Of the two leaching requirements the larger one should be used as this protects the soil from the effects of both salinity and sodicity.

Ameliorating the effects of poor quality water

Where good quality water is in short supply, poor quality water may have to be used. The following management procedures will reduce the hazards (Section 14.6):

- The salinity of water can be reduced by mixing with good quality water.
- The sodicity of water can be reduced in the same way: just as the SAR of soil solution is increased as water evaporates, so the SAR of poor quality irrigation water can be reduced by dilution with good quality water.
- The application of gypsum to soil will raise the calcium concentration in the soil solution and lower its SAR.

SALINITY IN GLASSHOUSES

Salinity may occur in soils or composts in glasshouses in the form of potassium, nitrate and chloride resulting directly from the application of fertilizers or from the accumulation of residues of fertilizers and liquid feeds (nutrient solutions) in excess of crop needs. The measurement and the use of an EC index is described in Section 14.7.

FURTHER STUDIES

Ideas for projects are in Section 14.8 and calculations are in Section 14.9.

Section 14.1 The determination of the composition of irrigation water

The water sample

To ensure that a representative sample is obtained, mix together several portions collected at different times. Samples from wells should be collected after the pump has been running for some time, and samples from streams should be taken from running water.

The carbonate plus bicarbonate concentration should be measured as soon as possible to prevent changes due to absorption or loss of CO_2 and precipitation of calcium carbonate. The latter also decreases the calcium concentration.

The carbonate plus bicarbonate concentration

These are determined by titration with standard acid. Very little carbonate is present unless the pH value is high. Carbonate is determined by titrating down to pH 8.2 using a pH meter or phenolphthalein indicator and the bicarbonate by further titration to pH 4.5 or using methyl orange (Nuffield Advanced Chemistry II, Topic 12.5). For most purposes the combined concentration only is required: the normal range of concentration is 0–$10 \, mmol_c \, l^{-1}$ (Note 1).

Reagents

Hydrochloric acid, 0.01 mM.
Methyl orange indicator. Dissolve 0.1 g of methyl orange in 100 ml of water.

Method

Pipette 10 ml of irrigation water into a 250 ml conical flask, dilute to about 50 ml with distilled water, add a few drops of indicator, and titrate with 0.01 M HCl to the end point (yellow → red).

Calculation

The reaction is

$$HCO_3^- + HCl = H_2O + CO_2 + Cl^-$$
$$CO_3^{2-} + 2HCl = H_2O + CO_2 + 2Cl^-$$

In both cases $1 \, mol_c$ of CO_3^{2-} or HCO_3^- reacts with 1 mol of HCl (1 mol CO_3^{2-} contains $2 \, mol_c$). Thus the number of mol_c of $CO_3^{2-} + HCO_3^-$ in 10 ml of water is equal to the number of mol of HCl used in the titration. If the volume of acid used is for example 5.2 ml, then the number of mol of acid is

$$0.01 \, mol \, l^{-1} \times 5.2/1000 \, l = 5.2 \times 10^{-5}$$

or $\qquad\qquad 0.052 \, mmol$

Thus there is $0.052 \, mmol_c \, CO_3^{2-} + HCO_3^-$ in 10 ml of irrigation water, or $5.2 \, mmol_c \, l^{-1}$.

The calculation simplifies to

$$mmol_c(CO_3^{2-} + HCO_3^-)l^{-1} = \text{titration volume (ml)}$$

and is often referred to simply as a HCO_3^- concentration or as the *alkalinity* of the water. The latter is probably the best term because very small concentrations of hydroxyl ions are also titrated. Note that the $CO_3^{2-} + HCO_3^-$ concentration is broadly related to pH which can be used to check for errors in the concentration measurements. The concentration seldom exceeds $10 \, mmol_c \, l^{-1}$ if pH is at or below 8.5 and seldom exceeds 3–$4 \, mmol_c \, l^{-1}$ at pH 7 or below.

Electrical conductivity

The ability of solutions to conduct electricity depends on the concentration of ions present and their electrical charge. A conductivity meter measures the current passing through a solution between two electrodes (normally platinum surfaces) in a conductivity cell.

Definitions and relationships

Conductance This is the ability of a solution to carry electricity. It is the reciprocal of resistance and has the unit siemens $(S) = ohm^{-1}$

Electrical conductivity (EC). This is the conductance of a solution filling the space between two metal surfaces 1 m apart, each with an area of $1\,m^2$. It has the symbol K and the unit $S\,m^{-1}$. Most of the salinity literature uses the old unit, $mmho\,cm^{-1}$ (pronounced milli mow), for electrical conductivity, where mho = S. For convenience therefore data are often expressed as $dS\,m^{-1}$ (decisiemens per metre) because $1\,dS\,m^{-1} = 1\,mmho\,cm^{-1}$. The unit dS is $10^{-1}\,S$.

Concentration For a range of mixed salt solutions, Fig. 14.6 shows the approximate relationship between concentration and EC at 25 °C. As a rough guide

the concentration of cations or anions
$$(mmol_c\,l^{-1})\ (Note\ 1) \simeq 10 \times EC\ (dS\,m^{-1})$$

The unit $mmol_c\,l^{-1}$ has to be used because conductivity depends on both charge and the number of ions. The fit of data in Table 14.1 to the graphs can be checked: for example, the River Pecos water has 17.3, 9.2 and $11.5\,mmol_c\,l^{-1}$ of Ca^{2+}, Mg^{2+} and Na^+ respectively, giving a total of 38 for an EC of $3.2\,dS\,m^{-1}$. Converting numbers to logarithms gives 1.58 and 0.51 which is plotted on Fig. 14.6 and which falls within the predicted range.

Temperature The EC of a solution is temperature dependent. Values are given at 25 °C. To convert values measured at 20 °C to values at 25 °C, multiply by 1.112. ADAS conductivity measurements (Section 14.7) are given at 20 °C.

Equipment and reagents

Electrical conductivity meter.
Conductivity cell.
Conductivity standards. Dry NaCl at 105 °C for 1 h and cool in a desiccator. Dissolve 2.922 g in water and make up to 1 l. This is 50 mM NaCl. Dilute this by 10 and by 100 to give 5 and 0.5 mM solutions. These have EC values of 5.550, 0.604 and $0.0625\,dS\,m^{-1}$ at 25 °C or 4.995, 0.543 and $0.0562\,dS\,m^{-1}$ at 20 °C.

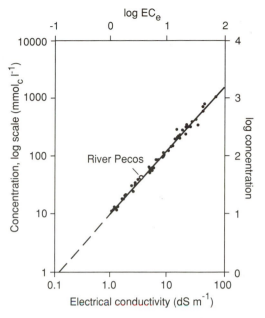

Figure 14.6 The relationship between solution concentration (saturation extracts) and electrical conductivity at 25 °C. The line was determined for saturation extracts from 50 soils from the western USA. It also applies to irrigation waters since both are mixed salt solutions with a similar range of ion concentrations. The graph has been extrapolated back to $0.1\,dS\,m^{-1}$ to include the range for good quality waters. The solution concentration is the total concentration of cations or anions (Section 14.1, Note 1). From Richards (1954).

Method

Many conductivity cells have a cell factor supplied with them. The meter can be set to this factor and the cell used directly for measurement. It is advisable, however, to calibrate the meter with the standard NaCl solutions. Either fill the cell with the solution, or dip the cell into the solution depending on the type of cell being used. Use the 5 mM solution to calibrate the meter, and check that it reads correctly on the other two solutions. Work at either 20 or 25 °C. Measure the EC of the sample, and record the value at 25 °C, using the conversion factor if necessary.

The meter will probably give values in $mS\,cm^{-1}$ or $\mu S\,cm^{-1}$ as these units are still commonly used. There is a direct conversion because $1\,mS\,cm^{-1} = 1\,dS\,m^{-1}$.

Sodium, calcium and magnesium

The usual ranges of concentration $(mmol\,l^{-1})$ in irrigation water are 0–50 Na^+, 0–20 Ca^{2+} and 0–10 Mg^{2+}, which are approximately $(mg\,l^{-1})$ 0–1000 Na^+, 0–800 Ca^{2+} and 0–250 Mg^{2+}. An atomic absorptiometer is normally used for their measurement. Details of its use are not given here, but the normal calibration ranges for these ions are $(mg\,l^{-1})$ 0–1 Na^+, 0–5 Ca^{2+} and 0–0.5 Mg^{2+}. The irrigation water may have to be diluted to bring it into the range for measurement. Possible dilutions are ×500 for Na^+ (by emission), ×100 for Ca^{2+} and ×250 for Mg^{2+}, but trial and error may be needed to obtain satisfactory readings.

If an atomic absorptiometer is not available, the following methods may be used.

1. Sodium by flame photometry

Equipment and reagents

Flame photometer and sodium filter.
Sodium chloride standards. Dry NaCl at 105 °C for 1 h and cool in a desiccator. Dissolve 0.254 g in water and make up to 100 ml. This contains $1\,g\,Na^+\,l^{-1}$. Pipette 10 ml into a 100 ml volumetric flask and make up to the mark. Into 100 ml volumetric flasks, pipette 0, 2, 4, 7 and 10 ml of this diluted standard and make up to the mark. These contain 0, 2, 4, 7 and $10\,mg\,Na^+\,l^{-1}$. Alternatively a volumetric standard containing 1 g $Na^+\,l^{-1}$ may be available from the manufacturers.

Method

Calibrate the flame photometer following the methods of Section 7.2.

The irrigation water will probably have to be diluted ×50 to bring it into the range for measurement: pipette 1 ml into a 50 ml volumetric flask and make up to the mark. Measure the concentration of this diluted irrigation water.

Calculation

Method summary

water sample $(?\,mmol\,Na^+\,l^{-1})$
$$\downarrow$$
1 ml → 50 ml, measured concentration $(mg\,l^{-1})$

Example The measured concentration in the diluted sample is $6.5\,mg\,Na^+\,l^{-1}$ and so the concentration of the undiluted sample is

$$6.5 \times 50 = 325\,mg\,l^{-1} \text{ or } 0.325\,g\,l^{-1}$$

Its molarity is $0.325\,g\,l^{-1}/23\,g\,mol^{-1} = 14\,mmol\,l^{-1}$.

2. Calcium plus magnesium by titration

The method is given in Section 7.2. Titrate 10 ml of the irrigation water against 0.005 M EDTANa$_2$.

Calculation

Method summary

water sample $(?\,mmol\,Ca^{2+} + Mg^{2+}\,l^{-1})$
$$\downarrow$$
10 ml titrated against 0.005 M EDTANa$_2$

Example The titration volume is 7.5 ml. The number of mol of EDTANa$_2$ used is

$$0.005\,mol\,l^{-1} \times 7.5/1000\,l = 3.75 \times 10^{-5}\,mol$$
or
$$0.0375\,mmol$$

The same number of mol of $Ca^{2+} + Mg^{2+}$ is present in 10 ml of water. The concentration is therefore

$$0.0375 \times 1000/10 = 3.75\,mmol\,Ca^{2+} + Mg^{2+}\,l^{-1}$$

The calculation simplifies to

$$mmol\,Ca^{2+} + Mg^{2+}\,l^{-1} = \text{titration volume (ml)}/2$$

3. Check your values

The relationship between concentration and EC in Fig. 14.6 allows a check to be made on your measurements. Using the above examples, the concentration of the irrigation water is

$$14\,mmol_c\,Na^+\,l^{-1} + (3.75 \times 2)\,mmol_c\,Ca^{2+} + Mg^{2+}\,l^{-1}$$
$$= 21.5\,mmol_c\,l^{-1}$$

Assuming only small amounts of other cations (K^+, NH_4^+), the EC should be about $2\,dS\,m^{-1}$.

4. Other measurements

Calcium and magnesium can be determined separately by titration (Section 7.2) and sulphate and chloride by the methods of Sections 10.5 and 7.6 respectively.

5. Sodium adsorption ratio

Calculate from 1. and 2. above:

$$SAR = [Na^+]/([Ca^{2+} + Mg^{2+}])^{1/2}$$

where [ ] is the concentration in $mmol^{-1}$.

Note 1 The use of the unit mole of charge (mol_c) for exchangeable cations has been explained in Section 7.2. In salinity studies the same unit is often used for the total concentration of a mixture of ions with different valency *in solution* (normally $mmol_c\,l^{-1}$), principally because the EC is related to the total charge carried on the ions. Also the measurement of $HCO_3^- + CO_3^{2-}$ concentration by titration is conveniently expressed by $mol_c\,l^{-1}$, as is $H^+ + Al^{3+}$ (Section 7.3). Note that in this chapter both mol_c and mol are used depending on the purpose, and that the number of moles of charge = the number of moles × the charge on the ion. Note also that the total concentration of a mixed salt solution expressed as $mmol_c\,l^{-1}$ is determined by summing either the cation or the anion concentrations, and that $mmol_c$ cations = $mmol_c$ anions.

Section 14.2 The preparation of a saturation extract and the analysis of soil salinity and sodicity

Soil salinity is of significance as a property of the soil solution. This can be studied by displacing solution from a field moist soil (Kinniburgh and Miles, 1983), but a simpler method is needed for routine purposes. Solution is normally extracted from a saturated paste or from a soil suspension (1:5, soil:water) and in the solution EC and the concentrations of sodium, calcium and magnesium are measured.

Soil sodicity is measured through the ESP. The methods of Section 7.2 are modified to take account of the presence of soluble salts and of gypsum and calcite which are slightly soluble.

SOIL SALINITY

The saturated paste

The soil sample should be air dried and passed through a 2 mm sieve. Determine its water content as $g\,H_2O$ per 100 g oven-dry soil following the methods of Section 3.3.

Weigh about 300 g (±0.1) of air-dry soil into a weighed 500 ml plastic beaker. Add distilled water with stirring until nearly saturated. Allow the paste to stand for several hours to wet thoroughly, and then add more water to form a saturated paste. In this state the paste glistens as it reflects light, flows slightly when the beaker is tipped, slides freely and cleanly off a spatula, and is consolidated easily by tapping the beaker after a trench has been formed in the paste with a spatula. Cover the beaker and stand overnight and then check the condition of the paste. If necessary add more distilled water or more dry soil. Weigh the paste + beaker to determine the mass of water added to the soil.

Soil pH

If required measure pH in the paste at this stage (Section 8.1).

The saturation percentage

This is the mass of water in the saturated paste expressed as a percentage of the oven-dry mass of soil.

Calculation

The mass of water in the air-dry soil plus the mass of water added to the paste is the total water content and should be expressed as mass of water per 100 g of oven-dry soil.

The saturation extract

Reagent

Sodium hexametaphosphate, 0.1 per cent. Dissolve 0.1 g of $(NaPO_3)_6$ in 100 ml of water.

Method

Fit a Buchner funnel with a Whatman No. 50 filter paper. Transfer the paste to the funnel and apply a vacuum. If the initial filtrate is turbid, refilter through the same paper. Terminate the filtration when air begins to pass through the paper. Normally between 1/4 to 1/3 of the water in the paste is recovered (i.e. between 20 and 60 ml).

Add sodium hexametaphosphate solution to the filtrate at the rate of one drop per 25 ml to prevent precipitation of calcium carbonate. It adds about $0.02\,mmol\,Na^+\,l^{-1}$ which is insignificant in relation to $soil\text{-}Na^+$.

Measurements on the saturation extract

1. Electrical conductivity

Follow the method given for irrigation water (Section 14.1).

2. Sodium, calcium and magnesium

The methods of Section 14.1 can also be used. For atomic absorption greater dilution may be required than for irrigation water because of the larger concentrations found in the extracts, the ranges being $0\text{--}150\,Na^+$, $0\text{--}20\,Ca^{2+}$ and $0\text{--}20\,Mg^{2+}$ ($mmol\,l^{-1}$). Suggested dilutions are ×1000 for Na^+, ×100 for Ca^{2+} and ×500 for Mg^{2+}.

3. Sodium by flame photometry

Dilute 1 ml of extract to 100 ml before measurement and adjust the calculation accordingly (Section 14.1).

4. Calcium plus magnesium by titration

Titrate 5 ml of extract and adjust the calculation accordingly (Section 14.1).

5. Sodium adsorption rates

Calculate as in Section 4.1.

Table 14.5 The saturation percentages of soils

Texture	Saturation percentage (g H_2O per 100 g oven-dry soil)
Coarse	32† (16–43)
Medium	43 (26–60)
Fine	60 (42–79)
Organic	142 (81–255)

† An average is given followed by the range. Sixty two soils were examined (Richards, 1954).

Values and relationships

Saturation percentages vary with texture as shown in Table 14.5 and are greater than the water contents in field moist soils. Soil solution salinities are therefore greater than those measured in saturation extracts. However, water contents in field moist soils also vary with texture (Table 5.2), and for all except sands, the saturation percentage is about four times the water content at the permanent wilting point and about twice the water content at field capacity. Thus although the saturation extract is two or four times more dilute than the soil solution at field capacity and the wilting point respectively the constancy of these factors means that the EC of the saturation extract can be used as an index of the effect of salinity on plant growth.

For sands the saturation percentage is between two and four times the water content at field capacity because of the large volume of transmission pores which are water-filled in the saturated paste but empty at field capacity. Thus the EC of the saturation extract underestimates the effect of soil salinity and the method below should be used for these soils.

A modified extraction procedure for sands

If a small volume of a sandy soil is wetted, but remains in contact with dry soil, water spreads until the regions of moist soil are approximately at field capacity. This water content is measured. Addition of 4 times this water content allows an extract to be obtained by vacuum filtration. The method above for heavier textured soils gives a saturation extract which is 2 times the water content at field capacity. Thus the EC of the modified extract for sands is multiplied by 2 to give a soil salinity value on the same scale as those obtained from saturation extracts of other soils.

Equipment

A soil container, 10–12 cm diameter with a loosely fitting basket made from a mesh with approximately 6 mm openings.

Method

Place the basket in the container . Add <2 mm air-dry soil to about 3 cm depth. Pipette 2 ml of water on to the soil surface wetting several spots but leaving the rest of the soil dry. Leave to stand for 15 min. Lift out the basket and gently sieve out the dry soil. Weigh the pellets of moist soil.

Calculation

The moist soil contains 2 ml of water (2 g). Therefore the mass of air-dry soil in the pellets = (mass of moist pellets − 2) g. The water content of the pellets is

$$100[2/(\text{mass of moist pellets} - 2)] \text{ g } H_2O \text{ per 100 g air-dry soil}$$

The water content of air-dry sands is small and so this value is approximately equal to g H_2O per oven-dry soil.

The modified extract

To 300 g of air dry-soil add water to bring the water content to 4 times the water content of the pellets. Mix and stand overnight. Extract and measure following the methods for the saturation extract. Multiply the EC by 2 to obtain a value which is used directly as an index of salinity (Table 14.3). The method assumes that gypsum and other minerals are not dissolving to any extent in the extract.

Direct measurements on the saturated paste

The electrical conductivity measured in the paste can be used to predict EC values in saturation extracts by the equation of Rhoades *et al.* (1989a) so avoiding the need for vacuum filtration.

The use of a 1:5 extract

If vacuum filtration is not available, the EC of any soil can be determined using a 1:5 extract: to 20 g of air-dry soil add 100 ml of water. Shake for 1 h. Allow the suspension to settle for 30 min and pipette 20 ml of the supernatant suspension into a beaker for the measurement of EC.

The EC of the saturation extract can be calculated as follows:

$$EC_{\text{saturation extract}} \simeq 6.4 \times EC_{\text{1:5 extract}}$$

The factor 6.4 is not simply the dilution factor (which is between 8 and 15) because clay in the suspension contributes to the EC of the 1:5 extract.

The concentrations of ions in the extracts

Although dilution of the soil solution to produce an extract has a predictable effect on EC, the effect on individual ions is less predictable. The concentrations of non-adsorbed Cl^- and NO_3^- change in proportion to the extra water added. Cations, however, are subject to exchange: dilution favours the adsorption of Ca^{2+} and Mg^{2+} with the release of exchangeable Na^+. Thus cation ratios in the extracts are not the same as in soil solution. The ratios in the saturation extract can, however, be used as a measure of sodicity as shown in the following sections.

SOIL SODICITY

The ESP can be measured directly or predicted from the composition of the saturation extract. Direct measurement involves either

- the displacement of soluble and exchangeable cations together, with exchangeable cations determined by difference using the saturation extract as a measure of soluble cations, or
- the use of a prewash to remove soluble cations before displacing the exchangeable cations.

Direct measurements are subject to uncertainties resulting from the presence of salts, and there is no agreement regarding the best method. That given here is a simple modification of the method in Section 7.2: the leaching solution contains 60 per cent ethanol which allows soluble salts to dissolve but largely prevents the dissolution of calcite and gypsum. Ammonium chloride instead of ethanoate (acetate) is used because a strongly buffered solution is not required, and the pH is adjusted to 8.5 which is close to the soil pH and helps to prevent dissolution of calcite.

Determination of exchangeable sodium and cation exchange capacity

1. Extraction

Reagent *Alcoholic ammonium chloride, 1 M.* Dissolve 53.5 g NH_4Cl in about 300 ml of water and add 600 ml of ethanol (95% *v/v*). Adjust the pH to 8.5–8.6 with concentrated ammonia solution and make up to 1 l. Store in a well-stoppered bottle to prevent access of CO_2.

Method Follow the leaching procedure of Section 7.2.

2. Measurement of leached sodium

Make up a set of standards for a flame photometer following the methods of Section 14.1. The concentration range is 0–$10\,mg\,Na^+\,l^{-1}$ but they should be made up in 1 M alcoholic ammonium chloride.

Dilute the extract by 10 (pipette 10 ml into a 100 ml volumetric flask and make up to the mark with alcoholic ammonium chloride) and measure its concentration using a flame photometer.

Method summary

$$<2\,mm \text{ air-dry soil } (?\,mmol\,Na^+\,kg^{-1} \text{ oven-dry soil})$$
$$\downarrow$$
$$5\,g \rightarrow 250\,ml$$
$$\downarrow$$
$$10\,ml \rightarrow 100\,ml, \text{ concentration measured as } mg\,Na^+\,l^{-1}$$

Example The flame photometer measurement was $2.0\,mg\,l^{-1}$. Therefore in 250 ml of leachate the concentration was $20\,mg\,l^{-1}$ and the amount present was

$$20\,mg\,l^{-1} \times 250/1000\,l = 5.0\,mg$$

This came from 5 g of soil and so there was $5.0 \times 1000/5 = 1000\,mg\,kg^{-1}$ which is

$$1000\,mg\,kg^{-1}/23\,g\,mol^{-1} = 43.5\,mmol\,Na^+\,kg^{-1} \text{ air-dry soil}$$

Express in terms of oven-dry soil (Section 3.3), e.g. $45.68\,mmol\,kg^{-1}$ oven-dry soil if the air-dry soil contained 5 g H_2O per 100 g oven-dry soil.

3. The amount of soluble sodium

This is measured in the saturation extract.

Example The saturation percentage = 48.5 g H_2O per 100 g oven-dry soil. The saturation extract has a concentration of $14.1\,mmol\,Na^+\,l^{-1}$. In the saturation extract there is

$$14.1 \times 48.5/1000 = 0.684\,mmol\,Na^+ \text{ per } 100\,g \text{ oven-dry soil}$$

or $6.84\,mmol\,kg^{-1}$ oven-dry soil.

4. The exchangeable sodium

This is found by difference. Exchangeable Na^+ is

$$\text{leached } Na^+ - \text{soluble } Na^+ = 45.68 - 6.84 = 38.84\,mmol\,kg^{-1}$$

or in the normal units of exchangeable cations, $3.9\,cmol_c\,kg^{-1}$.

5. Measurement of cation exchange capacity

After leaching with alcoholic ammonium chloride

(1 above), continue as in Section 7.2 with ethanol and KCl leachings and determine CEC.

6. Exchangeable sodium percentage

ESP = 100 (exchangeable Na^+/CEC), both expressed as $cmol_c\,kg^{-1}$.

7. An alternative method for exchangeable sodium

If salinity data from a saturation extract are not required, soluble salts can be removed from the soil by a preliminary leaching with glycol-ethanol which does not displace exchangeable cations. The procedure is then as above, but the Na^+ leached out with alcoholic ammonium chloride is now the exchangeable Na^+.

Reagent *Glycol-ethanol* Mix 100 ml of ethylene glycol (ethanediol) with 900 ml of ethanol (95% *v/v*).

Method To 5 g of soil in a beaker add 20 ml of glycol-ethanol, stir and allow to stand for 15 min. Transfer the suspension with washings to a funnel fitted with a Whatman No. 44 filter paper. Leach the soil with two further 20 ml volumes of glycol-ethanol running the leachate to waste.

Leach with alcoholic ammonium chloride into a 250 ml flask as in (1) above.

8. An alternative method for cation exchange capacity

The exchange capacity of saline and sodic soils is dominated by Na^+, Ca^{2+} and Mg^{2+} with small amounts of K^+. If in the above methods, exchangeable Ca^{2+} and Mg^{2+} are also measured, then CEC is approximately equal to exchangeable $Na^+ + Ca^{2+} + Mg^{2+}$. Section 7.2 gives the analytical methods for calcium and magnesium.

The determination of exchangeable sodium percentage from the sodium adsorption ratio of the saturation extract

Measurements show that there is a simple relationship between the ratios of exchangeable cations and of cations in saturation extracts. This is to be expected from the physical chemistry of ion exchange systems. In this context, Ca^{2+} and Mg^{2+} behave almost identically and can be considered together giving

$$Na^+_{ex}/(Ca^{2+}_{ex} + Mg^{2+}_{ex}) = G[Na^+]/[Ca^{2+} + Mg^{2+}]^{1/2}$$
$$[14.1]$$

where subscript ex indicates the amount of the

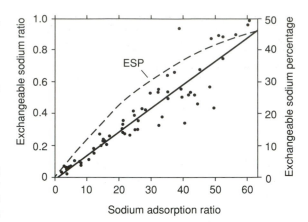

Figure 14.7 The relationship between ESR, ESP and SAR for the saturation extracts of 59 soils from the USA. From Richards (1954).

exchangeable cation in $cmol_c\,kg^{-1}$, [ ] indicates concentration in the saturation extract in $mmol\,l^{-1}$ and G is an exchange coefficient (Note 1). Eq. 14.1 is an example of the *Gapon exchange relationship*. The term on the left-hand side of the equation is known as the *exchangeable sodium ratio* (ESR) and that on the right hand side is the SAR.

Figure 14.7 illustrates the relationship. Fitting a straight line to the data by regression analysis (Section 9.6) gives

$$ESR = -0.013 + 0.015SAR$$

Other data are more scattered around the line, indicating a greater variability of exchange coefficients than is shown here.

Note 1 For some purposes the exchange relationship is given with the solution concentrations in $mol\,l^{-1}$. However, the value of SAR is calculated using $mmol\,l^{-1}$. The two are not interchangeable: because of the square root term, different values are obtained. For example, if $[Na^+] = 79$ and $[Ca^{2+} + Mg^{2+}] = 35.5\,mmol\,l^{-1}$, $SAR = 79/(35.5)^{1/2} = 13.25$. Using $mol\,l^{-1}$, $[Na^+] = 0.079$ and $[Ca^{2+} + Mg^{2+}] = 0.0355$ and the concentration ratio $= 0.079/(0.0355)^{1/2} = 0.419$.

The ESP can be calculated from ESR using the following equation:

$$ESP = 100ESR/(1 + ESR)$$

where $ESP = 100(Na^+_{ex}/CEC)$ and CEC is assumed to be equal to $Na^+_{ex} + Ca^{2+}_{ex} + Mg^{2+}_{ex}$.

The relationship between ESP and SAR is shown in Fig. 14.7. Over the range of practical interest ESP is almost equal to SAR. One of the reasons for calculating SAR using $mmol\,l^{-1}$ is to give this simple relationship.

Section 14.3 The measurement of plant tolerance to salinity in pot experiments

SOIL EXPERIMENTS

Preparation of saline soils

Using Table 14.3 and Fig. 14.1 choose the range of salinity required for treatment in terms of the EC of the saturation extract. Fig. 14.6 gives the total salt concentration for the EC values chosen. *Example* EC = $10 \, dS \, m^{-1}$ and the salt concentration = $120 \, mmol_c \, l^{-1}$.

Determine the saturation percentage of the soil to be treated (Section 14.2) or use the guidelines in Table 14.5 if an approximate value is adequate. *Example* 50 g H_2O per 100 g oven-dry soil.

Determine the water content of the air-dry soil (Section 3.3) or use the guidelines in Table 5.2 if an approximate value is adequate. *Example* 6.4 g H_2O per 100 g oven-dry soil which is 6.0 g H_2O per 100 g air-dry soil.

To calculate the salt treatment for 1 kg of air-dry soil, proceed as follows. It is assumed that before treatment the amounts of soluble salts present in the soil are insignificant compared to the treatments being applied. If this is not the case, measure the EC of the saturation extract, subtract from the treatment EC value and determine the required salt addition from Fig. 14.6.

In 1 kg of the above air-dry soil there is 60 g of water and 940 g of oven-dry soil. Made into a saturated paste this would contain

$$50 \times 940/100 = 470 \, g \, H_2O$$

in which there must be $120 \, mmol_c$ salt l^{-1}. The amount of salt to be added is therefore

$$120 \times 470/1000 = 56.4 \, mmol_c \, salt \, kg^{-1} \, air\text{-}dry \, soil$$

Choice of salts

The predominant ions in saline soils are Na^+, Ca^{2+}, Mg^{2+}, SO_4^{2-} and Cl^-. The proportions of each vary between soils and so an arbitrary choice has to be made. A mixture of $NaCl$, $CaCl_2$ and $MgCl_2$ can be used with equal amounts of Na^+, Ca^{2+} and Mg^{2+} ($mmol_c \, l^{-1}$). Some sulphate can be substituted if required. For this example the requirements are shown in Table 14.6(a). These amounts of salt can be mixed directly with the soil. However, if a range of treatments is required it is convenient to make up a concentrated solution and add known volumes to the soil. For example, dissolve 11.0 g $NaCl$, 13.8 g $CaCl_2.H_2O$ and 19.1 g $MgCl_2.6H_2O$ in 1 l

Table 14.6 Examples of the salt requirements for preparing (a) a saline soil and (b) a nutrient solution

(a) EC of the saturation extract at 25 °C = $10 \, dS \, m^{-1}$, total salt = $56.4 \, mmol_c \, kg^{-1}$ air−dry soil

	Na^+	Ca^{2+}	Mg^{2+}
Amount required			
($mmol_c \, kg^{-1}$)	18.8	18.8	18.8
($mmol \, kg^{-1}$)	18.8	9.4	9.4
Salt	$NaCl$	$CaCl_2.2H_2O$	$MgCl_2.6H_2O$
Molar mass ($g \, mol^{-1}$)	58.44	147.02	203.3
Mass of salt required			
($g \, kg^{-1}$)	1.10	1.38	1.91

(b) EC of the nutrient solution = $120 \, dS \, m^{-1}$

Required salt concentration ($mmol_c \, l^{-1}$)	143.5
Required ion ratio (m/m)	1 Na^+ : 2 Ca^{2+}
Molar mass ($g \, mol^{-1}$)	22.99 Na^+ 40.08 Ca^{2+}
($mol_c \, g^{-1}$)	0.0435 Na^+ 0.0499 Ca^{2+}
Required ion ratio ($mol_c : mol_c$)	1 Na^+ : 2.3 Ca^{2+}
Required ion concentrations	
($mmol_c \, l^{-1}$)	43.5 Na^+ 100 Ca^{2+}
($g \, l^{-1}$)	2.54 $NaCl$ 7.34 $CaCl_2.2H_2O$

of water. Using a pipette, sprinkle on to 1 kg of soil 100 ml of this solution with mixing. This gives the above treatment. Other volumes can be chosen to give the required range of treatments.

Note that this mixture of salts dissolved in the saturation extract (470 ml $H_2O \, kg^{-1}$ air-dry soil) gives an SAR of $40/(20 + 20)^{1/2} = 6.3$ if cation exchange does not alter the concentrations significantly. The ratios of ions are chosen so that soil structure is unlikely to be damaged by sodicity.

Add nutrients as required (Section 9.2).

The watering of pots or seed trays

Soluble salts move in water applied to soil. Stand the pots in saucers to prevent loss of drainage water containing salts. If germination tests are being conducted, seed trays should not have drainage holes. A uniform distribution of salt cannot be maintained even though losses can be prevented. Watering on to the surface of the soil will move salts down, but these will move up again as evaporation occurs.

In non-saline soils water stress is considered in terms of a lowering of the matric potential as water content decreases. In saline soils the osmotic potential has to be added to the matric potential to give a total water potential (Section 5.4). However, as water content decreases, the salt concentration in soil solution increases, so lowering the osmotic potential. Glasshouse experiments can be maintained with water contents near to field capacity, or the combined effects can be studied. The change in water potential caused by salts can be calculated as follows:

$$\text{osmotic potential (MPa)} = -0.04 \times EC \, (dS \, m^{-1})$$
[14.2]

Thus the osmotic potential in the saturation extract for the above soil is $-0.04 \times 10 = -0.4 \, MPa$. If the saturation percentage is 2 times the water content at field capacity, then the osmotic potential of the soil solution is $-0.8 \, MPa$. This can be compared to the water potential in a non-saline soil at field capacity ($-0.01 \, MPa$) and at wilting point ($-1.5 \, MPa$). Thus the treated soil must be maintained near to field capacity if water stress is to be avoided. Note that sea water (Table 14.1) has an osmotic potential of $-2.0 \, MPa$, thus preventing the growth of plants which are not adapted to saline environments. Artificial sea water can be prepared by dissolving the following salts in water containing a raised concentration of CO_2. This can be prepared by bubbling CO_2 through distilled water and prevents the precipitation of $CaCO_3$ when the salts dissolve. To 965 ml of water add the following salts (g): NaCl 27.2, $MgCl_2$ 3.8, $MgSO_4$ 1.6, $CaSO_4$ 1.3, K_2SO_4 0.9, $CaCO_3$ 0.1, $MgBr_2$ 0.1. The properties of sea water are discussed further by the Open University Course Team (1989).

Note that the osmotic potential can be determined from the composition of the solution using Fig. 14.6 or approximately as follows:

$$\text{osmotic potential (MPa)} \simeq -0.004 \times \text{concentration} \, (mmol_c l^{-1})$$

(Section 14.1, Note 1). Section 5.6, Project 3 brings together the effects of osmotic and matric potentials in a study of seed germination.

SAND CULTURE EXPERIMENTS

Plant tolerance is often studied in solution or sand culture in order to maintain more uniform salinity around the roots. Methods are in Hewitt (1966). To the nutrient culture solutions, NaCl and $CaCl_2$ are added to give the required EC. The ratio of $Na^+:Ca^{2+}$ (*m/m*) is often 1:2. An example is given in Table 14.6(b). The required salt concentration is obtained from Table 14.3 and Fig. 14.6. The required ion ratio (*m/m*) is then converted into a ratio expressed as $mol_c:mol_c$ to allow the salt to be partitioned between NaCl and $CaCl_2$.

FIELD PLOTS

Salinity studies in field plots are beyond the scope of this book. In treated plots salinity varies both with depth and time depending on irrigation and evapotranspiration and plants respond to a constantly changing heterogeneous system.

Section 14.4 The swelling and dispersion of clays

Clay layers and crystals (Figs 2.3 and 2.4) bind together because of electrical interactions between their surfaces, and are held together when humus, sesquioxides and calcium carbonate act as cements. Only the interactions between 'clean' clay surfaces will be considered here. Clays with their CEC values (cation exchange capacities) dominated by calcium and magnesium (neutral and calcareous soils) or by aluminium (acidic soils) are strongly bound together. With an increasing proportion of exchangeable sodium, 2:1 lattice clays swell causing the layers to separate and disperse. Even the presence of cements cannot completely protect soil structure from the damaging effect of swelling and dispersion.

Of the swelling clays, smectites are commonly found in arid and semi-arid regions. Of the non-swelling 1:1 lattice clays, kaolinite is commonly present. The explanation of swelling which follows is based on two parallel layers of smectite, but the same principles apply to the surfaces of two adjacent crystals. Section 2.2 and Table 2.3 give information on clay minerals.

A calcium smectite in air

A clay in which Ca^{2+} satisfies the CEC is known as a calcium clay. When completely dry the layers are close together with exchangeable Ca^{2+} sandwiched between them. When placed in the atmosphere, the calcium ions attract water to become hydrated. The layers have to move apart to make space for the water (Fig. 14.8(a)).

The forces operating between the layers are repulsion due to hydration, and attraction by electrostatic forces operating between the negatively charged layers and the positive cations which are now positioned midway between the layers. The cations apparently provide a link between the charged sites in the facing layers. There are also attractive forces operating directly between the layer surfaces known as van der Waals' forces.

The changes in the thickness of the film of water as the air becomes more humid are shown in Fig. 14.9(a). The water is sorbed in monomolecular layers causing stepwise expansion. The water molecule is 0.26 nm diameter. The third water layer is apparently less well

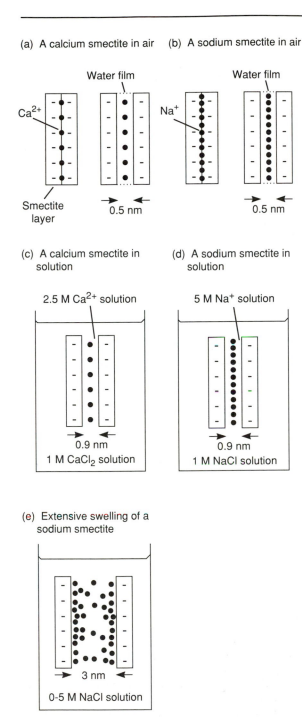

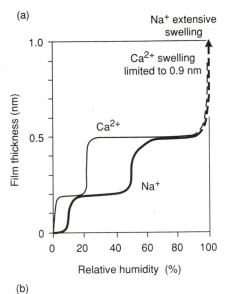

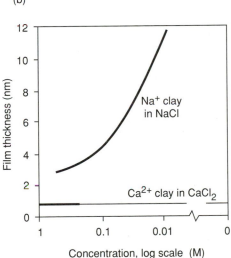

Figure 14.9 Expansion of an oriented flake of smectite in a humid atmosphere and in solutions. Smectites vary in their swelling characteristics. The lines drawn in (a) give typical results. Particularly between 90 and 100 per cent relative humidity the measurements are unstable. The line on the graph is broken to indicate this uncertainty. (a) From Suquet *et al.* (1975) and Brindley and Brown (1980); (b) from Norrish (1954).

Figure 14.8 The swelling of smectite.

two water layers predominates over an extensive range of humidities. In a saturated atmosphere, a water film of 0.9 nm thickness forms and the layers are still firmly bound together.

A sodium smectite in air

There are twice the number of exchangeable sodium ions compared to calcium because of their charge (Fig. 14.8(b)). The tendency for a sodium ion to hydrate is less than for a calcium ion, and so the thickness of the water film lags behind that for the calcium clay as humidity increases (Fig. 14.9(a)). In a saturated atmosphere extensive swelling occurs beyond 0.9 nm.

In moist soils, the humidity of the soil atmosphere is slightly less than saturated because of the attraction of water to the particle surfaces and to the dissolved salts. As a result, both calcium and sodium smectite will be held with a water film 0.5 nm in thickness. However, in wet soils, the clay is bathed in the soil solution and the tendency to swell must now be considered in this context.

A calcium smectite in solution

Imagine the calcium smectite with a 0.5 nm water film thickness (Fig. 14.8(a)) placed in a calcium chloride solution. Repulsion can now be considered as an osmotic phenomenon. The water film can be considered as a solution containing calcium ions. The concentration of calcium in the 0.5 nm film is about 4 M. If the concentration of the bathing solution is less than this, there is an osmotic pressure difference between the two solutions and water tends to move into the space between the layers to dilute this solution. The alternative, the movement of calcium ions out into the bathing solution, is prevented by the electrostatic attraction of the layer charge for the ions. The clay and its water are thus behaving like a plant cell: instead of a cell membrane holding the cell contents, the layer charge is holding the ions (Nuffield Advanced Biology I, Ch. 9). In 1 M $CaCl_2$ solution the film thickness expands to 0.9 nm (Figs 14.8(c) and 14.9(b)), the interlayer solution concentration is about 2.5 M and attractive and repulsive forces are balanced.

The maximum repulsion due to osmotic pressure occurs when the clay is placed in pure water. Even in this situation, the attractive forces balance the repulsion and the clay is stable with a 0.9 nm water film (Fig. 14.9(b)). This is the situation in all soils not containing appreciable quantities of sodium. Even with rain-water flowing into the soil, the clay system is stable.

A sodium smectite in solution

The attraction is of similar magnitude to that in the calcium clay, but the osmotic repulsion is approximately doubled due to the greater number of sodium ions. Note that osmosis is a colligative property, i.e. it is determined by the number of dissolved ions or molecules (Nuffield Advanced Biology I, Ch. 8.2). If the concentration of sodium in the bathing solution is more than about 0.5 M, the clay is stable with a water film of 0.9 nm and the sodium concentration in the water film is about 5 M (Figs 14.8(d) and 14.9(b)). Below 0.5 M, osmotic repulsion overcomes the attraction and the layers are forced further apart.

There is now a fundamental change in the balance of forces. The exchangeable sodium ions swarm against each layer, and the charge is satisfied by what is termed a *diffuse layer* of cations against each surface. There is no longer sufficient electrostatic attraction to hold the layers together. As the water film expands, its concentration of sodium ions decreases reducing the osmotic pressure difference between the film and the bathing solution. The repulsive force decreases until it is balanced by the weak van der Waals' forces, giving a film thickness of 3 nm or greater. Fig. 14.9(b) shows the expansion of the film as the concentration of the bathing solution decreases. At about 10 nm, the layers are so loosely held together that the clay is effectively dispersed.

The implications of swelling and dispersion

Figure 14.2 shows the uptake of water by flakes of clay in solution. The amount of water taken up is greater than would be expected from the changes in film thickness: the larger spaces between crystals expand more than the films within the crystals. Once expansion begins the strength of the clay rapidly decreases, and mechanical forces (raindrop impact, wheels, etc.) more easily cause damage to structure.

Mixed calcium–sodium clays have intermediate properties. An ESP of 10–15 is required to allow swelling to begin. At these ESP values, the exchangeable sodium apparently is present primarily on the outer faces of the crystals and not between layers in the crystals so that swelling and dispersion are primarily a separation of crystals. As ESP increases more sodium penetrates between the layers and swelling increases within the crystals.

The hydration expansion shown in Fig. 14.9(a) is reversible. However, although the extensive swelling of a single sodium smectite crystal (Fig. 14.9(b)) may be partly reversed if the solution concentration is increased, in soils this swelling is largely irreversible because the relative orientation of clay layers, crystals and other soil particles has been changed. Thus drying is necessary to reverse the effects of extensive swelling but even so the arrangement of the soil material is not the same as before swelling. Soil structure has been damaged: the soil has become less porous, is more plastic when wet and harder when dry. These changes are more extreme if the clay is dispersed and then dried. Additions to damaged soils of amendments such as gypsum which increase the salt concentration and reduce the amount of exchangeable sodium are therefore of little value until drying occurs. The reclamation of structure damaged by sodicity requires chemical amendment, drying and the structure-forming processes induced by plant roots and other organisms.

Other changes in sodic soils

The dilution of soil solution by rain or irrigation water has other important effects:

- Cation exchange causes the preferential adsorption of calcium and a reduction in ESP (Section 14.6). Only if the initial salinity is high does this confer significant protection from the effects of sodicity.
- Hydrolysis of the sodium-rich clays is a weathering process which involves desorption of exchangeable sodium, and an increase in alkalinity. The reactions are complex but involve exchange of Na^+ with H^+ from the water, and a subsequent attack of the clay by exchangeable H^+. Components of the clay are released, including magnesium and other cations which become exchangeable.

$$clay^{2-} Na_2^+ + 2H_2O = clay^{2-} H_2^+ + 2Na^+ + 2OH^-$$
$$\downarrow$$
$$clay^{2-} Mg^{2+}$$

The ESP of the clay is reduced, and the pH rises. The CEC is also reduced by the weathering process.

Differences between clay minerals

- Smectites have pronounced swelling properties because of their layer characteristics, charge and small particle size.
- Vermiculite also swells, but its layer characteristics and greater charge prevent extensive swelling of the sodium clay.
- Water does not penetrate into illite crystals: swelling is therefore limited to the separation of crystals which can damage the structure of illite-rich soils.
- Water does not penetrate into kaolinite crystals and there is little swelling between crystals. Their low charge makes them relatively insensitive to the effects of sodicity.

Section 14.5 Laboratory experiments to measure the effects of sodicity on clay swelling and soil hydraulic conductivity

Relatively simple experiments have proved useful in developing an understanding of the effects of sodicity on soils. Those described here involve bringing soils or clay flakes to known ESP values using concentrated solutions of known SAR. More dilute solutions of the same SAR allow changes to be measured which simulate those occurring under field conditions.

The preparation of solutions

For most purposes calcium and magnesium can be assumed to behave identically and solutions can be prepared using sodium and calcium salts only. In this case $SAR = [Na^+]/[Ca^{2+}]^{1/2}$ where $[\ ] = mmol\,l^{-1}$. However, total cation concentrations of 1000, 100 and $10\,mmol\,l^{-1}$ are often used as a suitable series for bringing soil or clay to a known ESP, and so the expression for SAR has to be in $mmol_c\,l^{-1}$. The use of $mmol_c\,l^{-1}$ for the total concentration means that a comparison of effects at different SAR values can be made at the same salinity. Note that salinity is rated according to the EC of the soil solution which is related to $mol_c\,l^{-1}$ in a mixed solution (Section 14.1).

Thus, using $mmol_c\,l^{-1}$ the total concentration $= \{Na^+\} + \{Ca^{2+}\}$ and $SAR = \{Na^+\}/(\{Ca^{2+}\}/2)^{1/2}$ where $\{\ \}$ is $mmol_c\,l^{-1}$. Taking as an example a concentration of $100\,mmol_c\,l^{-1}$ and SAR 15, the following method is used to calculate the concentrations of each cation:

$$\{Na^+\}/(\{Ca^{2+}\}/2)^{1/2} = 15$$
Squaring gives $\{Na^+\}^2/(\{Ca^{2+}\}/2) = 225$
and so $\{Ca^{2+}\} = \{Na^+\}^2/112.5$
Also $\{Na^+\} + \{Ca^{2+}\} = 100$

These simultaneous equations can be solved by substitution:

$$\{Na^+\} + \{Na^+\}^2/112.5 = 100$$
and so $\{Na^+\}^2 + 112.5\{Na^+\} - 11\,250 = 0$

This is a quadratic equation of the form

$$ax^2 + bx + c = 0$$

which can be solved using the equation

$$x = [-b \pm (b^2 - 4ac)^{1/2}]/2a$$

Thus substituting $a = 1$, $b = 112.5$ and $c = -11\,250$ gives $x = \{Na^+\} = 63.8$ and so $\{Ca^{2+}\} = 36.2$. Table 14.7 gives solution compositions for a range of concentrations and SAR values. Other combinations can be calculated as required. Note that to calculate the mass of $CaCl_2.2H_2O$ to give a concentration of $36.2\,mmol_c\,l^{-1}$, either convert to $mmol\,l^{-1}$ ($= 18.1$) and multiply by $147.02 \times 10^{-3}\,g\,mmol^{-1}$, or take $36.2\,mmol_c\,l^{-1}$ and multiply by $73.51\,g\,mmol_c^{-1}$.

Solution volumes required to bring soil to the required SAR value

In the following experiments, soil or clay is washed with solution until its exchange sites are in equilibrium with the solution. As an example, a soil column containing 80 g of soil (a column 15 cm long × 2.5 cm diameter, bulk density $1.1\,g\,cm^{-3}$) with an exchange

Table 14.7 Solution compositions having known SAR values and total concentrations

SAR	Total concentration	Na$^+$	Ca^{2+}	NaCl (g l^{-1})	CaCl$_2$.2H$_2$O (g l^{-1})
	(mmol$_c$ l^{-1})				
∞	1000	1000	0	58.44	0
(sodium	100	100	0	5.84	0
only)	10	10	0	0.58	0
25	1000	424.2	575.8	24.79	42.33
	100	79.7	20.3	4.66	1.49
	10	9.70	0.30	0.567	0.022
15	1000	283.8	716.2	16.59	52.65
	100	63.8	36.2	3.73	2.66
	10	9.24	0.76	0.540	0.056
5	1000	105.7	894.3	6.18	65.74
	100	29.7	70.3	1.74	5.17
	10	6.56	3.44	0.383	0.253
0	1000	0	1000	0	73.51
(calcium	100	0	100	0	7.35
only)	10	0	10	0	0.735

Molar mass (g mol$^-$): NaCl = 58.44, CaCl$_2$.2H$_2$O = 147.02, and 1 mol$_c$ of CaCl$_2$.2H$_2$O = 73.51 g.

capacity of 20 cmol$_c$ kg^{-1} has a column exchange capacity of 1.6 cmol$_c$. If a 1000 mmol$_c$ l^{-1} solution is passed through the column, 16 ml would contain 1.6 cmol$_c$ of Na$^+$ + Ca^{2+}. The pore volume of the column would be about 50 cm^3, and so saturation of the column with this solution provides about 3 times the amount of cations held on the soil. Displacement of one pore volume by another (allowing 50 ml to run through the column) will bring the soil close to equilibrium.

For the 100 mmol$_c$ l^{-1} solution, 10 times as much solution is required, and for the 10 mmol$_c$ l^{-1} solution 1000 times as much is required.

Clay flakes may contain about 0.1 g of clay with an exchange capacity of only 0.1 mmol$_c$ (100 cmol$_c$ kg^{-1} for smectite), and so only 1 ml of the 1000 mmol$_c$ l^{-1} solution contains 10 times the amount of cations held on the clay, and the flake is easily brought to equilibrium by soaking in a few ml of solution.

The relationship between ESP and SAR

To determine equilibrium ESP values produced with these solutions, use the relationship in Fig. 14.7. However, with large concentrations this relationship may not hold because the ratio of exchangeable ions depends strictly on a ratio of cation activities in solution (Section 7.1, Note 2) and not on concentrations (Eq. 14.1). Corrections for activities are not included in the calculation of SAR. The ESP values with large concentration may be greater than predicted from Fig. 14.7 for this reason. However, equilibrating soil with a series of solutions of fixed SAR but decreasing concentration rectifies this error because the relationship holds with small concentrations. If required, ESP can be measured using the methods of Section 14.2.

A simple experiment to measure the structural damage caused by sodicity

A simple leaching system is described in Section 5.5 (Fig. 5.16(a)). After placing a pad of absorbent cotton wool in the base of the tube add about 25 ml of <2 mm air-dry soil (light or medium texture) and cover with another pad of cotton wool. Place about 400 ml of a 1000 mmol$_c$ l^{-1} solution in the flask, close the clip, insert into the leaching tube and open the clip. The flask acts as a reservoir maintaining the head of solution in the leaching tube. If the flow rate is fast use a spatula to press down the cotton wool and soil to adjust the flow rate to about one drop per second.

When the soil is thoroughly wetted and the flow rate is steady, collect the effluent solution in a measuring cylinder for 1 min and record its volume. Close the clip, remove the flask, and refill with solution having the same SAR but a lower concentration. Place this in the leaching tube, and measure flow rate over 15 min. Continue with decreasing concentrations finally passing distilled water. Figure 14.3 gives a guide to the concentrations and ESP values at which swelling begins and flow rate changes may be recorded. The simplest experiment compares SAR ∞ and 0 which are sodium- and calcium-saturated soils respectively.

It is not necessary to measure hydraulic conductivity as in Section 5.5: a simple comparison of flow rates is adequate. Comparison of results from different treatments is complicated by the fact that the initial flow rates will be different because of variations in packing. It may be useful to present data as relative flow rates, i.e. measured flow rate/initial flow rate in the 1000 mmol$_c$ l^{-1} solution. In this way all experiments start with a relative flow rate of 1.

The measurement of hydraulic conductivity changes caused by sodicity

A glass tube 2.5 cm diameter and 25 cm long forms a suitable leaching column. In one end insert a bung drilled to hold a glass tube. Place a disc of nylon cloth on the bung and cover with a 1 cm layer of coarse sand. Pour <2 mm air-dry soil into the tube with tapping to

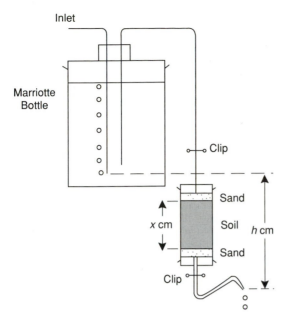

Figure 14.10 Equipment for measuring hydraulic conductivity.

consolidate the soil until the soil column is 15 cm long. Cover this with a 1 cm layer of coarse sand and a disc of nylon cloth. Into the top of the tube insert another drilled bung and glass tube. Heat a 10 cm length of soft glass tubing over a bunsen to narrow one end, and to form an S-bend in the centre as shown in Fig. 14.10. Attach with a polythene tube (and clip) to the outflow from the leaching column.

The reservoir is a *Marriotte bottle*. Two glass tubes are inserted through the bung to reach to within a few cm of the base of the bottle. To one is attached a polythene tube and clip to allow it to be linked to the leaching column.

Saturating the column

Place a concentrated solution in the Marriotte bottle and link it to the outflow tube of the column. Blow into the inlet tube of the bottle to force solution through to the leaching column. The soil will be soaked from the base upwards and this, to a large extent, avoids air being trapped in the soil. When the leaching tube is full, change the Marriotte supply to the top of the column (clip the tubes as required), and allow flow to begin.

A further means of preventing air bubbles being trapped in the soil is to pass CO_2 into the leaching tube before saturating with solution. Trapped bubbles of CO_2 will then dissolve in the leaching solution.

Measuring hydraulic conductivity

Hydraulic conductivity has been discussed in Section 5.5. Neglecting resistance to flow due to the equipment, flow is related to hydraulic conductivity as follows:

$$Q = -KAh/x$$

where Q is the flow ($cm^3 h^{-1}$), K is the hydraulic conductivity ($cm h^{-1}$), A is the cross-sectional area of the soil column (cm^2), h is the head of water (cm) and x is the length to the soil column (cm). The term h/x is the pressure gradient through the soil measured as $cm H_2O cm^{-1}$. The flow rate can be altered by adjusting h (both the height of the Marriotte bottle and the position of the outflow tube can be adjusted).

The reason for using a Marriotte bottle is that the head remains constant even though the water level is falling. Figure 14.10 shows that once the outlet tube is full, water siphons from the bottle and air is drawn in through the bottle inlet tube. The head is the distance between the bottom of the inlet tube where the bubbles escape, and the level of the end of the outlet tube from the leaching column where the water drops form.

Set the flow rate to about 1 drop s^{-1}. Allow the concentrated solution to flow until the soil has been brought to the required ESP, and flow rate is steady. Record Q, A, h and x.

Refill the reservoir with a more dilute solution (clip the tubes as required) making sure that no air enters the system. Allow the solution to flow, measuring rate of flow occasionally until it is constant. Continue with decreasing concentrations finally passing distilled water.

Plot changes in hydraulic conductivity against time or against cumulative volume. The latter is often preferred when comparison is being made between soils or treatments. In this case all the effluent solution has to be collected with flow rates measured occasionally as required. Figure 14.3 shows data obtained using a similar method: the hydraulic conductivities shown are those obtained when the soil had come to equilibrium with a given solution, i.e. the flow rate had become constant.

Note that the resistance of the equipment to water flow does not need to be taken into account when considering changes in hydraulic conductivity. Section 5.5 describes how the method can be adapted if required to give absolute values of hydraulic conductivity.

The swelling of clay flakes

Clay suspensions

Smectite is the clay which swells to the greatest extent.

It can be obtained from chemical reagent suppliers as bentonite powder. This is normally a sodium clay, but contains impurities of quartz and sesquioxides. On to 100 ml of water slowly sprinkle 6 g of bentonite with vigorous stirring. Shake to suspend and disperse the clay. This may take several days. Allow to stand overnight to settle the larger particles. Table 2.4 gives settling rates: if the suspension is placed in a beaker such that the depth is 10 cm, then after 8 h most of the <2 μm clay remains in suspension. Use this to prepare clay flakes.

Soil clays can be extracted using the methods of Section 2.3, but do not add sodium hexametaphosphate.

Clay flakes

Suitable supports for flakes can be made from wire and glass microscope cover slips. Bend the wire to make a frame on to which a cover slip can be glued with Araldite. The frame is arranged to sit on a 25 ml beaker and to support the cover slip horizontally in the beaker.

Weigh the frame plus cover slip. Pipette on to the cover slip a few drops of the clay suspension and allow to dry in air. Repeat with several more additions of clay until between 20 and 100 mg of clay is in the flake.

Measurements

Figure 14.2 shows the type of data that can be obtained. Place the flake in a beaker containing a concentrated solution of known SAR, and soak overnight. Remove, gently blot away the excess liquid and weigh. Dry the flake at 105 °C, cool in a desiccator containing dry silica gel and reweigh as quickly as possible to obtain the mass of water held per unit mass of clay.

For more dilute solutions, initially soak in a concentrated solution having the same SAR as that required in the final dilute solution. Remove the flake from the beaker, blot away excess liquid, rinse the flake in a little dilute solution and then place in a beaker of this solution. Measure the water content of the clay after overnight soaking. Note that when swelling begins, the flake may break and fragments may drift off the slide. Preliminary experiments are useful to gain experience in handling the slides.

Section 14.6 Leaching requirements, water mixing and the use of gypsum

LEACHING REQUIREMENTS

Salinity control

To maintain an unchanged salt content in a soil profile requires that salt removals must equal inputs. The components of the budget are irrigation water, drainage water, crop removals, fertilizer, precipitation and dissolution of salts, atmospheric deposition and groundwater. Only irrigation water and drainage water need to be considered in a budget provided the other components are small (Fig. 14.11). Thus to maintain a steady state, salt inputs in the irrigation water must be equal to salt removals in drainage. Amounts of salt are given by volume, V, × concentration, C, and so

$$V_{iw}C_{iw} = V_{dw}C_{dw}$$

where subscripts iw and dw are irrigation water and drainage water respectively. Rearranging,

$$V_{dw}/V_{iw} = C_{iw}/C_{dw}$$

For field use V is replaced by depth of irrigation water, D (mm), and C is replaced by the electrical conductivity, EC_w. Thus

$$D_{dw}/D_{iw} = EC_{iw}/EC_{dw} \qquad [14.3]$$

The depth ratio is the leaching fraction. It is equal to the ratio of the EC of the irrigation water to the EC of the drainage water. If EC_{iw} is known and a limit is set for EC_{dw}, then the leaching fraction is also a leaching

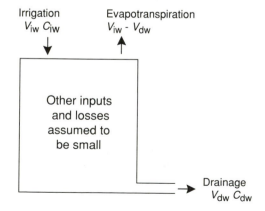

Figure 14.11 The salt budget of a soil profile.

requirement, LR, and can be calculated using Eqn. 14.3. The choice of the limit for EC_{dw} is based on crop tolerance data (Table 14.3), and is often set at a higher value than the EC of the saturation extract for 50 per cent yield reduction because the average EC within the profile is less than that in the drainage water (Fig. 14.5). Ayers and Westcot (1985) give methods for calculating the average root zone salinity which take account of the greater removal of water by roots in the upper part of the profile.

Example

$$EC_{iw} = 4\,dS\,m^{-1}$$
$$\text{Required } EC_{dw} = 40\,dS\,m^{-1}$$
$$LR = 4/40 = 0.1 \text{ (see Fig. 14.5)}$$

The water needs of the crop must now be considered. If for example the amount of water required to satisfy evapotranspiration needs during a growing season, D_{et}, is 1500 mm, then total water needs $D_{iw} = 1500 + D_{dw}$.

$$LR = 0.1 = D_{dw}/D_{iw}$$
$$D_{dw} = 0.1D_{iw} \quad\text{and}\quad D_{et} = 0.9D_{iw}$$
Thus $\quad 1500 = 0.9D_{iw}$
$$D_{iw} = 1667\,mm \quad\text{and}\quad D_{dw} = 167\,mm$$

The calculation simplifies to $D_{iw} = D_{et}/(1 - LR)$.

Sodicity control

When water moves through a soil profile its SAR is increased as a result of the increase in concentration of the ions as roots extract water. The factor by which the concentration of each ion is increased at the base of the profile is EC_{dw}/EC_{iw} which is equal to $1/LF$. The SAR changes from $[Na^+]/[Ca^{2+} + Mg^{2+}]^{1/2}$ to

$$([Na^+] \times 1/LF)/([Ca^{2+} + Mg^{2+}] \times 1/LF)^{1/2}$$

which is $([Na^+]/[Ca^{2+} + Mg^{2+}]^{1/2}) \times (1/LF)^{1/2}$. Thus

$$SAR_{dw} = SAR_{iw}/(LF)^{1/2} \qquad [14.4]$$

and when equilibrium is established, the ESP at the base of the profile is approximately equal to SAR_{dw}.

Example Figure 14.5 shows SAR values calculated using Eq. 14.4. If SAR_{iw} is 4 and LF is 0.1, then $SAR_{dw} = 4/0.1^{1/2} = 12.6$. Thus using the leaching requirement for salinity control, i.e. the leaching fraction required to maintain $EC_{dw} = 40\,dS\,m^{-1}$, the ESP is maintained below 12.6 in the profile. If it was decided that the ESP should be maintained below 10, then $10 = 4/(LR)^{1/2}$ and $LR = 0.16$. The amount of irrigation water would have to be increased to 1786 mm and the salinity, EC_{dw}, would be maintained below $25\,dS\,m^{-1}$ (Eq. 14.3).

AMENDMENT OF IRRIGATION WATERS

Mixing poor with good quality waters

Salinity is changed according to the equation

$$V_1C_1 + V_2C_2 = (V_1 + V_2)C_3$$

where V is volume, C concentration, subscripts 1, 2 and 3 are good quality water, poor quality water and mixture respectively. EC can be substituted for C.

The change in sodicity can be calculated using the above equation for each of the cations separately and substituting the new values in the SAR expression.

Addition of gypsum

Gypsum is normally applied to the soil surface where there is a sodicity hazard associated with the use of a poor quality water. Its effects are difficult to predict quantitatively because its slow dissolution depends on particle size and rate of water movement through the soil. When gypsum has come to equilibrium with soil solution, the calcium concentration is raised to about $15\,mmol\,l^{-1}$ depending on the sulphate concentration in the water. This value can be substituted in the SAR expression in place of the irrigation water value. It gives a new lower SAR value, which is the potential minimum but is unlikely to be established in practice.

Section 14.7 A conductivity index for use with soils or soil-based composts in glasshouses (ADAS method)

The principles are similar to those described in Section 14.2. Salinity is measured as the electrical conductivity in a 2:5 (m/v) suspension of soil in saturated calcium sulphate solution. The time-consuming preparation of a saturation extract is not worth while for this purpose where only a simple index of salinity is required. The calcium sulphate flocculates the clay in the suspension which will then give a clear filtrate for measurement of EC.

Equipment and reagents

Conductivity meter and cell.
Saturated calcium sulphate. Shake 7 g of $CaSO_4.2H_2O$ with 2 l of water at 20 °C for 2 h. Filter and store at 20 °C. Its EC should be $1.96\,dS\,m^{-1}$ at 20 °C.

Method

Preparation of the suspension Transfer 20 g (Note 1) of <2 mm air-dry soil or soil-based compost to a glass bottle and add 50 ml of saturated calcium sulphate solution. Shake for 15 min at 20 °C. Filter through a Whatman No. 2 paper and retain the filtrate at 20 °C for determination of EC.

Calibration and measurement Calibrate the electrical conductivity meter using the methods of Section 14.1. Ensure that the filtrate is at 20 °C (stand in a temperature-controlled water bath) and measure its EC value as dS m^{-1}. Subtract 1.96 from the EC reading to give the increase in EC above that of the calcium sulphate solution, and thus give the soil salinity value.

The conductivity index

Soils are classified as shown in Table 14.8.

Note 1 For routine purposes it is convenient to dispense 20 ml of soil using a standard measuring cup (28 mm internal diameter and 32 mm deep) as for nutrient index determinations (Section 9.1). However, the density of <2 mm air-dry soil in the scoop may vary greatly for composts. Thus it is necessary to record the mass of the 20 ml of soil or compost: tare the glass bottle (put it on a balance and set the reading to zero), add 20 ml of soil and note the mass. After subtracting 1.96 dS m^{-1} from the reading, the EC of a 20 g sample can be calculated as

$$EC_{20g} = EC_{20ml} \times 20/\text{mass of soil in 20 ml}$$

since EC is proportional to the mass of dissolved salt and therefore proportional to the mass of soil used.

Table 14.8 Conductivity indices for soils and soil−based composts (ADAS method)

Index	Electrical conductivity at 20 °C (dS m^{-1})	Index	Electrical conductivity at 20 °C (dS m^{-1})
0	0.00–0.30	5	0.91–1.10
1	0.31–0.50	6	1.11–1.40
2	0.51–0.70	7	1.41–1.80
3	0.71–0.80	8	1.81–2.10
4	0.81–0.90	9	Over 2.10

Note that the listed values are calculated from the measured values minus 1.96. MAFF (1988) lists the measured values.

Table 14.9 Guidelines on the effects of salinity on glasshouse and nursery crops (the numbers are conductivity indices)

Crop growth	All seedlings, bulbs and containerized nursery stock	All other soil grown vegetables and flowers	Carnations tomatoes, and peppers
No growth restrictions	0–2	0–3	0–4
Possible restrictions, especially with young plants	3 and 4	4 and 5	5 and 6
Severe damage likely	Over 4	Over 5	Over 6

From MAFF (1988).

Effects of salinity on crops

The guidelines in Table 14.9 can be used.

For valuable glasshouse crops, salinity should be maintained well below the level at which crop growth is affected. Comparison can be made with Table 14.3. A range of between 3.6 and 18 dS m^{-1} (saturation extract) is given for a 50 per cent reduction in yield. The 2:5 extract used here is about six times more dilute than a saturation extract (250 compared to about 45 g H$_2$O per 100 g soil, Section 14.2). The above range would therefore be about 0.6–3 dS m^{-1} in a 2:5 extract. In the ADAS guidelines restrictions are predicted between 0.5 and 1 dS m^{-1}.

Salinity management

Leaching will remove soluble salts. Guidelines are given in Table 14.10 to reduce the index to less than 3. The water should be applied in small amounts over a period of 4–5 d.

Leaching removes nitrate to give a N index 0 (Section 11.4).

Table 14.10 Leaching requirements for glasshouse soils

Soil type	Index						
	<3	3	4	5	6	7	>7
	(l m^{-2})						
Sandy soils	Nil	15	20	25	35	50	65
Other soils	Nil	20	25	35	50	70	90

From MAFF (1988).

Section 14.8 Projects

1. Use the methods of Section 14.5 to determine the effects of sodicity on the hydraulic conductivity of a soil. Work with a medium- or sandy-textured soil: heavy-textured soils give many practical difficulties in column experiments. Repeat one of your column experiments (SAR 25) but equilibrate the $10 \, mmol_c \, l^{-1}$ solution (Table 14.7) with calcium sulphate before use (shake 21 with 7 g $CaSO_4.2H_2O$ for 2 h and filter). The concentration of calcium in this solution can be measured, its effect on ESP predicted, and its protective effect on structure measured in the column.

2. Determine the effects of salinity on germination and growth for a crop of your choice (Section 14.3).

3. Prepare a solution with sodium, calcium and magnesium concentrations equal to those in the River Pecos water (Table 14.1). Use chloride salts. Leach this through a soil and then air dry it. Determine the salinity and sodicity of the soil using the methods of Section 14.2.

   Prepare a series of solutions to give a range of sodicities (0–20 SAR) at $40 \, mmol_c \, l^{-1}$. Carry out the above experiment using each solution and plot the relationship between ESP and SAR of the leaching solution and of the saturation extract. Compare your results to Fig. 14.7.

4. Obtain soil samples from a coastal marsh or tidal mudflats. Measure the salinity and sodicity of the soils. Do your results show the effects of sea water on the soils? Sea water composition is given in Table 14.1, and ESP–SAR relationships are in Fig. 14.7. Compare your data to those found for soils from the Kent and Essex marshes which are influenced by sea water at depth in the profile (Hazelden *et al.*, 1986).

5. Obtain soils with a range of texture. Determine the saturation percentage (Section 14.2) and the water content at field capacity (Section 5.3) for each soil. Plot the relationship between the two values. Does the slope approximate to 2 as suggested in Section 14.2?

Section 14.9 Calculations

1. For one of the waters in Table 14.1 calculate the concentrations of Ca^{2+}, Mg^{2+} and Na^+ in terms of $mmol_c \, l^{-1}$ and $mg \, l^{-1}$. Molar masses are in Appendix 2. (*Ans.* River Colombia. 0.9, 0.4, 0.2 and 18.0, 4.9 and 4.6)

2. Taking the saline soil in Table 14.2, calculate the EC in the soil solution at field capacity using the information in Section 14.2. (*Ans.* 24 dS m^{-1}) Calculate the osmotic potential of this solution, Section 14.3. (*Ans.* −0.96 MPa)

3. Using Fig. 14.1 and Table 14.3 predict the threshold and slope conditions for the effects of salinity on barley. (*Ans.* 8 dS m^{-1}, 5 %/dS m^{-1})

4. The composition of the irrigation water used at the Hofuf Experimental Station in Saudi Arabia is given in Table 14.11, along with that of the drainage water from one of the experimental plots. Calculate:

   - the EC of each water using Fig. 14.6. (*Ans.* 2.3 and 4.6 dS m^{-1})
   - the SAR of each water. (*Ans.* 5.2 and 7.9)
   - the ESP which would probably be present in soil brought to equilibrium with each water using Fig. 14.7. (*Ans.* 5.2 and 8.4)
   - the leaching fraction that had been used in the plot assuming no rainfall. (*Ans.* 0.5)

Table 14.11 Irrigation and drainage water composition

Ion	Irrigation water	Drainage water
	($mmol_c \, l^{-1}$)	
Ca^{2+}	7.55	15.25
Mg^{2+}	4.12	8.96
Na^+	12.65	27.32

From W. Abder−Rahman.

If the drainage water is reused as irrigation water, calculate the leaching requirement to maintain the EC below 8 dS m^{-1} in the profile. (*Ans.* 0.58) If irrigation water and drainage water are mixed in a ratio of 1:1 calculate again the leaching requirement to maintain the EC below 8 dS m^{-1} in the profile. (*Ans.* 0.43)

5. The high salt-water dilution method is occasionally used to reclaim saline sodic soils. This involves leaching with a highly saline water, e.g. sea water, then with a 1:1 mixture of sea water and fresh water, then with a 1:2 mixture and so on until both salinity and sodicity have been reduced to low levels. Find out how this method prevents structural damage as follows. Use the data in Table 14.1. Predict from Fig. 14.3 what the effect of sea water would be on the hydraulic conductivity of a smectite-rich soil. (*Ans.* 598 mmol$_c$ l^{-1}, SAR 59, no effect) Now calculate the composition of a 1:1 mixture of sea water and fresh water (use pure water for the purposes of the calculation), and predict again the effects using Fig. 14.3. (*Ans.* 299 mmol$_c$ l^{-1}, SAR 42, no effect) Continue with the 1:2 water mixture (*Ans.* 199 mmol$_c$ l^{-1}, SAR 34, no effect) You should see that for each SAR value, the salinity of the water

does not fall below the critical value at which damage begins

6. For the Nahal Oz groundwater (Table 14.1) calculate the possible effects of soil-applied gypsum as a means of controlling the development of sodicity. (*Ans.* Without gypsum SAR = 26, with gypsum SAR = 10 if the Ca^{2+} concentration rises to $15\,mmol\,l^{-1}$)

CHAPTER 15

Pesticides and Metals

It has been estimated that pests and diseases may be the cause of the loss of about one-third of the yield of crops grown in the world. Figure 13.1 shows the place of pests and diseases alongside other limiting factors in controlling yield. Traditionally crop rotations controlled pests and diseases, and cultivations, rotations and hand weeding controlled weeds. During the twentieth century the increasing world population and the movement of people from agriculture to urban living have necessitated the development of other methods of pest and disease control, and at present the only effective method on a world scale is the use of pesticides. These, together with adequate inputs of nutrients and water, allow intensive agriculture to produce yields close to the potential of the crop for a given climate. Progress in these aspects of crop management together with the breeding of crop varieties with increased yield potential have allowed total world food production to keep pace with the increasing population. However, even in intensive agriculture, average yields are only about half the potential yield, and in the poorer countries yields are only a fraction of the potential (see also Ch. 13). As a result the distribution of production does not match the distribution of population, resulting in both famine and excess.

Pesticides fall into three major categories: herbicides control weeds, fungicides control fungal diseases and insecticides control insect pests. As an example, the intensive production of winter wheat involves herbicides to control both broad-leaved and grass weeds, insecticides to control aphids and fungicides to control mildew. A detailed discussion of these is beyond the scope of this book. However the principles are developed in Section 15.1 where the effects of weeds on the yield of winter wheat are discussed and pot and plot techniques to study weed competition are described. Figure 15.1 shows examples of the depression of wheat yields caused by two weed species with differing competitiveness, showing the importance of weed control in management systems.

The environmental implications of pesticide use are discussed in this chapter within the context of soil science. There is clearly a need to reduce inputs of pesticides to the minimum required to control the

problems. This has led to steadily decreasing application rates through improved compounds and spray technology. There is also an increased understanding of the competitiveness of weeds, and a technique is described in Section 15.1 which allows yield depression to be predicted from weed counts early in the season. Using this information, pesticide management can be improved and 'unnecessary' applications avoided. However, the proverb, 'one year's seeding means seven years' weeding' applies in this situation and there is a need to prevent a build-up of weed seeds in the soil. In contrast to fertilizer use where an environmentally optimum application may simply be less than the economic optimum (Ch. 11) the management of herbicides requires a decision to apply or not to apply depending on weed populations and potential yield penalty, i.e. the likely depression resulting from weeds if the herbicide is not used. The profitability of the operation is obviously important and the environmental costs need to be assessed. The prediction of yield depression is still at the early stage of development and is complicated by the effects of variations in weather conditions from year to year. In this context there are similarities with the prediction of N availability (Ch. 11), and in both cases computer modelling is a tool of current research.

Those who believe that no pesticides should be used in crop production have reintroduced traditional organic farming techniques. Here competition from weeds and other pests with the associated yield penalty is an accepted part of the system. In the case of winter wheat an extended period of autumn cultivation is required to control weeds. There is a yield penalty associated with late sowing together with the penalty from surviving weeds. However, if fertilizers are also unacceptable in the production system, then the yield penalties from nutrient shortages, particularly N, are normally much greater than those from weeds.

THE ENVIRONMENTAL FATE OF PESTICIDES

Modern pesticides are nearly all neutral (uncharged) organic molecules. Because the majority are not found naturally in soil–plant systems, it is important that

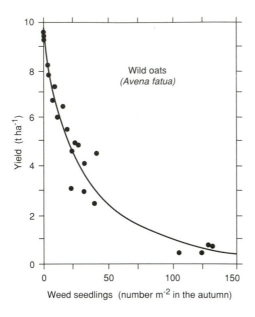

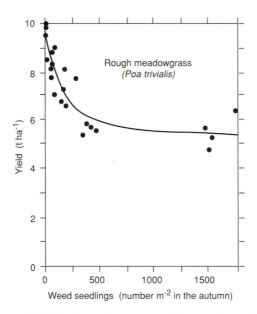

Figure 15.1 The effects of weed populations on the yield of wheat (Wilson and Wright, 1990).

their behaviour should be understood and their effects on the environment determined. Although in the early days of pesticide use controls were lax, agrochemical companies must now produce a rigorous package of data on the properties and environmental implications of the use of a compound to satisfy legal requirements before a licence is granted for its use. This is true for almost every country in the world. As a result more is known about the behaviour of pesticides in the environ-

ment than any other group of synthetic chemicals used by man. The pathway of pesticides through the soil is summarized in Fig. 15.2.

Development and testing a new pesticide involves first of all the determination, using a *biological dose response curve* (Section 15.2), of the rate of application required to control the target organism, i.e. the insect, fungus or weed causing damage to the crop. Rates of application vary greatly, the general range being from about $4\,kg$ of active ingredient ha^{-1} (e.g. simazine when used for total weed control) down to as low as $5\,g$ of active ingredient ha^{-1} (e.g. the sulphonyl urea herbicides). When incorporated into the top $20\,cm$ of soil, these rates give concentrations between about 2 parts per million and 2 parts per billion. The small inputs have to be taken into account in assessing their environmental impact. As the compounds disperse and break down, so the concentrations in soil and water become extremely small and require the most sophisticated analytical techniques for measurement. The influence of the pesticide on non-target organisms must also be determined during testing and development.

There are certain key questions which must be answered in order for risks to be assessed:

- How long will the chemical persist in the soil and what compounds are produced during its breakdown?
- What proportion of the chemical is free to affect non-target organisms?
- What proportion of the chemical is free to leach into groundwater and other watercourses?

Pesticide degradation

All organic compounds are subject to the normal decomposition processes of soil organic matter. Thus, the pesticide is 'decomposed' both by the microbial biomass and chemically. The overall process is normally termed *degradation*. The standard measurement of rate of degradation is the *half-life* of the compound and many pesticides have decay curves which follow approximately first-order kinetics (Section 3.5). An example is shown in Fig. 15.3. This is a simple concept, but often is a difficult measurement to make, especially because of the small initial concentrations in the soil. It normally requires both complicated procedures to extract the compound and produce a sample fit for analysis and sophisticated analytical equipment. Section 15.2 gives an outline of the methods used to measure half-lives of pesticides. The standard techniques are beyond the scope of this book, but there are other less accurate ways of estimating degradation rates: Section 15.2 describes a technique for monitoring the loss of phytotoxic effects of a herbicide using test plants.

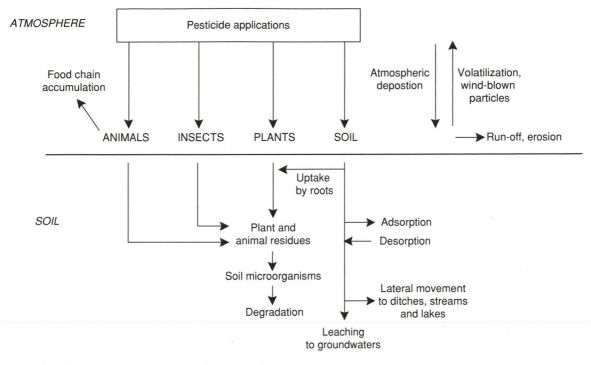

Figure 15.2 The pathway and reactions of pesticides in soils.

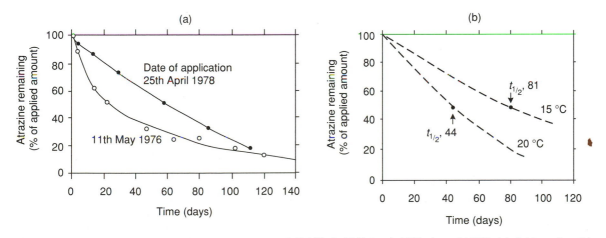

Figure 15.3 The decay of atrazine in a sandy loam soil (1.2% C, 70% sand, 18% clay, pH 7.0). (a) field studies, (b) laboratory incubation. From Walker (1978) and Nicholls *et al.* (1982).

Degradation rates of pesticides in soils vary enormously, and examples are shown in Table 15.1. The variations depend on a variety of factors:

- the affinity of micro-organisms for the chemical;
- the susceptibility of the chemical to attack by enzymes and other chemicals in the soil;
- the accessibility of the pesticide to attack by micro-organisms which is usually dependent on the extent

and strength of adsorption on to soil particle surfaces, particularly organic matter; and
- soil temperature, water content and other soil properties.

Metabolic pathways

In assessing the environmental impact of a pesticide it is also necessary to investigate the products of

Table 15.1 Degradation rates of pesticides given as half-lives, and adsorption coefficients, partition coefficients and leaching potentials rated by the GUS index

No. Pesticide	Use	Half life, $t_{1/2}$ at 20 °C (d)†	Adsorption coefficient, K_{oc} (cm^3g^{-1})†	Partition coefficient, K_{ow}†	GUS Index
1. Flauzifop-p-butyl	Herbicide	<2	2 000	32 000	0.21
2. 2,4–D	Herbicide	15	400	650	1.65
3. EPTC	Herbicide	6	200	1 600	1.33
4. Atrazine	Herbicide	70	100	220	3.27
5. Trifuralin	Herbicide	130	3500	120 000	0.97
6. Parathion	Insecticide	20	10 000	600	0.00
7. DDT	Insecticide	4 000	240 000	2 300 000	−4.97
8. Ethirimol	Fungicide	20	1200	200	1.20
9. Aminotriazole	Herbicide	14	100	100	2.30
10. Simazine	Herbicide	60	130	91	3.36
11. Mecoprop	Herbicide	21	20‡	1.3	3.56
12. Dichlobenil	Herbicide	60	400‡	1 100	2.49
13. Paraquat	Herbicide	1 000‡	10^6§	—*	−6.00
14. Glyphosate	Herbicide	47	24 000‡	—*	−0.64

† The values are only given to 2 significant figures. The factors which influence the values are as follows: $t_{1/2}$; climate (temperature, moisture content), soil composition, microbial biomass. K_{oc}; differences in soil pH which cause the charge on humus and on the pesticide molecule to vary, differences in the nature of the organic matter (Table 15.6). K_{ow}; provided the solutions are buffered (pH) and temperature is controlled an absolute value results: pH 6.8–7.0 and 20–25 °C (room temperature) are the normal conditions.
‡ Estimated value. Strong adsorption protects paraquat in soils: if it were free in soil solution its half-life would be about 14 d.
§ Paraquat is almost unique in its strength of adsorption, being held tightly by clay mineral surfaces. Only in peat soils does organic matter have a significant role in its adsorption. Its large K_{oc} value simply reflects its strong adsorption in soils.
* Ionic compounds have very small coefficients of no value in predicting adsorption in soils.
From Worthing (1991) and other sources.

degradation, since these may also have detrimental effects in the environment. The degradation of the pesticide often through a series of products is known as its *metabolic pathway*: its determination is both a complicated and a time-consuming process, and the methods are outlined in Section 15.2. An example is shown in Fig. 15.4. It can be seen that eventually simple compounds are produced which occur naturally in soils, and that the elements in the pesticide become incorporated in CO_2, water, humus and the biomass, entering the natural cycle in the soil, plant and atmosphere.

Pesticide adsorption

The extent to which a pesticide is free in a soil depends on the strength of adsorption. Because most pesticide molecules are uncharged and hydrophobic they have a stronger affinity for humus than for other soil particles. As a result the adsorption of most pesticides in soils depends on the nature of the adsorption reaction between the pesticide and humus, and the amount of humus present. There are exceptions, for example paraquat, which being a cationic compound adsorbs primarily on the negatively charged surfaces of clays. Figure 15.5 illustrates the main adsorption mechan-

isms. Soil pH also influences adsorption through its effect on both the properties of particle surfaces, particularly the charge on humus (Section 7.1), and the properties of the pesticide molecule.

Only those pesticide molecules which are free in soil solution can be easily taken up by plants and microorganisms, and so a measure of adsorption is a useful indicator of availability to non-target organisms. Potential leaching into streams and groundwaters also depends on how much of the pesticide resides in the soil solution (McCall *et al.*, 1980). To determine the extent of adsorption, the partitioning of the pesticide is normally measured between soil and water in a suspension which has been allowed to come to equilibrium. From this an *adsorption coefficient*, K_d, can be calculated and details are given in Section 15.3. However, the coefficient obtained only applies to the experimental soil. It is more useful to obtain one index of adsorption which can be used for a range of soils, and to this end the adsorption coefficient is often expressed in terms of soil organic carbon, K_{oc}, since organic matter is normally the predominant adsorbing component. Values are given in Table 15.1 for common pesticides for soils with pH values between 5 and 7. Different values may apply in acidic and calcareous soils. A knowledge of this coefi-C$_3$H$_7$NH   N   NH C$_2$H$_5$

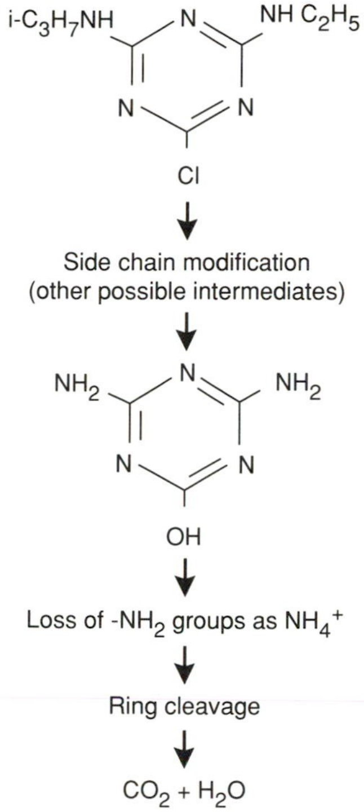

N   N

Cl

↓

Side chain modification
(other possible intermediates)

↓

NH$_2$ N   NH$_2$

N   N

OH

↓

Loss of -NH$_2$ groups as NH$_4^+$

↓

Ring cleavage

↓

CO$_2$ + H$_2$O

Figure 15.4 The metabolic pathway of atrazine in soils.

ficient and the organic C content of a soil then allows adsorption to be predicted without a direct measurement of K_d (Section 15.3).

The extent of adsorption of a pesticide on to organic matter is related to its hydrophobic character (hydrophobic = dislike of water) which can be measured through the partitioning of the compound between an organic solvent and water. Based on this principle, the *partition coefficient* of the pesticide between *n*-octanol and water, K_{ow}, has been used as an index of adsorption, being easier to measure than the coefficient for soil and water (Section 15.3) and values are given in Table 15.1. The relationship between K_{ow} and K_{oc} has been established for a wide range of soils (Section 15.3).

Pesticide mobility

Pesticide movement can be measured directly in the field by core sampling. This is a laborious, time-consuming process and provides data which only apply to that site and to the prevailing weather. Controlled experiments can be carried out using lysimeters of undisturbed soil: the effects of soil and climate can be studied in more detail. Future EC regulations may require data from lysimeters to be supplied for licensing purposes. Again this is laborious and time-consuming.

Mobility can be predicted using the principles which apply to chromatography discussed in Section 11.5. The retardation factor, R_f depends on the adsorption coefficient, K_d, with movement also depending on the pore characteristics and the amount of water leaching through the soil. Details are given in Section 15.4. Although relatively simple to apply to homogeneous porous materials, for field soils this approach requires an understanding of water flow through a structured material with a wide range of pore size. Thus the prediction of pesticide movement faces the same difficuties as for nitrate and computer-assisted modelling of the process is a primary tool of research.

An alternative way to determine the mobility of a pesticide is described in Section 15.4. It involves leaching the pesticide down a thin layer of saturated soil, followed by growing test plants on the soil to determine the distribution of pesticide from its toxicity.

The combined effects of adsorption and decay on leaching

It is a complex and time consuming process to determine in the field the potential of a pesticide to leach through soil and contaminate groundwater. However, for commonly used pesticides it may be necessary. Simpler methods are needed to produce an estimate of leaching potential during testing and development.

Leaching potential depends on two parameters:

1. The length of time the pesticide persists in the soil, measured by its half-life.
2. The strength of binding to the soil, measured by its partition coefficient.

In essence, a compound which degrades quickly will not be in the soil long enough to leach very far even if it is free in solution, e.g. 2,4–D, and a compound which is strongly adsorbed will not be free to leach even if it is persistent, e.g. DDT.

The most commonly used classification of leaching potential is the McCall *et al.* (1980) scale, which is based on K_{oc} values (Section 15.4). More useful is the classification developed by Gustafson (1989) known as the groundwater ubiquity score or GUS index, based on both the half-life and the K_{oc} value (Section 15.4). Table 15.1 gives examples. It can be seen that atrazine is moderately persistent, and has a potential to move into groundwaters. It has been in common use since 1957 and since then testing requirements have changed. Its leaching into groundwaters is currently being investigated in detail.

Any new chemical which in field trials is shown to be of value to the farmer but which, in laboratory tests, is predicted to have a high leaching potential, may require large-scale assessment trials to establish its actual leach-

(a) Hydrophobic bonds

(c) Ligand bonds

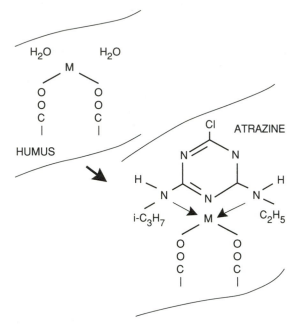

The energy change in displacing water from the humus surface favours the adsorption of the molecule : uncharged molecules, e.g. atrazine and other s-triazines.

(b) Hydrogen bonds, Van der Waals' forces and other weak intermolecular bonds

Form between uncharged or charged molecules and metals bound to humus. Atrazine binds in this way: the N has a free pair of electrons which is shared between the N and the metal as indicated by the arrows. Water hydrating the metal is displaced by the atrazine. (see Section 15.3 Note 4)

(d) Electrostatic bonds

Usually occur with molecules which are polar, i.e. they have uneven electron distributions in certain bonds so that, although the molecule as a whole is electrically neutral, certain parts have a finite charge (δ^+ in the diagram). Thus they tend to be attracted to other polar molecules: s-triazines are attracted to humus in this way. These bonds normally occur in conjunction with hydrophobic bonding.

Occurs between ionic (charged) compounds and charged soil particles, replacing exchangeable ions, e.g. paraquat and diquat.

Figure 15.5 Mechanisms of pesticide adsorption soils.

ing potential, before being released on to the market. Normally this involves:

- predictions using computer-assisted modelling (Section 15.4);
- measurements in lysimeters under controlled environmental conditions; and
- field trials in which movement of the chemical is measured through the soil and if necessary into watercourses and groundwaters. Colour Plate 20 illustrates a trial in progress.

It is not within the scope of this book to deal with the implications of pesticides in terms of food chains, wildlife and human health. Goring and Hamaker (1972) give an introduction to these topics. The influence of agriculture on the amounts of pesticides in rivers and lakes in England and Wales is discussed by NRA (1992).

POTENTIALLY TOXIC ELEMENTS IN SOILS

Toxicities

Unlike organic chemicals which are eventually broken down to simple non-toxic compounds, the addition of metals and other potentially toxic elements to soils leaves a residue which is permanent unless leaching occurs. They are therefore potentially much more damaging to crops and animals than are pesticides.

The elemental composition of soils varies greatly depending on the nature of their parent materials (Table 15.2). Industrial wastes, particularly from mining, have added metals locally in such large amounts that only tolerant plants survive. Sewage sludge added to soils is at present another source of potentially toxic metals. There is the likelihood of increased applications as dumping in the sea becomes progressively restricted,

Table 15.2 Recommended upper limits of total metals in soils

Metal	Typical total metal content, uncontaminated soils		Recommended upper limit	
	$(mg\,kg^{-1})$	$(kg\,ha^{-1})$†	$(mg\,kg^{-1})$	$(kg\,ha^{-1})$†
Zinc	80	160	300	600
Copper	20	40	135	270
Nickel	25	50	75	150
Cadmium	0.5	1	3	6
Lead	50	100	250	500

† Assuming 2000 t ha^{-1} to 15 cm depth.
From DOE/NWC (1981) and ADAS (1987).

although the metal content of sludges is decreasing as industry improves techniques for retrieving metals from water before discharge.

The potentially toxic elements added to soils can be divided into two groups:

1. Zinc, copper, nickel and boron can have direct effects on crop growth if the concentrations are high enough. Recommmended limits for the amounts in soils are based on their effects on crops and examples are given in Table 15.2.
2. Cadmium, lead, mercury, molybdenum, arsenic, selenium, chromium and fluorine are not normally toxic to crops but may affect animals feeding on crops grown on contaminated land. Cadmium can be toxic to crops but effects on animals occur at smaller concentrations and so recommended limits for all the elements in this group are based on their effects on animals (Table 15.2).

Of all these elements, only boron is lost from soil by leaching, the others remaining almost entirely in the soil as non-degradable contaminants. The metals usually present in sludges in the greatest amounts are zinc, copper and nickel.

Recommended metal limits are given as amounts soluble in a nitric acid–perchloric acid mixture (Section 15.5). This method is now being replaced by an aqua-regia extraction which gives similar results. A simpler and safer analytical procedure is to extract the metals in ethylenediaminetetra-acetic acid (EDTA) solution which gives an indication of the availability of the metals (Section 15.5). The amounts of acid-soluble metals can be approximately determined through correlation with amounts extractable in EDTA.

Sludge-contaminated soils normally contain raised concentrations of more than one metal but there is no agreed procedure for assessing their combined effects on crops and animals. The zinc equivalent method (ADAS, 1982) has been used to sum the effects of zinc, copper and nickel, taking account of their relative toxicities. However, this approach has been discontinued because it is unlikely that the effects are additive and no account is taken of possible interactions (Section 15.5).

The effects of metals on soil biomass are discussed in Ch. 3 and Section 6.9, Project 2.

Deficiencies

Crops can suffer from deficiencies of copper and zinc, particularly in sands, peats and chalky soils. Extraction in EDTA solution is a useful indication of the availability of copper (Section 15.5) but is less good for zinc. Many other methods have been used to extract micronutrients (MAFF, 1986a; Page, 1982) but their ability to predict deficiencies is limited.

FURTHER STUDIES

Ideas for projects are in Section 15.6, and calculations in Section 15.7.

Section 15.1 The effects of weeds on crop yield

Inputs of pesticides need to be reduced to a minimum both from a environmental viewpoint and because the cost of application can be a large part of the cost of crop production. Measurement and prediction of the yield loss associated with pests and diseases is the basis for sound management decisions. The principles of weed control are discussed by Attwood (1985) and Hance and Holly (1990).

Weeds and wheat

The competitiveness of weeds

The common weeds found in arable crops vary greatly in their size and pattern of growth. Thus their effect on yield also varies. Generally, cereals are most sensitive to competition in the early stages of growth, and weeds which emerge with the crop will be more competitive than those emerging later. Small, early flowering weeds have little effect, whereas larger weeds which grow in phase with the crop reaching their maximum size in July may seriously interfere with crop growth. The weeds are competing for light, water and nutrients. For wheat grown in the UK, the most competitive weeds are cleavers *Galium aparine* and wild oats *Avena fatua* which grow above and shade the crop.

Figure 15.1 shows the effects of wild oats and rough meadowgrass (*Poa trivialis*) on winter wheat yield measured in field trials at Long Ashton Research Station. Note the difference in the numbers of weeds plotted on the abscissa. A 20 per cent reduction in yield is caused by about 5 wild oats m^{-2} or by about 100 meadowgrass plants. Both grow in phase with the crop, but the latter is much less competitive because of its small size.

The initial slope of the lines in Fig. 15.1 is a measure of the yield loss per weed plant per square metre for small weed populations. Weed numbers can therefore be used to predict yield loss, and as a basis for making decisions regarding the need for a herbicide: cost of application relative to the value of the predicted yield loss allows the calculation of a *threshold weed population* below which spraying may not be justified. The following thresholds can be used based on an acceptable yield loss of 2 per cent: cleavers 1, poppy *Papaver rhoeas* 21, chickweed *Stellaria media* 13, field pansy *Viola arvensis*

100 seedlings present in the autumn per m^{2}. There is, however, much variability from year to year, and the problem of increase in weed seed numbers has to be considered.

The competitiveness of wheat

The condition of the crop is important in influencing its ability to compete with weeds. Figure 15.6 shows the results of an experiment on the influence of crop density (numbers of wheat plants m^{-2}) on the competitive effects of poppies. A normal crop density is about 250 plants m^{-2}. In this crop high densities of poppies reduced yield by about 15 per cent. However, a poor stand (low crop density) allowed the weeds to grow more vigorously and to reduce yield by up to 70 per cent. These data simply quantify the observation that a poor crop can be rapidly smothered by weeds and emphasize the need for good crop establishment particularly if herbicides are not to be used. Thus, to maintain yield losses below 2 per cent, lower threshold densities are needed for thinly established crops. The effect of wheat on the growth of poppies in this experiment was reflected in the number of seeds produced per plant: a plant growing in isolation produced about 500 000 seeds, but when grown in competition with wheat, only about 6000 seeds were produced.

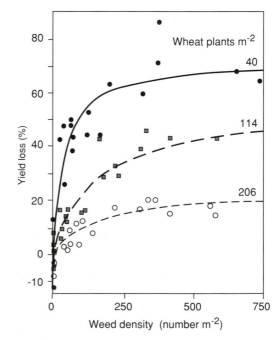

Figure 15.6 The effects of weed populations (poppies) on the yield of wheat, as influenced by the population density of wheat plants (Wilson, 1989).

Table 15.3 The effects of nitrogen and weeds on the yield of winter wheat

(a) Cleavers (Wilson, unpublished)

	Wheat yield (t ha^{-1})	
	With 52 weed plants m^{-2}	Weed free
No N	2.2	2.8
With 160 kg N ha^{-1}	4.2	7.3

(b) Wild oats (Wright and Wilson, 1992)

	Wheat yield (t ha^{-1})	
	With 45 weed plants m^{-2}	Weed free
No N	2.1	4.3
With 200 kg N ha^{-1}	0.7	6.7

Nitrogen availability and competition

The difficulty of predicting yield losses is shown in Table 15.3 in which a soil with N index 0 (Section 11.4) was used to compare the effects of weeds on wheat grown with and without added N fertilizer. In the experiment with cleavers, the crop responded to N both with and without weeds. Without N the weeds only had a small effect on yield: cleavers grows very slowly on N-deficient soils. With N, the growth of cleavers was rampant and seriously affected the ability of wheat to respond to the N. The data are another example of an interaction effect (Section 13.1). Without N and with weeds the yield was 2.2 t ha^{-1}. The response to N is 2.0 t ha^{-1} and to the removal of weeds 0.6 t ha^{-1}. Together these treatments produce a response of 5.1 t ha^{-1} which is greater than 2.0 + 0.6 = 2.6 t ha^{-1}, and illustrates the importance of integrated management of nutrients, pests and diseases.

In the second experiment, a mild winter allowed the wild oats to grow aggressively and to respond markedly to N. Without N the weeds reduced the yield of wheat by 50 per cent: with N the weeds reduced the yield by 90 per cent so that in the presence of the weeds, the yield of wheat was depressed by added N. The interaction effect is even more striking in this case. The response to added N and removal of weeds together is 4.6 t ha^{-1}, whereas the sum of the separate responses is −1.4 + 2.2 = 0.8 t ha^{-1}.

Organic farming systems which do not use herbicides normally have less available N, and so the weeds are probably less competetive than in an intensive production system.

Predicting the effects of mixed populations of weeds

Although experiments can be conducted using only one weed species, in practice a mixture of weeds will be present. Their combined effect can be predicted using the concept of a *crop equivalent*. The principle is that the competitiveness of each weed species can be rated according to the ratio, mass of one weed plant:mass of one crop plant, during the period of vigorous growth. This ratio is termed the *competitive index*. The crop equivalent of a given population of a species is the number of plants m^{-2} × the competitive index. It is then assumed that the yield depression is related to the ratio, mass of weeds:mass of crop plus weeds, based on the fact that the weeds are using a fraction of the total resources of light, water and nutrients.

Example The competitive index of poppies in winter wheat is about 0.6.

Thus, if there are 8 poppies m^{-2}, their crop equivalent = 0.6 × 8 = 4.8. If the density of wheat is 250 plants m^{-2}, then the yield depression is

$$4.8/(250 + 4.8) = 0.02, \text{ or about 2 per cent}$$

If there are also 2 cleavers plants m^{-2} with a competitive index of about 7, their crop equivalent is 7 × 2 = 14. The total crop equivalent of the weeds is 4.8 + 14 = 18.8 and the yield depression is

$$18.8/(250 + 18.8) = 0.07 \text{ or 7 per cent}$$

The concept of crop equivalents is still being developed but points the way forward in terms of quantifying and modelling effects on yields. Values for common weeds are shown in Table 15.4.

Table 15.4 The competitive index of some common arable weeds in cereal crops

		Competitive index
Strongly competitive	Cleavers	7†
	Wild oats	3
Moderately competitive	Poppy	0.6
	Chickweed	0.5
Weakly competitive	Red dead nettle,	
	Lamium purpureum	0.3
	Field pansy	0.1

† The competitive index of weeds which grow over and shade the crop is greater than the mass ratio. There is much variability in these values from year to year (Hance and Holly, 1990).

Methods for studying competition

Using pots

The methods of Section 9.2 can be followed. Suitable plants are barley and mustard *Brassica napis*, which simulate either the effect of a broad-leaved weed in a cereal crop, or the effects of barley growing as a weed in a crop of oilseed rape. The latter is a common situation where a rape crop follows barley in an arable rotation. Both are easy to grow.

Germinate the plants in a seed tray, and plant four seedlings of each species in each pot. Crop seeds are available from local agricultural merchants (Appendix 1).

Treatments Use barley alone, mustard alone, and barley and mustard together, with 3 replicates of each (9 pots). An N treatment could be imposed requiring 18 pots.

Harvest, separate the plant material and determine dry matter (Section 9.3).

Mildew tends to develop on cereals grown in glasshouses. Spray with Benlate as required following the manufacturer's instructions.

Using tanks

Growth conditions in pots are very different from field conditions. Large containers of soil allow a closer simulation of natural competition. Plastic cistern tanks available from building suppliers are suitable. The base should be drilled to allow drainage. For example the 'Ferham 16 litre' tank measures 30 × 43 cm and is 30 cm deep. It requires about 50 kg of soil when filled at a bulk density of $1.3\,g\,cm^{-3}$. Pass soil through a garden sieve and mix before filling the tanks. These containers are large enough to grow barley at its field spacing with and without weeds. Fertilization can be carried out by watering the soil with nutrient solutions following the methods of Section 9.2. The area of the tank is about 10 times the area of a 5 inch pot so apply 10 times the volume of solution to give the same application rate. Weed seeds are available from the address in Appendix 1.

Treatments Use barley with and without chickweed. An N treatment can also be used. Because a larger area of crop is being sampled than in pot experiments replication is not so important and for demonstration purposes, one tank per treatment can be used. If statistically reliable data are required, three replicates per treatment should be used. Broadcast the chickweed seed ($500\,m^{-2}$ or 65 per tank which has an area of $0.13\,m^2$: 2000 seeds weigh approximately 1 g). Sow the

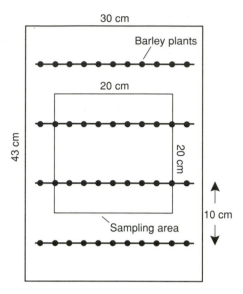

Figure 15.7 Tanks for weed competition studies.

barley in rows 10 cm apart with seeds 2.5 cm apart in the rows, and about 3 cm deep. This gives a seed density of approximately 340 seeds m^{-2} which is a normal seeding rate giving between 250 and 350 plants m^{-2} depending on germination and establishment (Fig. 15.7). Stand the tanks in the open and place the tanks together so that the crop spacing is continuous in and between the tanks. Apply water if required: about 6 l per tank brings a 'dry' soil back to field capacity (Table 12.1). Apply the same volume to all the tanks. Hand weed as required. Mildew and aphids may be a problem: if required spray with Benlate or Malathion respectively following the manufacturer's instructions.

Harvesting Vegetative growth can be harvested after a few weeks, particularly if a rapidly growing spring barley variety is used. If grain yield is required, the tanks should be in a bird-proof enclosure. Using a 20 × 20 cm frame, harvest the central part of the crop (Fig. 15.7). This leaves a *discard* area at the edge of the tank where *edge effects* alter growth and competition: the plants at the edge (the *guard rows*) have extra light and space (Section 13.1). The area sampled is 400 cm^2 or 0.04 m^2. Separate the plant material and determine dry matter.

Alternative crops Broad-leaved weeds are a problem in the establishment of grass leys. Ryegrass and chickweed can be grown in the above experiments. The seeding rate for ryegrass is about $30\,kg\,ha^{-1}$ or $3\,g\,m^{-2}$, and it should be lightly worked into the soil surface.

The use of herbicides

There are strict controls over the use of herbicides. Of those which are available for general use, Verdone, which contains 2,4–D and mecoprop, is a selective weedkiller used to remove broad-leaved weeds from lawns leaving the grasses relatively unaffected. It can be used on cereals which are also Gramineae, although at certain stages of growth it affects the crop. The above pot and tank experiments can be modified by planting all containers with both crop and weed seeds, and spraying as required to remove weeds. Follow the manufacturer's instructions and avoid spray drift onto containers which are not to be treated.

Using small plots

Section 13.1 describes small plot techniques. The grass experiment could be carried out using a spray treatment to remove broad-leaved weeds.

If an area of land can be cultivated, 1×1 m plots can be planted with barley and chickweed following the methods described for soil tanks above. Ten rows of barley should be planted, and the central 60×60 cm area harvested (six rows). Fertilizer applications and watering are described in Section 13.1.

Section 15.2 Phytotoxicity and herbicide persistence in soils

The phytotoxicity of herbicides

Herbicides enter target plants through leaves and stems or through roots. The amount required to kill plants varies, depending on the plant and its stage and rate of growth. When applied to leaves, the amounts required per hectare in the field also depend on formulation and spray technology. If the herbicide is applied to and is then taken up from the soil, required amounts also depend on soil type.

The effects of a herbicide can be studied by determining a biological dose response curve under a given set of conditions either in the glasshouse or in the field. A range of applications are made, and the effects on a test plant recorded. A glasshouse experiment is described below and a field experiment is illustrated in Colour Plate 20.

Decay curves and half-lives

Using the half-life of a pesticide in soil as a measure of persistence is a simple concept but often is a difficult assessment to make because of the low initial concen-

trations added to soils. The measurement of residual concentrations to obtain a decay curve and the half-life can be summarized as follows.

A subsample of soil taken at a given time interval after application is extracted with an organic solvent (e.g. methanol, acetone or acetonitrile) either by shaking the soil–solvent slurry in a tube or by refluxing. The methods and solvents used will vary depending on how strongly the pesticide is bound to the soil and how stable the chemical is in the solvents. Problems often arise at this stage as fractions of humus often co-extract with the pesticide, and the extract has to be 'cleaned up' before analysis. The clean-up process is normally by column chromatography or solvent partition. The cleaned sample is then concentrated to an appropriate volume for analysis. The most common analytical techniques are high performance liquid chromatography (HPLC), gas liquid chromatography and ultraviolet or visible spectroscopy.

Examples of decay curves for atrazine are in Fig. 15.3(a) and (b), and half-lives of common pesticides are in Table 15.1. The figure illustrates two important features of decay curves:

1. The rate of decay can be variable depending on soil conditions and climate. Thus although the half-life of atrazine is given as 70 d in Table 15.1, Fig. 15.3(a) shows half-lives of 20 and 60 d. The summer of 1976 was unusually warm which may have been the cause of the more rapid decay. The soil temperature at 10 cm depth rose from 7 to 17 °C over the experimental period in 1978. A similar soil at the National Vegetable Research Station, Wellesbourne, UK, was used in both cases. In laboratory incubation experiments (Fig. 15.3(b)), the half-life of atrazine in this soil varied between 16.5 and 194 d for temperatures between 30 and 5 °C respectively (Walker, 1978). The figure shows the results at 15 and 20 °C.

2. The decay curve does not fit to first-order kinetics (Section 3.5). For example, the 1976 data show the first half-life to be 22 d, the second 35 d and the third 68 d. In this context, the pesticide is behaving in a similar way to plant residues in soils (Fig. 3.6).

The direct measurement of the decay curve of a pesticide is beyond the scope of this book. However, an experiment is described below to measure the effects of time on the loss of toxicity from which a decay curve and the half-life can be determined indirectly.

Metabolic pathways

Even with a detailed knowledge of organic chemistry it is not possible to predict the metabolic pathway of a pesticide simply from a knowledge of the structure of the molecule. The products of degradation have to be identified as they are produced in the soil. The basic methods are similar to those outlined above for the pesticide itself, but with extra steps to separate the products before analysis.

The normal procedure is to synthesize the pesticide so that part of the molecule is labelled with a radioactive isotope, usually ^{14}C. Any metabolite containing the labelled portion of the molecule can then be detected. The pesticide is added to the soil which is then incubated. Production of volatile compounds such as CO_2 is monitored, and at given times subsamples are taken for analysis. Extraction is by organic solvents. The amount of radioactivity in the solvent extract is determined by liquid scintillation counting, and residual labelled chemical left in the soil is determined by combustion to produce CO_2 which is trapped and also counted. Metabolites in the extract are separated by thin layer chromatography or HPLC, counted and identified using a variety of techniques such as mass spectroscopy and nuclear magnetic resonance spectroscopy. At this stage the chemist becomes a detective: although it is not possible to predict the pathway, likely metabolites can be guessed from studies on compounds of similar structure, so aiding the identification process.

If it is found that a metabolite is persistent in the soil, its potential to harm non-target organisms or to leach is investigated as for the pesticide itself.

Obtaining pesticides

Apart from the determination of K_{ow} (Section 15.3) the experiments which follow have been designed to use formulations of pesticides which are readily available from horticultural suppliers. Pure pesticide reagents can be obtained from British Greyhound Chromatography and Allied Chemical Co. Ltd (Appendix 1) who also supply a useful catalogue of pesticide and environmental standard reagents.

An experiment to measure a biological dose response curve

Verdone-2 is a commonly available garden herbicide. It is sprayed on to lawns, entering plants through the leaves and stems. It is selective, killing only broad-leaved plants (dicotyledons). On entering the soil it is also taken up through roots, again with selective action. Its phytotoxicity can be determined by treating soil with varying amounts, growing a test crop and recording the effects on growth. The resulting *biological dose response curve* is a bioassay which can be used to determine concentrations of herbicide in soil from observed growth as in the experiment to measure the rate of breakdown of herbicides.

Reagents and equipment

Verdone-2. (Zeneca Agrochemicals)
Methanol.
Air-dry loam or sandy loam soil: 5 kg passed through a 5 mm sieve.
Plant pots. 24×3 inch pots
Mixing bowl.
Eppendorf (or similar) pipette, 0–100 µl.
Seeds. Lettuce is a suitable test crop. Alternatively, weed seeds are available from Herbiseeds, The Nurseries, Billingbear Park, Wokingham, Berks. RG11 5RY.

Method

The fact that common herbicides are available for garden use means that hazards from their use are minimal provided that manufacturers' instructions are read and followed. Thus, their use in the laboratory requires only that the normal precautions in handling chemicals should be observed. Wear gloves and a laboratory coat. A pipette filler or automatic pipette MUST be used. However, the use of these chemicals in the following experiments is not covered by their registration and therefore experiments are carried out at the operator's risk.

Choice of treatments The concentrations of added herbicide given below are for a sandy loam soil sampled from under grass (4–5% organic matter). Because the effectiveness of a soil-applied herbicide depends on its adsorption in the soil (its K_d value), suitable treatments depend primarily on the soil's organic matter content since this is the predominant adsorbing component. A sandy loam from a continuous arable site (2–3% organic matter) would require about half the amounts given below.

Preparation of treated soils Verdone-2 is a formulation containing 100 g mecoprop l^{-1} and 50 g 2,4–D l^{-1}. To prepare soils with graded amounts of the compound, treated and untreated soil are mixed in varying proportions.

Treated soil 1: using an Eppendorf pipette, add 30 µl of Verdone-2 to 10 ml of methanol and shake vigorously. Add this to 150 ml of water and shake. Place 1.5 kg of soil in a mixing bowl and slowly add the solution with mixing to ensure uniform distribution of the herbicide. The moist soil now contains 2 µg mecoprop and 1.0 µg 2,4–D g^{-1} air-dry soil.

Treated soil 2: follow the above procedure but use 60 µl of Verdone-2. The moist soil contains 4 µg mecoprop and 2 µg 2,4–D g^{-1} air-dry soil.

Untreated soil: to 2 kg of air-dry soil in the mixing bowl add 200 ml of water and mix thoroughly.

Table 15.5 Soil preparation and the resulting concentrations of mecoprop and 2,4–D

Mass of treated soil 1 (g)	Mass of treated soil 2 (g)	Mass of untreated soil (g)	Concentration of:	
			mecoprop	2, 4–D
			(μg g^{-1}air–dry soil)	
0	0	200	0	0 (control)
50	0	150	0.5	0.25
100	0	100	1.0	0.5
150	0	50	1.5	0.75
200	0	0	2.0	1.0
0	125	75	2.5	1.25
0	150	50	3.0	1.5
0	200	0	4.0	2.0

Thoroughly mix samples of treated and untreated soil in a polythene bag as indicated in Table 15.5. Prepare three replicates of each (200 g) and place in pots labelled with the treatment details.

Growth of a test crop Lightly sprinkle lettuce seed on each pot, and cover with a thin layer of untreated soil. Add water to the saucers to increase the water content to 20 per cent *m/m*. Weigh each pot and record the weights. Place the pots in a glasshouse and cover them all with a polythene sheet to prevent evaporation until the seedlings emerge. Remove the sheet. Add water as required to maintain the initial weights (Section 9.2). After 3 weeks, rate the growth on a scale 0–10 where 0 = death and 10 = growth similar to the control. Alternatively record the height of the plants (mm) or harvest the plants, dry at 100 °C and record the dry matter production. Plot the growth rating (or percentage of the control height or percentage of the control dry matter) against concentration as in Fig. 15.8(a). The concentration which reduces growth by 50 per cent can be used as a measure of the threshold concentration.

Alternative herbicides Pathclear (Zeneca Agrochemicals) is a formulation containing simazine, aminotriazole, paraquat and diquat. It is watered on to weeds and soil. It is non-selective killing all plants in the treated area both by leaf contact and root uptake. Paraquat and diquat are inactivated in the soil by strong adsorption and are not taken up by roots. Dissolve 10 mg (treated soil 1) and 20 mg (treated soil 2) of Pathclear in 10 ml of methanol with vigorous shaking. Add to 150 ml of water with shaking, mix with 1.5 kg of soil and proceed as above. The maximum treatment is 13 μg g^{-1}.

Deeweed (Arable and Bulb Chemicals Ltd) is a for-

mulation containing 48% atrazine and 38% aminotriazole. It is used in the same way as Pathclear. Dissolve 50 mg of Deeweed in 100 ml of methanol. Pipette 0.5 ml (treated soil 1) or 1.0 ml (treated soil 2) into 150 ml of water and mix with 1.5 kg of soil. The maximum treatment is 0.33 μg g^{-1}. Pure atrazine can be obtained from British Greyhound Chromatography and Allied Chemical Co. Ltd (Appendix 1). Follow the method given above for Deeweed but use 25 mg atrazine giving a maximum treatment of 0.165 μg g^{-1}.

An experiment to measure the rate of breakdown of a herbicide in soil

When a herbicide enters the soil, microbial and chemical breakdown begins, reducing the concentration of the active compound. The effects can be studied by growing a test plant in soil samples which have been incubated with the herbicide for varying periods of time. The concentration of herbicide can be determined indirectly by using a bioassay carried out under the same conditions, thus allowing a decay curve to be plotted and the half-life of the herbicide to be calculated.

It is important to check the biological dose response curve in a preliminary experiment before measuring the decay of the herbicide to ensure that suitable concentrations are used. The former takes about 2 weeks, whereas the latter may take several months and can be wasted time if the wrong concentration is chosen: see the previous experiment.

Reagents and equipment

Herbicides described in the previous experiments. Verdone-2 contains mecoprop and 2,4–D with half-lives of about 21 and 15 d respectively. These times are convenient for

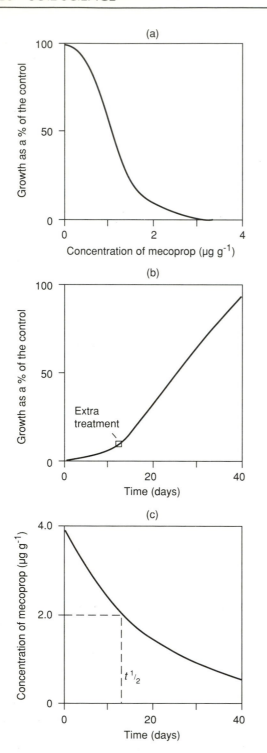

Figure 15.8 The effects of concentration and time of incubation on the phytotoxicity of Verdone-2 to lettuce growing in a sandy loam. (a) The biological dose response curve; (b) the decrease in phytotoxiciy; (c) the decay curve.

a glasshouse experiment. Details are also given for Pathclear, which contains simazine and aminotriazole with half-lives of 60 and 14 d respectively together with paraquat and diquat which are strongly adsorbed by soil and so are not measured in the bioassay. Deeweed or atrazine can also be used: atrazine has a half-life of 70 d (Table 15.1). The soil storage times are varied to suit the half-lives.

Methanol.

Air-dry loam or sandy loam soil as used in the biological dose response experiment. 4 kg passed through a 5 mm sieve.

Plant pots. 18 × 3 inch pots.

Eppendorf (or similar) pipette, 0–100 μl.

Mixing bowl.

Freezer.

Method

Preparation of the treated soil Using an Eppendorf pipette add 120 μl of Verdone-2 to 10 ml of methanol and shake vigorously. Add to this 300 ml of water and shake. Place 3 kg of soil in a mixing bowl and slowly add the solution with mixing. Transfer the treated soil to a polythene bag. The soil contains 4 μg mecoprop and 2 μg 2,4–D g^{-1} air-dry soil.

The control soil Place 1 kg of soil in a mixing bowl and add 100 ml of water with mixing. Transfer to a polythene bag.

Soil incubation Incubate the two bags of soil loosely folded at room temperature in the dark. Occasionally open the bags and shake to aerate the soil and weigh and add water as required to maintain their water content.

After preparing the treated soil, allow to stand for 2–3 h and then take a 600 g subsample into a polythene bag, label and store in a freezer. Remove further 600 g samples from the bag of incubated, treated soil after 5, 10 and 20 d, storing each in the freezer. No further breakdown of the herbicide occurs once the soils are frozen. After 40 d allow the frozen samples to thaw. You now have a control soil together with treated soils which have been incubated for 0, 5, 10, 20 and 40 d (a total of six samples).

The pot experiment From each of the samples prepare three pots each containing 200 g of moist soil. Label the pots and proceed as described under 'Growth of a test crop' in the previous experiment. At the same time, carry out a bioassay as described in the previous experiment. After 3 weeks, harvest and plot your biological dose response curve (Fig. 15.8(a)) and your measure of growth against time of incubation (Fig. 15.8(b)).

Results

The threshold concentration of Verdone-2 (50% of control growth in Fig. 15.8(a)) is about 1 μg mecoprop g^{-1} soil. At the threshold there is also about 0.5 μg 2,4–D g^{-1} soil, not marked on the graph. The incubation experiment shows that the threshold concentration is reached after 27 d with an initial treatment of 4 μg g^{-1}. A decay curve can be drawn (Fig. 15.8(c)) by taking the growth data from Fig. 15.8(b) and using the bioassay in Fig. 15.8(a) to obtain the concentration present in the soil. The decay curve shows that Verdone-2 has a half-life of about 13 d. Note that this is a 'combined' half-life for mecoprop and 2,4–D and that it is dependent on the incubation temperature (Fig. 15.3(b)).

A method to determine the half-life of Verdone-2 without a full bioassay

Follow the above procedure, but prepare an extra 300 g of treated soil: for convenience, add 140 μl of Verdone-2 to 3.5 kg soil. At time zero take an extra 300 g soil and store in the freezer until all the samples are ready for the pot experiment. After thawing, mix this 300 g sample with 300 g of untreated soil: it now contains 2 μg mecoprop and 1 μg 2,4–D g^{-1} air-dry soil which is half the initial concentration of the main experiment. Place in three pots and measure growth at the same time as the main experiment. This extra treatment is a bioassay indicating growth when the concentration of herbicide in the main experiment has decreased to half its initial value. It is marked on Fig. 15.8(b) as a square. The half-life is therefore about 12 d. This method is less reliable than the full bioassay, especially if the extra treatment gives a measure of growth which lies towards one end of the biological dose response curve.

Alternative herbicides

Pathclear Dissolve 40 mg in 10 ml of methanol. Add to 300 ml of water and mix with 3 kg of soil. Proceed as for Verdone but with sampling times of 0, 50, 100, 150 and 200 d.

Deeweed and atrazine Dissolve 50 mg of Deeweed or 25 mg of atrazine in 100 ml of methanol. Pipette 2 ml ino 300 ml of water and mix with 3 kg of soil. Proceed using the sampling times above for Pathclear.

Section 15.3 Adsorption of pesticides in soils

The partitioning of a pesticide when equilibrated between particle surfaces (the adsorbed phase) and the soil solution (the aqueous phase) is usually displayed as

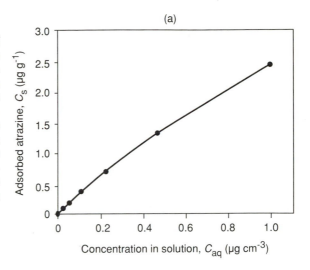

(a)

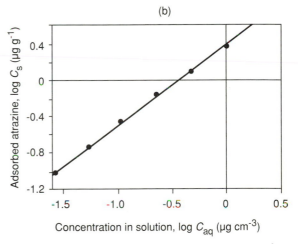

(b)

Figure 15.9 The adsorption isotherm of atrazine for a sandy loam soil (3.4% organic matter, pH 6.0, K_d = 3.5 cm^3 g^{-1}, K_{oc} = 177 cm^3 g^{-1}). Data by M. Lane, Zeneca Agrochemicals.

an adsorption isotherm (Section 9.4). Fig. 15.9(a) gives an example for atrazine. At the small concentrations involved in pesticide use, the isotherm is normally a straight line described by the equation

$$C_s = K_d \times C_{aq} \qquad [15.1]$$

where C_s is the concentration of the adsorbed pesticide (μg g^{-1}), C_{aq} is the concentration in the liquid phase (μg cm^{-3}) and K_d is the adsorption coefficient C_s/C_{aq} with the units cm^3 g^{-1}. The initial slope in Fig. 15.9(a) (up to 0.5 μg g^{-1}) is 3.5 cm^3 g^{-1} which is the value of K_d. With increased concentrations the isotherm becomes curved because the adsorption capacity of the soil becomes saturated. The curve can be described by the Freundlich equation $C_s = K_d \, C_{aq}^x$ where x is a constant. For convenience, the logarithmic form of this equation is used:

$$\log C_s = \log K_d + x \log C_{aq}$$

The atrazine data plotted in this form are shown in Fig. 15.9(b). The slope is 0.91 which is the value of x and the intercept on the $\log C_s$ axis ($\log C_{aq} = 0$) is $\log K_d$, giving $K_d = 2.6 \, cm^3 \, g^{-1}$ which is lower than the initial K_d value because of the decreasing slope in Fig. 15.9(a). Brouwer *et al.* (1990) give 0.91 as the mean value of x for atrazine. Values of K_d are normally reported at the expected environmental concentration of the pesticide.

The use of an adsorption coefficient to quantify the partitioning of a pesticide between solution and particle surfaces is similar to the use of *buffer power* for nutrient ions (Section 9.4).

Measurement of adsorption

K_d is measured in a similar way to that described for potassium (Section 9.4) and phosphorus (Section 10.2). Soil is shaken with a solution (often 10 mM CaCl$_2$) containing a known concentration of pesticide. Equilibration takes about 24 h, the solution is separated by filtration or centrifugation and the concentration measured. By difference the adsorbed pesticide is determined, and K_d calculated.

Methods of analysis are those mentioned in Section 15.2. The experiment has to take into account (a) the stability of the pesticide in the soil:water suspension for the duration of the shaking period, (b) the expected adsorption which will affect the chosen ratio of soil to water and (c) the expected environmental concentration depending on its application rate and mode of use.

Adsorption on organic matter

Organic matter is the predominant adsorbing component in soils. Figure 15.10 shows the values of K_d for atrazine measured for soils with a wide range of organic matter contents. There is considerable scatter indicating the importance of other soil properties also. Because of its importance, adsorption coefficients are often expressed in terms of organic matter, K_{om}, or organic C, K_{oc}, where

$$K_{om} = (\mu g \text{ adsorbed pesticide g}^{-1} \text{ organic matter})/ (\mu g \text{ absorbed pesticide cm}^{-3} \text{ solution})$$

Both coefficients have, like K_d, the units cm^3 g^{-1}. For a given soil,

$$K_d = K_{om} \times OM \text{ per cent}/100 = K_{oc} \times C \text{ per cent}/100$$

The factors OM per cent/100 and C per cent/100 are the fractional contents in the soil (g OM g^{-1} soil or g C g^{-1} soil).

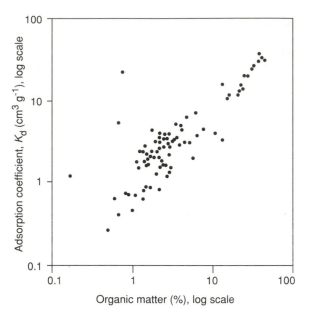

Figure 15.10 The relationship for atrazine between K_d and organic matter content of soils. From Brouwer *et al.* (1990).

To determine K_{oc} or K_{om} for a given pesticide, data for K_d and organic matter content (Fig. 15.10) are plotted. The slope of the line fitted through the data is $K_{om}/100$, allowing a mean value of K_{om} to be found. Since organic matter contents are determined from measurements of organic C using the factor 1.724 (58% C in organic matter, Section 3.4), then $K_{om} = 0.58K_{oc}$. The value of K_{oc} for atrazine is about 100 cm^3 g^{-1}. Using the above relationships, $K_{om} = 58 \, cm^3 \, g^{-1}$ and for a soil with 3.4 per cent organic matter (Fig. 15.9), $K_d = 2.0 \, cm^3 \, g^{-1}$. This value is smaller than that calculated from the isotherm (3.5 cm^3 g^{-1}) reflecting the variability between soils in the relationship between K_d and organic matter content as shown by the scatter of data in Fig. 15.10. Variations in the value of K_{oc} for a given compound in a range of soils are caused primarily by

Table 15.6 The effects of soil properties on the adsorption of atrazine

Soil No.	pH	Organic C (%)	Texture	K_{oc} (cm^3 g^{-1})
1	4.6	10.9	Loamy sand	434
2	5.8	3.7	Loam	79
3	5.9	1.5	Clay	636
4	7.0	1.7	Silt loam	57
5	8.0	10.9	Silt loam	60
6	8.3	0.3	Silt	80

Data from M. Lane, Zeneca Agrochemicals.

differences in pH and the nature of the organic matter as shown by Table 15.6. Thus there will be significant errors in using a single value of K_{oc} for a pesticide as a means of predicting a soil's K_d value from its organic C content. The errors are, however, small in comparison both to other errors in predicting behaviour in the field and to the large differences between the K_{oc} values of different pesticides (Table 15.1).

Values of K_{oc} are normally determined for agricultural soils with pH values between 5 and 7. For calcareous or acidic soils an adjusted K_{oc} value may be needed depending on the adsorption properties of the pesticide.

An index of adsorption

Because of the difficulties of measuring K_{oc} and K_d values, methods have been devised to estimate them, based on the principle that organic molecules adsorb because they are to varying degrees hydrophobic. Generally, the more hydrophobic an organic molecule is, the more it will adsorb, and the less it will reside in the soil solution. The most commonly used index of adsorption is K_{ow}, the *octanol–water partition coefficient*. The coefficient is measured by shaking a solution of pesticide in water with the solvent octanol until equilibrium is attained. The water is separated from the solvent and the concentration measured. By difference the concentration in the solvent is calculated, and K_{ow} is expressed as

(μg pesticide cm^{-3} octanol)/(μg pesticide cm^{-3} water)

and is thus dimensionless. Values for a range of pesticides are given in Table 15.1 along with their K_{oc} values. The relationship between the two parameters has been quantified through empirical equations, the most commonly used being the Briggs equation

$$\log K_{oc} = 0.52 \log K_{ow} + 0.86$$

or

$$\log K_{om} = 0.52 \log K_{ow} + 0.62$$

There can be large errors in estimates of K_{oc} using this equation, as evidenced by the values in Table 15.1, and its validity depends on the extent to which a chemical's hydrophobic properties are similar to those of the compounds initially used to produce the equation.

Measurement of K_{ow}

Reagents and equipment

UV spectrophotometer and 1 cm cells.
Buffer solution. Dissolve 1.39 g of potassium dihydrogen orthophosphate, KH_2PO_4, and 1.74 g of potassium hydrogen orthophosphate, K_2HPO_4, in water and make up to 1 l. This has a pH of 6.8.

Octan–1–ol, spectroscopy grade.
Methanol.
Magnetic stirrer and follower.
Atrazine. The reagent (Note 1) can be obtained from British Greyhound Chromatography and Allied Chemical Co. Ltd. (Appendix 1).

Method

Shake vigorously in a separating funnel about 50 ml of octanol and 300 ml of the buffer solution. Allow the two phases to separate completely and run off the two liquids into separate containers. Alternatively the liquids can be efficiently separated by centrifuging. The buffer solution is now saturated with the solvent.

Preparation of the stock atrazine solution Shake 10 mg of atrazine reagent with 10 ml of methanol. This contains about 1000 μg atrazine ml^{-1}. This solution is stable for a few days when kept in a refrigerator. Pipette 1 ml into a 250 ml volumetric flask and blow air into the flask until the methanol has evaporated. Make up to 250 ml with the buffer solution and shake until the atrazine has dissolved to give a solution containing about 4 μg ml^{-1} (Note 2).

The partition experiment Pipette 100 cm^3 (Note 3) of the atrazine solution into a conical flask. Add by pipette 1 cm^3 of buffer-saturated octanol. Stir with a magnetic stirrer for 2 h. Allow the solution to stand until the octanol has formed a discrete drop on the surface of the water. Using a pipette take about 10 cm^3 of the buffer solution avoiding the octanol. Allow the solution to stay in the pipette and tap occasionally so that any stray drops of octanol rise to the surface. Run the necessary amount of solution into a spectrophotometer cell. Repeat the experiment to obtain duplicate samples.

Measurement of atrazine concentrations Place octanol-saturated buffer solution in the spectrophotometer reference cell (Section 10.1). Measure the absorbance of the stock atrazine solution, A_1, and of the atrazine solution after partition, A_2, at a wavelength of 220 nm. If a scanning spectrophotometer is available scan each solution between 190 and 290 nm, and from the print-out measure the absorbance at the peak wavelength.

Calculation

By definition $K_{ow} = (\mu\text{g cm}^{-3}$ octanol)/($\mu\text{g cm}^{-3}$ water). The relationship between concentration, C, and absorbance is $C = kA$ (the Beer–Lambert Law, Section 10.1). Thus, the equilibrium concentration in the water is kA_2. The concentration in the octanol can be calculated from the amount of atrazine which has been lost from the water which is

$k(A_1 - A_2)\,\mu\text{g cm}^{-3} \times 100\,\text{cm}^{-3} = 100k(A_1 - A_2)\,\mu\text{g}$

This has moved into $1\,\text{cm}^3$ of octanol and so the concentration is $100k(A_1 - A_2)\,\mu\text{g cm}^{-3}$. Thus $K_{\text{ow}} = 100\,(A_1 - A_2)/A_2$. If as an example $A_1 = 0.8$ and $A_2 = 0.25$, then $K_{\text{ow}} = 220$.

Note 1 Although commercial formulations of pesticides have been used for other experiments in this book, the measurement of K_{ow} requires the reagent alone. Formulations may contain a mixture of pesticides with wetting agents and oils. Thus they are not suitable for spectrophometric analysis without careful separation and cleaning.

Note 2 The value of K_{ow} does not depend on the concentration of atrazine in the stock solution, nor is this value needed for the calculation.

However, a suitable concentration is required for measurement using the spectrophotometer. A saturated solution of atrazine in water contains about $30\,\mu\text{g cm}^{-3}$ at $20\,^\circ\text{C}$. The prepared solution contains about $4\,\mu\text{g cm}^{-3}$ giving an absorbance reading of about 0.8 using a 1 cm cell at 220 nm wavelength.

Note 3 Solution volumes here and in Section 15.4 are given in cm^3 rather than ml because the units $\text{cm}^3\,\text{g}^{-1}$ are commonly used for partition coefficients.

Note 4 The main types of chemical (Fig. 15.5) bond are as follows:

1. *Ionic or electrostatic.* Electrons are transferred from one atom or molecule to another: each is a (charged) ion. They are attracted to each other but can separate in solution, e.g. $\text{Na}^+\,\text{Cl}^-$ and exchangeable ions.
2. *Covalent.* Electrons from two atoms are shared to fill the electron shells of both atoms which generally do not separate to any extent in solution, e.g. H_2O.
3. *Ligand (chelation).* Electrons (typically a pair) from one atom are donated to complete the electron shell of another and the atoms cannot easily separate, e.g. a metal bound to humus. Usually the binding occurs with more than one site on the surface. These bonds are shown as arrows in Fig. 15.5(b).
4. *Hydrogen bonds.* These are explained in Fig. 15.5(c) and are shown as dotted lines in the figure.

Section 15.4 The mobility of pesticides and their leaching into groundwaters

ADSORPTION AND PESTICIDE MOBILITY

A simple basis for studying pesticide mobility is to consider the soil as a chromatography column through which water is flowing. A pulse of applied pesticide is adsorbed in the surface layer, the extent of adsorption depending on the adsorption coefficient, K_d. The pesticide remaining in solution moves down through the surface layer to be adsorbed below. The water entering the surface layer causes desorption, moving more pesticide down the column. The result is a movement of the pulse at a rate depending on K_d and the rate of water movement.

The principles governing leaching have been developed in Section 11.5 for the movement of nitrate in soils and Section 11.6, Project 5 considers the effect of adsorption of nitrate on leaching in positively charged soils. The relationship between movement and adsorption can be derived as follows.

Consider a soil packed at a bulk density of $\rho\,(\text{g cm}^{-3})$ with a volumetric water content of $\theta\,(\text{cm}^3\,\text{cm}^{-3})$. The total amount of pesticide per cm^3 of soil is the sum of that from the solid phase, $C_s \times \rho\,(\mu\text{g g}^{-1}\,\text{soil} \times \text{g cm}^{-3}\,\text{soil} = \mu\text{g cm}^{-3}\,\text{soil})$, and the liquid phase, $C_{\text{aq}} \times \theta$ $(\mu\text{g cm}^{-3}\,\text{solution} \times \text{cm}^3\,\text{solution cm}^{-3}\,\text{soil} = \mu\text{g cm}^{-3}\,\text{solution})$. Thus the ratio of the amount of pesticide adsorbed to that in solution is $(C_s \times \rho)/(C_{\text{aq}} \times \theta)$ which simplifies to $K_d\rho/\theta$ because $K_d = C_s/C_{\text{aq}}$ (Eqn. 15.1), and is dimensionless. The pesticide molecules are in dynamic equilibrium, i.e. they are constantly being adsorbed and desorbed but with no net change in concentration in either phase. The ratio also indicates the amount of time on average each molecule spends in each phase. Thus time adsorbed on the surface:time in solution $= K_d\rho/\theta{:}1$ and a molecule at equilibrium spends 1 s in solution and $K_d\rho/\theta$ seconds on the surface out of a total of $1 + K_d\rho/\theta$ seconds.

Imagine now that water is moving through the soil, and that an adsorbed pesticide molecule cannot move. If the water moves for $1 + (K_d\rho/\theta)$ seconds, the molecule only moves in the water for 1 s. Distance of movement is directly proportional to time of movement, and so the water moves $1 + (K_d\rho/\theta)$ cm for each 1 cm movement of the molecule. The ratio of these two distances is the *retardation factor*, R_f, where

$$R_f = \text{distance moved by the pesticide}/$$
$$\text{distance moved by the water} = 1/[1 + (K_d\rho/\theta)]$$

The term $K_d\rho/\theta$ is known as the *delay*, i.e. the distance by which the pesticide lags behind the water when the pesticide moves 1 cm.

The retardation factor can also be used as a volume ratio. In the above example the volume of water in the pores (the pore volume) is $\theta\,cm^3$. Thus $\theta\,cm^3$ water entering a face of the cube would move each water molecule on average to the opposite face, a distance of 1 cm. The pesticide, however, will only have moved a fraction of this distance (R_f). To move the pesticide right through the cube requires $1/R_f = 1 + K_d\rho/\theta$ pore volumes of water.

The relationship between K_d and R_f in the above equation is the basis for understanding the mobility of pesticides in soils. A measurement of K_d allows mobility to be predicted for an idealized situation: an example is given below. An experiment follows in which mobility is measured and K_d calculated.

Example: the prediction of mobility using a K_d value Atrazine has a K_{oc} value of about $100\,cm^3\,g^{-1}$, and so a soil with 2 per cent organic C has a K_d value of about $2.0\,cm^3\,g^{-1}$. If the bulk density of a field soil is $1.3\,g\,cm^{-3}$, and water is moving through the soil at just above the field capacity with $\theta = 0.4$ (Table 4.3), then $K_d\rho/\theta = 6.5$ and $R_f = 1/7.5$ which means that leaching water moves 7.5 times as far as a pulse of atrazine. Taking the example in Section 9.4 for the Reading area where about 30 cm $H_2O\,a^{-1}$ leaches through the soil, the distance of water movement is 30/0.4 = 75 cm, and the distance of movement of the pulse is 75/7.5 = 10 cm. This calculated value is for an idealized situation and does not take account of the effects of soil structure on water flow.

The measurement of pesticide mobility

A method developed by Gerber *et al.* (1970) uses the phytotoxicity of the pesticide as a means of determining the position of a pulse of pesticide in soils. Water is passed through a thin layer of soil on a sloping plate (Plate 15.1). A pulse of pesticide is added to the soil and is eluted by the water. Grass or lettuce is then grown on the plate and the position of maximum growth depression marks the position of the pulse.

Equipment and Reagents

Aluminium plates, 5 cm wide × 0.5 cm deep × 30 cm long.
Muslin cloth.
A seed propagator.
Loam or sandy loam soil. Air dry 1 kg of soil and pass through a 2 mm sieve.
Coarse sand.
Herbicides. Use Deeweed (which contains atrazine) or

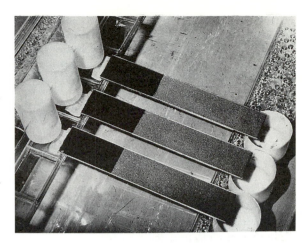

Plate 15.1 Plates containing thin layers of soil used to measure the mobility of pesticides during leaching. By M. Lane, Zeneca Agrochemicals.

pure atrazine or Casoron G4 (Vitax Ltd). Section 15.2 gives information about Deeweed and atrazine. Casoron G4 contains 4 per cent *m/m* dichlobenil in small limestone granules. It is sprinkled on to paths and shrubberies where it is washed into the soil and acts as a non-selective herbicide.
Calcium chloride, 10 mM. Dissolve 1.47 g of $CaCl_2.2H_2O$ in 1 l of water.
Seed. Perennial ryegrass or lettuce.
Methanol.
Eppendorf (or similar) pipette, 0–100 μl.

Method

Glue a muslin wick 5 × 10 cm in size inside the top end of the plate. Weigh the plate and fill it evenly with soil, tamping as necessary to compact to the volume of the plate (75 cm³). Packed in this way the bulk density is normally between 1 and 1.2 g cm⁻³ for a loam (75–90 g per plate). Weigh the plate plus soil. Place the plate in the propagator or in a suitable covered container to prevent evaporation losses. Support the plate at an angle of 5–10° (raise the upper end by 3–5 cm). Place the wick in a beaker of 10 mM $CaCl_2$ and allow solution to soak the soil and elute into a beaker under the lower end of the plate. Remove the wick from the supply beaker and allow solution to drain overnight.

Dissolve 250 mg of Deeweed or atrazine (or 500 mg Casoron G4) in 10 ml of methanol. Note that because Casoron G4 is formulated with limestone granules, these remain after the dichlobenil has dissolved. Using an Eppendorf pipette, apply 30 μl of herbicide solution in a band 2 cm from the top of the plate. Place the wick in the beaker again, place a clean beaker under the end of the plate and allow solution to elute. Collect about 100 ml of solution (or 200 ml if Casoron is used) over a period of 12–24 h. Remove the wick from the supply beaker and allow the plate to drain. Measure the volume of eluted solution.

Sprinkle seed liberally and evenly along the soil surface and cover with a thin layer of sand. Cut off the wick. Place the plate in the propagator in a glasshouse and allow the plants to grow. Spray with water as required to keep the soil moist. After about 1 week the plants should have grown sufficiently to show the peak of the phytotoxic effect. Measure the distance of the peak effect from the position where the pesticide was applied.

Calculation of R_f and K_d

The calculation follows that given for atrazine above.

Example

Mass of soil in the plate	$= 80\,g$
Distance of movement of the pulse	$= 10\,cm$
Volume of solution collected	$= 100\,ml$

The bulk density of the soil is $80/75 = 1.07\,g\,cm^{-3}$. Assuming a particle density of $2.6\,g\,cm^{-3}$, the porosity is

$$1 - 1.07/2.6 = 0.59\,cm^3\,cm^{-3} \quad \text{(Section 4.2)}$$

The soil is effectively saturated during elution and so the water-filled porosity is also $0.59\,cm^3\,cm^{-3}$.

The pesticide moved through 10 cm of the plate ($25\,cm^3$ soil) which contains $25 \times 0.59 = 14.75\,cm^3$ water. The total amount of water which moved through the plate was $100\,cm^3$ and so the retardation factor is $14.75/100 = 0.1475$ and equals $1/[1 + (K_d\rho/\theta)]$. Substituting $\rho = 1.07$ and $\theta = 0.59$ gives $K_d = 3.2$ with a delay of 5.8 pore volumes.

Limitations

The above approach does not take account of the complexities of water and solute movement in structured soils. However, the experiment can be used to rank the mobility of pesticides relative to one another. The technique always overestimates mobility in the field because degradation is not taken into account. Section 11.5 introduced computer-assisted models to predict flow of nitrate in field soils, and this approach is extended to account for adsorption below.

ADSORPTION, DEGRADATION AND PESTICIDE MOBILITY

Degradation during leaching reduces the concentration of a pesticide leaving the soil and entering groundwater. Various schemes have been introduced to classify the leaching potential of pesticides depending on their adsorption and persistence.

The McCall scale

This is based on K_{oc} values only. It takes direct account of adsorption, but only indirectly takes account of persistence through the broad relationship between persistence and adsorption. The scale is given in Table 15.7.

Table 15.7 The McCall scale of leaching potential

K_{oc} $(cm^3\,g^{-1})$	Leaching potential
0–50	Very high
50–150	High
150–500	Medium
500–2000	Low
2000–5000	Slight
>5000	Immobile

The groundwater ubiquity score, or GUS index

This takes account of both persistence and adsorption. Gustafson (1989) gathered together data for the half-lives and adsorption coefficients of pesticides, and plotted $\log t_{1/2}$ against $\log K_{oc}$. He then marked on the graph those pesticides which had been found in groundwaters. Fig. 15.11 is a similar graph for the compounds listed in Table 15.1. Gustafson found a pattern in the distribution, the leaching pesticides being in the upper left sector of the graph. They were separated from those which had not been found in groundwaters by fitting lines with the form $\log t_{1/2} = k/(4 - \log K_{oc})$ as shown in the figure. The equation allows a pesticide's tendency to leach to be rated: those likely to leach have a k value >2.8, those unlikely to leach have a k value <1.8 and those between 1.8 and 2.8 are transitional. When used in this way, the k value is called the *groundwater ubiquity score* (GUS) which is obtained by rearranging the equation to give

$$\text{GUS} = k = \log t_{1/2} \times (4 - \log K_{oc})$$

Most of the commonly used pesticides have a score of less than 2.8. Of those listed in Table 15.1 and plotted in Fig. 15.11, atrazine, simazine and mecoprop have a small tendency to leach. Values of GUS are given in Table 15.1.

It should be noted that a high score does not mean that the pesticide will leach in all soils since the risk

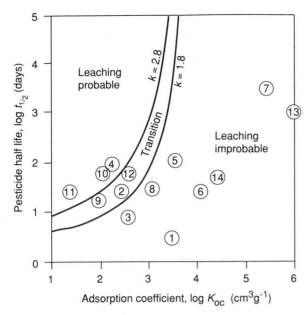

Figure 15.11 The Gustafson (1989) relationship between $t_{1/2}$, K_{oc} and occurrence of pesticides in groundwaters. The lines are plotted using the equation $\log t_{1/2} = k/(4 - \log K_{oc})$ where $k = 2.8$ and 1.8.

depends both on K_{oc} and organic matter content of the soil: a pesticide with a small K_{oc} value may be strongly adsorbed in an organic soil and unlikely to leach. Thus the score identifies those pesticides with a high leaching risk in soils of the type present in Gustafson's original investigation. In order for the pesticide to be detected in groundwater these soils must have been easily leached, mostly light textured with low organic matter contents. Thus the score identifies high risk pesticides in high risk soils. In normal agricultural use, the crop may intercept a large fraction of the applied chemical, significantly reducing the risk.

A MODEL TO SIMULATE PESTICIDE LEACHING

The movement of solutes in soils is discussed in Section 11.5 where the Wineglass model is used to simulate the leaching of nitrate. A similar approach is used for pesticides and other adsorbed solutes, but the partitioning of the pesticide between solution and particle surfaces has to be taken into account (Section 15.3). For this purpose a subroutine has been added to the programme. Similarly the decay of the pesticide can be modelled assuming first order kinetics (Section 3.5).

Example Following the example in Section 11.5 where

calculations are carried out manually to give Fig. 11.19, consider a column of soil $50 \times 10 \times 10$ cm divided into five layers each containing 1000 cm^3 of soil. The water content of each layer at field capacity is 300 cm^3, and assuming a bulk density of 1.3 g cm^{-3} there is 1300 g of soil per layer. Atrazine is now added at a rate of 4 kg ha^{-1} which is 4000 µg on the surface of the column. This is washed into the top layer with 150 cm^3 of water, where it mixes with the 300 cm^3 already there and equilibrates between the particle surfaces and solution to satisfy a K_d value of 3.4 µg g^{-1}/µg cm^{-3}.

It has been shown on page 320 that after equilibration the ratio of the amount of adsorbed pesticide to the amount in solution is $K_d \rho / \theta : 1$ and so the amount in solution as a fraction of the total amount in the soil is $1/(1 + K_d \rho / \theta)$. Substituting gives 0.0924 as the value of this fraction. The total amount in the soil is 4000 µg and so the amount in solution is 4000×0.0924 which is 370 µg. Of this one-third (123.3 µg) moves down into the second layer in 150 cm^3 of water to mix and equilibrate there. The process is repeated for each step in the leaching sequence, and for further sequences until the atrazine has been leached through the column.

Examples of the use of the Wineglass model are shown in Fig. 15.12(a) and (b). The inputs used for the simulation are based on the properties of the soils used by Smith et al. (1992) in atrazine leaching experiments so that comparison could be made with measurements. The soil was very light-textured, with a small amount of organic matter (about 0.3% C). The adsorption coefficients were calculated from the organic carbon contents assuming $K_{oc} = 100$ cm^3 g^{-1} (Section 15.3) and a half-life of 70 d was assumed.

Figure 15.12a shows, as expected, that without adsorption the peak appears after 1 pore volume. With adsorption the peak is delayed to about 3 pore volumes, and the pulse is much more dispersed. If the soil had 5 times the organic matter content ($K_d = 1.7$ cm^3 g^{-1}), the peak would be delayed to about 9.5 pore volumes. Decay reduces the concentrations to low values with the peak at about 2 pore volumes.

Figure 15.12b gives the data obtained by Smith et al. (1992). Water was added to four undisturbed soil columns at intervals during a 60 d period and so the leaching rate was not uniform. The columns gave variable results reflecting structural differences, but it can be seen that simulation which includes adsorption and decay (the same data is shown in Fig. 15.12a) gives results within the range of measured values. The programme together with an Atrazine.dat file which contains further details of the experiment and the data is available on a computer disk which can be obtained from the address in Appendix 3. More complex models and further data are discussed by Nicholls et al. (1982).

(a)

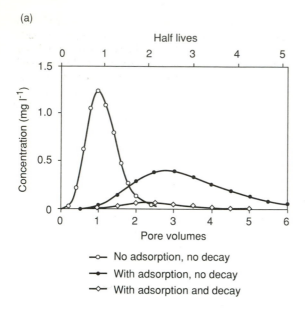

- —○— No adsorption, no decay
- —●— With adsorption, no decay
- —◇— With adsorption and decay

(b)

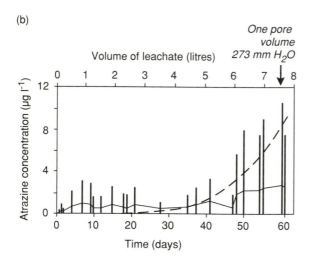

Figure 15.12 Breakthrough curves for atrazine leaching from columns of loamy sand (110 cm long, 19.1 cm diameter), after applying 329 mg atrazine m^{-2}. (a) Simulation of the concentration of atrazine in the leachate (K_d varies down the column with an average value of 0.33 cm^3 g^{-1}, $t_{1/2}$ = 70 d, one pore volume leached in 60 d, 22 soil layers, 30 mm water per sequence); (b) measured and simulated concentrations of atrazine. The vertical bars show the range of concentrations for four columns: in one column the concentration did not rise above the detection limit (0.3 µg l^{-1}). The full line is the average for the columns, and the dotted line shows the simulated concentration based on the average properties of the four columns.

Section 15.5 The determination of metals in soils: toxicities and deficiencies

TOXICITIES

Guidelines in terms of maximum allowable concentrations are given as totals in the soil determined by digestion of the soil in a mixture of nitric acid and perchloric acid (MAFF, 1986a). This is a potentially dangerous procedure and should only be carried out under supervision. It is being replaced by digestion in aqua-regia, which is the method described here.

Extraction of total metals

Soil is digested in a nitric–hydrochloric acid mixture and the metal concentration measured by atomic absorption spectrophotometry.

Reagents and equipment

Digestion acid mixture. Add 130 ml of HCl (approx. 36% *m/m* HCl) to 120 ml of water (CARE) and mix. Add 150 ml of this solution (CARE) to 50 ml of HNO$_3$ (approx. 70% *m/m* HNO$_3$) and mix.
Potassium chloride solution, 5 per cent m/v. Dissolve 5.0 g of potassium chloride, KCl, in water and make up to 100 ml.
Nitric acid solution, 8.8 per cent. Add 125 ml of HNO$_3$ (approx. 70% *m/m* HNO$_3$) to 40 ml of KCl solution (CARE) and make up to 1 l.
Boiling tubes, 60 ml graduated, borosilicate.
Heating block, thermostatically controlled.
Atomic absorption spectrophotometer and standard solutions.
Pestle and mortar.

Method

Grind some <2 mm air-dry soil in a pestle and mortar. Weigh 1.2 g (±0.01) of soil into a graduated boiling tube. Add 15 ml of digestion acid and swirl to wet the sample. Place a small glass funnel in the neck of the tube to aid with refluxing. Allow to stand overnight. Heat the tube at 50 °C for 30 min in the heating block and then increase the temperature to 120 °C for 2 h. Allow to cool. Remove the funnel and make up to 60 ml with 8.8 per cent HNO$_3$ solution. Filter through a Whatman No. 541 paper discarding the first few ml and retain the remainder for analysis.

Carry out a blank determination without soil.

Measurement by atomic absorption spectrophotometry

It is assumed that technical help will be available and full details are not given here. Standard solutions for each metal are needed for calibration. Lanthanum chloride must be added to extracts and standards to prevent interferences in the determination and background correction is needed.

Standard solutions The following calibration ranges apply (μg ml^{-1}): cadmium 0–0.2, copper 0–2, lead 0–2, nickel 0–0.5, zinc 0–1.5. These solutions should be made up in 8.8 per cent HNO$_3$/KCl solution.

Lanthanum chloride solution Dissolve 2.68 g of LaCl$_3$.7H$_2$O in water and dilute to 100 ml.

To 20 ml of each standard or extract add 1 ml of lanthanum chloride solution and mix before measurement by atomic absorption. It may be necessary to dilute the extract with 8.8 per cent HNO$_3$/KCl solution before adding lanthanum chloride to obtain a reading within the calibration range.

Calculation

Method summary

air-dry soil (? mg copper kg^{-1})

     ↓

    1.20 g digested → 60 ml

             ↓

              concentration measured,

              y μg copper ml^{-1}

Example The concentration y (corrected for the blank reading) was 1.8 μg ml^{-1}. The amount in 60 ml of extract was $1.8 \times 60 = 108$ μg. This was extracted from 1.20 g of soil and so the extractable copper is

$$108 \times 1000/1.2 = 90\,000 \ \mu\text{g kg}^{-1}$$

or

$$90 \ \text{mg copper kg}^{-1} \ \text{air-dry soil}$$

Recommended maximum metal concentrations in soils

Table 15.2 gives typical total metal contents of uncontaminated soils and the maximum recommended by the DOE/NWC (1981). Soils vary greatly in their natural content of these metals. However, the difference between columns 2 and 4 is a guide to the total amount of metal that can be added to the soil. It is recommended that not more than 1/30 of this amount should be added in any one year. Initially the metals are relatively free but they become strongly bound in the soil with time.

The recommended maxima apply to arable soils with pH > 6.5 and grassland soils with pH > 6.0. In acidic soils the metals are more available but there is insufficient evidence on which to base modified recommendations. In calcareous soils the metals are less available, and the upper limits for zinc and nickel can be increased to 1.5 times those given in Table 15.2: at high pH, the increased charge on humus favours adsorption (Section 7.1) and metals are also adsorbed on the surface of calcite.

Differences between methods

Although the recommendations in Table 15.2 are based on nitric–perchloric extractions, the amounts removed by nitric–hydrochloric extraction differ by only a few per cent and so for most purposes the two methods are direct alternatives (Franks, 1984).

The combined effect of metals on crops

Zinc, copper and nickel are normally all present in sewage sludge, and will act together in affecting crop growth. There is no general agreement regarding additive effects but tentative guidelines (ADAS, 1982) have in the past been based on their relative toxicities: compared to zinc, copper is about twice as toxic and nickel is about eight times as toxic. Thus a *zinc equivalent* can be calculated based on the total amount of each metal present:

zinc equivalent =
$$\text{Zn}^{2+} + (2 \times \text{Cu}^{2+}) + (8 \times \text{Ni}^{2+}) \ \text{mg kg}^{-1}$$

Using Table 15.2 a typical natural zinc equivalent would be about 320 mg kg^{-1}. It has been recommended that this should not be increased by more than 560 mg kg^{-1}. Because of uncertainties about combined effects recommendations are now made based on each metal separately.

The approach here is similar to that used for the crop equivalent when considering the effects of weeds (Section 15.1).

Extraction of available metals by EDTA

Availability of metals to crops is determined by extraction in the complexing reagent EDTA (MAFF, 1986a) and is a basis for considering both toxicities and deficiencies. It complexes strongly with metal ions in solution (Section 7.2), reducing their concentrations to very low values and causing metal desorption from particle surfaces until all the readily released metals are complexed in solution. Measurement is by atomic absorption spectrophotometry.

Reagent

Ammonium-EDTA, approximately 0.05 M at pH 7. Dissolve 14.6 g of ethylenediaminetetra-acetic acid (EDTA) in approximately 950 ml of water containing 8 ml of ammonia solution (approx. 35% *m/m* NH_3). Adjust the pH to 7.0 by addition of approximately 1 M nitric acid or 1 M ammonia solution and dilute to 1 l. Concentrated nitric acid solution (1.42 g ml^{-1}, 70% HNO_3) is about 16 M and concentrated ammonia solution (0.88 g ml^{-1}, 35% NH_3) is about 18 M. Dilute 6 ml or 5.5 ml respectively to 100 ml to give 1 M solutions.

Extraction

Transfer 10 ml of <2 mm air-dry soil using a measuring cup (Section 9.1) (or 10 g) into a bottle. Add 50 ml of ammonium–EDTA solution and shake for 1 h at 20°C. Filter through a Whatman No. 40 paper and retain the filtrate for analysis. If necessary, dilute by pipetting 10 ml of filtrate into a 100 ml flask and making up to the mark with ammonium-EDTA solution.

Measurement by atomic absorption spectrophotometry

Follow the procedure given in the section 'Extraction of total metals', but make up the standard solutions in 0.05 M ammonium–EDTA.

Calculation

Method summary

air-dry soil (? mg Zn^{2+} l^{-1} soil)
↓
  10 ml + 50 ml
    ↓
    10 ml → 100 ml
      ↓
      concentration measured,
      y mg Zn^{2+} ml^{-1}

Example Ten ml of the extract was diluted to 100 ml. The concentration *y* was 1.2 μg ml^{-1}. The amount of Zn^{2+} in 100 ml of diluted extract was 100 × 1.2 = 120 μg. This was also in 10 ml of undiluted extract and so in 50 ml there was 600 μg Zn^{2+}. This came from 10 ml of soil. The extractable zinc is

$$600 \times 1000/10 = 60\,000\ \mu g\,l^{-1}$$

or

$$60\ mg\ Zn^{2+}\,l^{-1}\ soil$$

Use of the data: toxicities

Although the total metal content is used to monitor safe disposal of sewage on land, the amount of EDTA-

extractable metal is a better index of availability. ADAS (1987) advise that when the total metal content reaches one-half of the permitted maximum, and again at three-quarters, EDTA-extractable metals should be measured irrespective of the total metal concentration. If there is a risk that EDTA-extractable metal concentrations will exceed 120 mg Zn^{2+} l^{-1} soil, 70 mg Cu^{2+} l^{-1} or 20 mg Ni^{2+} l^{-1} before the total metal limit is reached then there may be problems.

Extraction with EDTA is a more pleasant procedure than with concentrated acid, and can be used to calculate approximate total concentrations. Williams and Unwin (1983) measured total (*y*, in mg kg^{-1}) and available (*x*, in mg l^{-1} soil) metals in English soils and found the following relationships:

Zinc:	*y* = 2.50*x*	*r* = 0.80	(Section 9.6)
Copper:	*y* = 1.93*x*	*r* = 0.92	
Nickel:	*y* = 3.75*x*	*r* = 0.86	

Use of the data: deficiencies The occurrence of copper deficiencies in cereals on soils in southern England is the basis for the guidelines shown in Table 15.8. Severe deficiency usually only occurs on peaty soils, shallow calcareous soils with a large organic matter content or sand-textured soils reclaimed from heathland. The normal treatment is a foliar spray of 1 kg copper oxychloride ha^{-1} (about 0.5 kg Cu^{2+} ha^{-1}) or a soil application of 20 kg copper sulphate ha^{-1} (5 kg Cu^{2+} ha^{-1}) which will prevent deficiencies for a number of years. In the context of English arable farming this is a cheap treatment and is worth while where 'likely' or 'possible' responses are indicated.

Table 15.8 Recommendations for copper applications to arable crops growing in southern England (ADAS, unpublished)

EDTA extractable copper (mg l^{-1} soil)†	Soil status	Response to copper application
0–1.0	Deficient	Likely
1.0–2.5	Low	Possible on soils with > 6% organic matter
2.5–4.0	Satisfactory	Unlikely
>4.0	Well supplied	No response

† The measurement is made using a 10 ml measuring cup of soil (Section 9.1).

Section 15.6 Projects

1. Sections 15.2, 15.3 and 15.4 include projects studying the effects and properties of pesticides in soils. If

compounds other than those described are to be examined, use values of K_{oc} and $t_{1/2}$ to estimate suitable conditions for the experiment (e.g. time, concentration). Where a compound's characteristics are not known, a preliminary simplified experiment to establish suitable conditions may save much time. An initial measurement of K_{ow} and the organic matter content of a soil may also be worth while so that K_d values can be estimated.

2. Following the methods in Section 6.9, Project 2 measure the amount of EDTA-extractable zinc and copper before and after treatment. Use soils with both small and large amounts of organic matter. What fractions of the added metals cannot be extracted?

3. Using the Wineglass model and the Atrazine.dat file check the simulations shown in Fig. 15.12. Check also the effects of increasing the amount of organic matter in the soil. Run the model also for other pesticides (Table 15.1).

Section 15.7 Calculations

1. Using the half-lives given in Table 15.1, plot the decay curves for atrazine, 2,4–D and DDT over a period of 150 d assuming first-order kinetics (Section 3.5). Compare your theoretical atrazine curve to the measured curves in Fig. 15.3

2. Using the K_{oc} values in Table 15.1, calculate K_d values for atrazine, mecoprop and ethirimol for a soil with 3.5 per cent organic matter. (*Ans.* 2, 0.4, 24 $cm^3 g^{-1}$) Plot a Freundlich isotherm for each compound assuming that the coefficient x in the Freundlich equation is 0.9. Work over the concentration range 0–1 $\mu g\ cm^{-3}$ as in Fig. 15.9(b). Using the equation or your graphs, plot the adsorption isotherms (linear axes) as in Fig. 15.9(a)

3. An example of the data obtained in the sloping plate experiment is given in Section 15.4. Using the same values of ρ and θ, calculate the expected delay in a saturated soil plate for atrazine, dichlobenil and ethirimol based on K_d values calculated for a soil with 3.5 per cent organic matter as in Calculation 2 above. (*Ans.* K_d = 2, 8, 24 $cm^3 g^{-1}$; delay = 4, 15, 44 pore volumes) For this soil in the field with a bulk density of 1.3 $g\ cm^{-3}$ and leaching occurring at a water content of $\theta = 0.4$, calculate the amount of leaching (cm H_2O) required to displace the pesticide through a 25 cm depth of topsoil. (*Ans.* 76, 274, 802)

4. A wheat crop typically contains 4 g copper t^{-1} grain and 2.5 g copper t^{-1} straw. Calculate the amount of copper in a crop of 6 t grain and 4.5 t straw ha^{-1} (Table 9.2). (*Ans.* 35 g). Compare this to (a) the amount of copper applied in a 1 kg ha^{-1} spray of copper oxychloride, $CuCl_2.3Cu(OH)_2$ and (b) the amount of EDTA-extractable copper ha^{-1}_{2500t} at the limit of 1 mg copper l^{-1} (Table 15.8). (*Ans.* (a) 0.6 kg, (b) 2.5 kg approx.)

Epilogue

Although the emphasis in this book is on measurement and interpretation of data, observation is also important in gaining an understanding of the processes occurring in soils and their effects on plant growth. How else could the parable of the sower, included as the Prologue, have been written?

During the last few decades our activities in using soils have rightly come under close scrutiny but the need to care for the land is not a recent development. The Genesis story includes instructions to people to 'cultivate and keep' the land. I understand that the word translated 'keep' has much more depth in the original Hebrew language: it also means to explore, examine, investigate, experiment and develop with care. I hope this book will help readers to follow more successfully the Maker's instructions, so that we also will be able to say:

You show care for the land by sending rain; you make it rich and fertile.

You fill the streams with water; you provide the earth with crops.

This is how you do it: you send abundant rain on the ploughed fields and soak them with water; you soften the soil with showers and cause the young plants to grow.

What a rich harvest your goodness provides! Wherever you go there is plenty.

The pastures are filled with flocks; the hillsides are full of joy.

The fields are covered with sheep; the valleys are full of wheat.

Everything shouts and sings for joy.

(David, King of Israel; *c.* 1000 BC)

328

APPENDIX I

Symbols, units and other matters

UNITS

g	gram
m	metre
s	second
h	hour
d	day
a	annum (year)
mol	mole
mol_c	mole of charge
l	litre (dm^3)
ha	hectare
t	metric tonne (1000 kg)

PREFIXES

M	mega	$\times 10^6$
k	kilo	$\times 10^3$
d	deci	$\times 10^{-1}$
c	centi	$\times 10^{-2}$
m	milli	$\times 10^{-3}$
μ	micro	$\times 10^{-6}$
n	nano	$\times 10^{-9}$

GREEK LETTERS AND THEIR USE

α	alpha, impedance factor
β	beta
γ	gamma, surface tension
Δ	delta, change in
ϵ	epsilon, air-filled porosity
η	eta, viscosity
θ	theta, volumetric water content
μ	mu
π	pi
ρ	rho, density
σ	sigma, standard deviation
Σ	sigma, sum of
ψ	psi, potential

SYMBOLS

m/m	used after the percentage symbol to indicate a mass expressed as a percentage of a mass
v/v	as above but for volumes
[ ]	concentration in solution ($mol\,l^{-1}$)
( )	activity in solution ($mol\,l^{-1}$)
{ }	concentration in solution ($mol_c\,l^{-1}$)
$\bar{x}$	$\bar{x}$ bar, the mean value of x

STANDARD SOLUTIONS OF ACIDS AND BASES

Methods have not been given in the text for the preparation of standard acids and bases. These are normally prepared from volumetric solutions available from the suppliers. For example, BDH supply a $5\,mol\,l^{-1}$ volumetric standard hydrochloric acid solution, which by suitable dilution will give the required solutions.

FILTER PAPERS

Throughout the book references to Whatman papers have been included in the methods. Apart from the filter paper method for measuring soil water suction given in Section 5.2, alternative papers can be used. Technical data relating to Whatman papers is given below.

Grade	Retention (μm)	Filtration speed	Wet strength
1	11	M	L
40	8	M	L
41	20–25	F	L
42	2.5	S	L
44	3	S	L
50	2.7	S	H
541	20–25	F	H

M, medium; F, fast; S, slow; L, suitable for gravity or low suction filtration; H, high wet strength suitable for Buchner funnels and suction filtration.

EQUIPMENT AND REAGENT SUPPLIERS

Aldrich Chemical Co. Ltd, The Old Brickyard, New Road, Gillingham, Dorset SP8 4JL. (Specialist and general chemicals).

BDH-Merck Ltd, Merck House, Poole, Dorset BH15 1TD. (Chemicals and equipment).

British Greyhound Chromatography and Allied Chemical Co., Grange Road, Birkenhead, Merseyside L43 4XF. (Pesticides and environmental chemicals and standards.)

Eijkelkamp Agrisearch Equipment, Van Walt Ltd., Prestwick Lane, Grayswood, Haslemere, Surrey GU27 2DU. (Soil sampling equipment).

ELE International Ltd., Eastman Way, Hemel Hempstead, Hertfordshire HP2 7HB. (The ELE Agronomics catalogue is a useful source of information on a wide range of products for soil, atmosphere and crop measurements.)

FSA Laboratory Supplies (Fisons), Bishop Meadow Road, Loughborough, Leicestershire LE11 0RG. (Chemicals and equipment).

AGRICULTURAL SUPPLIES

Herbiseeds, The Nurseries, Billingbear Park, Wokingham, Berks RG11 5RY. (Weed seeds).

Midland Shires Farmers Ltd., Detford Mill, Earls Croome, Worcester WR8 9DF. (Crop seeds, fertilizers and agricultural chemicals).

USEFUL ADDRESSES

The Geological Survey, Keyworth, Nottingham NG12 5GG

Her Majesty's Stationery Office, 49 High Holborn, London WC1V 6HB (HMSO publications)

Macaulay Land Use Research Institute, Craigiebuckler, Aberdeen AB9 2QJ. (The publications of the Scottish Soil Survey are available from this address.)

Ministry of Agriculture, Fisheries and Food (Publications), Lion House, Willowburn Estate, Alnwick, Northumberland NE66 2PF. (ADAS publications are available from this address).

The Ordnance Survey, Romsey Road, Maybush, Southampton SO9 4DH

Soil Survey and Land Research Centre, Cranfield Rural Institute, Silsoe, Bedfordshire MK45 4DT. (The publications of the Soil Survey of England and Wales are available from this address.)

COURSES IN SOIL SCIENCE

Details of courses taught in British universities are available from the registrars at the following universities and colleges:

Silsoe College, Bedford MK43 0AL

The University, Aberdeen AB9 1FX

University College of North Wales, Bangor, Gwynedd LL57 2DG

University College of Wales, Aberystwyth, Dyfed SY23 2AX

The University of London, Wye College, Wye, Ashford, Kent TN25 5AH

The University, Newcastle upon Tyne NE1 7RU

The University, Nottingham NG7 2RD

The University, Whiteknights, Reading, Berks. RG6 2AH

Molar masses of selected elements

Element	Symbol	$(g\,mol^{-1})$
Aluminium	Al	26.91
Boron	B	10.81
Bromine	Br	79.91
Cadmium	Cd	112.41
Caesium	Cs	132.91
Calcium	Ca	40.08
Carbon	C	12.01
Chlorine	Cl	35.45
Chromium	Cr	52.00
Cobalt	Co	58.93
Copper	Cu	63.55
Fluorine	F	19.00
Hydrogen	H	1.01
Iodine	I	126.90
Iron	Fe	55.85
Lead	Pb	207.19
Magnesium	Mg	24.31
Manganese	Mn	54.94
Mercury	Hg	200.59
Molybdenum	Mo	95.94
Nickel	Ni	58.71
Nitrogen	N	14.01
Oxygen	O	16.00
Phosphorus	P	30.97
Potassium	K	39.10
Selenium	Se	78.96
Silicon	Si	28.09
Sodium	Na	22.99
Strontium	Sr	87.62
Sulphur	S	32.06
Tin	Sn	118.69
Zinc	Zn	65.38

Elements commonly referred to in this book are Al, Ca, C, H, Mg, N, O, P, K, Na and S. Where the element is covalently bound as part of a compound, eg. in soil organic matter, plant material or manures, the symbol alone is used. Where the element is present as an ion the charge is inserted, e.g. Ca^{2+}, K^+, Cl^-. Thus exchangeable ions, ions in solution, or soil minerals, in fertilizers and in plant cells are shown with their charges. Where an element is present as part of an ion it is shown as, for example, NO_3^--N (nitrate–nitrogen), $SO_4^{2-}-S$ etc.

The program of the Bucket model for the soil water budget (Section 12.3)

Computing by Lester P. Simmonds

The program is written in BASIC. The program and data files are available on a 3.5 or 5.25 inch IBM PC compatible disk from the Secretary, Department of Soil Science, The University, Whiteknights, Reading RG6 2AH. A charge will be made to cover costs. The disk also holds the programs for the Carbon turnover model (Section 3.5) and the Wineglass model (Sections 11.5 and 15.4) together with the following files: Tritium.dat (Section 11.5), Water.dat (Section 12.1) and Atrazine.-dat (Section 15.4).

LISTING OF THE BASIC PROGRAM FOR CALCULATING THE SOIL WATER BALANCE

```
5     REM THE BUCKET MODEL OF THE SOIL WATER BALANCE – 9 JULY 1993
10    DIM CUMEPOT(13), CUMEVAP(13), CUMDRAIN(13), CUMRAIN(13), CUMSWD(13):REM monthly totals
49    REM
50    REM *********** Initialisation section ***************************************************************
51    REM
100   GOSUB 1000: REM Input soil characteristics and limiting soil water deficit
200   SWD=RZAWC: REM set initial water content at the permanent wilting point
300   GOSUB 1500: REM open data file
323   REM *******************************************************************************************
324   REM The program now works out the water balance week by week. The annual
325   REM cycle is repeated 10 times (as if there were a run of 10 years with
326   REM identical weather. The purpose of this repetition is to remove the
327   REM influence of the initial soil water content that was chosen in line 200)
328   REM *******************************************************************************************
329   PRINT:PRINT"week  Epot  Evap  Rain  Evap/Epot  SWD  Drainage":PRINT
330   FOR YEAR = 1 TO 10
332     CLOSE #1: REM The data file is closed and reopened at the start of each year
333     OPEN DATAFILE$ FOR INPUT AS #1:LINE INPUT#1, TITLE$ :REM first line is title
336     REM
337     REM ********** Calculations for each week  ****************************************************
338     REM
339     MONTH=0
340     FOR WEEK=1 TO 52
345       IF INT(WEEK/4)=WEEK/4 THEN MONTH=MONTH+1:CUMRAIN(MONTH)=0
            :CUMEVAP(MONTH)=0:CUMDRAIN(MONTH)=0:CUMEPOT(MONTH)=0:CUMSWD(MONTH)=0
350       INPUT#1, WEEKNO, EPOT, RAIN: REM input current week's data from file
352       OLDSWD=SWD
355       IF SWD<LIMDEF THEN DRYFACT=1 ELSE ERATIO=(RZAWC–SWD)/(RZAWC–LIMDEF)
357       FOR ATTEMPT=1 TO 3
360         SWD=OLDSWD+EPOT*ERATIO*CROPFACT–RAIN: DRAINAGE=0: IF SWD>RZAWC THEN
              SWD=RZAWC–.1
362         IF SWD<LIMDEF THEN ERATIO1=1 ELSE ERATIO1=(RZAWC–SWD)/(RZAWC–LIMDEF)
363         ERATIO=(ERATIO+ERATIO1)/2
365         IF SWD<0 THEN DRAINAGE=–SWD: SWD=0: REM water > field capacity drains
366       NEXT ATTEMPT
367       EVAP=ABS(SWD–OLDSWD+RAIN–DRAINAGE)
370       CUMRAIN(MONTH)=CUMRAIN(MONTH)+RAIN: CUMEVAP(MONTH)=CUMEVAP(MONTH)+EVAP
371       CUMEPOT(MONTH)=CUMEPOT(MONTH)+EPOT
372       CUMDRAIN(MONTH)=CUMDRAIN(MONTH)+DRAINAGE
373       CUMSWD(MONTH)=CUMSWD(MONTH)+SWD :REM this is divided later by 4 to calc the mean SWD
400       IF YEAR=10 THEN GOSUB 3500: REM Finish calculations and print results
450     NEXT WEEK
```

```
500   NEXT YEAR
600   REM ***********************************************************************************
601   REM PRINT OUT MONTHLY AND ANNUAL TOTALS FOR WATER BALANCE COMPONENTS
602   REM ***********************************************************************************
604   INPUT"PRESS ENTER FOR MONTHLY TOTALS";DUM$
605   CLS
606   TOTRAIN=0: TOTEPOT=0: TOTEVAP=0: TOTDRAIN=0
607   PRINT TITLE$:PRINT
608   PRINT"         Available water content of soil = ";AWC;         "% water content"
609   PRINT"      Soil depth = ";DEPTH;"cm";"   root zone water hold. cap. = ";         RZAWC;"mm"
610   PRINT"        land use : ";LANDUSE$;"            limiting water deficit = ";         LIMDEF;         "mm"
612   PRINT:PRINT"MONTH  RAIN  EPOT  EVAP  EVAP/EPOT  DRAIN  MEAN  SWD"
615   FOR MONTH=1 TO 13
620   PRINT USING"##     ####    ####    ####    #.##    ####    ####";
              MONTH,CUMRAIN(MONTH),CUMEPOT(MONTH),CUMEVAP(MONTH),
              CUMEVAP(MONTH)/CUMEPOT(MONTH),CUMDRAIN(MONTH),CUMSWD(MONTH)/4
640   TOTRAIN=TOTRAIN+CUMRAIN(MONTH):TOTEPOT=CUMEPOT(MONTH)+TOTEPOT
641   TOTEVAP=TOTEVAP+CUMEVAP(MONTH):TOTDRAIN=CUMDRAIN(MONTH)+TOTDRAIN
650   NEXT MONTH
651   PRINT
655   PRINT USING"T_O_T_A_L  ####   ####   ####  #.##   ####";
              TOTRAIN,TOTEPOT,TOTEVAP,TOTEVAP/TOTEPOT,TOTDRAIN:PRINT
800   REM ***********************************************************************************
801   REM REPEAT CALCULATIONS IN ORDER TO RE-DISPLAY RESULTS
802   REM ***********************************************************************************
850   INPUT "DO YOU WANT TO DISPLAY RESULTS AGAIN (Y/N)";DUMS
855   IF (DUM$="Y") OR (DUM$="y") THEN :CLS: GOTO 325
899   CLOSE
900   STOP
949   REM
950   REM ********** SUBROUTINES ************************************************************
951   REM
999   REM ***********************************************************************************
1000  REM Input soil characteristics and limiting water deficit
1001  REM ***********************************************************************************
1010  CLS
1100  INPUT "Available water content of soil (%water content – range 8 to 30)   "          ;AWC:PRINT
1105  IF (AWC>30) OR (AWC<8) THEN PRINT:PRINT"outside range – try again"          :PRINT:GOTO 1100
1110  INPUT "soil depth (cm) ";DEPTH:PRINT
1115  RZAWC=(AWC/100)*(DEPTH*10):REM water holding capacity of profile
1122  LANDUSE$="none":INPUT "Bare soil or Crop (B or C) ";CHOICE$
1124  IF (CHOICE$="B") OR (CHOICE$="b") THEN LANDUSE$="bare": LIMDEF=RZAWC/4: IF LIMDEF>20
        THEN LIMDEF=20
1125  IF (CHOICES="C") OR (CHOICES="c") THEN LANDUSES="crop": LIMDEF=RZAWC/2: IF LIMDEF>60
        THEN LIMDEF=60
1130  IF LANDUSES="none" THEN PRINT"Try again: GOTO 1122
1137  IF LANDUSE$="bare" THEN IF DEPTH>40 THEN RZAWC=(AWC/100)*400
1138  IF LANDUSE$="crop" THEN IF DEPTH>80 THEN RZAWC=(AWC/100)*800
1139  CLS: PRINT "Input a value for crop factor: Typical values might be" :PRINT
1140  PRINT"Ground        soil          sand          loam          clay"
1141  PRINT"Cover         surface":PRINT
1142  PRINT"bare soil    mostly dry     0.2           0.3           0.4"
1143  PRINT"bare soil    mostly wet     0.4           0.5           0.6"
1144  PRINT"50% cover    mostly dry     0.6           0.6           0.6"
1145  PRINT"50% cover    mostly wet     0.7           0.8           0.8"
1146  PRINT"95% cover    wet or dry     0.9           0.9           0.9":PRINT
1147  PRINT"For simplicity, the program assumes the crop factor stays the same"
1148  PRINT"all year. See the notes at the end of the program about how to modify"
1149  PRINT"the program to allow different values for each month":PRINT
1150  INPUT"Crop Factor";CROPFACT
1151  PRINT:PRINT"water holding capacity of extraction zone is "; RZAWC; "mm":PRINT
1155  PRINT"Limiting soil water deficit is ";LIMDEF;"mm":PRINT
1160  RETURN
1499  REM **************
```

```
1500 REM open data file
1501 REM **************
1505 GOSUB 4000: REM list data files available
1510 INPUT "name of file containing weather data"; DATAFILE$
1520 OPEN DATAFILE$ FOR INPUT AS #1
1525 PRINT:LINE INPUT#1, TITLE$: PRINT TITLE$
1530 RETURN
3499 REM ******************************
3500 REM     print weekly results
3501 REM ******************************
3510 PRINT USING "##   ###.#   ###.#   ###.#   #.##   ###   ###";
                 WEEKNO,EPOT,EVAP,RAIN,EVAP/EPOT,SWD,DRAINAGE
3520 IF (WEEK=17) OR (WEEK=34) THEN INPUT"PRESS ENTER TO CONTINUE";DUM$
                 : PRINT:PRINT"week  Epot  Evap  Rain  Evap/Epot  SWD  Drainage":PRINT
3530 RETURN
3999 REM ******************************
4000 REM     print list of data files
4001 REM ******************************
4010 DIM FILELIST$(15), FILEDESC$(15)
4020 DATA "FILENAME","DESCRIPTION"
4021 DATA ********,***********
4022 DATA "READING.83","Reading, UK, 1983 (dry year)"
4023 DATA "READING.85","Reading, UK, 1985 (wet year)"
4024 DATA "CAIRNGOR.83","Cairngorm, UK, 1983 (dry year)"
4025 DATA "CAIRNGOR.85","Cairngorm, UK, 1985 (wet year)"
4026 DATA "SYRIA.83","Aleppo, Syria, 1983"
4027 DATA "NIAMEY.AV","Average data for Niamey, Niger"
4028 DATA "KADUNA.AV","Average data for Kaduna, N Nigeria"
4029 DATA "PORTHARC.83","Port Harcourt, SE Nigeria, 1983"
4030 DATA " ", " ": REM lines 4030 to 4034 can be completed as above for new data
4031 DATA " ", " "
4032 DATA " ", " "
4033 DATA " ", " "
4034 DATA " ", " "
4038 FOR FILENUM=1 TO 15
4040    READ FILELIST$(FILENUM),FILEDESC$(FILENUM)
4050    PRINT FILELIST$(FILENUM),FILEDESC$(FILENUM)
4060 NEXT FILENUM
4090 RETURN
5000 REM *********************************************************************************************
5001 REM       List of variables used in the program
5002 REM *********************************************************************************************
5009 REM variables with soil, vegetation or weather significance
5010 REM
5011 REM CUMEPOT(I)      Monthly totals for cumulative potential evaporation
5012 REM CUMEVAP(I)      Monthly totals for cumulative actual evaporation
5013 REM CUMRAIN(I)      Monthly totals for cumulative rainfall
5014 REM CUMDRAIN(I)     Monthly totals for cumulative drainage
5015 REM EPOT            weekly potential evaporation (mm) read in from datafile
5016 REM DRAINAGE        weekly drainage (mm) calculated from soil water balance
5017 REM EVAP            weekly evaporation (mm) calc'd from soil water balance
5018 REM ERATIO          ratio of actual to potential evap'n, depending on SWD
5019 REM ERATIO1         revised ERATIO, in light of revised estimate of SWD
5020 REM TOTEPOT         Annual potential evaporation (mm)
5021 REM TOTEVAP         Annual actual evaporation (mm)
5022 REM TOTRAIN         Annual rainfall (mm)
5023 REM TOTDRAIN        Annual drainage (mm)
5100 REM AWC             Available water content of soil (percent by volume)
5101 REM DEPTH           Depth of soil profile (cm)
5102 REM RZAWC           Available water capacity of root zone (mm)
5103 REM                 (in the case of bare soil, the root zone means the depth
5104 REM                 from which water is lost through evaporation)
5105 REM SWD             the soil water deficit (i.e. mm water required to bring
5106 REM                 the soil profile back to field capacity
```

```
5107 REM LIMDEF          the SWD above which evaporation is reduced through
5108 REM                 an inadequate supply of water
5109 REM CROPFACT        Crop factor
5198 REM
5199 REM
5200 REM variables used in program control
5201 REM
5202 REM WEEK            week of the year
5203 REM WEEKNO          the value listed in the first col of the datafile
5204 REM MONTH           month (4 weeks per month, 13 months per year!!)
5205 REM YEAR            there are 10 annual cycles of calculation (year=1 to
5206 REM                 year=10) to remove effect of arbitrary initial SWD
5207 REM DATAFILE$       string variable containing name of datafile
5208 REM TITLE$          string variable containing first line of datafile
5209 REM FILEDESC$(i)    list of descriptions of datafiles
5210 REM ATTEMPT         there are three attempts (iterations) to calculate the
5211 REM                 midweek SWD (which is used to calculate EVAP) from the
5212 REM                 known SWD at the start of the week, and the latest
5213 REM                 estimate of the SWD at the end of the week.
6000 REM *******************************************************************************************
6001 SUGGESTED MODIFICATION TO ALLOW DIFFERENT CROP FACTORS EACH MONTH
6002 REM *******************************************************************************************
6003 REM                 Line 360 – change CROPFACT to CROPFACT(MONTH)–1
6004 REM                 Line 10 – add ,CROPFACT(13) to list of arrays declared in the DIM statement
6005 REM                 line 1150 should be
6006 1150               FOR I=1 TO 13:PRINT"Month ";I:INPUT"Crop factor";CROPFACT(I):NEXT I
```

Footnote: The data files (READING.83, READING.85, CAIRNGOR.83, CAIRNGOR.85, SYRIA.83, NIAMEY.AV, KADUNA.AV and PORTHARC.83) are required by the above program. These files contain a first line of text, which is a description of the data set that is read into the string variable FILEDESC$ (line 4040). The next 52 lines contain the weekly data (the week number, the weekly potential evaporation and the weekly rainfall), that is read into the variables WEEKNO, EPOT and RAIN in line 350. For example, the file READING.83 would contain the following:

```
READING, UK, 1983 DATA
1       2.8        17.3
2       4.1         6.6
3       5.3         6.2
```

and so on, through to

```
51      3.5        31.3
52      4.3        10.4
```

The full data set for this file is enclosed by the dashed box in the following table. The data files for the other years and locations can be constructed similarly. If you have any data from other sources they can also be used. Lines 4030 to 4034 are 'spares' that can be completed for new data sets.

WEATHER RECORDS – UK

Week of year	Reading, UK 1983 (dry)		1985 (wet)		Cairngorm, UK 1983 (dry)		1985 (wet)	
	EPOT	Rain	EPOT	Rain	EPOT	Rain	EPOT	Rain
1	2.8	17.3	1.0	6.4	3.2	12.9	1.6	14.7
2	4.1	6.6	2.1	4.5	4.0	34.3	1.1	57.9
3	5.3	6.2	1.6	1.3	4.1	27.8	0.1	12.8
4	2.9	4.0	2.6	19.2	5.0	7.8	1.2	43.7
5	5.2	21.3	2.7	26.5	4.6	44.4	1.6	2.6
6	4.1	11.2	3.9	1.5	3.4	51.8	3.7	22.0
7	3.7	4.3	3.8	30.1	1.5	16.1	3.1	1.2
8	4.6	0.0	3.7	0.0	1.9	0.7	2.4	3.2
9	4.9	13.5	3.7	0.2	4.2	7.1	5.1	1.5
10	5.2	2.0	4.8	14.2	7.1	3.2	3.4	5.3
11	5.9	4.9	5.9	3.0	6.4	23.6	7.0	11.7
12	7.6	13.4	5.9	3.8	8.0	24.2	6.0	26.8
13	8.1	11.6	6.3	17.5	6.9	31.6	3.7	26.7
14	10.7	24.6	10.9	5.6	7.4	23.0	5.3	28.6
15	12.4	15.5	11.8	13.1	9.2	21.7	9.0	23.7
16	10.0	15.6	14.6	9.3	11.5	10.3	9.5	16.9
17	15.0	31.9	15.2	5.1	9.1	36.0	12.1	19.3
18	12.2	21.9	13.8	6.3	6.3	26.3	10.2	32.3
19	16.8	10.5	18.3	0.3	11.0	42.4	11.1	6.0
20	19.7	21.0	12.0	36.7	14.2	12.6	13.2	30.0
21	13.6	11.8	16.2	27.4	11.1	33.5	9.5	18.8
22	16.4	30.5	17.2	18.5	11.4	34.6	14.9	42.9
23	20.8	7.5	29.9	16.4	9.9	26.6	22.8	0.2
24	23.1	0.7	16.5	51.6	20.1	11.0	13.7	35.9
25	23.8	0.0	19.0	4.5	18.1	1.6	14.7	39.8
26	18.9	22.0	17.3	46.7	16.2	21.7	15.3	57.9
27	26.6	3.4	19.6	1.7	14.1	11.4	14.5	31.7
28	26.8	0.4	27.4	1.7	18.0	9.4	13.3	23.1
29	30.7	1.5	21.8	6.4	17.5	3.3	18.4	13.9
30	24.8	12.3	23.4	15.1	18.6	1.6	17.4	9.3
31	25.9	6.9	20.7	23.3	17.3	8.7	12.3	35.5
32	23.3	0.0	20.2	27.9	13.7	2.3	17.0	30.5
33	24.2	0.1	19.8	11.9	17.3	7.2	15.0	27.9
34	20.9	7.5	17.6	9.0	11.4	12.5	11.2	30.2
35	19.5	0.0	18.5	15.9	14.4	2.1	14.2	54.2
36	18.1	3.6	17.3	18.2	15.4	22.9	10.4	41.9
37	12.9	18.5	14.0	2.2	9.1	74.1	9.7	24.2
38	15.1	20.5	14.4	1.2	12.0	24.0	11.6	19.2
39	9.7	6.0	10.8	2.2	10.4	12.7	7.7	40.1
40	8.5	6.4	11.4	0.0	8.3	29.4	6.6	8.5
41	10.4	12.3	12.4	20.5	7.4	16.7	9.7	7.9
42	10.7	32.6	7.6	0.0	10.4	48.7	5.7	2.6
43	6.4	0.0	4.7	0.0	7.1	15.8	3.2	1.4
44	4.0	0.1	4.9	0.0	8.6	8.8	2.8	1.5
45	3.6	6.3	5.4	1.7	5.0	5.4	3.7	39.0
46	4.2	0.0	6.2	13.6	1.7	2.0	3.7	53.2
47	2.5	0.1	2.9	16.2	3.6	3.9	2.9	15.1
48	3.7	44.0	2.5	1.0	1.7	22.8	1.3	40.4
49	0.8	1.3	3.8	20.4	4.0	1.6	2.8	32.2
50	2.5	16.5	2.2	38.5	3.6	22.7	0.8	28.9
51	3.5	31.3	3.0	4.4	1.4	42.6	3.2	25.1
52	4.3	10.4	4.7	23.3	1.9	48.7	2.5	19.5

EPOT (mm each week) = the potential rate of evaporation from short, well-watered grass, calculated from the amount of radiant energy absorbed by the surface, the humidity of the air and the windspeed.
Rain = the weekly rainfall (mm).

WEATHER RECORDS – SYRIA and WEST AFRICA

Week of year	Aleppo, Syria (1983)		Niamey, Niger		Kaduna, N. Nigeria		P. Harcourt, SE Nigeria	
			(averages of a number of years)					
	EPOT	Rain	EPOT	Rain	EPOT	Rain	EPOT	Rain
1	7	1.6	39	0	48	0	55	0
2	5.6	2.6	41	0	50	0	57	0
3	8.4	12.5	35	0	47	0	53	0
4	14	0.4	42	0	46	0	51	0
5	14	15.3	40	0	45	0	56	0
6	11.9	4.6	38	0	45	0	52	0
7	11.9	37.2	41	0	41	0	54	3
8	11.2	9.4	45	0	45	3	52	0
9	9.1	32.6	48	0	44	0	55	0
10	25.9	0	49	1	44	0	55	0
11	31.5	0.8	53	0	41	15	51	15
12	23.8	28.4	55	0	45	0	55	0
13	43.4	2.2	50	1	42	0	52	2
14	25.2	28.6	49	0	40	10	51	7
15	36.4	2.8	53	2	42	0	48	48
16	31.5	15.4	55	0	40	10	55	21
17	46.9	0.5	53	0	39	24	47	52
18	61.6	0.5	58	5	40	40	42	27
19	63	14.8	49	1	37	24	33	81
20	39.2	3.9	52	15	41	5	27	56
21	74.2	0.8	51	25	36	85	22	140
22	81.2	0.4	48	18	36	35	24	85
23	84	1.6	49	12	31	41	29	33
24	95.2	0	52	28	33	23	22	137
25	98	0	47	36	28	103	21	73
26	99.4	0	44	15	31	21	22	97
27	98	0	35	47	25	107	20	67
28	96.6	0	38	35	26	58	24	71
29	105	0	41	80	22	23	21	120
30	95.9	0	36	29	24	35	19	58
31	112.7	0	31	95	21	15	21	60
32	98	0	29	42	25	114	18	49
33	95.9	0	29	35	21	79	17	75
34	97.3	0	31	10	24	101	18	110
35	81.9	0.2	29	25	26	12	19	35
36	93.1	0	29	40	26	58	17	61
37	76.3	0	31	5	30	38	13	145
38	68.6	1.2	32	25	27	136	14	94
39	65.1	0.2	28	31	32	21	17	56
40	44.8	4.2	30	2	30	42	16	34
41	47.6	0.6	35	0	34	20	17	60
42	49.7	0	37	15	35	15	21	85
43	37.1	0.8	38	4	33	5	22	34
44	35.7	13	39	0	35	46	20	90
45	17.5	10.2	38	0	37	0	26	27
46	18.2	8.6	36	0	35	3	23	36
47	14.7	5.8	39	0	38	2	24	0
48	8.4	10.8	35	0	35	0	23	47
49	7	20.5	38	0	39	0	34	0
50	9.1	0	34	0	41	0	23	21
51	7.7	2.8	35	0	38	0	36	0
52	6.3	29.0	36	0	45	0	37	44

EPOT (mm each week) = the potential rate of evaporation from short, well-watered grass, calculated from the amount of radiant energy absorbed by the surface, the humidity of the air and the windspeed.
Rain = the weekly rainfall (mm).

REFERENCES

A GENERAL INTRODUCTION TO SOIL SCIENCE

Brady, N. C. (1990). *The Nature and Properties of Soils*. 10th edn. Macmillan, New York.

Wild, A. (ed.) (1988) *Russell's Soil Conditions and Plant Growth*. 11th edn. Longman, Harlow.

BACKGROUND SCIENCE

Nuffield Advanced Science (1984) *Book of Data*. Longman, Harlow.

Nuffield Advanced Science (1986) *Biology Study Guides I and II*. Longman, Harlow.

Nuffield Advanced Science (1988) *Chemistry Students Book I and II*. Longman, Harlow.

Nuffield Advanced Science (1989) *Physics Students' Guides I and II*. Longman, Harlow.

Nuffield Coordinated Sciences (1988) *Physics, Chemistry, Biology*. Longman, Harlow.

METHODS

Avery, B.W. and Bascomb, C.L. (1974). *Soil Survey Laboratory Methods*. Soil Survey Technical Monograph No. 6, Harpenden.

Landon, J. R. (ed.) (1984) *Booker Tropical Soil Manual*. Booker Agriculture International, Longman, Harlow.

MAFF (1986a) *The Analysis of Agricultural Materials*. Reference Book 427. HMSO, London.

Page, A. L. (ed.) (1982) *Methods of Soil Analysis, Parts I and II*. Agronomy No. 9. American Society of Agronomy, Madison.

Smith, K. A. and Mullins, C. E. (eds) (1991) *Soil Analysis. Physical Methods*. Marcel Dekker, New York.

and for older methods

Piper, C. S. (1947) *Soil and Plant Analysis*. University of Adelaide.

OTHER REFERENCES

ADAS (1982) *The Use of Sewage Sludge on Agricultural Land*. Booklet 2409. MAFF Publications, Alnwick.

ADAS (1987) *The Use of Sewage Sludge on Agricultural Land*. Booklet 2409. MAFF Publications, Alnwick.

Addiscott, T. M. and Wagenet, R. J. (1985) Concepts of solute leaching in soils: a review of modelling approaches. *Journal of Soil Science* **36**, 411–24.

Addiscott, T. M., Whitmore, A. P. and Powlson, D. S. (1991) *Farming, Fertilizers and the Nitrate Problem*. CAB International, Wallingford.

Aitken, R. L., Moody, P. W. and McKinley, P. G. (1990) Lime requirement of acidic Queensland soils. *Australian Journal of Soil Research* **28**, 695-701 and 703–15.

Amato, M. and Ladd, J. N. (1988) Assay for microbial biomass based on ninhydrin-reactive nitrogen in extracts of fumigated soils. *Soil Biology and Biochemistry* **20**, 107–14.

Anderson, J. P. E. and Domsch, K. H. (1980) Quantities of plant nutrients in the microbial biomass of selected soils. *Soil Science* **130**, 211–16.

Archer, J. R. (1975) Soil consistency. In *Soil Physical Conditions and Crop Production*. Technical Bulletin 29, Ministry of Agriculture, Fisheries and Food. HMSO, London, pp. 289–97.

Archer, J. (1988) *Crop Nutrition and Fertilizer Use*. Farming Press, Ipswich.

Arnold, P. W. and Close, B. M. (1961) Release of non-exchangeable potassium from some British soils cropped in the glasshouse. *Journal of Agricultural Science* **57**, 295–304.

Attwood, P. J. (1985) *Crop Protection Handbook – Cereals*. BCPC Publications, Croydon.

Ayers, R. S. and Westcot, D. W. (1985) *Water Quality For Agriculture*. Irrigation and Drainage Paper 29 Rev. 1. FAO, Rome.

Banks, L. and Stanley, C. (1990) *The Thames. A History from the Air*. Oxford University Press.

Barraclough, D., Jarvis, S. C., Davies, G. P. and Williams, J. (1992) The relation between fertilizer nitrogen applications and nitrate leaching from grazed grassland. *Soil Use and Management* **8**, 51–6.

Batey, T. (1971) *Soil Field Handbook*. ADAS Advisory Papers No. 9. Ministry of Agriculture, Fisheries and Food, London.

Batey, T. (1988) *Soil Husbandry*. Soil and Land Use Consultants, Aberdeen.

Bhat, K. K. S. and Nye, P. H. (1973). Diffusion of phosphate to plant roots in soil. I. Quantitative autoradiography of the depletion zone. *Plant and Soil* **38**, 161–75.

Bhat, K. K. S. and Nye, P. H. (1974). Diffusion of phosphate to plant roots in soil. III. Depletion around onion roots without root hairs. *Plant and Soil* **41**, 383–94.

Birch, S. P. and Moss, B. (1990) *Nitrogen and Eutrophication in the U.K.* Report to the Fertilizer Manufacturers Association. University of Liverpool.

Böhm, W. (1979) *Methods of Studying Root Systems.* Springer-Verlag, Berlin.

Bolton, J. (1972) Changes in soil pH and exchangeable calcium in two liming experiments on contrasting soils over 12 years. *Journal of Agricultural Science* **79**, 217–22.

Bridges, E. M. (1978) *World Soils.* 2nd edn. Cambridge University Press.

Brindley, G. W. and Brown, G. (1980) *Crystal Structure of Clay Minerals and their X-Ray Identification.* Monograph No. 5. The Mineralogical Society, London.

Brookes, P. C. and McGrath, S. P. (1984) Effects of metal toxicity on the size of the soil microbial biomass. *Journal of Soil Science* **35**, 341–6.

Brookes, P. C., Powlson, D. S. and Jenkinson, D. S. (1984) Phosphorus in the soil microbial biomass. *Soil Biology and Biochemistry* **16**, 169–75.

Brouwer, W. W. M., Boesten, J. J. T. I. and Siegers, W. G. (1990). Adsorption and transformation products of atrazine by soil. *Weed Research* **30**, 123–8.

Bryant, C. (1971) *The Biology of Respiration.* Studies in Biology No. 28. Edward Arnold, London.

Bullock, P. (1971) Soils of the Malham Tarn area. *Field Studies* **3**, 381–408.

Bullock, P. *et al.* (1985) *Handbook for Soil Thin Section Description.* Waine Research Publications, Wolverhampton.

Bullock, P. and Gregory, P. J. (1991) *Soils in the Urban Environment.* Blackwell, Oxford.

Burns, I. G. (1974) A model for predicting the redistribution of salts applied to fallow soils after excess rainfall or evaporation. *Journal of Soil Science* **25**, 165–78.

Burton, R. G. O. and Hodgson, J. M. (1987) *Lowland Peat of England and Wales.* Soil Survey Special Survey No. 15. Soil Survey of England and Wales, Harpenden.

Cameron, K. C. and Wild, A. (1984) Potential aquifer pollution from nitrate leaching following the ploughing of temporary grassland. *Journal of Environmental Quality* **13**, 274–8.

Campbell, D. J., Kinniburgh, D. G. and Beckett, P. H. T. (1989) The soil solution chemistry of some Oxfordshire soils: temporal and spatial variability. *Journal of Soil Science* **40**, 321–40.

Campbell, G. S. (1985) *Soil Physics with Basic.* Elsevier, Amsterdam.

Campbell, G. W., Devenish, M., Heyes, C. J. and Stone, B. H. (1987) *Acid Rain in the United Kingdom: Spatial Distributions in 1987.* Warren Spring Laboratory, Stevenage.

Carter, M. R. (1991) Ninhydrin–reactive N released by the fumigation-extraction method as a measure of biomass under field conditions. *Soil Biology and Biochemistry* **23**, 139–43.

Caudle, N. (1991) *Groundworks 1. Managing Soil Acidity.* Tropsoils Publications, Box 7113, North Carolina State University, Raleigh, NC.

Chander, K. and Brookes, P. C. (1991a) Microbial biomass dynamics during the decomposition of glucose and maize in metal-contaminated and non-contaminated soils. *Soil Biology and Biochemistry* **23**, 917–25.

Chander, K. and Brookes, P. C. (1991b) Effects of heavy metals from past application of sewage sludge on microbial biomass and organic matter accumulation in a sandy loam and silty loam U.K. soil. *Soil Biology and Biochemistry* **23**, 927–32.

Church, B. M. and Skinner, R. J. (1986) The pH and nutrient status of agricultural soils in England and Wales 1969–83. *Journal of Agricultural Science* **107**, 21–8.

Cochrane, T. T., Salinas, J. G. Sanchez, P. A. (1980) An equation for liming acid mineral soils to compensate aluminium tolerance. *Tropical Agriculture* **57**, 133–40.

Cooke, G. W. (1982) *Fertilizing for Maximum Yield.* Granada, London.

Court, M. N., Stephen, R. C. and Waid, J. S. (1964) Toxicity as a cause of the inefficiency of urea as a fertilizer. II. Experimental. *Journal of Soil Science* **15**, 49–65.

Currie, J. A. (1970) Movement of gases in soil respiration. In *Sorption and Transport Processes in Soils.* Society of Chemical Industry Monograph No. 37, pp. 152–69.

Cuttle, S. P. *et al.* (1992) Nitrate leaching from sheep-grazed grass/clover and fertilized grass pastures. *Journal of Agricultural Science* **119**, 335–43.

Davies, B., Eagle, D. and Finney, B. (1972) *Soil Management.* Farming Press, Ipswich.

de Datta, S. K. (1981) *Principles and Practice of Rice Production.* Wiley Interscience, New York.

De Gee, J. C. (1950). Preliminary oxidation potential determinations in a 'Sawah' profile near Bogor (Java). *Transactions of the 4th International Congress of Soil Science* **1**, 300–3.

Denmead, O. T. and Shaw, R. H. (1962) Availability of soil water to plants as affected by soil moisture content and meteorological conditions. *Agronomy Journal* 54, 385–90.

Dent, D. and Young, A. (1981) *Soil Survey and Land Evaluation*. George Allen and Unwin, London.

Dixon, J. B. and Weed, S. B. (eds) (1989) *Minerals in Soil Environments*. 2nd edn. Soil Science Society of America, Madison.

DOE (1986) *Nitrate in Water*. Pollution Paper No. 26. Department of the Environment Central Directorate of Environmental Protection. HMSO, London.

DOE (1989) *Digest of Environmental Pollution and Water Statistics No. 11, 1988*. HMSO, London.

DOE (1990) *Acid Deposition in the United Kingdom 1986–1988*. Third Report of the United Kingdom Review Group on Acid Rain (Chairman J. G. Irwin). Department of the Environment, Warren Spring Laboratory.

DOE (1991) *Acid Rain – Critical and Target Loads Maps for the United Kingdom*. Air Quality Division, Department of the Environment.

DOE/NWC (1981) *Report of the Sub-committee on the Disposal of Sewage Sludge to Land*. Standing Technical Committee Report No. 20. Department of the Environment, London.

Doorenbos, J. and Pruitt, W. O. (1977). *Guidelines for Predicting Crop Water Requirements*. FAO Irrigation and Drainage Paper 24. FAO, Rome.

Dyke, G. (1974) *Comparative Experiments with Field Crops*. Butterworths, London.

Edmeades, D. C., Wheeler, D. M. and Clinton, O. E. (1985) The chemical composition and ionic strength of soil solutions from New Zealand topsoils. *Australian Journal of Soil Research* 23, 151–65.

Edwards, C. A. and Lofty, J. R. (1977) *Biology of Earthworms*. Chapman and Hall, London.

Elkhatib, E. A., Hern, J. L. and Staley, T. E. (1987) A rapid centrifugation method for obtaining soil solution. *Soil Science Society of America Journal* 51, 578–83.

Emerson, W. W. (1967) A classification of soil aggregates based on their coherence in water. *Australian Journal of Soil Research* 5, 47–57.

FAO (1978) *Report on the Agro-Ecological Zones Project*. Vol. 1. *Methodology and Results for Africa*. Food and Agriculture Organization of the United Nations, Rome.

Fitter, A. H. and Hay, R. K. M. (1987) *Environmental Physiology of Plants*. Academic Press, London.

Fitzpatrick, E.A. (1980) *The Micromorphology of Soils*. Department of Soil Science, University of Aberdeen.

FMA (1981) *Fertilizer Statistics*. Fertilizer Manufacturers' Association, London.

Foster, R. C., Rovira, A. D. and Cock, T. W. (1983) *Ultrastructure of the Root–Soil Interface*. The American Phytopathological Society, St Paul.

Franks, C. L. (1984) The use of alternative extractants for determination of total heavy metals in soils. Unpublished ADAS paper, Wolverhampton.

Gardner, C. M. K. *et al.* (1990) Hydrology of the saturated zone of the chalk of south-east England. In *Chalk* (ed. J.B. Burland), Thomas Telford, London, pp. 611–18.

Garvin, J. W. (1986) *Skills in Advanced Biology*. Vol. 1. *Dealing with Data*. Stanley Thornes, Cheltenham.

Gasser, J. K. R. (1973) An assessment of the importance of some factors causing lime loss from agricultural soils. *Experimental Husbandry* 25, 86–95.

Gerber, H. R., Ziegler, P. and Dubach, P. (1970) Leaching as a tool in the evaluation of herbicides. *Proceedings of the 10th British Weed Conference* 1, 188–225.

Giller, K. and McGrath, S. (1989) Muck, metals and microbes. *New Scientist* 4 November, 31–2.

Goring, C. A. I. and Hamaker, J. W. (eds) (1972) *Organic Chemicals in the Soil Environment*. Vols 1 and 2. Marcel Dekker, New York.

Goulding, K. W. T., McGrath, S. P. and Johnston, A. E. (1989) Predicting the lime requirement of soils under permanent grassland and arable crops. *Soil Use and Management* 5, 54–8.

Green, W. H. and Ampt, G. A. (1911) Studies on soil physics. I The flow of air and water through soils. *Journal of Agricultural Science* 4, 1–24.

Gregory, P. J., Crawford, D. V. and McGowan, M. (1979) Nutrient relations of winter wheat. I. Accumulation and distribution of Na, K, Ca, Mg, P, S and N. *Journal of Agricultural Science* 93, 485–94.

Gustafson, D. I. (1989) Groundwater ubiquity score: a simple method for assessing pesticide leachability. *Environmental Toxicology and Chemistry* 8, 339–57.

Hamblin, A. P. (1981) Filter-paper method for routine measurement of field water potential. *Journal of Hydrology* 53, 355–60.

Hance, R. J. and Holly, K. (1990) *Weed Control Handbook, Principles*. 8th edn. Blackwell Scientific Publications, Oxford.

Hawkesworth, D. L. and Hill, D. J. (1984) *The Lichen-forming Fungi*. Blackie, Glasgow.

Haynes, R. J. and Williams, P. H. (1992) Changes in soil solution composition and pH in urine-affected areas of pasture. *Journal of Soil Science* 43, 323–34.

Hazelden, J., Loveland, P. J. and Sturdy, R. G. (1986) *Saline Soils in North Kent*. Special Survey No. 14. Soil Survey of England and Wales, Harpenden.

Helyar, K. R., Cregan, P. D. and Godyn, D. L. (1990) Soil acidity in New South Wales – current pH values and estimates of acidification rates. *Australian Journal of Soil Research* 28, 523–37.

Henkens, C. H. (1986) Changing production targets and techniques and their effect on the potassium balance sheet. *Proceedings of the 13th Congress of the International Potash Institute*, Berne, pp. 129–39.

Hewitt, E. J. (1966) *Sand and Water Culture Methods Used in the Study of Plant Nutrition*. Technical Communication No. 22. Commonwealth Bureaux, Farnham Royal.

Hillel, D. (1982) *Introduction to Soil Physics*. Academic Press, San Diego.

Hillel, D. (1992) *Out of the Earth: Civilization and the Life of the Soil*. Aurum Press, London.

Hodges, R. D. (1992) Soil organic matter: its central position in organic farming. In *Advances in Soil Organic Matter Research: The Impact on Agriculture and the Environment* (ed. W. S. Wilson), The Royal Society of Chemistry, Cambridge, pp. 355–64.

Hodgson, J. M. (1974) *Soil Survey Field Handbook*. Technical Monograph No. 5. Soil Survey of England and Wales, Harpenden.

Hoogmoed, W. B. and Klaij, M. C. (1990) Soil management for crop production in the West African Sahel I. Soil and crop parameters. *Soil and Tillage Research* 16, 85–103.

House of Lords (1989) *Nitrate in Water*. A report of the Select Committee on the European Communities. HMSO, London.

Howells, G. and Dalziel, T. R. K. (1992) *Restoring Acid Waters: Loch Fleet 1984–1990*. Elsevier Applied Science, London.

IRRI (1988) *Rice Facts, 1988*. World Rice Statistics. International Rice Research Institute, Manila.

Isaac, R. A. and Jones, J. B. (1972) Effects of various dry ashing temperatures on the determination of 13 different elements in five plant tissues. *Communications in Soil Science* 3, 261–9.

Jackson, R. M. and Raw, F. (1966) *Life in the Soil*. Studies in Biology No. 2. Edward Arnold, London.

Jarvis, R. A. (1968) *Soils of the Reading District*. Soil Survey of England and Wales, Harpenden.

Jarvis, S. C. (1986) Forms of aluminium in some acid permanent grassland soils. *Journal of Soil Science* 37, 211–22.

Jarvis, S. C. (1992) Grazed grassland management and nitrogen losses: an overview. *Aspects of Applied Biology* 30, 207–14.

Jarvis, S. C. and Robson, A. D. (1983) A comparison of the cation/anion balance of ten cultivars of *Trifolium subterraneum* L., and their effects on soil acidity. *Plant and Soil* 75, 235–43.

Jenkinson, D. S. (1981) In *The Chemistry of Soil Processes*, (eds D. J. Greenland and M. H. B. Hayes) John Wiley, Chichester, pp. 505–61.

Jenkinson, D. S. and Ladd, J. N. (1981) In *Soil Biochemistry* (eds E. A. Paul and J. N. Ladd) Marcel Dekker, New York, Vol. 5, pp. 415–71.

Jenkinson, D. S. and Powlson, D. S. (1976) The effects of biocidal treatment on metabolism in soil. V. A method for measuring soil biomass. *Soil Biology and Biochemistry* 8, 209–13.

Joergensen, R. G. and Brookes, P. C. (1990) Ninhydrin-reactive nitrogen measurements of microbial biomass in 0.5 M K_2SO_4 soil extracts. *Soil Biology and Biochemistry* 22, 1023–7.

Johnston, A. E. (1986) Soil organic matter, effects on soils and crops. *Soil Use and Management* 2, 97–104.

Johnston, A. E. (1992) Soil fertility and soil organic matter. In *Advances in Soil Organic Matter Research: The Impact on Agriculture and the Environment* (ed. W. S. Wilson), The Royal Society of Chemistry, Cambridge, pp. 299–314.

Johnston, A. E., Warren, R. G. and Penny, A. (1970) *Rothamsted Annual Report for 1969*, Part 2, 39–68.

Julien, J. L. (1989) Détermination de normes d'interprétation d'analyse de terre en vue de la fertilisation potassique. *Science du Sol* 27, 131–44.

Kamprath, E. J. (1970) Exchangeable aluminium as a criterion for liming leached mineral soils. *Soil Science Society of America Proceedings* 34, 252–4.

Kaye, G. W. C. and Labey, T. H. (1973) *Tables of Physical and Chemical Constants*. Longman, London.

Kinniburgh, D. G. and Miles, D. L. (1983) Extraction and chemical analysis of interstitial water from soils and rocks. *Environmental Science and Technology* 17, 362–8.

Kutschera, L. (1960) *Wurzelatlas mitteleuropäischer Ackerunkraüter und Kulturpflanzen*. DLG-Verlags-GMBH, Frankfurt.

Lal, R., Sanchez, P. A. and Cummings, R. W. 1986. *Land Clearing and Development in the Tropics*. Balkema, Rotterdam.

Landon, J.R. (ed.) (1984) *Booker Tropical Soil Manual*. Booker Agriculture International. Longman, Harlow.

Lathwell, D. J. (1979) *Crop Response to Liming of Ultisols and Oxisols*. Cornell International Agriculture Bulletin 35. Cornell University, Ithaca.

Lawrence, G. P., Payne, D. and Greenland, D. J. (1979) Pore size distribution in critical point and freeze dried aggregates from clay subsoils. *Journal of Soil Science* 30, 499–516.

Likens, G. E., Borman, F. H., Johnson, N. M., Fisher, D. W. and Pierce, R. S. (1970) Effects of forest cutting and herbicide treatment on nutrient budgets in the Hubbard Brook watershed-ecosystem. *Ecological Monographs* 40, 23–47.

Lindsay, W. L. (1979) *Chemical Equilibria in Soils*. Wiley, New York.

Maas, E. V. (1986) Salt tolerance of plants. *Applied Agricultural Research* 1, 12–26.

McCall, P. J., Laskowski, D. A., Swann, R. L. and Dishburger, H.J. (1980) Measurement of sorption coefficients of organic chemicals and their use in environmental fate analysis. In: *Test Protocols for Environmental Fate and Movement of Toxicants*, Proceedings of a Symposium of the Association of Official

Analytical Chemists, 94th Annual Meeting, October 1980 Washington DC, pp. 89–109.

McKeague, C., Wong, C. and Topp, G. C. (1982) Estimating saturated hydraulic conductivity from soil morphology. *Soil Science Society of America Journal* **46**, 1239–44.

McNeal, B. L. and Coleman, N. T. (1966). Effect of solution composition on soil hydraulic conductivity. *Soil Science Society of America Proceedings* **30**, 308–12.

McNeill, A.M. and Wood, M. (1990) ^{15}N estimates of nitrogen fixation by white clover (*Trifolium repens* L.) growing in a mixture with ryegrass (*Lolium perenne* L.). *Plant and Soil* **128**, 265–73.

MAFF (1970) *Modern Farming and the Soil*. Report of the Advisory Council on Soil Structure and Soil Fertility. HMSO, London.

MAFF (1976) *Agriculture and Water Quality*. Technical Bulletin 32. HMSO, London.

MAFF (1981) *Lime and Liming*. Ministry of Agriculture, Fisheries and Food Reference Book 35. HMSO, London.

MAFF (1986b) *Changes in ADAS Lime Recommendations*. Technical Bulletin SS/R/86/8. Internal Publication.

MAFF (1988) *Fertilizer Recommendations*. Reference Book 209. HMSO, London.

Marshall, T. J. (1958) A relation between permeability and size distribution of pores. *Journal of Soil Science* **9**, 1–8.

Marshall, T. J. and Holmes, J. W. (1988) *Soil Physics*. Cambridge University Press.

Mead, R. and Curnow, R. N. (1983) *Statistical Methods in Agriculture and Experimental Biology*. Chapman and Hall, London.

Molloy, L. (1988) *Soils in the New Zealand Landscape. The Living Mantle*. Mallinson Rendel, Wellington.

Monteith, J. L., Szeicz, G. and Yabuki, K. (1964) Crop photosynthesis and the flux of carbon dioxide below the canopy. *Journal of Applied Ecology* **1**, 321–37.

Moore, R. E. (1939) Water conduction from shallow water tables. *Hilgardia* **12**, 383–426.

Nicholls, P. H., Walker, A. and Baker, R. J. (1982) Measurement and simulation of the movement and degradation of Atrazine and Metribuzin in a fallow soil. *Pesticide Science* **12**, 484–94.

Norrish, K. (1954) The swelling of montmorillonite. *Discussions of the Faraday Society* **18**, 120–34.

NRA (1992) *The Influence of Agriculture on the Quality of Natural Waters in England and Wales*. A report by the National Rivers Authority, Bristol.

Nye, P. H. and Greenland, D. J. (1960) *The Soil under Shifting Cultivation*. Technical Communication No. 51. Commonwealth Bureau of Soils, Harpenden.

Nye, P. H. and Tinker, P. B. (1977). *Solute Movement in the Soil–Root System*. Blackwell Scientific Publications, Oxford.

Ocio, J. A. and Brookes, P. C. (1990) An evaluation of methods for measuring the microbial biomass in soils following recent additions of wheat straw and characterization of the biomass that develops. *Soil Biology and Biochemistry* **22**, 685–94.

Ogunkunle, A. O. and Beckett, P. H. T. (1988) The efficiency of pot trials, or trials on undisturbed cores, as predictors of crop behaviour in the field. *Plant and Soil* **107**, 85–93.

Open University Course Team (1989) *Seawater: Its Composition, Properties and Behaviour*. Pergamon Press, Oxford.

Parsons, A. J. *et al.* (1990) Uptake, cycling and fate of nitrogen in grass–clover and grass swards continuously grazed by sheep. *Journal of Agricultural Science* **116**, 47–61.

Petersen, L. (1986) Effects of acid deposition on soil and sensitivity of the soil to acidification. *Experientia* **42**, 340–4.

Powlson, D. S., Pruden, G., Johnston, A. E. and Jenkinson, D. S. (1986) The nitrogen cycle in the Broadbalk Wheat Experiment: recovery and losses of ^{15}N-labelled fertilizer applied in spring and inputs of nitrogen from the atmosphere. *Journal of Agricultural Science* **107**, 591–609.

Pritchard, D. T. (1969) An osmotic method for studying the suction/moisture content relationships of porous materials. *Journal of Soil Science* **20**, 374–83.

Ramig, R. E. and Rhoades, H. F. (1962) Interrelationships of soil moisture level at planting time and nitrogen fertilization on winter wheat production. *Agronomy Journal* **55**, 123–7.

Reith, J. W. S. (1962) Long term effects of various liming materials. *Empire Journal of Experimental Agriculture* **30**, 27–41.

Reuss, J. O. and Johnson, D. W. (1986) *Acid Deposition and the Acidification of Soils and Water*. Ecological Studies Volume 59. Springer-Verlag, New York.

Rhoades, J. D., Manteghi, N. A., Shouse, P. J. and Alves, W. J. (1989a) Estimating soil salinity from saturated soil-paste electrical conductivity. *Soil Science Society of America Journal* **53**, 428–33.

Rhoades, J. D., Manteghi, N. A., Shouse, P. J. and Alves, W. J. (1989b) Soil electrical conductivity and soil salinity. New formulations and calibrations. *Soil Science Society of America Journal* **53**, 433–9.

Richards, L. A. (1954) *Diagnosis and Improvement of Saline and Alkali Soils*. Agriculture Handbook No. 60. USDA, Washington.

Roberts, T. M., Skeffington, R. A. and Blank, L. W. (1989). Causes of Type 1 Spruce decline in Europe. *Forestry* **62**, 179–222.

Robson, J. D. and Thomasson, A. J. (1977) *Soil Water Regimes*. Technical Monograph No. 11. Soil Survey of England and Wales; Harpenden.

Robson, M. J., Parsons, A. J. and Williams, T. E. (1989) In *Grass, its Utilization and Production* (ed. W. Holmes). Blackwell Scientific Publications, Oxford, pp. 7–88.

Rothamsted Experimental Station (1969) *Report for 1968, Part II*. Harpenden.

Rothamsted Experimental Station (1970) *Details of the Classical and Long-term Experiments to 1967*. Harpenden.

Rothamsted Experimental Station. (1977a) *Report for 1976, Part II*. Harpenden.

Rothamsted Experimental Station (1977b) *Details of the Classical and Long-term Experiments 1968–73*. Harpenden.

Rothamsted Experimental Station (1983) *Report for 1982, Part II*. Harpenden, pp. 5–44.

Rothamsted Experimental Station (1991) *Guide to the Classical Field Experiments*. Harpenden.

Rowell, D. L (1963) Effect of electrolyte concentration on the swelling of orientated aggregates of montmorillonite. *Soil Science* 96, 368–74.

Rowntree, D. (1981) *Statistics Without Tears*. Penguin Books, London.

Rowse, H. R. (1975) Simulation of the water balance of soil columns and fallow soils. *Journal of Soil Science* 26, 337–49.

Rowse, H. R. and Stone, D. A. (1978) Simulation of the water distribution in soil. I. Measurement of soil hydraulic properties and the model for an uncropped soil. *Plant and Soil* 49, 517–31.

Russell, R. S. (1977) *Plant Root Systems: Their Function and Interaction with Soil*. McGraw-Hill, London.

Salter, P. J. and Williams, J. B. (1969) The moisture characteristics of some Rothamsted, Woburn and Saxmundham soils. *Journal of Agricultural Science* 73, 155–8.

Sanchez, P. A. (1976) *Properties and Management of Soils in the Tropics*. Wiley, Chichester.

Schinas, S. and Rowell, D. L. 1977. Lime induced chlorosis. *Journal of Soil Science* 28, 351–68.

Shainberg, I. (1985) The effect of exchangeable sodium and electrolyte concentration on crust formation. *Advances in Soil Science* 1, 101–22.

Shainberg, I. and Letey, J. (1984) Response of soils to sodic and saline conditions. *Hilgardia* 52, 1–57.

Shoemaker, H. E., McLean, E. O. and Pratt, P. F. (1961) Buffer methods for determining lime requirements of soils with appreciable amounts of extractable aluminium. *Soil Science Society of America Proceedings* 25, 274–7.

Smart, P. and Tovey, N. K. (1981) *Electron Microscopy of Soils and Sediments: Examples*. Clarendon Press, Oxford.

Smith, D. L. O. (1987) Measurement, interpretation and modelling of soil compaction. *Soil Use and Management* 3, 87–93.

Smith, W. N., Prasher, S. O., Khan, S. U. and Barthakur, N. W. (1992) Leaching of ^{14}C-labelled atrazine in long, intact soil columns. *Transactions of the American Society of Agricultural Engineers* 35, 1213–20.

Soane, B. D. (1975) Studies on some physical properties in relation to cultivations and traffic. In *Soil Physical Conditions and Crop Productivity*, Technical Bulletin 29, Ministry of Agriculture, Fisheries and Food. HMSO, London, pp. 160–82.

Soil Survey Staff (1975) *Soil Taxonomy: A Basic System of Soil Classification for Making and Interpreting Soil Surveys*. USDA Soil Conservation Service, Washington DC.

Stevens, P. A., Hornung, M. and Hughes, S. (1989) Solute concentrations, fluxes and major nutrient cycles in a mature Sitka spruce plantation in Beddgelent Forest, North Wales. *Forest Ecology and Management* 27, 1–20.

Stevenson, F. J. (1982) *Humus Chemistry. Genesis, Composition, Reactions*. John Wiley, New York.

Stribley, D. P., Tinker, P. B. and Snellgrove, R. C. (1980) Effect of vesicular–arbuscular mycorrhizal fungi on the relations of plant growth, internal phosphate concentration and soil phosphorus analysis. *Journal of Soil Science* 31, 655–72.

Suquet, H., Calle, C. de la and Pezerat, H. (1975) Swelling and structural organization of saponite. *Clays and Clay Minerals* 23, 1–9.

Sylvester-Bradley, R. *et al.* (1987) *Nitrogen Advice for Cereals: Present Realities and Future Possibilities*. Proceedings No. 263. The Fertiliser Society of London.

Tennant, D. (1975) A test of a modified line intersect method of estimating root length. *Journal of Ecology* 63, 995–1001.

Thomas, M. D. (1924) Aqueous vapour pressure of soils. II. Studies in dry soils. *Soil Science* 17, 1–18.

Tipping, E. and Hurley, M. A. (1988) A model of solid–solution interactions in acid organic soils, based on the complexation properties of humic substances. *Journal of Soil Science* 39, 505–19.

van Raij, B. and Peech, M. (1972) Electrochemical properties of some oxisols and alfisols of the tropics. *Soil Science Society of America Proceedings* 36, 587–93.

van Schilfgaarde, J. (1974) *Drainage for Agriculture*. Agronomy No. 17. American Society of Agronomy, Madison.

Walker, A. (1978) Simulation of the persistence of eight soil applied herbicides. *Weed Research* 18, 305–13.

Waring, S. A. and Bremner, J. M. (1964) Ammonium production in soil under waterlogged conditions as an index of nitrogen availability. *Nature (London)* 201, 951–2.

Webster, R. and Oliver, M. A. (1990) *Statistical Methods in Soil and Land Resource Survey*. Oxford University Press.

West, L. T. *et al.* (1984) *Soil Survey of the ICRISAT Sahelian Centre, Niger, West Africa*. Soil and Crop Sciences Department/Tropsoils, Texas A and M.

White, R. E. and Beckett, P. H. T. (1964) Studies on the phosphate potential of soils. *Plant and Soil* **20**, 1–16 and **21**, 253–82.

Whittaker, R. H. and Likens, C. E. (1975) In *Primary Productivity of the Biosphere* (eds H. Leith and R. H. Whittaker). Springer-Verlag, Berlin, pp. 305–28.

Wild, A. (ed.) (1988) *Russell's Soil Conditions and Plant Growth*. 11th edn. Longman Group UK Ltd, Harlow.

Wild, A. (1993) *Soils and the Environment: An Introduction*. Cambridge University Press.

Williams, J. H. and Unwin, R. J. (1983) The relationship between total and extractable metal contents of contaminated soils and the implications for the addition of sewage sludge to land. Unpublished ADAS paper, Wolverhampton.

Williams, M. L. *et al.* (1989) *A Preliminary Assessment of the Air Pollution Climate of the UK*. Warren Spring Laboratory, Stevenage.

Williams, R. J. B. (1976) In *Agriculture and Water Quality* (eds W. Dermott *et al.*). Technical Bulletin 32. Ministry of Agriculture, Fisheries and Food. HMSO, London, pp. 174–200.

Wilson, B. J. (1989) *Predicting Cereal Yield Loss from Weeds*. Technical Report 89/4, Long Ashton Research Station.

Wilson, B. J. and Wright, K. J. (1990) Predicting the growth and competitive effects of annual weeds in wheat. *Weed Research* **30**, 201–11.

Wilson, M.J. (ed.) (1987) *A Handbook of Determinative Methods in Clay Mineralogy*. Blackie, Glasgow.

Wong, M. T. F., Hughes, R. and Rowell, D. L. (1990) Retarded leaching of nitrate in acid soils from the tropics: measurement of the effective anion exchange capacity. *Journal of Soil Science* **41**, 655–63.

Woodruff, C. M. (1948) Testing soils for lime requirement by means of a buffered solution and the glass electrode. *Soil Science* **66**, 53–63.

Worthing, C.R. (ed.) (1991) *The Pesticide Manual. A World Compendium*. 9th edn. British Crop Protection Council, Farnham.

Wright, K. J. and Wilson, B. J. (1992) The effects of fertilizer on competition and seed production of *Avena fatua* and *Galium aparine* in winter wheat. *Aspects of Applied Biology* **30**, 381–6.

Young, C. P., Hall, E. S. and Oakes, D. B. (1976) *Nitrate in Groundwater – Studies on the Chalk near Winchester, Hampshire*. Technical Report TR 31. Water Research Centre, Medmenham.

INDEX